The Force Concept in Chemistry

The Force Concept in Chemistry

Edited by

B. M. Deb
Department of Chemistry
Indian Institute of Technology
Bombay, India

VNR VAN NOSTRAND REINHOLD COMPANY
NEW YORK CINCINNATI ATLANTA DALLAS SAN FRANCISCO
LONDON TORONTO MELBOURNE

Van Nostrand Reinhold Company Regional Offices:
New York Cincinnati Atlanta Dallas San Francisco

Van Nostrand Reinhold Company International Offices:
London Toronto Melbourne

Library of Congress Catalog Card Number: 80-11263
ISBN: 0-442-26106-3

Manufactured in the United States of America

Published by Van Nostrand Reinhold Company
135 West 50th Street, New York, N.Y. 10020

Published simultaneously in Canada by Van Nostrand Reinhold Ltd.

15 14 13 12 11 10 9 8 7 6 5 4 3 2 1

Library of Congress Cataloging in Publication Data

Main entry under title:
The Force concept in chemistry.

Includes bibliographies and indexes.
1. Quantum chemistry. I. Deb, B. M.
QD462.F67 541.2'8 80-11263
ISBN 0-442-26106-3

Dedication

People who met Charles Coulson during his life sometimes came away wondering who is greater: Coulson the scientist, or Coulson the man. That there has never been a simple answer to this question is itself a tribute to the rigor with which Charles Coulson developed and practiced these qualities, for it is a tragic sign of our times that the scientist in many of us far outshines the human being. There is hardly any area of applied quantum mechanics which has not been enriched by Coulson's pioneering contributions (see S. L. Altmann and E. J. Bowen, *Biographical Memoirs of Fellows of the Royal Society*, Vol. 20, pp. 75–134, December 1974). His many-sided humanitarian activities were guided by a simple principle: to understand and to serve. He had a special concern for science and social development in developing countries. Indeed, the growth of the present small body of theoretical chemists in India has been due largely to the stimulation and encouragement provided directly or indirectly by Charles Coulson. Accordingly, this book is dedicated with respect and affection to

CHARLES ALFRED COULSON (1910–1974)
SCIENTIST AND MAN

List of Contributors

R. F. W. Bader, Department of Chemistry, McMaster University, Ontario Canada

A. J. Coleman, Department of Mathematics and Statistics, Queen's University, Kingston, Canada.

K. C. Daiker, Department of Chemistry, University of New Orleans, New Orleans, Louisiana

B. M. Deb, Department of Chemistry, Indian Institute of Technology, Bombay, India

S. T. Epstein, Department of Physics, University of Wisconsin, Madison, Wisconsin

J. Goodisman, Department of Chemistry, Syracuse University, Syracuse, New York

T. Koga, Department of Hydrocarbon Chemistry, Faculty of Engineering, Kyoto University, Kyoto, Japan

M. T. Marron, Department of Chemistry, University of Wisconsin-Parkside, Kenosha, Wisconsin

H. Nakatsuji, Department of Hydrocarbon Chemistry, Faculty of Engineering, Kyoto University, Kyoto, Japan

P. Politzer, Department of Chemistry, University of New Orleans, New Orleans, Louisiana

P. Pulay, Department of General and Inorganic Chemistry, Eötvös Loránd University, Múzeum Körút 6-8, Budapest, Hungary

Preface

The object of this book is to draw scientists' attention to the simple but rigorous concept of electrostatic forces in molecules. This approach views chemical processes in three-dimensional space through the net forces in a molecular system, or where detailed physical insight into a chemical phenomenon is desired, through appropriately partitioned components of these net forces. The forces themselves can be derived from a knowledge of the quantum mechanical electron density, by means of the Hellmann–Feynman (H–F) theorem. Although this theorem was proposed forty years ago, a fruitful realization of its implications and potential applications had to wait until the sixties. This occurred partly because the apparent simplicity of the H–F theorem had evoked some skepticism and suspicion that too much was probably being read into too little, and partly because of the delayed availability of reliable electron densities for systems of interest. However, it must also be said that the initial disbelief at results derived from the H–F theorem did contribute significantly toward the growth of the force concept in quantum mechanics.

It has been recognized for some time that a considerable disadvantage of many highly accurate, conventional quantum-chemical calculations is the fact that simple concepts and models highly useful for chemists are extremely difficult to retrieve from the resulting maze of high-precision numbers. The difficulties are compounded by the fact that the theoretical foundations of a number of earlier concepts and pictures in current chemical use have been shown by these accurate calculations to be either nonexistent or extremely weak. But the calculations themselves could hardly provide alternative simple concepts that might be applied easily by all chemists, and so the use of earlier pictures, even if somewhat unjustified, continued unabated. As Professor C. A. Coulson once remarked, the generation of accurate numbers, whether by computation or by experiment, is a somewhat futile exercise unless these numbers can provide us with simple and useful chemical concepts: if the numbers themselves were of the essence, one might as well be interested in a telephone directory.

The force concept provides a very powerful tool for developing these much needed chemical models, which will be based on firm quantum-mechanical foundations and yet retain an essential simplicity of character. This approach has already yielded outstanding successes in understanding chemical binding, molecular geometry, and chemical reactivity, among other subjects. These successes are due, first, to the fact that, within the Born–Oppenheimer approximation, the force is expressible merely as a balance of two opposing terms, one coming from nuclear–nuclear repulsions and the other from electron–nuclear attractions. (Interelectronic repulsions do not appear explicitly in the expression.) Secondly, the electron–nuclear attractive force is expressed in terms of the three-dimensional electron density. These two features provide a comfortable visualization of molecular behavior in three-dimensional space, a great advantage that was not accessible before. Still, the tremendous possibilities of the force concept, which also has considerable pedagogic value, have not yet been fully realized.

This book sets down in detail the theoretical machinery developed so far within the force concept. It then deals in depth with numerous applications of the concept to investigate molecular phenomena. The topics covered range from the H–F theorem to works which have been inspired directly or indirectly by the theorem, as well as other works related to the force concept. The presentation is at a level which should generally be comprehensible to a graduate student in chemistry or physics who has been exposed to quantum mechanics at the undergraduate level. Throughout the book, emphasis has been upon highlighting how new concepts and insights spring from the force concept, and how the reader can apply these concepts in dealing with molecular (and solid state) behavior. Considerable effort has been made to include all relevant work done so far in this area, thus making the book as comprehensive and exhaustive as possible. The editor would be very grateful to have his attention drawn to any omission of related work.

The idea for this book arose from the unexpected and very gratifying response, from all over the world, to a review article by the editor on the force concept.* Those people who showed an interest included mathematicians, physicists, chemists, biologists, geologists, medical researchers, chemical engineers, materials scientists, among others. While it is too much to expect that the present book would be of service to all these people, it is the editor's fond hope that this expectation may be realized at least in part.

The organization of the book is described below.

In Chapter 1, Epstein provides a thorough and elegant account of the general H–F theorem, and various other related theorems such as the electrostatic and time-dependent H–F theorems, integral and integrated H–F theorems, etc. Since the form of the force operator (the derivative of the Hamiltonian) depends on the choice of coordinate system, the effects of coordinate transformations on the general and electrostatic H–F theorems are discussed. Conditions under which optimal variational wavefunctions satisfy the various theorems are then examined;

Rev. Mod. Phys., *45*, 22 (1973).

these conditions may be regarded as virtues to be acquired by such approximate wavefunctions. The relation between perturbation theory and the H–F theorems, which is important for the treatment of, e.g., long-range intermolecular forces, as well as the connection between the H–F theorem and hypervirial theorems has also been dealt with. Finally, it is also possible to define a regional H–F theorem where the integration is performed over a subspace of the total space. (The implications of such quantum subspaces are brought out in Chapter 2). Chapter 1 thus lays down most of the theoretical foundations for the whole book.

Applications of the various H–F theorems to chemical problems begin with Chapter 2. In this chapter, Bader presents a rigorous and comprehensive discussion of chemical binding based on the properties of molecular electronic charge distributions, and the forces which these distributions exert on the nuclei. Chemical binding is examined in detail through density difference maps, which represent pictorially the redistribution of charge accompanying the formation of a chemical bond. A partitioning of the electron–nuclear attractive force vis-à-vis the density difference map reveals that in covalent binding the nuclei are bound by the electron density shared between them, while in ionic binding the nuclei are bound by the electron density localized about a single nucleus, thereby confirming the old hypothesis of G. N. Lewis. A combined use of the virial and electrostatic H–F theorems enables one to derive certain general constraints that might be placed on the signs and relative magnitudes of the changes in kinetic and potential energy over the complete range of internuclear distances involved in molecule formation. Bader discusses in considerable detail the changes in kinetic energy accompanying the formation of a molecule, and points to the existence of local regions of space in which the kinetic energy attains rather low values. The locations of these regions in different systems may be predicted from the charge distributions depicted in density difference maps. Finally, Bader highlights the role of the charge density in quantum mechanics, and discusses some very interesting properties of quantum subspaces.

From the theories of chemical binding it is but one step to enquire about the relative spatial arrangements of the bound atoms in a molecule. It is well known that many physical, chemical, and biological properties of substances can be correlated with their characteristic molecular shapes. Unfortunately, the broad problem of molecular geometry has defied solution for a long time. Nakatsuji and Koga open Chapter 3 with a brief discussion on several earlier models of molecular geometry, and then proceed to give an exhaustive account of the various phenomena governing molecular geometry, based on two force models. Of these two models, one is more pictorial, employing the HOMO (highest occupied molecular orbital) postulate, and provides a consistent qualitative view of all static aspects of molecular geometry. The other model is more quantitative in character, and can also deal with dynamic changes in molecular geometry occurring during chemical reactions, as well as with long-range intermolecular forces (see Chapter 7). In all such discussions the paramount importance of the three-dimensional electron density becomes quite obvious. The origin of the internal force which favors or disfavors the passage of a molecule from one configuration to another lies in the relaxation of the electron cloud during such processes. Obviously, the stretching and bending force constants of molecules (see Chapter 5) can be interpreted on the basis of such relaxation processes, some features of which have been described by Nakatsuji and Koga.

As Epstein explains in Chapter 1, the forces can be integrated to give energies and energy differences. However, here a problem arises, in that (except in some special cases) the resulting energy quantities are generally less accurate than the corresponding quantities obtained by a variational or perturbational calculation. Further, with the integral H-F theorem, for example, the value of the integral may sometimes depend on the choice of the path of integration. Indeed, by a careful choice of path one can obtain an energy difference superior to that given by the difference in expectation values! These and other related issues are discussed at length by Marron in Chapter 4, where he also explains how one can obtain useful physical insights into the mechanisms of internal motions.

The topics of chemical binding, molecular geometry, internal motions, and force constants are all intricately linked. The last-mentioned topic forms the subject matter of Chapter 5. Here Goodisman analyzes various ways of calculating and interpreting force constants of diatomic and polyatomic molecules. There are certain practical advantages in calculating force constants from forces over the conventional double differentiation of the energy, although the vexing question of accuracy again crops up here. However, there are situations where even wavefunctions of less than Hartree-Fock accuracy can provide good estimates of stretching and bending force constants via forces.

By means of the three-dimensional electron density one can define the electrostatic potential at any point in the space around a molecule; the negative derivative of the potential, multiplied by the charge, gives the force at any point in the three-dimensional space. Like the H-F force, the electrostatic potential is given by a balance of two opposing contributions, one a nuclear-nuclear repulsion term and the other an electron-nuclear attraction term. Using a number of interesting examples, which include nucleic acid bases as well as drug-receptor interactions, Politzer and Daiker explain in Chapter 6 how the electrostatic potential in large molecules serves as an attractive guide to their reactivity. For example, it can reveal the regions in a molecule that are most susceptible to electrophilic attack and provide a measure of their relative reactivities toward electrophiles. Politzer and Daiker also discuss other force approaches to chemical reactivity.

Chapter 7 gathers together a number of other interesting applications of the H-F theorem to atomic and molecular physics, solid state physics, statistical thermodynamics, etc. Topics as diverse as the calculation of three-center force integrals, inner-shell binding energies (ESCA), binding energy of biexcitons, relativistic and nonrelativistic classical analogs of the H-F theorem, the stress concept as a generalization of the force concept, etc., are discussed. Chapter 7 also includes an account of the long-range forces between atoms in their ground and excited or ionized states, as well as intermolecular forces in other systems. Feynman's original conjecture about the origin of the van der Waals attraction between molecules is verified.

All these applications of the force concept repeatedly highlight the role of the three-dimensional electron density, indicating that a great deal of chemistry can be dealt with simply in three-dimensional space. Since it is generally believed that all information of physical significance about a system is encoded in the single-particle and two-particle reduced density matrices, quantum mechanists have been trying for some time to devise algorithms for direct calculation of these density matrices, i.e., without first calculating the wavefunction. Unfortunately, these efforts have been beset with a very serious problem, that of N-representability of the 1-matrix and 2-matrix. In Chapter 8, Coleman provides an erudite account of these efforts and indicates how close we are now to this goal. He summarizes the basic properties of the 1- and the 2-matrix, and reviews the efforts to solve what is denoted as the ensemble N-representability problem, after which he includes a discussion of the numerical results achieved so far.

The final chapter concerns itself with the direct analytical evaluation of energy gradients ("exact" forces). Its author and main protagonist Pulay points out that direct analytical evaluation, wherever applicable, is computationally superior to numerical determination of gradients from calculated energy values. Further, with approximate wavefunctions the error in the "exact" force is always of second order in the error parameter, unlike the H–F forces. Pulay looks into the calculation of energy gradients from various types of SCF wavefunctions, and discusses in detail the determination of molecular geometries, saddle points on potential energy surfaces, and different types of force constants via these "exact," or non-H–F forces.

The editor is deeply appreciative of the enthusiastic participation of all the other authors of this book, and their great patience when the work was passing through difficult phases. If this book has any merits, it is because of their efforts to provide accounts of their areas which would be as comprehensive and readable as possible. If the book has any defects, in spite of its rich material, the editor assumes responsibility for them. This effort to produce the first book on the force concept in chemistry has received warm encouragement from many scientists in India and the world at large. Limitations of space prevent listing their names, but the editor is very grateful for their kindness. It is a pleasure to thank Dr. (Mrs.) Geeta Mahajan, Dr. (Mrs.) Anjuli Bamzai, Mr. S. K. Ghosh and Mr. G. B. Pattanaik for their help in various ways, and the editor's colleagues at I.I.T. Bombay for their encouragement and support. Finally, words of gratitude are hardly adequate for the efforts of the one particular individual who, besides helping with the author index, nursed her husband through continued ill health and kept him on his feet during the course of this project. Without her this book might not have been possible.

B. M. Deb
Editor

Acknowledgments

For permission to reproduce certain copyright material, including diagrams and tables, grateful acknowledgment is made to:

The American Chemical Society; The Chemical Society, London; Springer-Verlag, Germany; The North-Holland Publishing Company; Taylor and Francis Limited, Great Britain; Societes Chimique Belges, Brussels; John Wiley & Sons, Inc.; The National Research Council of Canada; The American Institute of Physics; Plenum Publishing Corporation.

Acknowledgments

Contents

The Force Concept in Chemistry

1
The Hellmann-Feynman Theorem*

Saul T. Epstein
Department of Physics
University of Wisconsin
Madison, Wisconsin

Contents

*Research supported by National Science Foundation Grant MPS74-17494.

1-1. INTRODUCTION

The electrostatic Hellmann-Feynman theorem will play an important role in many of the chapters of this book. Our major task in this first chapter is therefore to provide a reasonably precise and thorough introduction to that theorem. The method we will use will be to view the electrostatic theorem as "simply" a special case of a more general theorem–the Hellmann-Feynman (H–F) theorem. Our reasons for this approach are twofold. First, a more general point of view presumably enhances appreciation of special cases; and, second, many other special cases of the general theorem are also of interest (see Section 1-5). At the end of our discussion we will also include some brief comments on other general theorems of a related sort, and on some other matters as well.

We will be concerned with the normalizable (i.e., bound-state) eigenfunctions and corresponding eigenvalues of a Hamiltonian H. At first we will confine attention to the ideal situation in which we have an exact eigenfunction and eigenvalue; later we will turn to the more practical situation in which we have only approximations to these quantities.

1-2. STATEMENT OF THE HELLMANN-FEYNMAN THEOREM

Let ψ be a normalizable eigenfunction of H with eigenvalue E, thus

$$H\psi = E\psi \tag{1-1}$$

and let σ be a real parameter in H. Then the theorem[1] states that

$$\frac{\partial E}{\partial \sigma} = \frac{\left\langle \psi \left| \frac{\partial H}{\partial \sigma} \right| \psi \right\rangle}{\langle \psi | \psi \rangle}. \tag{1-2}$$

Below we offer three "different" proofs of this theorem.[2] It is true that one would do; however, each proof is instructive in its own way.

1-3. THREE PROOFS OF THE HELLMANN-FEYNMAN THEOREM

1-3-1. First Proof (*94, 95*)

In this first proof we simply appeal to well-known results (the second and third proofs are more self-contained). Namely, suppose that we change the parameter in H from σ to $\sigma + \delta\sigma$. Then H will change by δH, where

$$\delta H = \frac{\partial H}{\partial \sigma}\,\delta\sigma + O((\delta\sigma)^2), \tag{1-3}$$

whence, from first-order perturbation theory,[3] the energy will change by

$$\delta E = \frac{\left\langle \psi \left| \frac{\partial H}{\partial \sigma} \right| \psi \right\rangle}{\langle \psi | \psi \rangle}\,\delta\sigma + O((\delta\sigma)^2). \tag{1-4}$$

If now we divide both sides of (1-4) by $\delta\sigma$, and then let $\delta\sigma$ tend to zero, we evidently arrive at (1-2).[4] Q.E.D.

1-3-2. Second Proof (*25, 39*)

Here we first differentiate the identity

$$\langle \psi | H - E | \psi \rangle = 0 \tag{1-5}$$

with respect to σ to find

$$\left\langle \frac{\partial \psi}{\partial \sigma} \middle| H - E \middle| \psi \right\rangle + \left\langle \psi \left| \frac{\partial H}{\partial \sigma} - \frac{\partial E}{\partial \sigma} \right| \psi \right\rangle + \left\langle \psi \middle| H - E \middle| \frac{\partial \psi}{\partial \sigma} \right\rangle = 0. \tag{1-6}$$

However[5]

$$(H - E)|\psi\rangle = 0 \text{ and } \langle \psi|(H - E) = 0, \tag{1-7}$$

whence we have

$$\left\langle \psi \left| \frac{\partial H}{\partial \sigma} - \frac{\partial E}{\partial \sigma} \right| \psi \right\rangle = 0, \tag{1-8}$$

which of course is just (1-2) again[6,7].

1-3-3. Third Proof (*67*)

Instead of differentiating the scalar product (1-5), we differentiate (1-1) directly. Then we find

$$\left(\frac{\partial H}{\partial \sigma} - \frac{\partial E}{\partial \sigma} \right) \psi + (H - E) \frac{\partial \psi}{\partial \sigma} = 0, \tag{1-9}$$

and if we take the scalar product of this with ψ and use (1-7) again, we of course recover (1-2). However, now we can also (following Pauli) do more, as we could also have done in connection with the "first proof," there by invoking other familiar formulas from perturbation theory, to provide other interesting and useful results.

Namely, if ψ' is another eigenfunction of H with eigenvalue $E' \neq E$, then taking the scalar product of (1-9) with ψ' and using the fact that

$$\langle \psi' | \psi \rangle = 0 \tag{1-10}$$

we find

$$\left\langle \psi' \left| \frac{\partial \psi}{\partial \sigma} \right\rangle = - \frac{\left\langle \psi' \left| \frac{\partial H}{\partial \sigma} \right| \psi \right\rangle}{E' - E}. \right. \tag{1-11}$$

Hence, assuming no complications due to possible degeneracy of E, and assuming for convenience, that the $\langle \psi' |$ are normalized, we have that[8] (using the conventional discrete notation)

$$\frac{\partial \psi}{\partial \sigma} = - \sum \psi' \frac{\left\langle \psi' \left| \frac{\partial H}{\partial \sigma} \right| \psi \right\rangle}{E' - E}. \tag{1-12}$$

Equation (1-12), of course, looks just like another formula from first-order perturbation theory, and indeed, as we indicated above, we could have derived it on that basis. Also, by further differentiation of (1-2) and of (1-12) one can derive potentially useful formulas for $\partial^2 E/\partial\sigma^2$, $\partial^2\psi/\partial\sigma^2$, $\partial^2 E/\partial\sigma\partial\mu$, where μ is another parameter, etc.,[9] all of which, again, could also be derived from standard perturbation-theory formulas. Conversely, if one identifies σ with a perturbation parameter λ, and sets $\lambda = 0$ in these formulas, one can derive from them the usual equations of perturbation theory for $\psi^{(n)} \equiv (1/n!)\,(\partial^n\psi/\partial\lambda^n)_{\lambda=0}$ and $E^{(n)} \equiv (1/n!)\,(\partial^n E/\partial\lambda^n)_{\lambda=0}$, and hence for ψ and E (*15, 52*).

$$\psi = \sum_{n=0}^{\infty} \lambda^n \psi^{(n)}, \quad E = \sum_{n=0}^{\infty} \lambda^n E^{(n)}. \tag{1-13}$$

1-4. COORDINATE DEPENDENCE OF THE THEOREM

In the differentiation of E, ψ, and H in the second and third proofs, and in the variation of H in the first proof, it was implicit in the discussion that we were, to put it most abstractly, to keep the coordinates and their conjugate momenta fixed. More concretely, if we think in terms of a wavefunction in, say, coordinate or momentum space with the scalar product representing an integration, then we were to keep the integration variables fixed.[10]

Now clearly these implicit requirements have no effect on $\partial E/\partial\sigma$; E is E whatever coordinate system or canonical variables one may use. However, equally clearly $\partial H/\partial\sigma$ will in general be different in coordinate systems which differ from one another in a σ-dependent way, and hence the interpretation, though not the numerical value, of the right-hand side of (1-2) will change if one makes a σ-dependent coordinate transformation.

To illustrate what is involved we will first give a very simple example of a linear harmonic oscillator. (More physically interesting examples will be found in Section 1-6.) Consider, then, a linear harmonic oscillator of unit mass. Using an ordinary Cartesian coordinate system, one has

$$H = \frac{p^2}{2} + \frac{1}{2}\omega^2 x^2. \tag{1-14}$$

If now we identify σ with ω, (1-2) yields

$$\frac{\partial E}{\partial\omega} = \omega\,\frac{\langle\psi|x^2|\psi\rangle}{\langle\psi|\psi\rangle}. \tag{1-15}$$

On the other hand, suppose we make an ω-dependent change of variable (which in this case is actually a unitary transformation) according to

$$x = \frac{x'}{\sqrt{\omega}}, \quad p = \sqrt{\omega}\,p'. \tag{1-16}$$

Then in terms of the new variable we have the Hamiltonian

$$H' = \omega\left(\frac{p'^2}{2} + \frac{x'^2}{2}\right) \tag{1-17}$$

and hence from (1-2), but now with primes on the right-hand side,

$$\frac{\partial E}{\partial \omega} = \frac{\left\langle \psi' \middle| \frac{p'^2}{2} + \frac{x'^2}{2} \middle| \psi' \right\rangle}{\langle \psi' | \psi' \rangle} = \frac{E}{\omega}, \tag{1-18}$$

where ψ' is the wavefunction expressed in terms of x'. Thus in this coordinate system the H–F theorem for $\sigma = \omega$ is not (1-15) but rather $\partial E/\partial\omega = E/\omega$, which clearly exhibits the coordinate dependence of the theorem.

Moreover the nature of this coordinate dependence is quite predictable. Let us first note that, since (1-15) and (1-18) are both true, we can combine them to find

$$E = \omega^2 \frac{\langle \psi | x^2 | \psi \rangle}{\langle \psi | \psi \rangle}, \tag{1-19}$$

which will be recognized as one form of the virial theorem for a harmonic oscillator. Then the point we would make is that this is quite a general phenomenon. Namely, one can show (*16*) that if one can implement a change of coordinates by a σ-dependent unitary transformation U (or by such a transformation followed by any sort of σ-independent coordinate transformation) then equating the two right-hand sides of (1-2) will yield the hypervirial theorem[11] for $U^{-1}\partial U/\partial\sigma$. Thus, in the example just considered, the transformation involves a scaling. Therefore,[12] $U^{-1}\partial U/\partial\sigma$ is essentially the operator $(px + xp)$, and hence the hypervirial in question is the virial theorem,[13] in agreement with what we have just found by explicit calculation.

Similarly, to give another example, suppose we make a translation $x = x' + \omega$, $p = p'$. Then

$$H' = \frac{p'^2}{2} + \frac{1}{2}\omega^2(x' + \omega)^2 \tag{1-20}$$

and hence

$$\frac{\partial E}{\partial \omega} = \omega \frac{\langle \psi' | (x' + \omega)^2 | \psi' \rangle}{\langle \psi' | \psi' \rangle} + \omega^2 \frac{\langle \psi' | x' + \omega | \psi' \rangle}{\langle \psi' | \psi' \rangle}, \tag{1-21}$$

or, changing variables back again,

$$\frac{\partial E}{\partial \omega} = \omega \frac{\langle \psi | x^2 | \psi \rangle}{\langle \psi | \psi \rangle} + \omega^2 \frac{\langle \psi | x | \psi \rangle}{\langle \psi | \psi \rangle}. \tag{1-22}$$

Comparison with (1-14) then yields

$$\frac{\langle \psi | x | \psi \rangle}{\langle \psi | \psi \rangle} = 0, \tag{1-23}$$

i.e., the average force on the particle is zero, and indeed[14] this "force theorem" is the hypervirial theorem for the operator p which generates translations.

1-5. SOME EXAMPLES OF THE HELLMANN-FEYNMAN THEOREM[15]

1-5-1. Electric and Magnetic Examples

These were probably the first examples to which the theorem was applied explicitly *(94)*. Most simply, if one has a nonrelativistic system in a uniform electric field $\boldsymbol{E}$, then the electric field dependence of the Hamiltonian is contained in a term

$$-\boldsymbol{E} \cdot \sum_i q_i \boldsymbol{r}_i,$$

where q_i is the charge of the ith particle and $\boldsymbol{r}_i$ is its position vector. Thus in this case (1-2) tells us that the negative derivative of E with respect to $\boldsymbol{E}$ is the average of $\Sigma q_i \boldsymbol{r}_i$, i.e., the average electric dipole moment of the system. Thus, just as one would expect classically, polarizabilities and hyperpolarizabilities arise from an induced dipole moment. Further, for a system of electrons, since the electric dipole moment operator is a one-electron, momentum-independent operator, one can write its average as simply the electric dipole moment of the electronic charge density, and one thereby has a very classical and picturesque formula for $-\partial E/\partial \boldsymbol{E}$.

Similarly, in the presence of a uniform magnetic field (and using the gauge in which the vector potential is $\frac{1}{2}\boldsymbol{B} \times \boldsymbol{r}$) one finds that $-\partial H/\partial \boldsymbol{B}$ is the magnetic dipole operator and hence that, just as one would expect classically, $-\partial E/\partial \boldsymbol{B}$ is the average magnetic dipole moment of the system. Further, for a system of electrons one can write this as the magnetic dipole moment produced by the electronic current and magnetization (spin) densities, again yielding a very classical and picturesque result.

Further, if one also introduces an external magnetic dipole moment $\boldsymbol{\kappa}$ into the system then, with the usual choice of its vector potential, one finds that $-\partial H/\partial \boldsymbol{\kappa}$

is the operator representing the magnetic field produced by the charges at the position of the dipole. If then we continue to denote by $\boldsymbol{B}$ the external uniform field, it follows from (1-2) that one can calculate the "shielding tensor"[16]

$$-\left[\frac{\partial^2 E}{\partial \boldsymbol{\kappa} \partial \boldsymbol{B}}\right]_{\kappa=0, B=0}$$

using any one of three different approaches (*73, 86*). First, one can calculate it as the term in the average magnetic dipole moment at zero $\boldsymbol{B}$, $-(\partial E/\partial \boldsymbol{B})_{B=0}$, which is linear in $\boldsymbol{\kappa}$ (the moment induced by $\boldsymbol{\kappa}$ in zero external field). Second, one can calculate it as the term in the magnetic field at the position of $\boldsymbol{\kappa}$ in the absence of $\boldsymbol{\kappa}$, $-(\partial E/\partial \boldsymbol{\kappa})_{\kappa=0}$, which is linear in $\boldsymbol{B}$ (the field induced by $\boldsymbol{B}$ at the position $\boldsymbol{\kappa}$ is to occupy). Third, one can simply calculate the second derivative of E directly and not worry about pictures.

Also, if in general static electric and magnetic fields one introduces formal ordering parameters ν_e and ν_m, each with numerical value 1, to order the scalar and vector potential (in whatever gauge), respectively (i.e., if one replaces the scalar potential ϕ by $\nu_e \phi$, and the vector potential A by $\nu_m A$) then, ignoring spin-dependent interactions, one finds that the H-F theorems for $\sigma = \nu_e$ and $\sigma = \nu_m$ yield the classical results

$$\partial E/\partial \nu_e = \int q \phi \, dr$$

$$\partial E/\partial \nu_m = -\frac{1}{c}\int \boldsymbol{j} \cdot \boldsymbol{A} \, dr,$$

where q is the charge density, i.e., the average of $\Sigma_i q_i \delta(\boldsymbol{r} - \boldsymbol{r}_i)$; and $\boldsymbol{j}$ is the current density, i.e., the average of

$$\sum_i \frac{1}{2} \frac{q_i}{m_i} \left\{ \left(\boldsymbol{p}_i - \frac{q_i \nu_m}{c} \boldsymbol{A} \right) \delta(\boldsymbol{r} - \boldsymbol{r}_i) + \delta(\boldsymbol{r} - \boldsymbol{r}_i) \left(\boldsymbol{p}_i - \frac{q_i \nu_m}{c} \boldsymbol{A} \right) \right\}.$$

Finally, differentiation with respect to masses yields interesting results for the binding energies of nonadiabatic systems, for example, electron-positron complexes, muonic molecules, and polarons. For details see Refs. 35 and 89.

1-5-2. The Electrostatic Theorem (*25, 39*)

We come now to the case which will be of prime interest in this book, the celebrated electrostatic H-F theorem. Here we are concerned with the electronic Hamiltonian (including the nuclear repulsions, and most simply, in the absence of external fields) within the Born–Oppenheimer approximation. Thus we have, in atomic units,

$$H = \sum_s \frac{p_s^2}{2} - \sum_s \sum_a \frac{Z_a}{|r_s - R_a|} + \sum_{s>t}\sum \frac{1}{r_{st}} + \sum_{a>b}\sum \frac{Z_a Z_b}{|R_a - R_b|}, \tag{1-24}$$

where s and t label the electrons, a and b the nuclei, and where for the moment we have not made any particular choice of origin or orientation of our coordinate system.

One now readily finds that

$$-\frac{\partial H}{\partial R_a} \equiv F_a \tag{1-25}$$

is the operator which represents the force on nucleus a due to the electrons and other nuclei, i.e.,

$$-\frac{\partial H}{\partial R_a} = -\sum_s \frac{Z_a}{|R_a - r_s|^3}(R_a - r_s) + \sum_{b \neq a} \frac{Z_a Z_b}{|R_a - R_b|^3}(R_a - R_b). \tag{1-26}$$

Thus, with σ identified with the components of the R_a, (1-2) yields

$$-\frac{\partial E}{\partial R_a} = \frac{\langle \psi | F_a | \psi \rangle}{\langle \psi | \psi \rangle} \tag{1-27}$$

or, in the notation which we will use quite frequently from now on,

$$-\frac{\partial E}{\partial R_a} = \langle F_a \rangle, \tag{1-28}$$

and this is the theorem in question.

Note that it has an essentially classical form. Thinking of the nuclear coordinates simply as external parameters, as one does in the present context, $-\partial E/\partial R_a$ would be expected to be the force on nucleus a, and the right-hand side of (1-28) is in perfect agreement with this. Further, within the simple Born–Oppenheimer approximation, where E plays the role of the potential energy for the actual nuclear motion, one can also read (1-28) to say that the Born–Oppenheimer force $-\partial E/\partial R_a$ is equal to the H–F force $\langle F_a \rangle$, which latter one might have written down a priori. Finally, since F_a is a one-electron, momentum-independent operator, one can write $\langle F_a \rangle$ as a sum of a classical nuclear contribution plus a classical contribution due to the electronic charge density, thus yielding a very simple and appealing picture of the origin of the Born–Oppenheimer force on nucleus a. Namely, in detail, one finds (25) that

$$\langle F_a \rangle = -Z_a \int \rho(r) \frac{R_a - r}{|R_a - r|^3} dr + Z_a \sum_{a \neq b} Z_b \frac{R_a - R_b}{|R_a - R_b|^3}, \tag{1-29}$$

where ρ is the negative of the electronic charge density, i.e., ρ is a positive quantity. Since the interpretation and use of the theorem will be thoroughly discussed in the chapters which follow, we will not pursue these points further here. Rather, we will now turn to certain coordinate-dependent aspects of the theorem, and for definiteness and simplicity we will confine attention to diatomic molecules.

1-6. COORDINATE DEPENDENCE OF THE ELECTROSTATIC THEOREM FOR A DIATOMIC MOLECULE

In actual calculations involving diatomic molecules one usually does not put the origin just anywhere, and orient the coordinate axes in some arbitrary way. Rather, one will usually place the origin on the internuclear axis in a way linked to the nuclear positions—at the molecular midpoint, at the nuclear center of mass, etc. Also, one will usually choose the internuclear axis as a coordinate axis, say the z-axis. If the coordinate origin is placed at a fractional distance $(1 + \alpha)^{-1}$ along the internuclear axis, then denoting the nuclei by 1 and 2 with, for definiteness, nucleus 2 to the right of nucleus 1, in this coordinate system R_1 and R_2 have the components

$$R_1 = \left(0, 0, -\frac{R}{\alpha + 1}\right), \; R_2 = \left(0, 0, \frac{\alpha}{\alpha + 1} R\right), \tag{1-30}$$

where

$$R \equiv |R_2 - R_1|$$

is the internuclear separation.

Using this coordinate system, one readily finds that

$$-\frac{\partial H}{\partial R_2} = \frac{\partial H}{\partial R_1} = \left(\frac{\alpha}{\alpha + 1} F_{2z} - \frac{1}{\alpha + 1} F_{1z}\right) k,$$

where k is a unit vector along the internuclear axis. Thus, in this coordinate system (1-2), with σ identified with the components of R_1 and R_2, yields not

(1-28) but

$$-\frac{\partial E}{\partial \boldsymbol{R}_2} = \frac{\partial E}{\partial \boldsymbol{R}_1} = \left(\frac{\alpha}{\alpha+1}\langle F_{2z}\rangle - \frac{1}{\alpha+1}\langle F_{1z}\rangle\right)\boldsymbol{k}. \tag{1-31}$$

Comparison with (1-28) then shows that

$$\langle \boldsymbol{F}_1\rangle + \langle \boldsymbol{F}_2\rangle = 0, \tag{1-32}$$

i.e., that the total average force on the nuclei is zero, and also that each average force is directed along the internuclear axis, i.e.,

$$\boldsymbol{k} \times \langle \boldsymbol{F}_a\rangle = 0, \qquad a = 1, 2. \tag{1-33}$$

Further, these results are in accord with the discussion in Section 1-4. Namely, the coordinate system used to derive (1-31) will be related to that used to derive (1-28) by some translation plus rotation, and indeed the force theorems (1-32) are[17] the hypervirial theorems for the total electron momentum operator—the operator which generates translations. Similarly the hypervirial theorems for the total electron angular momentum operator, the operator which generates rotations, are torque theorems[18] which, when coupled with the force theorems, can be shown to be equivalent to (1-33).[19]

Of course (1-32) and (1-33) also follow from (1-28) and the fact that E depends on $\boldsymbol{R}_1$ and $\boldsymbol{R}_2$ only through $|\boldsymbol{R}_1 - \boldsymbol{R}_2|$. In this connection it is interesting to remark that one can also give a formal derivation of this last fact using (1-2). Namely, if we identify σ with R, then we find with the present coordinate system that

$$\frac{\partial E}{\partial R} = -\left(\frac{\alpha}{\alpha+1}\langle F_{2z}\rangle - \frac{1}{\alpha+1}\langle F_{1z}\rangle\right), \tag{1-34}$$

which one might call a scalar electrostatic theorem. Comparison with (1-31) then yields

$$\frac{\partial E}{\partial \boldsymbol{R}_1} = \frac{\partial E}{\partial R}\frac{\partial R}{\partial \boldsymbol{R}_1} \quad \text{and} \quad \frac{\partial E}{\partial \boldsymbol{R}_2} = \frac{\partial E}{\partial R}\frac{\partial R}{\partial \boldsymbol{R}_2}, \tag{1-35}$$

which evidently implies that $E = E(R)$.

Before leaving this case, one further point should be made. We viewed (1-15), (1-18), and (1-22) as expressing the content of (1-2) for $\sigma = \omega$ in three different coordinate systems. Similarly, we have viewed (1-31) as expressing the content of (1-2) for σ the components of the $\boldsymbol{R}_a$ in coordinate systems differing by the parameter α. However, given that (1-15), (1-18), and (1-22) are all true, and given that the right-hand side of (1-31) is independent of α, we can alternatively view them as yielding formally different but numerically equivalent formulas for the energy derivative within a given coordinate system. (It should be kept in mind that these equations may well not be true when one deals with approximate wavefunctions; see Section 1-9.) In particular (*78*), depending on the circumstances, one choice of α in (1-31) may be more useful than another.

As a final example (*12, 28, 68*) of coordinate dependence, one in which the interpretation of the theorem is drastically altered, suppose that we scale all electronic coordinates according to $\boldsymbol{r}_s = R\boldsymbol{r}_s'$ (and note that one can then pass from these Cartesian $\boldsymbol{r}_s'$ coordinates to the usual elliptical coordinates by an R-independent transformation). Then in these coordinates one finds that H' takes the form

$$H' = \frac{t}{R^2} + \frac{v}{R} \equiv T + V, \tag{1-36}$$

where $t/R^2 \equiv T$ represents the kinetic energy with the operator t explicitly independent of R, and where $v/R \equiv V$ represents the potential energy with v explicitly independent of R. Then identifying σ with R, we find that in these coordinates

$$\frac{\partial E}{\partial R} = -2\left\langle \frac{t}{R^3} \right\rangle - \left\langle \frac{v}{R^2} \right\rangle,$$

which upon multiplication by R will be seen to be not (1-34) but the familiar virial theorem

$$2\langle T\rangle + \langle V\rangle + R\,\frac{dE}{dR} = 0. \tag{1-37}$$

Further, in accord with the discussion in Section 1-4, this differs from (1-34) by the hypervirial theorem for the operator $\Sigma_s(\boldsymbol{r}_s \cdot \boldsymbol{p}_s + \boldsymbol{p}_s \cdot \boldsymbol{r}_s)$, the operator which generates scaling transformations.[20,21]

1-7. PERTURBATION THEORY AND THE HELLMANN-FEYNMAN THEOREM

A question which has been of some interest[22] in connection with the H-F theorem is the following. Suppose that the Hamiltonian involves a perturbation parameter λ, and denote by ϕ the sum of the first n terms in the perturbation series for ψ. Thus, from (1-13),

$$\phi = \psi^{(0)} + \lambda\psi^{(1)} + \cdots + \lambda^n \psi^{(n)}. \tag{1-38}$$

Then the question is, to what order of accuracy does

$$\frac{\left\langle \phi \left| \frac{\partial H}{\partial \sigma} \right| \phi \right\rangle}{\langle \phi | \phi \rangle} \tag{1-39}$$

approximate $\partial E/\partial\sigma$?

Using (1-2) we can write this question in symbols as follows:

$$\frac{\left\langle \phi \left| \frac{\partial H}{\partial \sigma} \right| \phi \right\rangle}{\langle \phi | \phi \rangle} = \frac{\left\langle \psi \left| \frac{\partial H}{\partial \sigma} \right| \psi \right\rangle}{\langle \psi | \psi \rangle} + O(\lambda^s); \; s = ? \tag{1-40}$$

To answer this question, we write

$$\phi = \psi + \lambda^{n+1}\,\Delta, \tag{1-41}$$

whence one readily finds that

$$\frac{\left\langle \phi \left| \frac{\partial H}{\partial \sigma} \right| \phi \right\rangle}{\langle \phi | \phi \rangle} = \frac{\left\langle \psi \left| \frac{\partial H}{\partial \sigma} \right| \psi \right\rangle}{\langle \psi | \psi \rangle} + \frac{\lambda^{n+1}\left\{\left\langle \psi \left| \frac{\partial H}{\partial \sigma} - \frac{\partial E}{\partial \sigma} \right| \Delta \right\rangle + \text{c.c.}\right\}}{\langle \psi | \psi \rangle} + O(\lambda^{2n+2}). \tag{1-42}$$

Therefore, since there is no reason to suspect that the term in braces will vanish, it follows that if

$$\frac{\partial H}{\partial \sigma} \text{ is of order 1 then } s = n + 1, \tag{1-43}$$

while if $\partial H/\partial\sigma$ is of order λ then $s = n + 2$, etc.[22]

To give an example, consider the calculation of the long-range force between two atoms as a function of R, by means of the electrostatic theorem. Then the perturbation is the interaction of the electrons and nuclei of each atom with those of the other. In a coordinate system like (1-30), R is involved in both the zero-order Hamiltonian and the perturbation. Hence $\partial H/\partial R$ is of order 1, whence it follows from (1-43) that to calculate $\partial E/\partial R$ and hence from (1-35) the force correctly through second order using the electrostatic theorem one needs the wavefunction through second order.[22]

However, there are two further remarks to be made. First, continuing to use the long-range force example, since, as is well known,

$$\langle\phi|H|\phi\rangle/\langle\phi|\phi\rangle = E + O(\lambda^{2n+2}), \tag{1-44}$$

it follows that by direct differentiation of $\langle\phi|H|\phi\rangle/\langle\phi|\phi\rangle$ we can certainly calculate the force through second order from a knowledge of ψ (and its derivative) through first order only. Indeed, this is what one usually does, i.e., one uses second-order perturbation theory for the energy.[23] In this connection we may also note that by simply equating the formal expressions derived from the two approaches and/or by directly differentiating the Schrödinger equation, one can derive "curious identities" (*85*) relating integrals involving $\psi^{(2)}$ to integrals involving $\psi^{(0)}$, $\partial\psi^{(0)}/\partial R$, $\psi^{(1)}$, etc. Similarly, going on to higher orders will yield other identities.

The second remark is that, though we seem to have found $s = n + 1$ (for $\partial H/\partial\sigma$ of order 1), it would appear that theoretically at least one can have $s = 2n + 1$. The main point is that although we have been implying that ϕ is a well-defined quantity, actually it is not. Thus, consider first $n = 1$. Then it is well known that $\psi^{(1)}$ is arbitrary to within an additive multiple (say, β) of $\psi^{(0)}$, and so if $\psi^{(1)}$ is a particular first-order function, we write for $n = 1$ that

$$\phi = \psi^{(0)} + \lambda(\psi^{(1)'} + \beta\psi^{(0)})$$

or

$$\phi = (1 + \lambda\beta)\left[\psi^{(0)} + \lambda\psi^{(1)'} - \frac{\lambda^2\beta}{1 + \lambda\beta}\psi^{(1)'}\right]. \tag{1-45}$$

Therefore, since the numerical factor $(1 + \lambda\beta)$ will cancel out of (1-39), we see, assuming $\partial H/\partial\sigma$ to be of order 1, that (1-39) will first be affected by the β term in second order, and then through a term linear in β.

We now assume that the coefficient of β does not vanish.[24] Then since (1-44), which is unaffected by the ambiguity in ϕ, allows us to calculate $\partial E/\partial\sigma$ through second order from $\psi^{(0)}$ and a $\psi^{(1)'}$, it follows that given these quantities one can choose β in such a way that for $n = 1$, s in (1-40) becomes 3 rather than 2.

More generally, for arbitrary n, ϕ will contain n arbitrary constants, and hence it would appear that one could, in principle, choose these in such a way that s in (1-43) becomes $2n + 1$ instead of $n + 1$.[25] However, note that all of this applies to just a single σ. If one is interested in several parameters (e.g., if one is interested in the long-range forces between molecules), then the situation changes.

Confining attention to $n = 1$, since the "correct" value of β will presumably be different for different parameters, it follows that no single first-order wavefunction can do the whole job through the second order via H–F theorems.

1-8. THE HELLMANN–FEYNMAN THEOREM FOR VARIATIONAL WAVEFUNCTIONS

Thus far we have been assuming that we had an exact eigenfunction and eigenvalue of H. However, for most systems this is a very unrealistic assumption. Usually the best one has is some optimal variational wavefunction. Hence one is led to ask, since clearly H–F theorems are very "nice" theorems, lending themselves to classical pictures, etc., if one can still have (1-2), but now with ψ an optimal variational wavefunction, and with E the corresponding optimal energy, i.e.,

$$E = \frac{\langle\psi|H|\psi\rangle}{\langle\psi|\psi\rangle}. \tag{1-46}$$

Indeed, it should be emphasized that if (1-2) is not satisfied, then one is faced with two equally plausible but now inequivalent methods for calculating the "same" effect; e.g., to calculate a polarizability one might proceed either by calculating an average dipole moment or by calculating an energy derivative.

Since (1-46) is identical in form to (1-5), we will still have[10] (1-6). Hence what we want is a condition which will ensure[26] that

$$\left\langle\frac{\partial\psi}{\partial\sigma}|H - E|\psi\right\rangle + \left\langle\psi|H - E|\frac{\partial\psi}{\partial\sigma}\right\rangle = 0 \tag{1-47}$$

as a consequence of the variation method. Such a condition was first given by Hurley (*46*). (For analogous results in statistical mechanics see Refs. 2, 43, and 96.)

1-8-1. Hurley's Condition

Hurley's condition can be stated in the following rather elegant way. If the set of trial functions used in a variation calculation is invariant with respect to changes in a parameter σ, i.e., if the same set is used for all values of σ, then the optimal trial functions will satisfy the H–F theorem (1-2).[27]

Granting the correctness of this condition, it follows that one will satisfy the theorem either if the individual trial functions have no explicit σ dependence[28] or, when the individual trial functions do depend explicitly on σ, if changing σ simply turns one trial function into another.[29]

Before turning to a proof, we now give some examples to show the power of this simple condition.[30] In these examples we will be assuming that H is the electronic Hamiltonian. (For a discussion of the status of the H–F theorem for a molecule as a whole within the Born–Oppenheimer approximation, see Ref. 19).

(i) In any configuration interaction calculation in which the set of basis functions is independent of σ the condition will evidently be satisfied, and hence we will have (1-2). In particular, if one uses basis functions which have no explicit σ dependence, though they may contain nonlinear variational parameters whose optimal values depend on σ, then Hurley's condition will be met and (1-2) will be satisfied. In practice this is usually the case for all σ's except possibly for nuclear coordinates, and magnetic fields when one uses so-called GIAOs.[31] Also, recently (*76, 97*) there has been interest in using basis sets which depend explicitly on an external electric field.

(ii) In all Hartree–Fock approximations (as distinct from SCF approximations–see (iii) below), be they restricted, unrestricted, open shell, closed shell, multiconfigurational, or whatever, the individual trial functions (whether general Slater determinants, symmetry-restricted Slater determinants, linear combinations of Slater determinants, or whatever) are never assumed to depend explicitly on a parameter in the problem–a nuclear charge, an electric field, a nuclear coordinate, etc. Therefore it follows that in all such approximations all H–F theorems will be satisfied.[30]

(iii) In SCF approximations one limits the Slater determinants to those which one can form from finite, one-electron basis sets. However, as in (i), these sets are usually independent of any σ's except possibly for nuclear coordinates (and magnetic fields when one uses so-called GIAOs[31] or electric-field-dependent sets). Thus for all other σ's (1-2) will be satisfied.

1-8-2. Proof That Hurley's Condition is Sufficient

To make the discussion precise it is helpful to introduce some more notation. First we will write a trial function in more detail as $\psi(A, \sigma)$, where A stands for the variational parameters, whether arbitrary numbers, arbitrary functions, or both, and where we have allowed for an explicit σ-dependence of the trial functions. Secondly, we write an optimal trial function as $\psi(\hat{A}(\sigma), \sigma)$ where $\hat{A}$ denotes the optimal values of the variational parameters, values which, as our notation indicates, may be σ-dependent. Now for the proof. First we generate from an optimal function a whole new set of trial functions $\psi(\hat{A}(\eta), \eta)$ by replacing σ with a real numerical variational parameter η. However, from the assumption that the original set was closed under changes in σ (and of course under changes in the A's), it follows that this new set must be a subset of the old. Therefore, when used in the variation method, the new set must yield (possibly among others) $\hat{\eta} = \sigma$ as a solution, with the same optimal energy as before.

Now the basic equation of the variation method is

$$\langle \delta\psi | H - E | \psi \rangle + \langle \psi | H - E | \delta\psi \rangle = 0. \tag{1-48}$$

Therefore, from what we have just said, this must be satisfied by

$$\delta\psi = \left(\frac{\partial \psi}{\partial \eta} (\hat{A}(\eta), \eta) \right)_{\eta = \sigma} \delta\eta, \tag{1-49}$$

which yields, upon canceling a factor of $\delta\eta$,

$$\left\langle \left(\frac{\partial \psi}{\partial \eta} (\hat{A}(\eta), \eta) \right)_{\eta = \sigma} | H - E | \psi \right\rangle + \left\langle \psi | H - E | \left(\frac{\partial \psi}{\partial \eta} (\hat{A}(\eta), \eta) \right)_{\eta = \sigma} \right\rangle = 0. \tag{1-50}$$

However $(\partial/\partial\eta\psi(\hat{A}(\eta), \eta))_{\eta = \sigma}$ will be recognized as what we previously denoted by simply $\partial\psi/\partial\sigma$. Therefore (1-50) is identical to (1-47), and the sufficiency of Hurley's condition is proven.[32]

In conclusion, an apparent objection to some of our applications of Hurley's condition should be mentioned. In our discussion we have implicitly assumed that the A's are independent variational parameters. However, in some of the examples one usually introduces constraints on the parameters, constraints which are then enforced by use of Lagrange multipliers. Nevertheless, the applications were fully justified: the use of Lagrange multipliers is, after all, only a convenience. In principle one could, without changing the results, use the constraint equations to eliminate dependent parameters. Then the results follow as claimed, because if the original set is independent of σ, the set derived by elimination will be independent of σ (provided of course, as in the examples, that the constraints do not involve σ explicitly).[33]

1-9. COORDINATE DEPENDENCE OF THE ELECTROSTATIC THEOREM FOR VARIATIONAL WAVEFUNCTIONS

Again we will confine attention to diatomic molecules. When, as is customary, one employs a coordinate system of the sort exemplified by (1-30), Hurley's condition tells us that if one uses a set of trial functions which, in this set of coordinates, is independent of $\boldsymbol{R}_1$ and $\boldsymbol{R}_2$, then (1-31) will be satisfied. The question then arises, will (1-28) also be satisfied?

First, we remark that usually symmetry alone will ensure that $\langle F_1 \rangle$ and $\langle F_2 \rangle$ are axial and hence (1-31) is, in effect, equivalent to

$$-\frac{\partial E}{\partial R_2} = \frac{\partial E}{\partial R_1} = \frac{\alpha}{\alpha + 1} \langle F_2 \rangle - \frac{1}{\alpha + 1} \langle F_1 \rangle. \tag{1-51}$$

Thus the only question remaining is under what circumstances do we have

$$\langle F_{1z} \rangle + \langle F_{2z} \rangle = 0 \tag{1-52}$$

and hence by symmetry

$$\langle F_1 \rangle + \langle F_2 \rangle = 0. \tag{1-53}$$

As we have remarked before, (1-52) is equivalent to the hypervirial theorem for the z-component of the total electronic momentum. Now it is known[34] that a sufficient condition to satisfy the hypervirial theorem for a Hermitian operator G is that one use a set of trial functions which is invariant to the unitary transformations generated by G, and therefore, in this case, to translation along the internuclear axis.

In particular, in the unrestricted Hartree-Fock approximation the set of trial functions, which is the set of all Slater determinants, is invariant to the unitary transformations generated by all Hermitian one-electron G's.[35] Therefore it follows that in this approximation not only, as noted earlier, are all H-F theorems satisfied, but also all one-electron hypervirial theorems[36] are satisfied, and hence as a special case (1-53).

Another example of a set of trial functions which will satisfiy both (1-51) and (1-53) is provided by what are called (*46*) "floating wavefunctions." To be specific, consider a one-electron two-center problem. Then trial functions of the form

$$A \exp\left[-a(x^2 + y^2 + (z - \gamma)^2)^{1/2}\right] + B \exp\left[-b(x^2 + y^2 + (z - \delta)^2)^{1/2}\right] \tag{1-54}$$

are said to be floating functions. Here γ and δ are numerical variational parameters and the numbers A, B, a, and b may or may not be variational parameters, but if not, they should be strictly independent of R_1 and R_2. Evidently these are described as floating functions because the individual atomic orbitals are not centered on the nuclei a priori. Rather the centers are allowed to float, the variation method then determining the optimal centering. To see that these floating functions will do the job is trivial. The individual trial functions, and hence the set, are independent of $\boldsymbol{R}_1$ and $\boldsymbol{R}_2$ and so (since symmetry does apply here) (1-51) is guaranteed. Further the set is invariant to translation along the z-axis since replacing z by $z + d$ is equivalent to replacing γ and δ by $\gamma - d$, and $\delta - d$ respectively, thus sending one member of the set into the other. Hence (1-53) is also guaranteed.

However, if one uses a set which, though independent of $\boldsymbol{R}_1$ and $\boldsymbol{R}_2$, is not translationally invariant, then one is guaranteed (1-51), but one will probably not have (1-52). Similarly if one uses a set which is translationally invariant, but not independent of $\boldsymbol{R}_1$ and $\boldsymbol{R}_2$ then one will have (1-52) but probably not (1-31).

1-10. THE INTEGRAL HELLMANN-FEYNMAN THEOREM

As indicated in our introductory remarks, there are other general theorems with some similarities to the H-F theorem, and which are also of interest in connection with the topics to be discussed in the remainder of this text. Most prominent among these is probably the so-called integral H-F theorem. This theorem, though without a name, had been used in various connections, but seemingly not in quantum chemistry before the work of Parr (*66*), who incidentally gave it its name, and of Richardson and Pack (*74*). Since then, however, it has been applied extensively, principally by Parr and coworkers.[37]

The theorem can be stated as follows. Let H_A and H_B be two different Hamiltonians, and let ψ_A and E_A be an eigenfunction, and corresponding eigenvalue of H_A, and similarly let ψ_B and E_B be an eigenfunction and corresponding eigenvalue of H_B. Then the identity (we assume $\langle\psi_A|\psi_B\rangle \neq 0$)

$$E_B - E_A = \frac{\langle\psi_B|H_B - H_A|\psi_A\rangle}{\langle\psi_B|\psi_A\rangle} \tag{1-55}$$

is called the integral H-F theorem.[38]

That it represents a generalization of the H-F theorem can be seen in the following way. Suppose that $H_A = H(\sigma)$, $H_B = H(\sigma + \delta\sigma)$, $\psi_A = \psi(\sigma)$, $\psi_B = \psi(\sigma + \delta\sigma)$, and therefore, that $E_A = E(\sigma)$ and $E_B = E(\sigma + \delta\sigma)$. Then writing

$$H_B - H_A = \frac{\partial H}{\partial \sigma}\delta\sigma + \cdots$$

$$\psi_B = \psi(\sigma) + \frac{\partial \psi}{\partial \sigma}\delta\sigma + \cdots \tag{1-56}$$

etc., dividing both sides of (1-55) by $\delta\sigma$, and taking the limit $\delta\sigma \to 0$ one is evidently led back to (1-2).

One charm of (1-55) is that it is a formula which directly involves an energy difference[39]—an energy difference due to a change in nuclear configuration, in nuclear charge or whatever. Note also that though in our discussion in the previous paragraph we assumed that ψ_B was the eigenfunction of H_B which correspond in some sense to ψ_A, no such requirement is implied in general.

Further (1-55), like (1-2), lends itself to pictures. For example, if H_A and H_B are Hamiltonians referring to different values of a uniform electric field then for a system of electrons, the scalar product on the right can be written as the electric dipole moment of a certain transition charge density. Similarly if H_A and H_B are electronic Hamiltonians which differ in nuclear configuration, then the right-hand side can be expressed as a nuclear difference potential plus a difference potential produced by the transition charge density. More precisely, the last statement has to be qualified by the remark that one is assumed to be using coordinates like those of (1-24) or (1-29). If one uses coordinates like those of (1-36) then $H_B - H_A$ will also involve kinetic energy terms.

Thus far we have been assuming that we have exact eigenfunctions. However, in practice one would like to know the status of (1-55) when one has only approximate wavefunctions. First of all, one can of course use the right-hand side of (1-55) to define an energy difference. A formal difficulty[40] here is that in general the error in the resultant energy difference will be of first order in the error in the wavefunctions.[41] Moreover there are often coordinate ambiguities. Thus, for example, the same physical change in the nuclear framework of a molecule can often be produced in several equally plausible ways, differing by an overall translation or rotation of the molecule. Such gross motions have no effect on exact energies; however, the energy differences calculated with approximate wavefunctions will usually depend on the particular path used to produce the change.[42] Nevertheless, some impressive successes have been recorded through the use of the integral H-F theorem to define an energy difference.[37] (See Chapter 4 for a survey of these results).

Also, when one uses approximate wavefunctions, one can ask whether the calculated energy difference will equal

$$\frac{\langle\psi_B|H_B|\psi_B\rangle}{\langle\psi_B|\psi_B\rangle} - \frac{\langle\psi_A|H_A|\psi_A\rangle}{\langle\psi_A|\psi_A\rangle}.$$

Since this difference is certainly of second order in the error in the wavefunctions, evidently from what we have just said this must in general not be true. The only known general exception[43] is when ψ_A and ψ_B are derived from linear variation calculations (one using H_A and the other H_B, of course) using the same basis set.[44]

1-11. MISCELLANEOUS TOPICS

1-11-1. A General Connection between the Hellmann-Feynman Theorem and Hypervirial Theorems

As we have noted, if a coordinate transformation can be implemented by a unitary transformation, then by comparing the H–F theorems for the same σ in the two coordinate systems one is led to the hypervirial theorem for the operator which generated the transformation. Here we wish to point out that this sort of discussion can be generalized so as to produce all hypervirial theorems from the H–F theorem. One approach is to note that, although in the discussion of Epstein (*16*) U was supposed to generate a coordinate transformation, formally the argument would be the same whatever U did. (Thus, for example, just as in Section 1-3, 1-6, and 1-9 we discussed the coordinate dependence of the H–F theorem, we could in a similar way discuss the gauge dependence of the theorem). Here however we will use a somewhat different approach (*64*).

Consider the family of Hamiltonians

$$H(\alpha) = e^{i\alpha G} H e^{-i\alpha G}, \tag{1-57}$$

where G is a given Hermitian operator, and where α is a real-number parameter not contained in H. Since $H(\alpha)$ is related to H by a unitary transformation, its eigenvalues are the same as those of H and hence independent of α. Therefore the H–F theorem for $\sigma = \alpha$ applied to $H(\alpha)$ (we assume we have an exact eigenfunction and eigenvalue) yields

$$0 = \left\langle \psi(\alpha) \left| \frac{\partial H(\alpha)}{\partial \alpha} \right| \psi(\alpha) \right\rangle \Big/ \langle \psi(\alpha) | \psi(\alpha) \rangle. \tag{1-58}$$

However from (1-57),

$$\frac{\partial H(\alpha)}{\partial \alpha} = i\,[G, H(\alpha)].$$

If now we insert this into (1-58) and set $\alpha = 0$, we evidently have the hypervirial theorem for the (arbitrary) Hermitian operator G and therefore, since the theorem is linear in G, for any operator G.

1-11-2. Accuracy of $\partial E/\partial \sigma$

Returning to variational calculations, it is well known that the error in E is of second order in the error in ψ. That is, if we now denote the exact eigenfunction and eigenvalue by ψ' and E', respectively, then writing

$$\psi = \psi' + \delta \tag{1-59}$$

we have that

$$E = E' + O(\delta^2). \tag{1-60}$$

Now quite often, either explicitly or implicitly, it is suggested that energy derivatives also automatically have a second-order error (this quite apart from whether or not the H–F theorem is satisfied).

However, we clearly have that

$$\frac{\partial E}{\partial \sigma} = \frac{\partial E'}{\partial \sigma} + O\left(\delta \frac{\partial \delta}{\partial \sigma}\right), \tag{1-61}$$

and therefore only if $\partial\delta/\partial\sigma$ is again of order δ do we have a second-order error. Thus, to give a trivial example, if σ is a perturbation parameter λ, and if δ is of order λ^{n+1}, then δ^2 is of order λ^{2n+2} while $\delta(\partial\delta/\partial\lambda)$ is of lower order, namely λ^{2n+1}.

To give more interesting examples, let us consider the situation at $\sigma = 0$. In particular, let us confine attention to $(\partial E/\partial\sigma)_{\sigma=0}$, though similar arguments can be given for the terms of higher order in σ, i.e., for higher derivatives of E evaluated at $\sigma = 0$. Then if we denote by γ the error in ψ at $\sigma = 0$ there are two different possibilities. The first is that if γ were equal to zero, i.e., if whatever it is in the Hamiltonian for $\sigma = 0$ which makes ψ inaccurate for $\sigma = 0$ were removed, then introducing the σ terms would still leave ψ exact. In this case clearly

$$\delta \sim \gamma(1 + A\sigma + \cdots), \tag{1-62}$$

where A is some number. Therefore

$$\left[\delta \frac{\partial \delta}{\partial \sigma}\right]_{\sigma=0} \sim \gamma^2 \tag{1-63}$$

and we have a second-order error.

The second possibility is that introducing the σ terms would make ψ inexact even if γ were equal to zero. Therefore in such a case

$$\delta \sim \gamma + A\sigma, \tag{1-64}$$

and hence

$$\left[\delta \frac{\partial \delta}{\partial \sigma}\right]_{\sigma=0} \sim \gamma \tag{1-65}$$

and we have a first-order error.

Each of these situations can be realized within the unrestricted Hartree-Fock approximation already referred to. For simplicity suppose that H involves σ linearly in a term σP. Then, since in this approximation all H–F theorems are satisfied, we have

$$\left(\frac{\partial E}{\partial \sigma}\right)_{\sigma=0} = \frac{\langle \psi(\sigma=0)|P|\psi(\sigma=0)\rangle}{\langle \psi(\sigma=0)|\psi(\sigma=0)\rangle}. \tag{1-66}$$

However, it is well known that only if P is a one-electron operator is the expectation value of P, and hence $(\partial E/\partial \sigma)_{\sigma=0}$, given correctly through first order; otherwise there is a first-order error. To connect up with our earlier remarks we need only note that if $H(\sigma=0)$ were replaced by a single-particle Hamiltonian such that $\psi(\sigma=0)$ would become exact (i.e., if γ were made equal to zero) then adding a single-particle operator would still leave unrestricted Hartree-Fock exact, while adding a multiparticle operator would not.[45]

1-11-3. The Integrated Hellmann–Feynman Theorem

An obvious difference between the integral H–F theorem (1-55) and the H–F theorem (1-2) is that the former can deal with finite energy differences, while the latter has to do with infinitesimal differences. However, one can of course compound infinitesimal differences to produce finite ones according to

$$E(\sigma) - E(\sigma_0) = \int_{\sigma_0}^{\sigma} \frac{\left\langle \psi(\sigma') \middle| \dfrac{\partial H(\sigma')}{\partial \sigma'} \middle| \psi(\sigma') \right\rangle}{\langle \psi(\sigma')|\psi(\sigma')\rangle}\, d\sigma', \tag{1-67}$$

a formula which has been dubbed (*23*) the integrated H–F theorem. For optimal variational wavefunctions which satisfy the H–F theorem, (1-67) will be satisfied with $E(\sigma)$ and $E(\sigma_0)$ the optimal variational energies $\langle\psi(\sigma)|H(\sigma)|\psi(\sigma)\rangle/\langle\psi(\sigma)|\psi(\sigma)\rangle$ and $\langle\psi(\sigma_0)|H(\sigma_0)|\psi(\sigma_0)\rangle/\langle\psi(\sigma_0)|\psi(\sigma_0)\rangle$, respectively. On the other hand, if one uses approximate wavefunctions which do not satisfy the H–F theorem, then one can use the right-hand side of (1-67) to define an energy difference, in general different from the difference $\langle\psi(\sigma)|H(\sigma)|\psi(\sigma)\rangle/\langle\psi(\sigma)|\psi(\sigma)\rangle - \langle\psi(\sigma_0)|H(\sigma_0)|\psi(\sigma_0)\rangle/\langle\psi(\sigma_0)|\psi(\sigma_0)\rangle$, and in general different from that provided by the integral H–F theorem. (Some simple examples are given in Ref. 23). However, note that to use the integrated formula one needs a wavefunction for a continuous range of σ values, whereas to use the integral formula or to subtract variational energies one needs only two wavefunctions, one for the "beginning" and one for the "end." Of course, if one is actually interested in a range of σ (e.g., a continuous range of nuclear geometries), then this distinction becomes irrelevant.

In our discussion of the integral H–F theorem we mentioned the possibility of coordinate dependence when one uses the theorem to define an energy difference. A similar situation can occur when one uses the integrated theorem in a similar way, if there is more than one parameter involved. When one has several parameters σ_i, and if (1-2) is satisfied for each parameter, then evidently, just as one can recover a potential energy from a field, one will have that the energy difference $E(\sigma) - E(\sigma_0)$ is given by the line integral

$$E(\sigma) - E(\sigma_0) = \int_{\sigma_0}^{\sigma} \sum_i \frac{\left\langle \psi(\sigma') \left| \frac{\partial H}{\partial \sigma_i'} \right| \psi(\sigma') \right\rangle}{\langle \psi(\sigma') | \psi(\sigma') \rangle} \, d\sigma_i', \tag{1-68}$$

where the exact path in σ-space is irrelevant since the integrand is an exact differential. However, the point we would make is that if one employs this formula to define an energy difference, and uses approximate functions which do not obey (1-2), then in general the resultant energy difference will depend on the integration path (see Chapter 4, Section 4-5).

One interesting theoretical application of (1-67) has been to give a simple formula for the total electronic plus nuclear energy of a molecule solely in terms of the electron charge density (*27, 99*).[46] The derivation is as follows. Replace the nuclear charges Z_a by the charges ZZ_a, where Z is to be thought of as a continuous parameter ranging from zero to one. Assuming that one has binding for all nonzero values of Z, and no binding for $Z = 0$, one then has from (1-67), with σ identified as Z, that

$$E = \int_0^1 \frac{\left\langle \psi(Z) \left| \frac{\partial H}{\partial Z} \right| \psi(Z) \right\rangle}{\langle \psi(Z) | \psi(Z) \rangle} \, dZ. \tag{1-69}$$

Then, using H as given in (1-24), one readily sees that E can be written as a sum of the nuclear potential energy plus a classical electron–nucleus potential-energy term, the charge density for the latter being the electronic charge density averaged over Z.[47]

Similarly we can, following Hurley (*45*), replace the nuclear coordinates $\mathbf{R}_a$ by the coordinates SR_a, where S is now to be thought of as a continuous parameter ranging from zero to infinity. Then the integrated H–F theorem for $\sigma = S$, with the limits of integration extended from 1 to ∞, will yield a formula for the energy difference between the molecule and its atomic dissociation fragments in terms of the quantity $-\Sigma_a(1/S)\,\mathbf{R}_a \cdot \langle \mathbf{F}_a \rangle$ averaged over S. Alternatively one may, after deleting the nuclear potential-energy term, integrate from 0 to 1 and derive an analogous formula for the electronic energy difference between molecule and united atom.

1-11-4. Evaluation of Integrals

If either theoretically or empirically one knows the energy as a function of certain parameters, then one can evidently use (1-2) to evaluate the integral on the right-hand side. Thus, to give a trivial example, referring to (1-15) if we know that $E = n + \frac{1}{2}$ then we can use (1-15) to evaluate the integral $\langle \psi | x^2 | \psi \rangle / \langle \psi | \psi \rangle$. Namely, we then have

$$\frac{\langle \psi | x^2 | \psi \rangle}{\langle \psi | \psi \rangle} = \frac{1}{\omega}\left(n + \frac{1}{2}\right). \tag{1-70}$$

As a more interesting example, suppose that σ is a perturbation parameter and that the perturbation is of the form λP. Then one readily sees from (1-2) that if one writes $\langle P \rangle = \Sigma \lambda^n p^{(n)}$, then

$$p^{(n-1)} = nE^{(n)}. \tag{1-71}$$

This technique coupled with the use of hypervirial theorems has proven quite useful in many interesting problems. However, since it is in principle quite straightforward, and also somewhat aside from our main interest, we content ourselves with giving some references together with some indication of what is done in each reference.

(i) Knowing the energy of a top as a function of moment of inertia parameters, one may use the theorem to determine the average values of the squares of the angular momentum components (*9*).

(ii) Various radial averages can be evaluated for the nonrelativistic hydrogen atom and harmonic oscillator, and for the Dirac hydrogen atom (*14*). (See also Refs. 101–103, 105, 106.)

(iii) Averages of x^n, and the $E^{(n)}$ are calculated for a general anharmonic oscillator (*8*, *88*, *90*). Killingbeck (*48*) does the same for a perturbed hydrogen atom.

In addition, similar techniques can be used to calculate off-diagonal matrix elements for, e.g., the nonrelativistic hydrogen atom and harmonic oscillator (*22*) and for the anharmonic oscillator (*91*, *92*).

1-11-5. Regional Theorems

Recently, mainly through the work of Bader,[48] interest has been aroused in the question as to whether or not theorems like (1-2) and the hypervirial theorem of note 11 hold if the integration is only over some finite region of configuration space. One can show (*18, 21*) that for H the electronic Hamiltonian the answer to this is yes provided that

$$\int_S [\nabla\psi^* w\psi - \psi^*\nabla(w\psi)] \cdot dS = 0, \tag{1-72}$$

where S is the surface of the region and ∇ is the gradient vector in configuration space, and where for the H–F theorem $w\psi = \partial\psi/\partial\sigma$ and for the hypervirial theorem $w\psi = G\psi$.

1-11-6. Time-Dependent Theorems

To this point we have been concerned only with stationary states. For completeness we now note that both the H–F theorem and the hypervirial theorem have time-dependent generalizations. However, since most of their applications to bound-state problems[49] have been in connection with the response of systems to periodic external fields (see, e.g., Appendix C of Ref. 18), a topic which is outside the scope of the present volume, we will content ourselves with simply quoting the generalizations and mentioning some applications, but will not go into details.

The time-dependent generalization of (1-2) was given only fairly recently (*37*), and is the statement that if

$$H\Psi = i\frac{\partial\Psi}{\partial t} \tag{1-73}$$

then[50]

$$i\frac{\partial}{\partial t}\left\langle\Psi\left|\frac{\partial\Psi}{\partial\sigma}\right.\right\rangle = \left\langle\Psi\left|\frac{\partial H}{\partial\sigma}\right|\Psi\right\rangle. \tag{1-74}$$

On the other hand, the generalization of the hypervirial theorem has been known and used since the inception of quantum mechanics. It is the familiar (generalized) Ehrenfest theorem[51]

$$\frac{d}{dt}\langle\Psi|G|\Psi\rangle = \left\langle\Psi\left|\frac{\partial G}{\partial t}\right|\Psi\right\rangle + i\langle\Psi|[H,G]|\Psi\rangle. \tag{1-75}$$

One application of (1-75) has been to derive (*24*) a relationship between the dipole shielding factor of an atom and its frequency-dependent dipole polarizability (see also Ref. 47). As an application of (1-74) we may mention that it can be used to show (see, e.g., Appendix C of Ref. 18) that in the steady state produced by a simple, harmonic, uniform electric field the time average of the time derivative of the phase of Ψ, a quantity for which there exists a variation principle (*38*), directly yields that part of the average electric dipole moment which is in phase with the electric field.

1-12. SUMMARY

In accord with our purpose as announced in Section 1-1, we have attempted to provide a soundly based general theoretical understanding of the H–F theorem and various of its near neighbors. Among other things, we have pointed out that the theorem often leads to very classical and intuitive pictures and in this connection there is a potentially important point. We have given conditions under which optimal variational wavefunctions satisfy the various theorems, and have by implication taken this to be a virtue of such a ψ. However, one might argue that, for example, with sufficient intuition one should be able to choose ψ so that

$$\frac{\langle\psi|\partial H/\partial\sigma|\psi\rangle}{\langle\psi|\psi\rangle}$$

is a better approximation to the exact $\partial E/\partial\sigma$ than is

$$\frac{\dfrac{\partial}{\partial\sigma}\langle\psi|H|\psi\rangle}{\langle\psi|\psi\rangle}.$$

However, although there have been isolated successes of this type, thus far there is no general theoretical understanding of the situation.

Notes

1. There are similar theorems in classical and semiclassical mechanics, and a related theorem for continuum eigenfunctions (see note 4). A similar theorem holds in Thomas–Fermi theory (*32*) and in the $X\alpha$ method (*20, 83*). Also, there are related theorems in statistical mechanics (see, e.g., Ref. 51). Finally, there is of course an analogous theorem for Hermitian matrices (see, e.g., Ref. 58): If $H\phi = ES\phi$, where H and S are Hermitian matrices, E is an eigenvalue, and ϕ is an eigenvector, then the theorem states that

$$\frac{\partial E}{\partial \sigma} = \left[\phi^+ \cdot \left(\frac{\partial H}{\partial \sigma} - E\frac{\partial S}{\partial \sigma}\right)\phi\right] \Big/ [\phi^+ \cdot \phi].$$

(For some applications to Hückel theory see Salem (*79*), Section 1-8 and references given therein. For some applications in molecular spectroscopy see Rowe and Wilson (*75*) and references given therein). Further generalizations can be found in Refs. 58, 101, 104, and 107) and in the references mentioned in note 9 below.

2. For some remarks on the history of this theorem, see Refs. 65 and 84. Sometimes the theorem is referred to as the generalized Hellmann–Feynman theorem, or simply Feynman's theorem.

3. Note that it is $\delta\sigma$ not σ which is to be identified as the perturbation parameter. Also, (1-4) holds whether or not E is degenerate provided that if E is degenerate then ψ is (in the sense of degenerate perturbation theory) a correct zero-order wavefunction at any particular value of σ. This requirement will also be implicit in the second and third derivations below, since we will there assume that ψ is a smooth differentiable function of σ.

4. In classical mechanics and in semiclassical mechanics (i.e., Bohr–Sommerfeld quantum mechanics) first-order perturbation theory yields the result that

$$\frac{\partial E}{\partial \sigma} = \overline{\frac{\partial H}{\partial \sigma}},$$

where the bar denotes time average, and where the differentiation on the left is with appropriate action variables (quantum numbers) held constant. For some applications of this theorem see Refs. 14 and 61.

When E is in the continuum then the first Born approximation is the analog of first-order perturbation theory. Correspondingly, the continuum version of the H–F theorem is a formula for the derivative of the phase shift (or related quantity) with respect to some parameter in the potential, in terms of the derivative of the potential (*4*, *54*, *62*, *87*).

5. At this point Hellmann (*39*) instead appealed to the variation principle. See notes 27 and 32.

6. If we differentiate (1-5) with respect to E, then we find

$$\left\langle \frac{\partial \psi}{\partial E} | H - E | \psi \right\rangle - \langle \psi | \psi \rangle + \left\langle \psi | H - E | \frac{\partial \psi}{\partial E} \right\rangle = 0, \qquad (*)$$

which upon use of (1-7) leads to the nonsensical result $\langle \psi | \psi \rangle = 0$. Actually, the flaw in this argument is concealed by our use of bracket notation, which is explicitly adapted to Hermitian operators. The point is, however, that $\partial\psi/\partial E$ is a very unnormalizable function (E is a discrete eigenvalue), and therefore one cannot successfully carry out the integration by parts needed to conclude that $\langle \psi | H - E | \partial\psi/\partial E \rangle = 0$. However, if one notes that (1-5) and hence (*) both hold even if one limits the implied integration to a finite region of configuration space, one can use (*) to yield a useful formula for $\langle\psi|\psi\rangle$ in terms of the asymptotic behavior of ψ and $\partial\psi/\partial E$ (*50*). Also, the continuum analog of this Kramers' theorem can be shown to yield Wigner's (*98*) theorem on the energy derivative of a phase shift.

7. Seemingly, general awareness of formula (1-2) came first through this rather formal derivation. Indeed, there was even some initial disbelief in the result. The essential point was the following. With $H = T + V$, where T is the kinetic energy operator and V the potential energy operator, suppose that σ occurs only in V. Then we have from (1-2) that

$$\frac{\partial E}{\partial \sigma}\delta\sigma = \left\langle \psi \left| \frac{\partial V}{\partial \sigma}\delta\sigma \right| \psi \right\rangle \Big/ \langle\psi|\psi\rangle, \qquad (**)$$

which was read to say that the first-order change in the total energy was equal to the first-order change in the potential energy alone, an obviously incorrect result, since usually the kinetic energy would change as well. Therefore, it was argued that there must be something wrong with the apparently straightforward differentiation procedure which we carried out.

However, the flaw of course is not with (**) but (7) with the interpretation of the right-hand side of (**) as the first-order change in potential energy. The latter is actually given by

$$\left(\frac{\partial}{\partial\sigma}\frac{\langle\psi|V|\psi\rangle}{\langle\psi|\psi\rangle}\right)\delta\sigma = \left\langle \psi \left| \frac{\partial V}{\partial \sigma}\delta\sigma \right| \psi \right\rangle + \frac{\left\langle \frac{\partial\psi}{\partial\sigma} | V - \bar{V} | \psi \right\rangle + \left\langle \psi | V - \bar{V} | \frac{\partial\psi}{\partial\sigma} \right\rangle}{\langle\psi|\psi\rangle}\delta\sigma,$$

where

$$\bar{V} \equiv \frac{\langle\psi|V|\psi\rangle}{\langle\psi|\psi\rangle},$$

and hence differs from the right-hand side of (**) by the second term. Further, as one can readily verify, this difference is numerically equal to the negative of $(\partial/\partial\sigma\langle\psi|T|\psi\rangle/\langle\psi|\psi\rangle)\,\delta\sigma$, the first-order change in kinetic energy.

8. One can, of course, add a multiple (say, α) of ψ to the right-hand side of (1-12). Our choice of $\alpha = 0$ simply serves to fix phases and normalization (see, e.g., Ref. 15).

9. There are a number of papers on this general subject (*1*, *10*, *15*, *52*, *54*, *64*, *77*). In several of these papers the formulas are viewed as providing "sum rules," i.e., they express certain sums in closed form as energy derivatives.

10. In our differentiation in proof 2 we evidently ignored the possibility that the volume element was σ-dependent. However, it is clear from proof 1, where the problem never arose, that this can have no effect on the theorem, and indeed one notes that any such term would have $(H - E)|\psi\rangle$ as a factor and hence would yield zero contribution. The situation is not quite so straightfor-

ward in the case of approximate wavefunctions (*17*), so for simplicity we will assume in our discussion in Section 1-8 that if the Jacobian of transformation from Cartesian coordinates does involve σ, it either does so in a multiplicative way, in which case this dependence cancels out, or if it involves σ in a more complicated way, that it has been absorbed into the definitions of ψ and H (as is automatically the case if, as below, the transformation can be viewed as a unitary one).

11. If $H\psi = E\psi$ then evidently $\langle\psi|[H, A]|\psi\rangle = 0$ where $[H, A] \equiv HA - AH$ is the commutator of H and A. This (*40*) is called the hypervirial theorem for the operator A.

12. In all of our examples U will be the form $e^{i\alpha G}$, where α is a σ-dependent parameter and where the Hermitian operator G which "generates" the transformation is independent of α. Thus in all our examples $U^{-1}(\partial U/\partial\sigma) = i(\partial\alpha/\partial\sigma)\,G$, and hence the hypervirial theorem for $U^{-1}(\partial U/\partial\sigma)$ is simply the hypervirial theorem for G.

13. See, e.g., Section 20 of Ref. 18.

14. See, e.g., Section 18 of Ref. 18.

15. McWeeny (*63*, Sections 5-6, 5-7) has recently given a collection of examples. Other examples will be mentioned in Section 1-11 below, in the discussion of evaluation of integrals.

16. This mixed second derivative is in general well defined only for a nondegenerate state (see, e.g., Section IV-E of Ref. 41.)

17. See, e.g., Section 18 of Ref. 18. The direct interpretation of the hypervirial theorems is that the total average force on the electrons vanishes. However, since the electron–electron contributions cancel anyway, it follows that we have 0 = average force on electrons due to nuclei = −average force on nuclei due to electrons. Then since the nucleus–nucleus contributions cancel, this in turn = −average force on nuclei, and we have (1-32). For the force theorems in the presence of external fields see, e.g., Section 18, Appendix D of Ref. 18 and references given therein.

18. See, e.g., Section 19 of Ref. 18.

19. Since the left-hand side of (1-31) must be independent of α, i.e., the choice of coordinates, the right-hand side must also be independent of α, which yields the z-component of (1-32), $\langle F_{1z}\rangle + \langle F_{2z}\rangle = 0$. This is again in accord with Section 1-4, since changing the value of α is equivalent to a translation in the z-direction.

20. Since (1-37) is already the virial theorem this may not be obvious. However, the hypervirial theorem for $\Sigma_s(r_s \cdot p_s + p_s \cdot r_s)$ is *not* (1-37); rather, it is (*53*)

$$2\langle T\rangle + \langle V\rangle + R\left\langle\frac{\partial H}{\partial R}\right\rangle = 0,$$

which in turn is precisely the difference between (1-34) and (1-37), since the former can, of course, also be written as $\partial E/\partial R = \langle\partial H/\partial R\rangle$.

21. As another example, Benston (*6*) considers just a scaling of the z-coordinate. This then yields a theorem differing from (1-34) by the hypervirial theorem for $\Sigma_s(z_s P_{zs} + P_{zs} z_s)$.

22. See, e.g., Refs. 3, 42, 80, 85, and 100.

23. In doing calculations of long-range forces (see Chapter 7), one usually uses a coordinate system in which the electrons of each atom are referred to the nucleus of that atom. In such coordinates, R appears only in the perturbation, i.e., $\partial H/\partial\sigma$ is of order λ, and hence, as has been frequently pointed out, in such coordinates one can calculate the force through the second order from the H–F theorem for $\sigma = R$ using only a first-order wavefunction. However, this theorem is not at all the electrostatic theorem, although of course $-\partial E/\partial R_1$ and $-\partial E/\partial R_2$ are still numerically equal to the H–F forces. For further discussion of these points, see Ref. 49.

24. It is easy to see that the coefficient of β will be

$$-\frac{\left\{\left\langle\psi^{(0)}\left|\frac{\partial H^{(0)}}{\partial\sigma}-\frac{\partial E^{(0)}}{\partial\sigma}\right|\psi^{(1)'}\right\rangle+\left\langle\psi^{(1)'}\left|\frac{\partial H^{(0)}}{\partial\sigma}-\frac{\partial E^{(0)}}{\partial\sigma}\right|\psi^{(0)}\right\rangle\right\}}{\langle\psi^{(0)}|\psi^{(0)}\rangle}. \qquad (\dagger)$$

This can then be written in a simpler form as follows. If one expands (1-2) in powers of λ, one finds through the first order that

$$\left\langle\psi^{(0)}\left|\frac{\partial H^{(0)}}{\partial\sigma}-\frac{\partial E^{(0)}}{\partial\sigma}\right|\psi^{(0)}\right\rangle = 0$$

and that

$$\left\langle\psi^{(0)}\left|\frac{\partial H^{(0)}}{\partial\sigma}-\frac{\partial E^{(0)}}{\partial\sigma}\right|\psi^{(1)'}\right\rangle+\left\langle\psi^{(1)'}\left|\frac{\partial H^{(0)}}{\partial\sigma}-\frac{\partial E^{(0)}}{\partial\sigma}\right|\psi^{(0)}\right\rangle + \left\langle\psi^{(0)}\left|\frac{\partial H^{(1)}}{\partial\sigma}-\frac{\partial E^{(1)}}{\partial\sigma}\right|\psi^{(0)}\right\rangle = 0.$$

The first of these is, of course, the H–F theorem for $H^{(0)}$, and indeed we used it already in writing (†). The second when used in (†) yields the desired formula

$$\left\langle\psi^{(0)}\left|\frac{\partial H^{(1)}}{\partial\sigma}-\frac{\partial E^{(1)}}{\partial\sigma}\right|\psi^{(0)}\right\rangle. \qquad (\dagger\dagger)$$

In particular, for the long-range force problem a short calculation shows that (††), though exponentially small, is nevertheless nonzero as desired.

25. The argument in the appendix of Yaris (*100*) would suggest that for what he calls factorable perturbations this should be true without special precautions.

(More accurately, since in such a case $\partial H/\partial\sigma$ is of order λ, one should always find $s = 2n + 2$ rather than $n + 2$). However, it would appear that he has drawn much too strong a conclusion from his equation (A4).

26. Note that, as implied by our wording, one cannot expect to do better than to give a sufficient condition. Thus, for example, the set of trial functions might consist of just a single function with no free parameters, and yet if that function is an eigenfunction, all theorems will be satisfied. However, in practice the conditions we give also seem to be necessary.

27. Since if the set of trial functions is completely unrestricted the optimal functions will be eigenfunctions, and since such a set is obviously independent of σ, Hurley's condition provides another proof (recall note 5) that for eigenfunctions (1-2) will be satisfied for any σ.

28. As a trivial example, if the set consists of a single σ-independent function then obviously (1-47) will be satisfied identically, since $\partial\psi/\partial\sigma \equiv 0$, and hence we will have (1-2).

29. Since the normalization of ψ is irrelevant to (1-46) and (1-2), one can relax this requirement and also relax our statement of Hurley's condition, and say that changing σ should change one trial function into another to within a numerical multiplicative factor. A trivial example of a set in which a change in σ changes one trial function into another is provided by the set

$$e^{-\gamma\sigma r}$$

where γ is a variational parameter and r is a coordinate. Namely, a change in σ is evidently equivalent to a change in γ and hence, as claimed, does send one member of the set into another.

30. Subsequent to Hurley's work there have been many detailed proofs that certain H–F theorems hold within certain variational approximations. Often such derivations (*11*, *82*) invoke Brillouin's Theorem, but since that theorem is essentially just a restatement of the basic equation (1-48) of the variation method (see, e.g., Section 10 of Ref. 18) it will be appreciated, especially upon comparison with the proof in note 32 below, that these authors have in effect been repeating Hurley's argument in each special case, and often in an indirect way. Also, such more involved considerations can lead to error (*33*; *44*, note 11). Further, the claim (*30*) that GFR does not satisfy the H–F theorem should evidently be qualified by the proviso (as in (iii) below) that this is the case only if the χ_μ depend explicitly on λ; otherwise the theorem *is* satisfied.

31. A considerable bibliography on the use of gauge-invariant atomic orbitals (GIAO) can be found in Ref. 13.

32. If there is no explicit σ-dependence a much simpler proof is possible (*46*). Namely, since now

$$\frac{\partial\psi}{\partial\sigma}\delta\sigma = \sum_i \frac{\partial\psi}{\partial\hat{A}_i}\frac{\partial\hat{A}_i}{\partial\sigma}\delta\sigma,$$

it immediately follows that $\partial\psi/\partial\sigma$ has the form

$$\sum_i \frac{\partial\psi}{\partial\hat{A}_i}\,\delta\hat{A}_i.$$

That is, it has the form of a $\delta\psi$, and therefore it must satisfy (1-48). However, if we replace $\delta\psi$ in (1-48) by $(\partial\psi/\partial\sigma)\delta\sigma$ and cancel the factor $\delta\sigma$ we have (1-47) again, which proves the point. Note, referring back to note 6, that a similar "proof" that $\langle\psi|\psi\rangle = 0$ fails in the variation case because the $\partial\hat{A}_i/\partial E$ are infinite.

Note also that if one has a set in which the individual trial functions depend on σ but the set itself does not, then one can always replace it by an equivalent set in which the individual trial functions do *not* depend on σ. Sometimes one can do this by redefining the variational parameters (thus, referring to note 29, one can use the equivalent set $e^{-\beta r}$), but in any case one can do it just by putting σ equal to some specific value (thus, again referring to note 29, one can use for example the equivalent set $e^{-3\gamma r}$).

33. If there *is* explicit σ-dependence then even if the set is not invariant to changes in σ, still (see note 32) $\Sigma(\partial\psi/\partial\hat{A}_i)(\partial\hat{A}_i/\partial\sigma)\delta\sigma$ will satisfy (1-49). Therefore in such a case we have that

$$\frac{\partial E}{\partial\sigma} = \frac{\left\langle\psi\left|\frac{\partial H}{\partial\sigma}\right|\psi\right\rangle}{\langle\psi|\psi\rangle} + \frac{\left\langle\frac{\eth\psi}{\partial\sigma}|H - E|\psi\right\rangle + \left\langle\psi|H - E|\frac{\eth\psi}{\partial\sigma}\right\rangle}{\langle\psi|\psi\rangle}, \tag{§}$$

where $\eth/\partial\sigma$ means that one need differentiate only the explicit σ-dependence (see, e.g., Section 2 of Ref. 71). In deriving (§) we have, of course, again assumed the A_i to be independent. If there are constraints but they have no physical consequences (like the orthonormality constraints of unrestricted SCF) then one can show that (§) remains valid. More generally, however, it is easy to see (*31*, *72*) that (§) is replaced by

$$\frac{\partial E}{\partial\sigma} = \frac{\left\langle\psi\left|\frac{\partial H}{\partial\sigma}\right|\psi\right\rangle}{\langle\psi|\psi\rangle} + \frac{\eth E}{\partial\sigma} + \sum_\sigma l_\alpha \frac{\eth C_\alpha}{\partial\sigma},$$

where $C_\alpha = 0$ are the constraints and the l_α are the Lagrange multipliers.

34. See Section 16 of Ref. 18.

35. The operator $\exp\,(i\alpha\Sigma_{s=1}^N g_s)$ applied to the Slater determinant $|\psi_1, \psi_2, \ldots, \psi_N|$ yields $|e^{i\alpha g}\psi_1, e^{i\alpha g}\psi_2, \ldots, e^{i\alpha g}\psi_N|$, i.e., another Slater determinant.

36. Note that, since the hypervirial theorem is linear in G, if it is satisfied for all Hermitian one-electron G's it is satisfied for all one-electron G's.

37. For an extensive bibliography see Refs. 70 and 81.

38. If ψ_B represents not a single function but a complete set of functions, then writing (1-55) as $\langle\psi_B|H_A|\psi_A\rangle = E_A\langle\psi_B|\psi_A\rangle$ we see that it is the Schrödinger equation for $|\psi_A\rangle$ in the B representation.

39. At the expense of introducing some auxiliary functions, Marron and Weare (*60*) have produced a formula for $E_B - E_A$ which is stationary around the result (1-55).

40. For some practical aspects, see Section IX of Ref. 23.

41. Even if the variational functional mentioned in note 38 is used the error, unless special precautions are taken, is not truly second order but rather is bilinear. For some calculations of this sort, see Ref. 93.

42. Lowe and Mazziotti (*59*, especially Section IV.b) give suggestions concerning a best choice of path (see Chapter 4).

43. See Section IV of Ref. 23.

44. If one has a pair ψ_A and ψ_B which do not have this property, then by using them as basis functions in two further linear variation calculations, one can evidently produce a new pair of functions with the desired properties.

45. Some time ago Hall (*36*) introduced the notion of stable wavefunctions. These are wavefunctions ψ such that if a perturbation λP were added to the Hamiltonian, they would change with λ in such a way that the energy would change in accord with the H–F theorem, i.e.,

$$\frac{\partial E}{\partial \lambda} = \frac{\langle\psi(\lambda)|P|\psi(\lambda)\rangle}{\langle\psi(\lambda)|\psi(\lambda)\rangle}. \qquad (\S\S)$$

(In some of the general equations of Hall's paper, e.g., in his equation (2.06), it appears that λ has been set equal to zero. However, in other places he seems to allow λ to be arbitrary, as we do.) We would like to make two comments on this concept. First, from our discussion in Section 1-8 of the various sorts of Hartree–Fock wavefunctions it follows that any of these is stable to *any* sort of perturbation, and not just to one-electron perturbations as is sometimes implied. Second, it should be noted that in and of itself the property of being stable is no guarantee of goodness. Thus, to give an extreme example, if we choose $\psi(\lambda) = \psi(0)$ then (see note 28) whatever the merits of $\psi(0)$, (§§) will be trivially satisfied for any P.

46. See Ref. 26 for an application to atoms, Ref. 34 for an application to molecules and Refs. 29 and 69 for related material.

47. We could, of course, let all charges (and in the example below all nuclear coordinates) vary independently and using (1-68) derive a formula involving a line integral in a multidimensional space. However, as noted above, for a wavefunction satisfying the requisite H–F theorems one path is as good as another, while for approximate wavefunctions not satisfying the theorem, certainly a one-parameter path is simplest.

48. See, e.g., Ref. 5.

49. For applications of (1-75) below to continuum problems see Refs. 55–57.

50. To justify our claim that (1-74) is a generalization of (1-2), let us verify that it reduces to (1-2) when $\Psi = \psi e^{-iEt}$. First we note that then $\langle\Psi|\partial\Psi/\partial\sigma\rangle =$

$\langle\psi|\partial\psi/\partial\sigma\rangle - i(\partial E/\partial\sigma)\, t\langle\psi|\psi\rangle$. Therefore,

$$i\frac{\partial}{\partial t}\left\langle\Psi\left|\frac{\partial\Psi}{\partial\sigma}\right.\right\rangle = \frac{\partial E}{\partial\sigma}\langle\psi|\psi\rangle,$$

and since $\langle\Psi|[\partial H/\partial\sigma]|\Psi\rangle = \langle\psi|[\partial H/\partial\sigma]|\psi\rangle$ the point is proven.

51. If $\Psi = \psi e^{-iEt}$ then $\langle\Psi|G|\Psi\rangle = \langle\psi|G|\psi\rangle$ and therefore $d\langle\Psi|G|\Psi\rangle/dt = \langle\psi|\partial G|\partial t|\psi\rangle$. Then, since further $\langle\Psi|[H, G]|\Psi\rangle = \langle\psi|[H, G]|\psi\rangle$ we evidently have that in a stationary state (1-75) becomes $0 = \langle\psi|[H, G]|\psi\rangle$, i.e., the hypervirial theorem for G.

References

1. Aizu, K., 1963, *J. Math. Phys.*, **4**, 762.
2. Argyres, P. N., Kaplan, T. A., and Silva, N. P., 1974, *Phys. Rev.*, **A9**, 1716.
3. Van der Avoird, A. and Wormer, C. E. S., 1977, *Mol. Phys.*, **33**, 1367.
4. Azuma, S., 1961, *Prog. Theoret. Phys.* (Japan), **26**, 861.
5. Bader, R. F. W. and Runtz, G. R., 1975, *Mol. Phys.*, **30**, 117.
6. Benston, M. L., 1966, *J. Chem. Phys.*, **44**, 1300.
7. Berlin, T., 1951, *J. Chem. Phys.*, **19**, 208.
8. Bonham, R. A. and Lu, S., 1966, *J. Chem. Phys.*, **45**, 2827.
9. Bragg, J. K. and Golden, S., 1949, *Phys. Rev.*, **75**, 735.
10. Brown, W. B., 1958, *Proc. Cambridge Phil. Soc.*, **54**, 251.
11. Coulson, C. A., 1971, *Mol. Phys.*, **20**, 687.
12. —— and Hurley, A. C., 1962, *J. Chem. Phys.*, **37**, 448.
13. Ditchfield, R., 1972, *J. Chem. Phys.*, **56**, 5688.
14. Epstein, J. H. and Epstein, S. T., 1962, *Am. J. Phys.*, **30**, 266.
15. Epstein, S. T., 1954, *Am. J. Phys.*, **22**, 613.
16. ——, 1965, *J. Chem. Phys.*, **42**, 3813.
17. ——, 1967, *J. Chem. Phys.*, **46**, 571.
18. ——, 1974, *The Variation Method in Quantum Chemistry* (Academic Press, New York).
19. ——, 1974, *J. Chem. Phys.*, **60**, 147.
20. ——, 1974, *J. Chem. Phys.*, **60**, 3328.
21. ——, 1974, *J. Chem. Phys.*, **60**, 3351.
22. ——, Epstein, J. H., and Kennedy, B., 1967, *J. Math. Phys.*, **8**, 1747.
23. ——, Hurley, A. C., Wyatt, R. E., and Parr, R. G., 1967, *J. Chem. Phys.*, **47**, 1275.
24. —— and Johnson, R. E., 1969, *J. Chem. Phys.*, **51**, 188.
25. Feynman, R. P., 1939, *Phys. Rev.*, **56**, 340.
26. Foldy, L. L., 1951, *Phys. Rev.*, **83**, 397.
27. Frost, A. A., 1962, *J. Chem. Phys.*, **39**, 1147.
28. —— and Lykos, P. G., 1956, *J. Chem. Phys.*, **25**, 1299.
29. Garcia-Sucre, M., 1976, *J. Chem. Phys.*, **65**, 280.
30. Goddard, W. A., III, 1968, *J. Chem. Phys.*, **48**, 5337.

31. Golebiewski, A., 1976, *Mol. Phys.*, **32**, 1529.
32. Goodisman, J., 1970, *Phys. Rev.*, **A2**, 1.
33. ——, 1973, *Diatomic Interaction Potential Theory 2* (Academic Press, New York), p. 231.
34. Goscinski, O. and Siegbahn, H., 1977, *Chem. Phys. Lett.*, **48**, 568.
35. Gutlyanskii, E. D. and Khartsiev, V. E., 1976, *Sov. Phys.–J.E.T.P.*, **44**, 248.
36. Hall, G. G., 1961, *Phil. Mag.*, **6**, 249.
37. Hayes, E. F. and Parr, R. G., 1965, *J. Chem. Phys.*, **43**, 1831.
38. Heinrichs, J., 1968, Phys. Rev., **172**, 1315; and **176**, 2167.
39. Hellmann, H., 1937, *Einführung in die Quantenchemie* (Franz Deuticke, Leipzig and Vienna), Section 54.
40. Hirschfelder, J. O., 1960, *J. Chem. Phys.*, **33**, 1462.
41. ——, Brown, W. B., and Epstein, S. T., 1964, in *Advances in Quantum Chemistry*, vol. 1, ed. P.-O. Löwdin (Academic Press, New York).
42. —— and Eliason, M. A., 1967, *J. Chem. Phys.*, **47**, 1164.
43. Huber, A., 1970, in *Methods and Problems of Theoretical Physics*, ed. J. E. Bowcock (North-Holland, Amsterdam).
44. Huo, W., 1968, *J. Chem. Phys.*, **49**, 1482.
45. Hurley, A. C., 1954, *Proc. Roy. Soc.* (London), **A226**, 170.
46. ——, 1954, *Proc. Roy. Soc.* (London), **A226**, 179.
47. Kaveeshwar, V. G., Dalgarno, A., and Hurst, R. P., 1969, *Proc. Phys. Soc.* (London), 2, 984.
48. Killingbeck, J., 1978, *Phys. Lett.*, **65A**, 87.
49. Koga, T. and Nakatsuji, H., 1976, *Theoret. Chim. Acta* (Berlin), **41**, 119.
50. Kramers, H. A., 1957, *Quantum Mechanics* (North-Holland, Amsterdam), p. 306.
51. Landau, L. D. and Lifshitz, E. M., 1969, *Statistical Physics*, 2nd rev. ed. (Addison-Wesley, Reading, Mass.), p. 105.
52. Landsberg, P. T. and Morgan, D. J., 1966, *J. Math. Phys.*, **7**, 2271.
53. Laurenzi, B. J. and Fitts, D. D., 1965, *J. Chem. Phys.*, **43**, 317.
54. Levine, R. D., 1966, *Proc. Roy. Soc.* (London), **A294**, 467.
55. ——, 1969, *Quantum Mechanics of Molecular Rate Processes* (Oxford University Press, London), Section 2-4.
56. Lippmann, B. A., 1965, *Phys. Rev. Lett.*, **15**, 11.
57. ——, 1966, *Phys. Rev. Lett.*, **16**, 135.
58. Löwdin, P.-O., 1959, *J. Mol. Spec.*, **3**, 46.
59. Lowe, J. P. and Mazziotti, A., 1968, *J. Chem. Phys.*, **48**, 877.
60. Marron, M. T. and Weare, J. H., 1968, *Int. J. Quantum Chem.*, **2**, 729.
61. McKinley, W. A., 1971, *Am. J. Phys.*, **39**, 905.
62. —— and Macek, J. H., 1964, *Phys. Lett.*, **10**, 210.
63. McWeeny, R., 1964, in *Orbital Theories of Molecules and Solids*, ed. N. H. March (Clarendon Press, Oxford).
64. Morgan, D. J. and Landsberg, P. T., 1965, *Proc. Phys. Soc.* (London), **86**, 261.
65. Musher, J. I., 1966, *Am. J. Phys.*, **34**, 267.
66. Parr, R. G., 1964, *J. Chem. Phys.*, **40**, 3726.

67. Pauli, W., Jr., 1933, in *Handbuch der Physik* (Springer-Verlag, Berlin), p. 511.
68. Phillipson, P., 1963, *J. Chem. Phys.*, **39**, 3010.
69. Politzer, P., 1976, *J. Chem. Phys.*, **64**, 4239.
70. —— and Daiker, K. C., 1978, *J. Chem. Phys.*, **68**, 5289.
71. Pulay, P., 1969, *Mol. Phys.*, **17**, 197.
72. ——, 1977, in *Applications of Electronic Structure Theory*, ed. H. F. Schaefer III (Plenum Press, New York).
73. Ramsey, N. F., 1952, *Phys. Rev.*, **86**, 243.
74. Richardson, J. W. and Pack, A. K., 1964, *J. Chem. Phys.*, **41**, 897.
75. Rowe, W. F. and Wilson, E. B., 1975, *J. Mol. Spec.*, **56**, 163.
76. Sadlej, A., 1977, *Chem. Phys. Lett.*, **47**, 50.
77. Salem, L., 1962, *Phys. Rev.*, **125**, 1788.
78. ——, 1963, *J. Chem. Phys.*, **38**, 1227.
79. ——, 1966, *The Molecular Orbital Theory of Conjugated Systems* (W. A. Benjamin, Reading, Mass.)
80. —— and Wilson, E. B., Jr., 1962, *J. Chem. Phys.*, **36**, 3421.
81. Simons, G., 1975, *J. Chem. Phys.*, **63**, 2206.
82. Stanton, R. E., 1962, *J. Chem. Phys.*, **36**, 1298.
83. Slater, J. C., 1972, *J. Chem. Phys.*, **57**, 2389.
84. ——, 1975, *Solid-State and Molecular Theory: A Scientific Biography* (John Wiley & Sons, New York).
85. Steiner, E., 1973, *J. Chem. Phys.*, **59**, 2427.
86. Stevens, R. M., Pitzer, R. M., and Lipscomb, W. N., 1963, *J. Chem. Phys.*, **38**, 550.
87. Sugar, R. and Blanckenbecler, R., 1964, *Phys. Rev.*, **136**, 8472.
88. Swenson, R. J. and Danforth, S. H., 1972, *J. Chem. Phys.*, **57**, 1734.
89. Thomchick, J., Lemmens L. F., and Devreese J. T., 1976, *Phys. Rev.*, **B14**, 1777.
90. Tipping, R. H., 1973, *J. Chem. Phys.*, **59**, 6433.
91. ——, 1973, *J. Chem. Phys.*, **59**, 6443.
92. ——, 1976, *J. Mol. Spec.*, **59**, 8.
93. Trindle, C. and George, J. K., 1975, *Theoret. Chim. Acta.* (Berlin), **40**, 119.
94. Van Vleck, J. H., 1928, *Phys. Rev.*, **31**, 587.
95. ——, 1932, *The Theory of Electric and Magnetic Susceptibilities* (Oxford University Press, London), Section 36.
96. Wagner, F. and Koppe, H., 1965, *Z. Naturforsch.*, **A20**, 1553.
97. Werner, H. J. and Meyer, W., 1976, *Mol. Phys.*, **31**, 855.
98. Wigner, E. P., 1955, *Phys. Rev.*, **98**, 145.
99. Wilson, E. B., Jr., 1962, *J. Chem. Phys.*, **36**, 2232.
100. Yaris, R., 1963, *J. Chem. Phys.*, **39**, 863.

Supplementary References

101. Banerjee, K., 1978, *Proc. Roy. Soc.* (London), **A363**, 147.
101a. Brown, C. W., 1980, *Phys. Rev. Lett.*, **44**, 1054.

102. Laurenzi, B. J., 1969, *Theoret. Chim. Acta* (Berlin), **13**, 106.
103. —— and Saturno, A. F., 1970, *J. Chem. Phys.*, **53**, 579.
104. Nakatsuji, H., 1977, *J. Chem. Phys.*, **67**, 1312.
105. Politzer, P. and Daiker, K. C., 1978, *Int. J. Quantum Chem.*, **14**, 245.
106. Scherr, C. W. and Knight, R. E., 1963, *Rev. Mod. Phys.*, **35**, 436.
107. Tachibana, A., Yamashita, K., Yamabe, T., and Fukui, K., 1978, *Chem. Phys. Lett.*, **59**, 249.

2
The Nature of Chemical Binding

R. F. W. Bader
Department of Chemistry, McMaster University
Hamilton, Ontario, Canada

Contents

2-1. INTRODUCTION

"The force on any nucleus (considered fixed) in any system of nuclei and electrons is just the classical electrostatic attraction exerted on the nucleus in question by the other nuclei and by the electron charge density for all electrons." The quotation is from Feynman (*49*), from a paper which introduced the Hellmann-Feynman theorem in a way which illustrated and emphasized the potential use of the theorem in providing a classical electrostatic basis for the discussion of chemical problems. In particular, his further statement that "It now becomes quite clear why the strongest and most important attractive forces arise when there is a concentration of [electronic] charge between two nuclei" is the electrostatic explanation of the origin of binding between atoms. What remained for further workers was to exemplify and categorize molecular charge distributions by their ability and manner of concentrating electronic charge in internuclear regions. Such studies constitute a portion of this chapter. Thus, chemical *binding*, as interpreted in terms of the forces exerted on the nuclei, is well documented and understood. Chemical *bonding* remains less well understood.

Berlin (*31*) was the first to distinguish between the use of the words *binding* and *bonding* in reference to the formation of a molecule, the former relating to the forces acting on the nuclei in such a process and the latter to the corresponding changes in energy. A discussion of chemical binding via the electrostatic approach presupposes that one has a knowledge of the electronic charge density for the system under study. The manner in which the electronic charge is distributed throughout space is determined by quantum mechanics. A theory of chemical binding can thus never be a complete theory in the predictive sense. Given a charge distribution, the electrostatic theorem allows for a rationalization, in terms of a well-defined physical model, of the binding or lack of it predicted by the distribution. The theorem also provides a physical basis for the classification of charge distributions. One can envisage many possible distributions which would satisfy the electrostatic requirement of zero forces on the nuclei. Not all are realized in nature. Those that are may be classified, in a manner to be outlined below, by means of a spatial partitioning of the charge density and the forces it exerts on the nuclei. The point to be emphasized, however, is that the criteria of some net forces acting on the nuclei, zero for some configuration of the nuclei, attractive or repulsive for others, cannot be put forth as the sole physical reason for the particular distribution of charge found in a system. Instead, the variational principle, applied in such a way as to minimize the energy, is the *only* operational principle directly determining the properties of the system. In recent years, however, it has been shown that the total energy of a molecular system is a functional of the charge density, and a variational principle for this energy functional has been proposed (*58,63,103*). The density functional approach is being applied to problems of chemical interest (*53, 86, 97*). The charge density plays a central role in the $X\alpha$ method of calculating molecular energies (*83*). Variational procedures for the energy functional utilizing the Hellmann-Feynman (H-F) or the virial theorems have been proposed (*71*). Finally, it has been shown that the topographical property of a molecular charge distribution determines the boundaries of a quantum subspace, thereby yielding a unique partitioning of the real space and properties of a molecular system (*88, 102*).

The application of the H-F theorem to chemical problems, particularly to the area of chemical binding (see e.g., Ref. 2), contributed to the focusing of attention on the charge density. By the very nature of the electrostatic theorem as discussed above, its application in a quantitative manner was forced to await the time, beginning in the early sixties, when good representations of the charge density were made available through variational calculations of molecular wavefunctions by the self-consistent-field method. Only then was a systematic study of the charge density and its properties possible (see e.g., Ref. 24). Molecular charge distributions are now frequently determined and related to many properties other than the forces they exert on the nuclei (*21, 97*).

The conceptual simplicity of the electrostatic approach to chemical binding is primarily the result of the charge density, a measurable quantity, playing the central role in this approach. Because of the physical reality of $\rho(r)$, what modeling is necessary in the electrostatic approach is clearly and easily related to physical principles. It is the purpose of this chapter to show how studies of chemical binding have contributed to the proposal that the charge density may be used as the physical vehicle for the development of the theory of chemical bonding, as well as of binding.

2-2. THE ELECTROSTATIC MODEL

As discussed in Section 1-5, the electrostatic statement of the H-F theorem, the form of the theorem employed in a discussion of chemical binding, requires the separation of nuclear and electronic motions as contained in the Born-Oppenheimer approximation. This approximation is totally adequate for a discussion of chemical binding. Within the framework of this approximation it requires no further assumption to equate the total force exerted on a nucleus $F_\alpha(R)$ to a sum of nuclear and electronic contributions,

$$F_\alpha(R) = F_\alpha^n(R) + F_\alpha^e(R), \tag{2-1}$$

where (R) denotes the dependence of the forces on the complete set of nuclear coordinates. The electronic contribution $F_\alpha^e(R)$ is given by the quantum mechanical average of the electric field operator at nucleus α,

$$F_\alpha^e(R) = Z_\alpha \int dr \cdots \int dr_N \, \psi^* \left[\sum_{i=1}^{N} \nabla_\alpha (|r_i - R_\alpha|)^{-1} \right] \psi. \tag{2-2}$$

Because of the equivalence of the electrons, (2-2) for $F_\alpha^e(R)$ is equivalent to N times the average force exerted on nucleus α by one electron, or

$$F_\alpha^e(R) = Z_\alpha \int dr \, \nabla_\alpha (|r - R_\alpha|)^{-1} \cdot \rho(r), \tag{2-3}$$

where $\rho(r)$, the electronic charge density, is

$$\rho(r) = N \int ds_1 \int dx_2 \cdots \int dx_N \, \psi^*(x_1, x_2, \ldots, x_N) \, \psi(x_1, x_2, \ldots, x_N) \tag{2-4}$$

and x_i denotes a product of the space coordinate r_i and spin coordinate s_i of the ith electron. In a stationary state $\rho(r)$ may be interpreted as a "number or charge" density and $\rho(r)\, dr$ as giving the amount of electronic charge in the volume element dr. The interpretation of $\rho(r)$ as the density of a static distribution of charge spread over the total space of the system is consistent with its method of measurement and with its use in the theoretical calculation of the electrical moments of a system.

Thus $F_\alpha^e(R)$ is determined by the electric field exerted at the position of nucleus α by a static distribution of negative charge spread throughout real space. However, the application of the electrostatic theorem requires that this interpretation of $\rho(r)$ be taken as a physical model of the electrons so as to include the possibility of ascribing a value to the electric field exerted at nucleus α by each and every element $\rho(r)\, dr$. Quantum mechanics provides a procedure for the calculation of the average value of an observable, as exemplified in (2-2) or (2-3). There is no quantum mechanical nor experimental procedure, however, which justifies ascribing a physical significance to the value of the electric field generated by each of the individual volume elements which contribute to the total spatial average. This assignment of physical properties to $\rho(r)$ at each point in space is the essential step which transforms the H-F theorem into an electrostatic *model* of the real system.

2-2-1. Berlin's Binding and Antibinding Regions

This model is well illustrated in the definition of binding and antibinding regions proposed by Berlin (*31*). In terms of the electrostatic model of a diatomic molecule Berlin defined a quantity $f(r)$.

$$f(r) = (Z_\alpha/r_\alpha^2)\cos\theta_\alpha + (Z_\beta/r_\beta^2)\cos\theta_\beta,$$

which is the component along the internuclear axis of the total force exerted on both nuclei by a unit negative charge at the point $\boldsymbol{r}$. With the coordinate system used to define $f(r)$, a right-handed one centered on nucleus α and a left-handed one centered on nucleus β, the total forces exerted on the nuclei are equal in sign and magnitude. Thus the electronic contribution to the force on either nucleus is (R is the internuclear separation)

$$F_\alpha^e(R) = F_\beta^e(R) = \tfrac{1}{2}(F_\alpha^e(R) + F_\beta^e(R)) = -\tfrac{1}{2}\int f(r)\,\rho(r)\,dr \qquad (2\text{-}5)$$

and one-half of the total of the forces exerted on both nuclei in the system is

$$F(R) = (Z_\alpha Z_\beta/R^2) - \tfrac{1}{2}\int f(r)\,\rho(r)\,dr. \qquad (2\text{-}6)$$

For a system at its equilibrium internuclear separation (R_e) the force $F(R)$ will equal zero.

The charge density $\rho(r)$ is everywhere positive and thus the sign of the contribution to $F(R)$ of the charge in each volume element, $f(r)\,\rho(r)\,dr$, is determined by the sign of $f(r)$. The negative charge in regions where $f(r) > 0$ reduces the value of $F(R)$, or *binds* the nuclei, while negative charge in regions where $f(r) < 0$ increases the value of $F(R)$. Thus Berlin defined binding as follows: Negative charge in a region of space where $f(r)$ is positive is binding; negative charge in a region of space where $f(r)$ is negative is antibinding. This definition leads immediately to a physical partitioning of the total space of a diatomic system into a binding region, where $f(r) > 0$, and antibinding regions, where $f(r) < 0$, the regions being separated by the two surfaces of revolution $f(r) = 0$.

The binding and antibinding regions for the molecules LiF and CO are illustrated in Fig. 2-1, superimposed on contour maps of their ground-state electronic charge distributions. For a heteronuclear system, with $Z_\beta > Z_\alpha$, the anti-

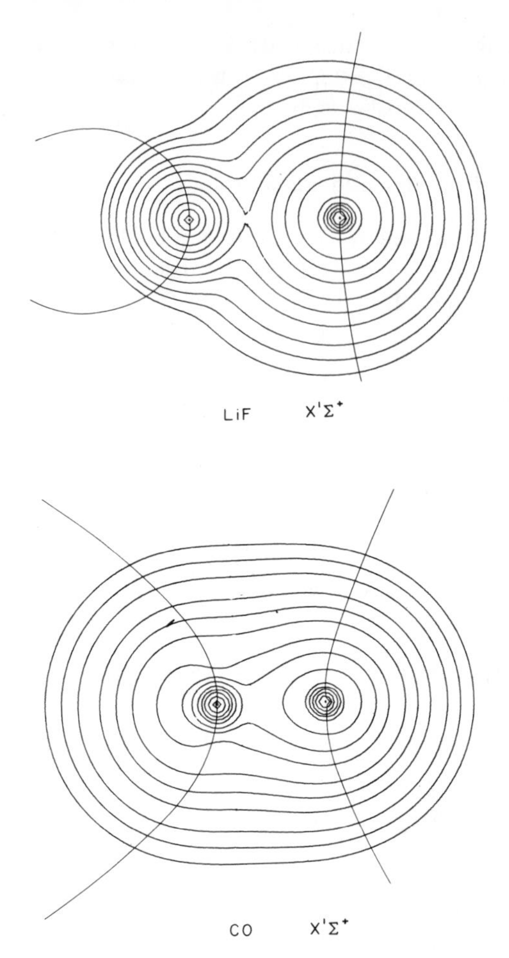

Fig. 2-1. Contour maps of the total electronic charge distributions showing the boundaries which define Berlin's binding and antibinding regions for LiF$(X^1\Sigma^+)$ and CO$(X^1\Sigma^+)$. In this and following figures, in which the values of the charge density contours are not indicated on the diagram, the contour lines have the following values: starting from the outermost contour, the values of the contour lines increase inwardly with values $2 \times 10^n, 4 \times 10^n$, 8×10^n, beginning with $n = -3$ up to $n = +1$ in steps of unity.

binding region for nucleus α is closed, while that for nucleus β in the limit $Z_\beta >> Z_\alpha$ approaches a plane perpendicular to the internuclear axis. The equations for the surfaces $f(r) = 0$ are given by Berlin.

The definition of binding and antibinding regions may be extended to polyatomic systems (*3, 5*). Since the charge distribution must possess the same symmetry as the nuclear framework, one may consider the forces exerted on the nuclei by a symmetrically equivalent set of point charges in deriving the boundaries of the binding region. Taking the ammonia molecule (C_{3v} symmetry) as an example, the set of symmetrically equivalent point charges reduces to a single point for charge along the threefold axis, to a set of three points for charge in the symmetry planes, and to a set of six points for charge placed in other regions. One may then define a surface enclosing a binding region which has the following significance: Any set of symmetrically equivalent point charges placed within the boundary surface (in the binding region) will exert forces on the nuclei such as to decrease all internuclear separations. A similar set of charges placed outside this boundary surface will tend to separate the molecule totally into atoms or into separate groups of atoms. Such a binding region is illustrated in Fig. 2-2 for the water molecule.

One may also define a binding region in a polyatomic molecule by demanding that the force exerted by any single element of charge in that region exert a binding force on all nuclei. This yields a more restricted binding region than the definition given above and is obtained by the superposition of all the possible diatomic boundary surfaces. The binding region for the polyatomic system is then that region of space which is common to the binding regions of all possible diatomics in the system. This more restricted definition of a binding region is also illustrated in Fig. 2-2 for the water molecule. Any single element of charge density within the boundary surface binds all three nuclei. Charge outside the region tends to separate the system into three atoms or an atom and a diatomic molecule (see also Ref. 98).

If for some range of decreasing R values $F(R) < 0$ in (2-6), then the force exerted by the charge density in the binding region exceeds in magnitude both the nuclear force of repulsion and the antibinding force exerted by the remainder of the charge density with the result that the nuclei are drawn together. Eventually for a further decrease in R to some value $R_e, F(R)$ is reduced to zero. If for simplicity we assume $F(R)$ is binding for all $R_e < R < \infty$, then the resulting state of electrostatic equilibrium is known to correspond to an energy minimum and the molecule is bound with respect to the separated atoms. Thus a necessary condition for chemical binding between atoms is that the molecular charge density be distributed so that the forces exerted on the nuclei by the electronic charge in the binding region exceed the antibinding forces exerted by the remaining charge density and by the nuclei for $R > R_e$, and that the binding forces balance the antibinding forces for $R = R_e$.[1]

Without making further extensions in the model, this is the only condition which the electrostatic theorem imposes on chemical binding. Additional questions which come immediately to mind, such as how much of the total charge is present in the binding region and how it is distributed to achieve electrostatic equilibrium may be answered only through a study of known molecular charge distributions.

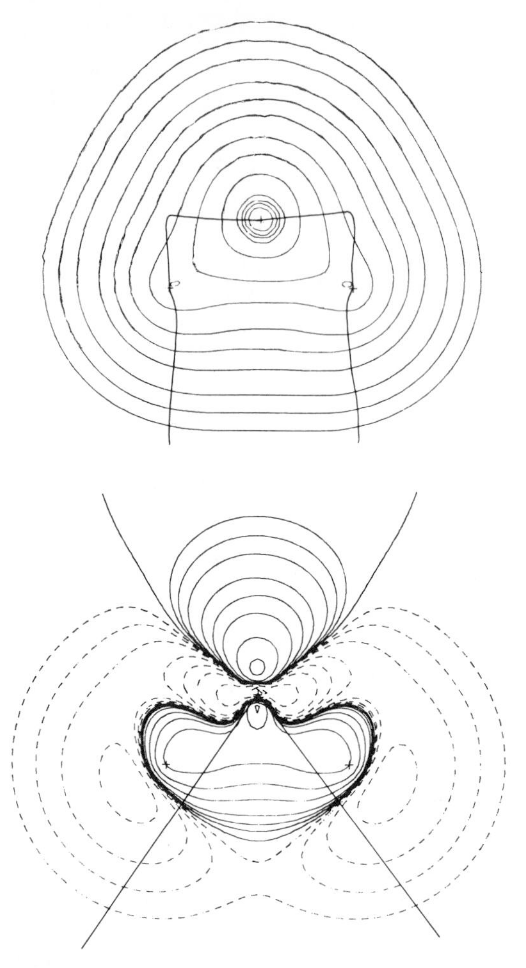

Fig. 2-2. Contour maps of the total electronic charge distribution and of the density difference distribution $\Delta\rho(r)$ $[\Delta\rho(r) = \rho(\text{molecule}) - \rho(\text{atoms})]$ for $H_2O(X^1A_1)$. The boundary line of the extended binding region for a polyatomic system (for a symmetrically related set of charge points) is shown on the contour map for the total density distribution. The more restricted definition of a polyatomic binding region is shown superimposed on the density difference map. The oxygen-atom density used in the density difference map is derived from the configuration $1s^2 2s^2 2p_x^{5/3} 2p_y^{5/3} 2p_z^{2/3}$, with $2p_z$ directed along the C_2 symmetry axis. This corresponds to the atomic state of oxygen, which in a C_{2v} field interacts with two hydrogen atoms to yield a ground-state $^1A_1$ water molecule; see Section 2-3 for further details. The geometry is: $\angle HOH = 106.14°$ and $R(O—H) = 1.7960$ au. The molecular density is calculated from a wavefunction close to the Hartree–Fock limit using (10, 6, 2/4, 2) Gaussian set, yielding an energy of −76.05964 au and zero forces on the protons. The atomic densities used in the calculation of $\Delta\rho(r)$ are of corresponding accuracy and basis-set composition.

2-3. THE ELECTROSTATIC THEOREM AND DENSITY DIFFERENCE MAPS

The results of such a study are easier to understand and categorize if a molecular charge density is compared to that of some reference system so that, as is frequently done in chemistry, one considers changes rather than the absolute values of a property. A theory of chemical binding must account for the electrostatic stability of a molecule relative to the separated atoms, and hence a density related to the appropriate states of the separate atoms is an obvious choice for the definition of a reference density.

The system of separate atoms is an illuminating choice of reference state in its own right for the discussion of the properties of molecular charge distributions. Daudel and Roux had earlier introduced the definition of the charge density difference function

$$\Delta\rho(r) = \rho_m(r) - \rho_a(r)$$

as the difference in the distribution of electronic charge between that found in the molecule $\rho_m(r)$ and that obtained by the overlap of the unperturbed atomic densities $\rho_a(r)$. The atomic distribution $\rho_a(r)$ is expressed as a sum of contributions from each of the separated atom densities, the nucleus of each atom occupying a position identical to that found in the molecular system (*75–77*). A contour map of the density difference distribution function $\Delta\rho(r)$ provides a display of the changes which occur in the atomic densities in the formation of a molecule. By combining the electrostatic theorem with the concept of a density difference distribution one obtains a physical model for the discussion of chemical binding (*3*).

The atomic distribution $\rho_a(r)$ is, of course, never realized in nature. A $\Delta\rho(r)$ distribution instead describes how a true molecular density differs from the distribution which would be obtained if the molecule were formed by the simple overlap of the atomic densities. While $\rho_a(r)$ is a hypothetical function, it nevertheless is defined in terms of measurable quantities. In fact, density difference functions are now obtained from experimental X-ray diffraction studies (*30*, *38*).[2] The use of $\rho_a(r)$ as the reference density offers a distinct advantage for the use of electrostatic theorem to interpret the $\Delta\rho(r)$ distributions; the density $\rho_a(r)$ happens to be one which places insufficient charge density in the binding region (in a diatomic or polyatomic system) to ever exceed or balance the forces of repulsion (*60*). Thus one has an immediate test of a *necessary* (but not sufficient) requirement for binding: The $\Delta\rho(r)$ function must exceed zero within the binding region.

Consider the forces exerted on the nuclei by a reference atomic charge distribution $\rho_a(r)$ in which the atomic densities are considered to be spherical.

Since each atomic distribution in $\rho_a(r)$ is unperturbed and centrosymmetric, it does not exert a force on its own nucleus. When the (neutral) atoms are initially infinitely separated from one another the force exerted on each nucleus by $\rho_a(r)$ is zero, as each nuclear charge is effectively screened from the other nuclei by its own electronic charge distribution. However, for finite internuclear separations, the nuclei will experience net forces of repulsion when the molecular charge distribution is described by $\rho_a(r)$.[3] This is a simple result of Gauss's electrostatic theorem, which states that the electric field exerted by a spherical continuous distribution of charge at a point R from the centroid of the distribution is identical to the field which would be exerted by a charge equal to the total charge contained in a sphere of radius R placed at the centroid of the distribution. For finite separations, each nucleus will have partially penetrated the charge density of all the other atoms in the system and hence will experience a net repulsive force proportional to the magnitude of the unshielded nuclear charge of each of the penetrated atoms.

Therefore, the reference distribution $\rho_a(r)$ does not place sufficient charge density in the binding region to balance the forces of repulsion and yield a state of electrostatic equilibrium. This simple electrostatic argument shows that the electronic charge distribution of a molecule, which is stable with respect to the separated atoms, cannot be obtained by the overlap of the undistorted atomic densities. Instead, the atomic densities must be changed in such a way that charge density is concentrated in the binding region so as to exert larger attractive forces on the nuclei than those obtained by the mere overlap of atomic densities.

Thus, the density difference distribution not only shows the redistribution of the atomic charge distributions accompanying the formation of a molecule, it also indicates how the molecular density differs from one which is known to contain insufficient charge density in the binding region to balance the forces of repulsion. It follows that a molecule whose charge distribution yields negative values for $\Delta\rho(r)$ in the binding region must necessarily be in a repulsive state, since it contains a smaller charge density in the binding region than does the reference distribution, which is known to be insufficient in this regard. A molecule for which $\Delta\rho(r) > 0$ in the binding region can be in a state which is stable relative to the separated atoms. In particular, a display of a $\Delta\rho(r)$ distribution for a molecule known to be in bound state at electrostatic equilibrium can be viewed as the change in the superimposed atomic densities required to balance the force of nuclear repulsion and hence yield a stable molecule. With this electrostatic interpretation, a contour map of $\Delta\rho(r)$ provides one with a display of the bond density (*6, 7, 12*).

2-3-1. Density Difference Maps for H_2 and He_2

The simplest and most direct manner in which $\rho_a(r)$ may rearrange to yield a state of electrostatic equilibrium is for electronic charge to be transferred from the antibinding to the binding region as the atoms approach to form a molecule. Correspondingly, a transfer of charge from the binding to the antibinding regions relative to $\rho_a(r)$ will yield an unstable system. Density difference maps (DDMs) exhibiting such simple charge rearrangements are found to be atypical, and are found only for the simplest systems, those involving H and He. Figures 2-3 and 2-5 illustrate the DDMs (*8*), for various stages in the formation of the ground-state charge distributions of $H_2(^1\Sigma_g^+)$ and $He_2(^1\Sigma_g^+)$. The charge distributions for H_2 were calculated from wavefunctions (*41*) which include a major portion of the correlation energy and yield the correct description of the system for infinite separation. Thus the calculated potential energy (PE) curve closely parallels the experimental one over a large range of internuclear separa-

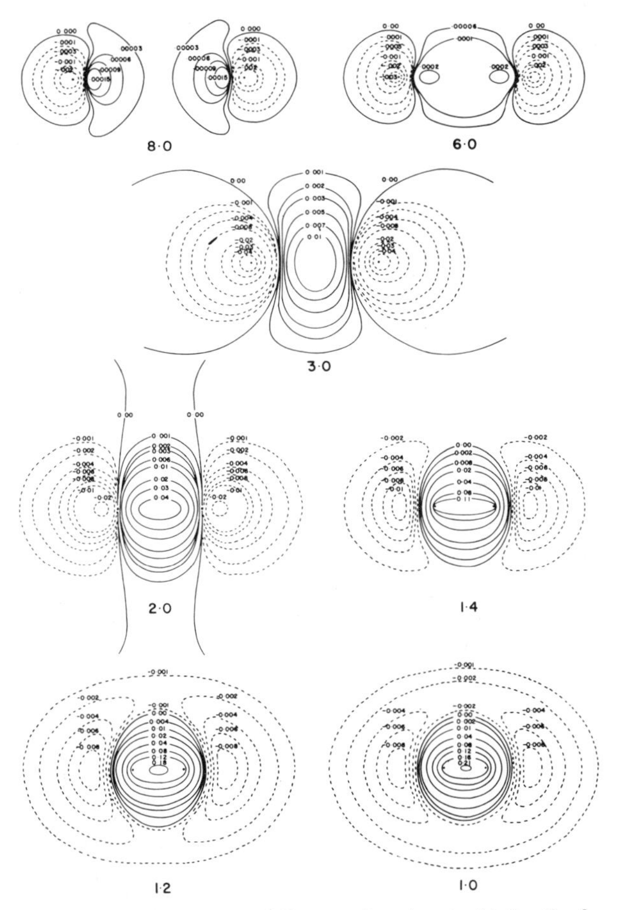

Fig. 2-3. Contour maps of the density difference function $\Delta\rho(r)$ for the formation of the ground-state hydrogen molecule. The maps for R = 8.0 and 6.0 au are drawn to one-half scale relative to the maps at the remaining values of R (Reproduced from Ref. 8, courtesy the National Research Council of Canada.)

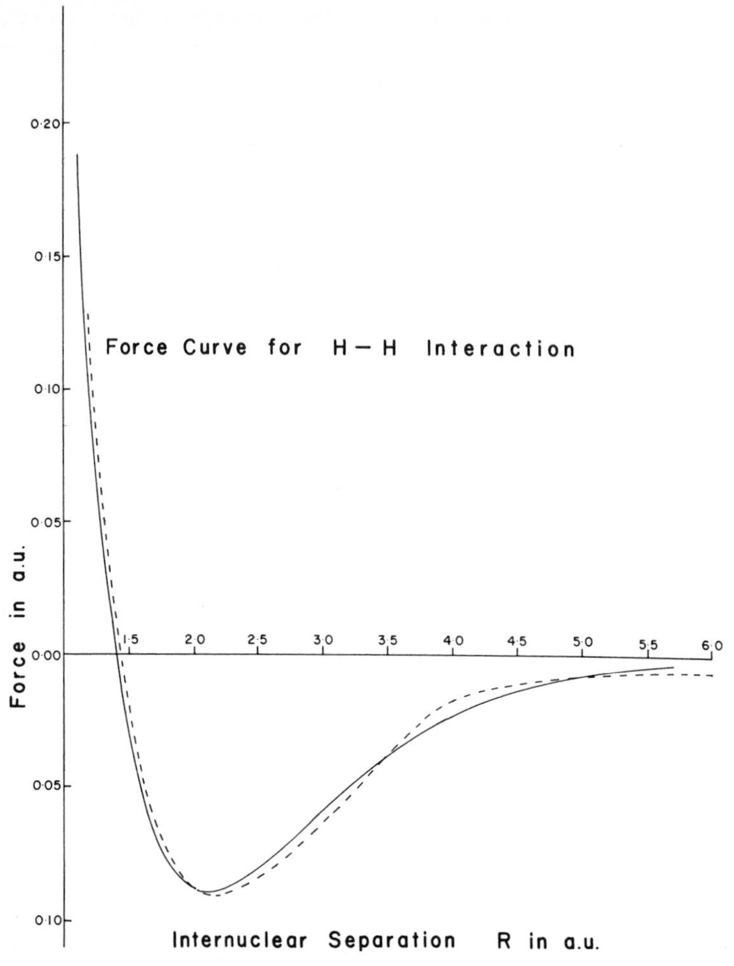

Fig. 2-4. The experimental (solid line) and theoretical force curves for the hydrogen molecule (1 au of force = e^2/a_0^2 = 8.2378 × 10^{-3} dyn). This is the general form of $F(R)$ for a bound (relative to the atoms) state of a diatomic molecule. It is the derivative of the usual potential curve $E(R)$. (Reproduced from Ref. 8, courtesy the National Research Council of Canada.)

tions, including the region of large separation where the interaction between the atoms is described by van der Waals or London dispersion forces.

The contour maps of $\Delta\rho(r)$ in Fig. 2-3 indicate that even for large R, e.g., R = 8.0 or 6.0 au, where the overlap of the atomic orbitals or densities is essentially zero, the atomic densities are inwardly polarized. The result of this polarization is that each nucleus experiences a binding force because of the

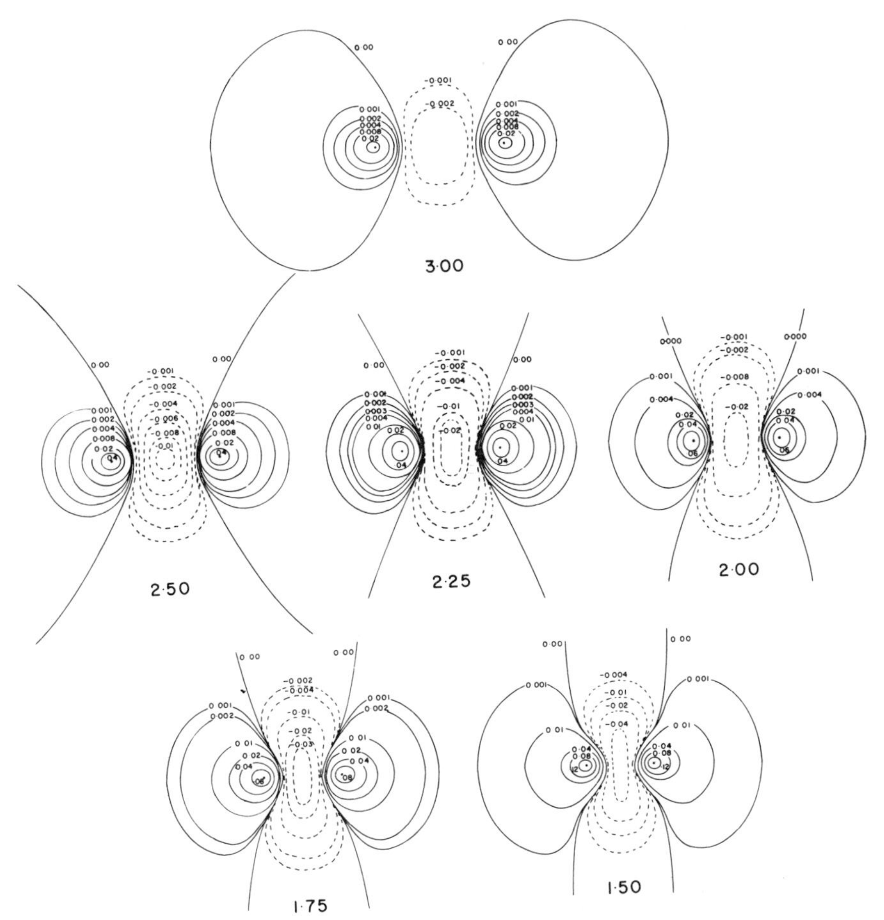

Fig. 2-5. Contour maps of the density difference function $\Delta\rho(r)$ for $He_2(X^1\Sigma_g^+)$. (Reproduced from Ref. 8, courtesy the National Research Council of Canada.)

polarization of its own charge density. This is the electrostatic description of van der Waals forces and was foreseen by Feynman (*49*):

> The Schrödinger perturbation theory for two interacting atoms at a separation R, large compared to the radii of the atoms, leads to the result that the charge distribution of each is distorted from central symmetry, a dipole moment of order $1/R^7$ being induced in each atom. The negative charge distribution of each atom has its center of gravity moved slightly toward the other. It is not the interaction of these dipoles which leads to van der Waals' force, but rather the attraction of each nucleus for the distorted charge distribution of its *own* electrons that give the attractive $1/R^7$ force.

It is important to realize that the polarizations of the atomic densities in the $\Delta\rho(\mathbf{r})$ map for $R = 8.0$ au are the result of correlative interactions between the electrons on the two centers. In terms of a configuration interaction wavefunction, the long-range polarizations are found to be primarily the result of the so-called left–right correlation of the electrons on the two atoms (*8*).

Hirschfelder and Eliason (*57*) have used the H–F theorem in a direct calculation of the coefficient C_6 of the leading term $-6C_6/R^7$ in the expression for the long-range interaction between two hydrogen atoms. Perturbation theory was employed to obtain the necessary second-order corrections to the wavefunction. Their results substantiate Feynman's suggestion that the long-range force on each nucleus is a result of its attraction to the centroid of its own charge cloud (see Section 7-2).

Nakatsuji and Koga (*69*) have given a detailed physical and perturbative analysis of the forms of the electrostatic interactions contributing to the leading terms of long-range forces between molecules. For two well-separated spherical atoms, the only contribution to the leading terms arises from the atomlike polarizations, as described above. If one atom is charged, or in a state other than an S state, then the extent to which the nucleus is over- or underscreened by its own charge density also contributes to the leading terms in the expression for the force.

Figure 2-3 indicates that for values of R where the overlap of the atomic distributions becomes appreciable an increasing amount of electronic charge is transferred from the antibinding to the binding region. Figure 2-4 illustrates the force curve for this system, a curve which indicates that the attractive force exerted on either nucleus is a maximum at $R = 2$ au. The $\Delta\rho(\mathbf{r})$ plots indicate that the transfer of charge from the antibinding to the binding region is maximized at this value of R, the surface separating the regions $\Delta\rho > 0$ and $\Delta\rho < 0$ passing through the nuclei. A continued decrease in R causes the region of $\Delta\rho > 0$ to expand into the antibinding region and the attractive forces on the nuclei decrease rapidly to yield the equilibrium distribution at $R = 1.4$ au.[4]

The contour maps of $\Delta\rho(\mathbf{r})$ for the He_2 system shown in Fig. 2-5 are calculated from Hartree–Fock wavefunctions (*51*). Since He_2 is a closed-shell system, the Hartree–Fock wavefunction provides the proper limiting description of the separated atoms. The functions for large R will not, however, predict the correlatively induced atomic dipoles. The $\Delta\rho(\mathbf{r})$ maps do predict with only small error, the alterations in $\rho_m(\mathbf{r})$ when the atomic densities overlap. The charge migrations relative to $\rho_a(\mathbf{r})$ found for the approach of two He atoms are the reverse of those found for the formation of H_2, with $\Delta\rho(\mathbf{r})$ becoming increasingly negative in the binding region and increasingly positive in the antibinding regions as R decreases. In H_2, the $\Delta\rho(\mathbf{r})$ maps indicate a change in charge density which corresponds to a loss of the individual atomic identities, i.e., a delocalization of the charge distribution to encompass both nuclei, while the approach of two He atoms leads to an accentuation of each atomic density with the charge being increasingly localized in two separate distributions. These differences in the redistribution of the atomic densities in the formation of H_2 and He_2 are made very clear in Figs. 2-6 and 2-7.

Since the spatial wavefunction for the ground state of H_2 is totally symmetric, there is no Fermi-type correlation operative in this system. In the He_2

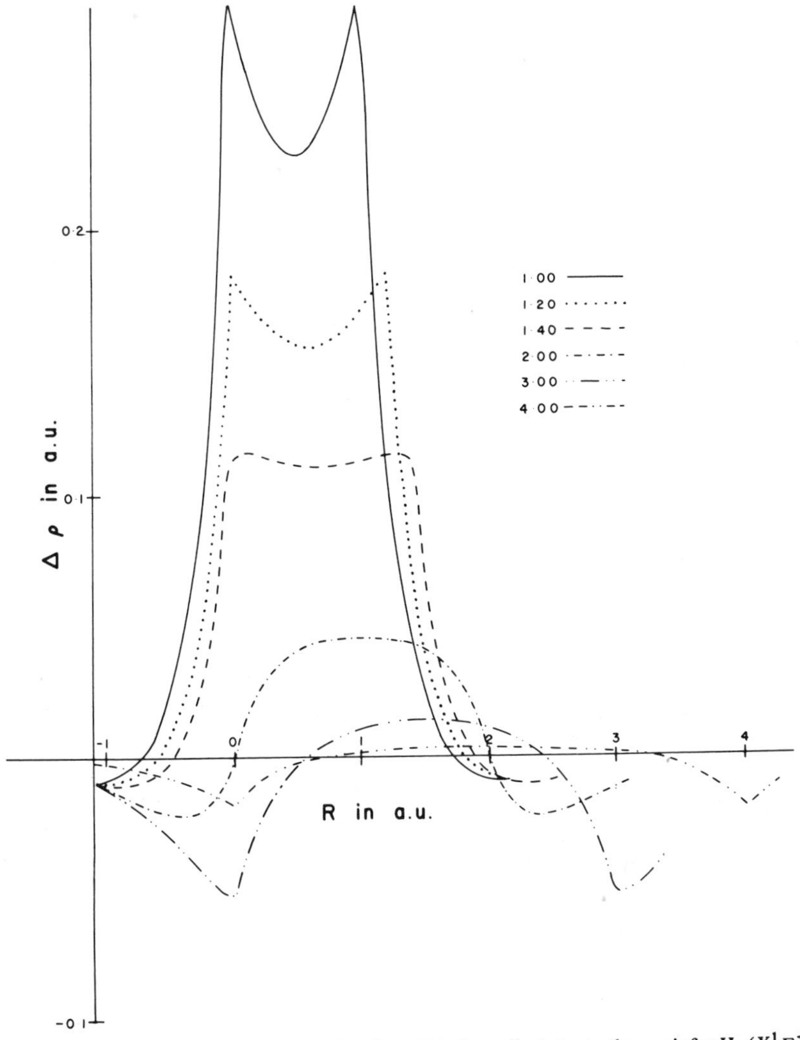

Fig. 2-6. Profiles of the $\Delta\rho(r)$ distribution function along the internuclear axis for $H_2(X^1\Sigma_g^+)$. One nucleus is fixed at the origin (*12*) and the other is placed at the distances indicated on the figure. (Reproduced from Ref. 8, courtesy the National Research Council of Canada.)

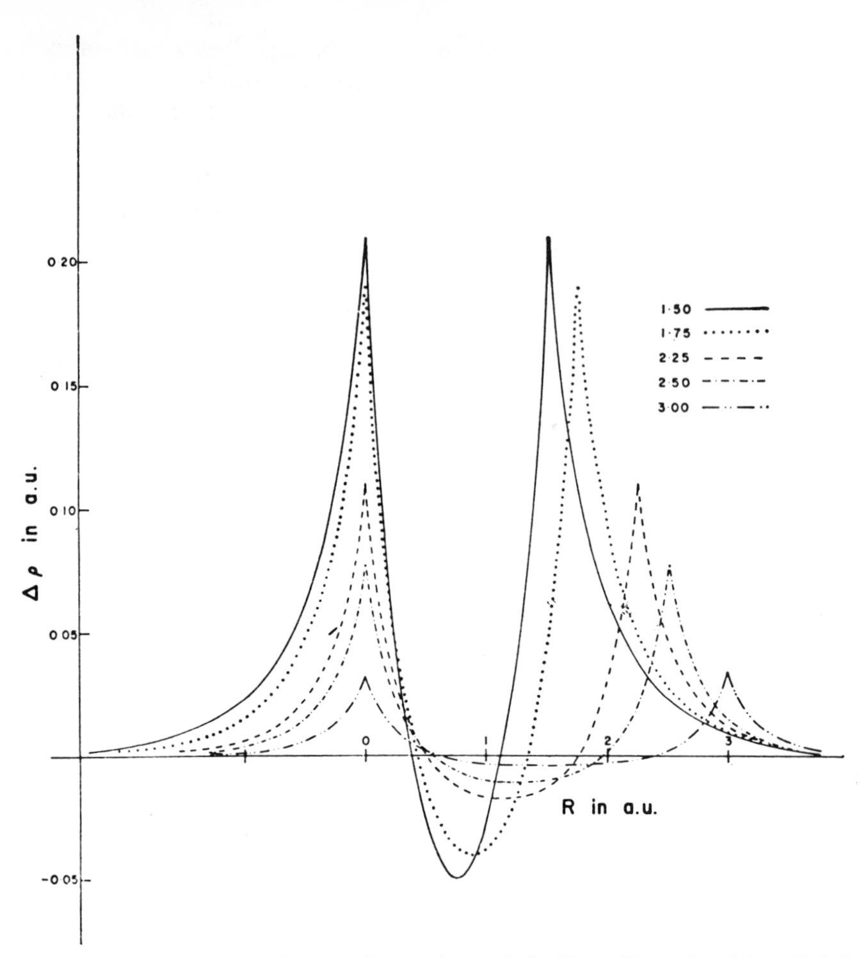

Fig. 2-7. Profiles of $\Delta\rho(r)$ along the internuclear axis for He_2. (Reproduced from Ref. 8, courtesy the National Research Council of Canada.)

system the requirement of the Pauli exclusion principle is the dominant factor responsible for the redistribution of charge illustrated in the $\Delta\rho(r)$ maps. Specifically, the extreme localization of the molecular density into two atomlike distributions is a consequence of a corresponding spatial localization of the Fermi-type correlative interactions between the two almost distinct sets of atomic electrons (*23*).

2-3-2. Density Difference Maps for N_2 and Ne_2

DMs (density maps) and DDMs are shown in Fig. 2-8 for $N_2(^1\Sigma_g^+)$ at R_e and in Fig. 2-9(a) for $Ne_2(^1\Sigma_g^+)$ at $R = 3.0$ au. These charge redistributions, repre-

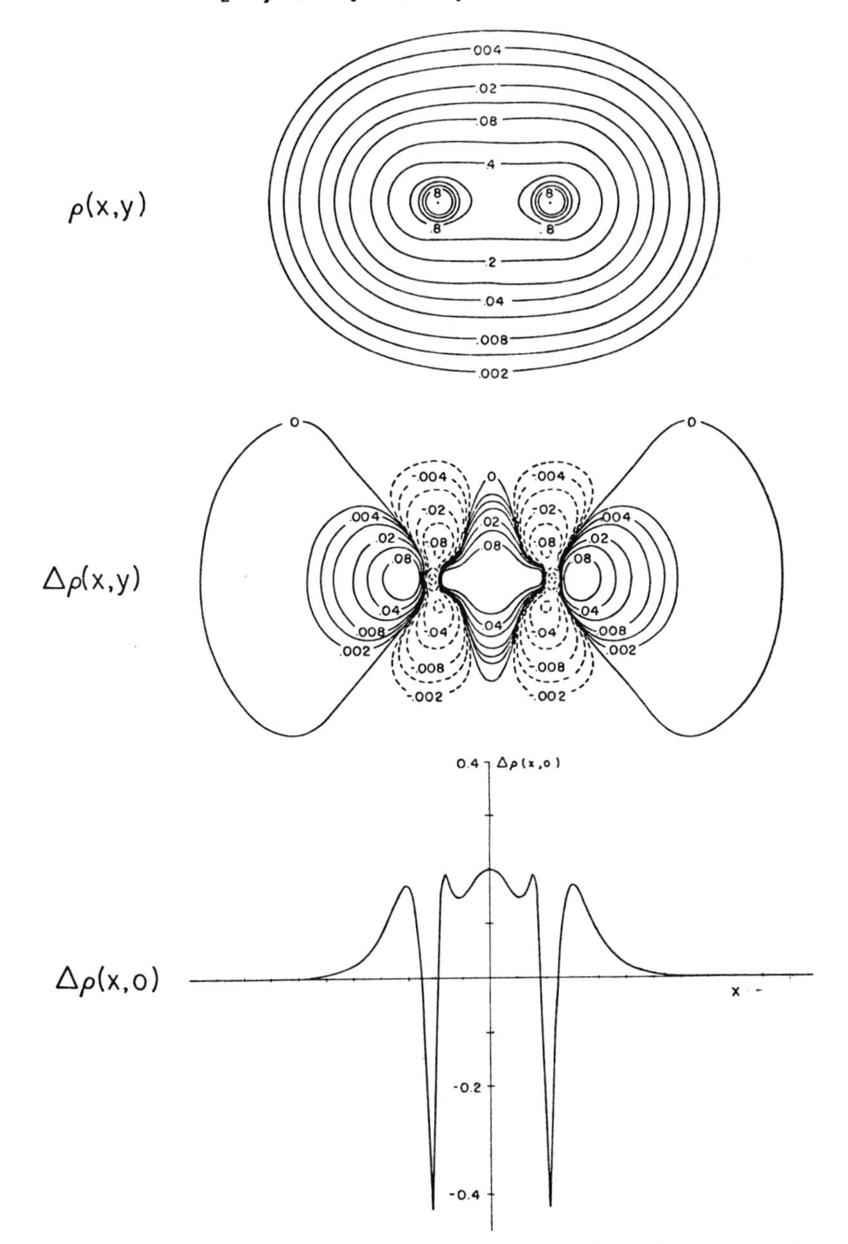

Fig. 2-8. Contour maps of the total molecular charge distribution and of the bond density function (referenced to N atoms in *S* states) for N_2 at the experimental equilibrium bond length.

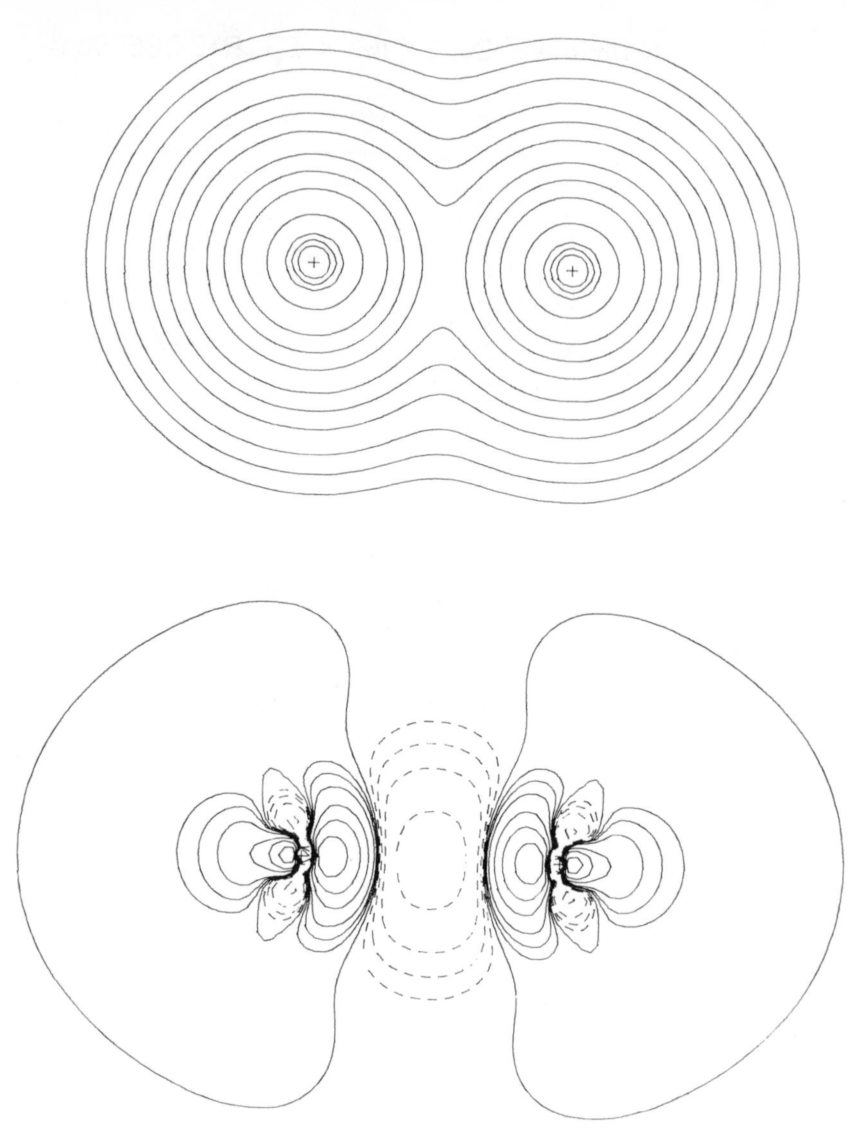

Fig. 2-9 (a). Density and density difference plots for $Ne_2(X^1\Sigma_g^+)$ calculated from Hartree-Fock wavefunctions for the atom (*27*) and the molecule (*51*) at an internuclear separation of 3 au. The contour values shown for the total density and density difference maps are of the same value as those employed in the corresponding maps for N_2 (Fig. 2-8) and Be_2 [Fig. 2-9 (b)].

sentative of the formation of molecules, respectively stable and unstable with respect to the separated atoms, are typical of those found for atoms from the second and (aside from more complicated core polarizations) third rows of the periodic table. In an orbital description of the electronic structure of the elements beyond He, *p* orbitals are either occupied or are energetically accessible and their participation in the charge rearrangements accompanying the formation of molecules from such elements is clearly evident in the $\Delta\rho(r)$ maps. Thus the general redistribution of charge in the field of any of the second-row nuclei Be → Ne or the third-row nuclei Mg → Ar in the formation of a ground-state molecular density is *quadrupolar* in form; charge density is concentrated along a bond axis in both the binding and antibinding regions and removed from a toruslike region perpendicular to the bond axis at the position of the nucleus (*14*).

A comparison of the bond density maps (BDMs) for the stable molecules H_2 and N_2 illustrates a (necessary) common characteristic, namely, an accumulation of charge in the binding region. Similarly, the $\Delta\rho(r)$ maps for the unstable systems He_2 and Ne_2 both exhibit a depletion of the charge density in the binding region. The similarities, however, end there. In N_2 there are separate and substantial accumulations of charge in the antibinding regions. In Ne_2 the effect of the quadrupolar polarization being superposed on the simple He_2-type $\Delta\rho$ pattern results in separate accumulations of charge in the binding region of each neon nucleus.

An integration of $\Delta\rho(r)$ over the volume of space in which $\Delta\rho(r) > 0$ in N_2 shows that the total amount of charge accumulated in these volumes in the antibinding regions ($\sim 0.26e^-$) is approximately the same as that accumulated in the binding region ($\sim 0.25e^-$) (*6, 7*). However, the charge accumulated in the binding region is much more contracted along the internuclear axis, along which it exerts the maximum force on the nuclei, than that accumulated in the antibinding regions. This aspect of $\Delta\rho(r)$ is particularly evident in the profile plot of $\Delta\rho(r)$. The result is that the force exerted by the charge increase in the binding region is sufficient to balance both the increase in the antibinding force from the charge placed in the antibinding regions and the force of nuclear repulsion.

Figure 2-9(b) illustrates the $\Delta\rho(r)$ distribution for the $Be_2(X^1\Sigma_g^+)$ system for a separation of 4 au. The Be_2 system is predicted to be unstable in the Hartree-Fock approximation. The pattern of the charge redistribution for the formation of this molecule is different in degree but not in kind from that found for the stable $N_2(X^1\Sigma_g^+)$ molecule. It does differ in kind from the $\Delta\rho(r)$ map for $Ne_2(X^1\Sigma_g^+)$. It appears that the observation of $\Delta\rho(r) < 0$ over a central portion of the binding region is characteristic only of molecules formed by the interaction of atoms which are true closed-shell systems, i.e., in which all the valence orbitals of a given principal number are doubly occupied. In $Be_2(X^1\Sigma_g^+)$ which

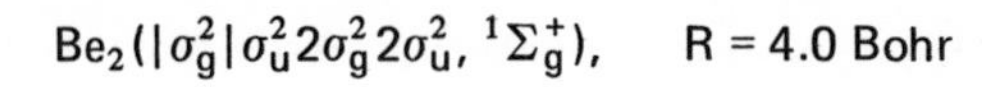

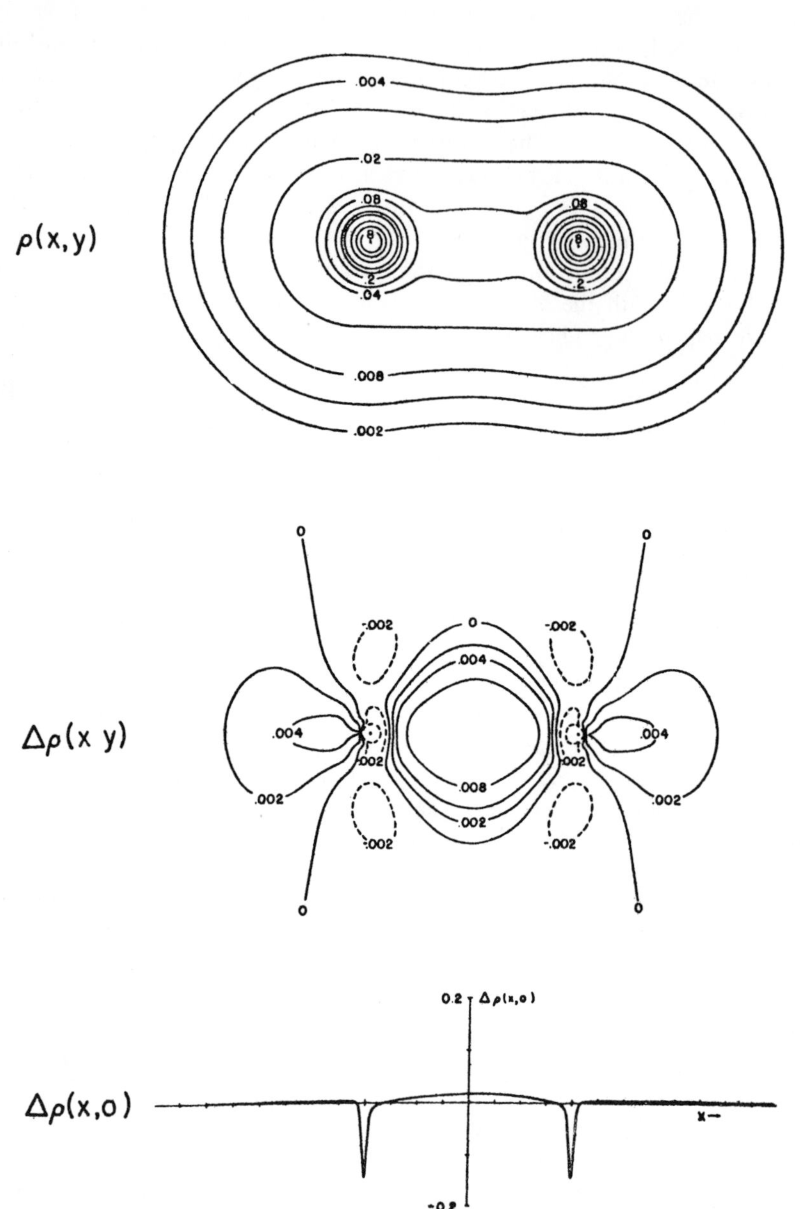

Fig. 2-9(b). Contour maps of the total molecular charge density and the density difference function $\Delta\rho(r)$ (referenced to Be atoms in S states) for Be_2.

is derived from $Be(^1S)$ atoms with the configuration $1s^2 2s^2$, the vacant $2p$ valence orbitals are energetically accessible, and from the appearance of the $\Delta\rho(r)$ map (and also from the LCAO-MOs) the $2p$ orbitals are heavily involved in the orbital description of the interaction between Be atoms. In an AO analysis of Be_2, the $\Delta\rho(r)$ map represents a clear case of considerable electron promotion. Because of orbital vacancies in the valence shell of Be, the characteristic depletion of charge from the region of overlap of two (truly) closed-shell systems, as demanded by the Pauli principle, is not observed in Be_2. It is interesting to note that the corresponding system Mg_2 has been observed experimentally and is found to have a very shallow potential well with a $D_0 \sim 399\ \text{cm}^{-1}$ (*28*). It is clear from the $\Delta\rho(r)$ map for Be_2 that whether a molecule formed by the union of two atoms with an outer configuration of ns^2 is stable or unstable is the result of a very delicate balance of forces exerted by the charges accumulated in the binding and antibinding regions. In Be_2 the integral of $\Delta\rho(r)\,dr > 0$ in the binding region yields $\sim 0.17e^-$, while the corresponding integrals evaluated for the antibinding regions yield a total of $\sim 0.22e^-$. In addition, the charge increase in the binding region is much more diffuse than that for N_2 and the result is an unbalanced repulsive force on the Be nuclei.

Before proceeding with the electrostatic interpretation of the charge rearrangements depicted in the DDMs, particularly those for a bound state at equilibrium (the BDMs), a discussion of the form chosen for the atomic distribution $\rho_a(r)$ is necessary.

2-3-3. Choice of Atomic Reference States in $\Delta\rho(r)$

All the molecules considered so far (H_2, He_2, N_2, Be_2) dissociate into atoms in S states. Since there is no spatial degeneracy, the definition of $\rho_a(r)$ is unambiguous. For systems which dissociate into atomic states with spatial degeneracy, $\rho_a(r)$ has been chosen in two ways; (a) sphericalize the separated atom densities before summing them to obtain $\rho_a(r)$; (b) employ those atomic states, or their linear combinations, which correctly describe the dissociated atoms in the limit of vanishing interactions. We illustrate the difference between these two choices of $\rho_a(r)$ through the formation of $H_2O(X^1A_1)$ from the separated atoms, $O(^1D)$ and $2H(^2S)$.

We pose the question, how does the ground-state density of H_2O differ from that of the superimposed atoms derived from states consistent with the C_{2v} symmetry of the equilibrium system? Of the five ^1D states of oxygen arising from $2p$ AOs, two have A_1 symmetry in a C_{2v} field. Taking the z-axis as C_2, the ^1D component state which interacts with H atoms to yield the ground $^1A_1$ state of H_2O is given by the following linear combination of determinants.

$$\psi(^1A_1) = (1/\sqrt{6})(2|p_x^2 p_y^2| - |p_x^2 p_z^2| - |p_y^2 p_z^2|).$$

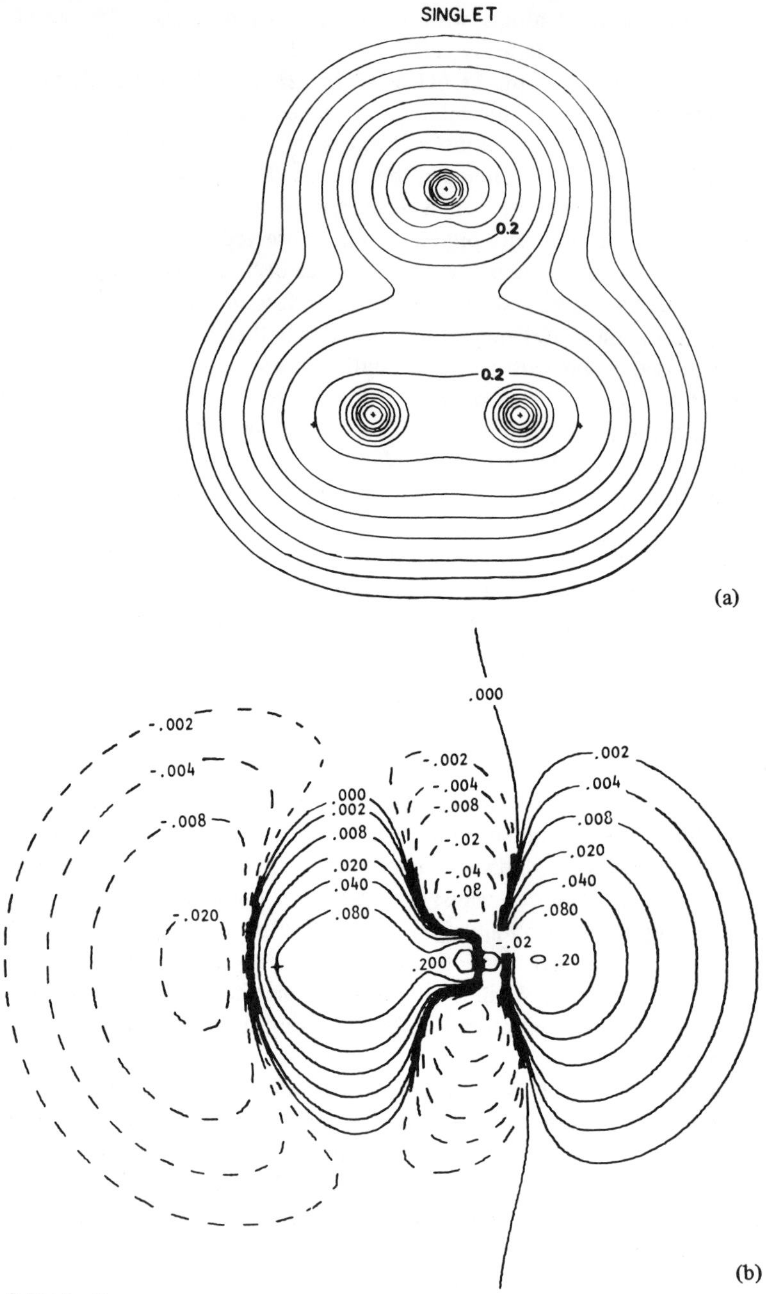

Fig. 2-10(a). Contour map of the total charge density for the formation of ground-state water for a separation of 3 au between the oxygen atom and the bond midpoint of H_2 (*15*). Note the deficit of charge along the C_{2v} axis in the region of the oxygen.

The corresponding oxygen density (with AO occupation numbers)

$$\rho(^1A_1):\ 1s^2 2s^2 2p_x^{5/3} 2p_y^{5/3} 2p_z^{2/3}$$

is not spherical but has a change deficit along the C_2-axis. This is clearly apparent from the DM in Fig. 2-10, where no significant bonding has occurred (the energy change at this distance of approach is ~10 kcal/mole relative to separated reactants) and is more evident at larger separations. The five states of oxygen ^1D level are not equivalent with respect to the C_{2v} symmetry of the field. In other words, a spherically averaged oxygen atom does not react with two H atoms (or, equivalently an H_2 molecule); rather, in the limit of vanishing interactions, one particular oxygen state of $^1A_1$ symmetry (see above) interacts with the H atoms to produce ground-state H_2O. This description is equivalent to applying perturbation theory to a degenerate level. One finds the correct linear combination of some arbitrary set of zero-order functions which is diagonal with respect to the perturbed Hamiltonian. In the limit of vanishing perturbation (i.e., infinite separation of atoms), the system is left in this particular linear combination and *reapplication of the same field finds the oxygen atom in the same state.* In the present example, taking a sphericalized density for oxygen yields a wrong description of the system; i.e., the necessary linear combination

$$\psi(^1A_1, {}^1S) = \frac{1}{\sqrt{3}}\left(|p_x^2 p_y^2| + |p_x^2 p_z^2| + |p_y^2 p_z^2|\right)$$

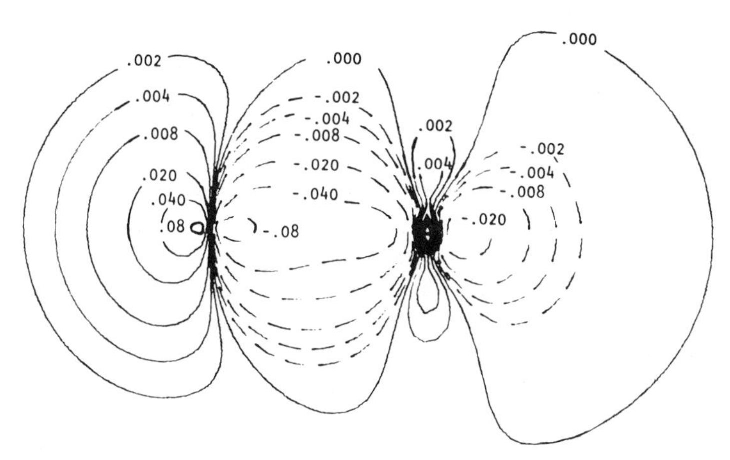

Fig. 2-10(b). Contour maps of the bond density and of the relaxation density for OH ($X^2\Pi$). The relaxation density is calculated (from Hartree–Fock wavefunctions) for the OH molecule when the bond length is increased by 0.166 au. The oxygen nucleus is held stationary in the subtraction of the two molecular density maps; $\rho(r)$ (equilibrium) $-\rho(r)$ (bond extended). (Reproduced from Ref. 14, courtesy the National Research Council of Canada.)

actually describes the ^1S state of oxygen which is *not* involved in forming the ground state of water, but instead correlates with an excited state (*15*). Figure 2-2 employs the correct atomic density $\rho(^1A_1)$ for oxygen.

Study of a large number of examples reveals that BDMs referenced to correct atomic densities exhibit only two basic patterns of charge polarization: dipolar and quadrupolar, as found for H and N, respectively. For $H_2O(X^1A_1)$, oxygen shows a quadrupolar pattern (Fig. 2-2) with a charge increase along the C_2-axis in its binding and antibinding regions together with a toruslike region of charge removal perpendicular to this axis; each hydrogen shows a simple dipolar polarization, antibinding to binding.

Thus, when referenced to appropriate atomic-state densities, BDMs exhibit the same quadrupolar patterns whether the systems dissociate into atomic S states or not. With sphericalized $\rho_a(r)$, however, different patterns for $\Delta\rho(r)$ are obtained for systems which do not dissociate into atomic S states. For example, the $\Delta\rho(r)$ plot for F_2 using a sphericalized $\rho_a(r)$ is totally different from the quadrupolar pattern (like N_2) obtained by referencing $\rho_a(r)$ to the appropriate $M_L = 0$ component of each separated F atom (*74*, *91*). We now give a number of observations which suggest that the dipolar and quadrupolar polarizations exhibited by BDMs represent the general primary response (*14*) of a charge distribution to a field (internal or external).

By plotting $\Delta\rho(r)$ for two slightly different values of R, one obtains a diagram of charge relaxation accompanying a displacement of the nuclei (*9*, *11*). Such a $\Delta\rho$ plot for, e.g., a bond extension corresponds to a slight reversal of polarizations found in the BDMs which represent the sum of all polarizations occurring in the formation of or molecule if and only if the BDMs are related to correct separated atomic-state densities. Thus, BDMs offer the added advantage that the dominant characteristic of change relaxation, whether it opposes or facilitates nuclear displacements, may be inferred from the polarizations relative to $\rho_a(r)$. For example, in $OH(X^2\Pi)$ bond *extension* causes a polarization of $\rho_m(r)$ which is quadrupolar in the oxygen region and dipolar in the H region (Fig. 2-10); both polarizations are reversed in sign relative to the BDM which represents bond formation (see also Refs. 9, 11, and 36).

The dipolar and quadrupolar components of charge relaxation determine, respectively, the quadratic and cubic force constants (*81*). For molecules containing first-column elements (e.g., LiH, NaH) the dominant polarization in the BDMs and relaxation maps is dipolar for both nuclei, whereas for molecules like HCl and HF quadrupolar polarization dominates in the regions of the A nuclei. For this reason, the electronic contributions to the cubic constant K_3 in HCl and HF are 10–40 times larger than those for LiH and NaH (*11*).

When a system is placed in an external electric field, the polarizations of the charge distribution again exhibit the same dipolar and quadrupolar rearrangements (*90*). The first-order change in the charge distributions of the HF and F_2 molecules, when they are perturbed by an external electric field defines a DDM $\Delta\rho(r) = \rho(\text{molecule in field}) - \rho(\text{unperturbed molecule})$. These maps are remarkably similar in appearance to the BDMs for the same molecules in the regions of the fluorine nuclei. The results for HF show that when the direction of the positive field is from the proton to the fluorine, charge is removed from a toruslike region perpendicular to the axis at the position of the F nucleus, and transferred to both the binding and antibinding regions of the F nucleus along the internuclear axis. Thus, the polarizations evident in the BDMs, which show the response of the atomic charge densities to a field resulting from the close approach of nuclei, are identical in form to the response of a system to an externally applied field. Similar polarizations are also observed during a chemical reaction (*47*) if one subtracts the reactant charge distributions from that of the reacting complex (see Section 3-5).

The principal features of quadrupolar polarization in homonuclear diatomics have been verified by gas-phase[5] electron diffraction experiments (*29*, *62*). The experimental BDMs for C_2, N_2, O_2, F_2, and CO compare favorably with the corresponding Hartree-Fock maps. The recent N_2 BDMs obtained from high-precision data for the scattering of 40-keV electrons by gas-phase N_2 molecules (*50*) also show good overall agreement with the Hartree-Fock map; charge is accumulated along the bond axis in both binding and antibinding regions and removed from toruslike rings around the bond axis at nuclear positions. The only serious difference between the theoretical and experimental maps is the presence of weak rings of charge increase within the rings (closer to nuclei) of charge removal (see Ref. 50 for possible reasons).

2-3-4. Characteristics of Bond Density Maps

All DMs and DDMs displayed in this chapter, except for H_2, are calculated from Hartree-Fock wavefunctions. Systematic studies of BDMs have been made for ground states of homonuclear diatomics $Li_2 \rightarrow F_2$ (Fig. 2-11; *6*), diatomic hydrides NaH $\rightarrow$ HCl (Figs. 2-12, 2-13; *33*) and LiH $\rightarrow$ HF (*7*), as well as two isoelectronic series N_2, CO, BF and C_2, BeO, LiF (*10*). Similar maps also illustrate charge reorganizations accompanying electronic excitation, ionization, and electron attachment (*34*).

In the present discussion we wish simply to summarize the important characteristics of those charge redistributions which yield bound molecular states with the nuclei in electrostatic equilibrium. This provides the necessary background for a later discussion which attempts to present a single perspective on the formation of a bound molecular state by complementing the theory of chemical

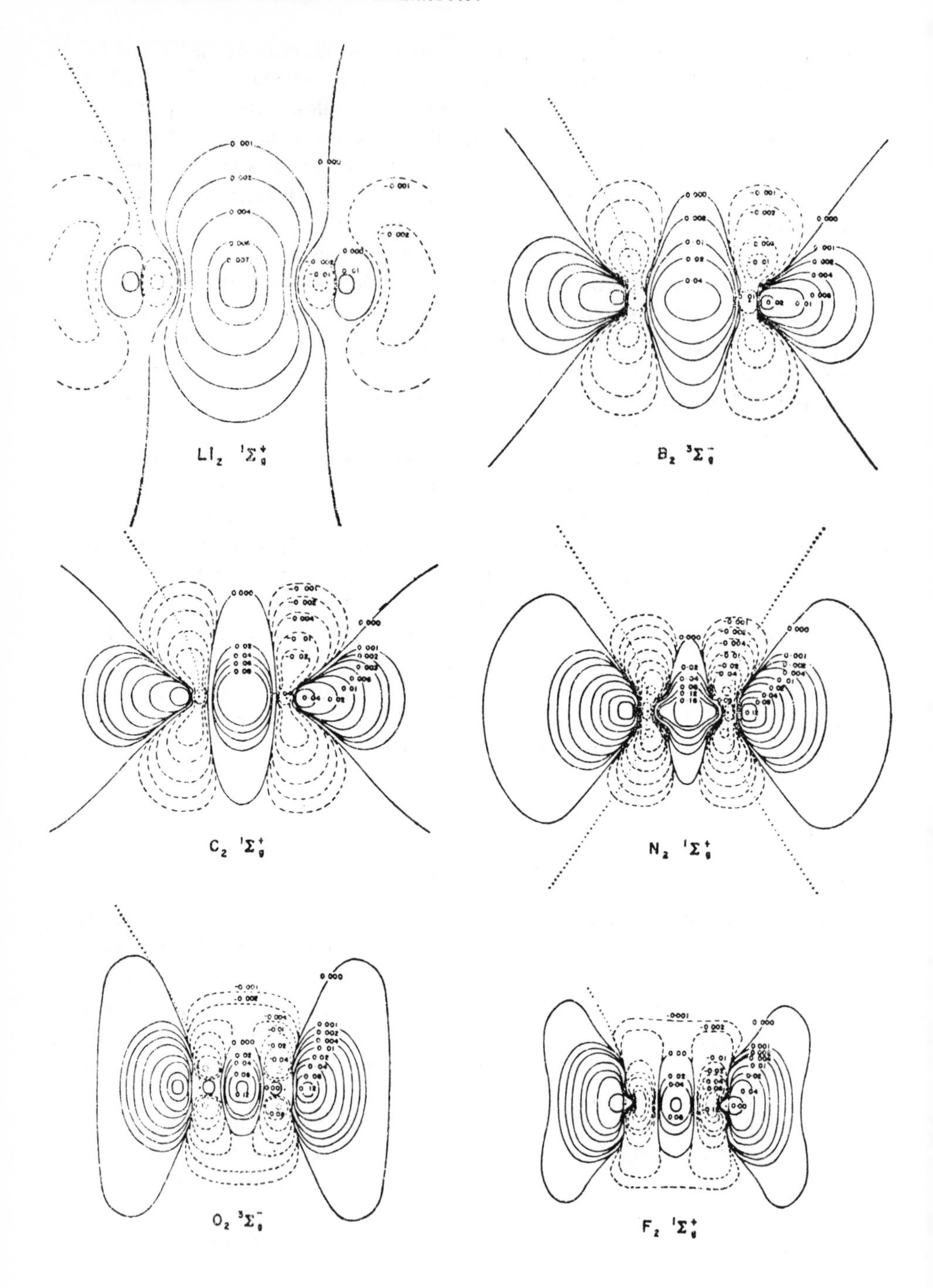

Li_2 $^1\Sigma_g^+$
B_2 $^3\Sigma_g^-$
C_2 $^1\Sigma_g^+$
N_2 $^1\Sigma_g^+$
O_2 $^3\Sigma_g^-$
F_2 $^1\Sigma_g^+$

binding (the electrostatic approach) with the energy analyses obtained from studies of chemical bonding.

All BDMs so far studied show that at equilibrium $\Delta\rho(r) > 0$ in some portion of the antibinding regions as well as in the binding region. This is understandable; the nuclear-electron attractive potential is most negative in the regions of the nuclei, including those portions of the nuclear potential wells that lie in the antibinding region. The gradient of this potential, which determines the forces acting on the nuclei, changes sign at the boundaries of the binding region, boundaries which approximately bisect the most negative portions of each nuclear potential well. Thus, charge accumulated in the immediate vicinities of the nuclei, in the binding or antibinding regions, will contribute to a lowering of the PE of the system. Care must be exercised in interpreting the PE changes ΔV associated with a DDM. The sign of ΔV depends both on the positioning relative to the nuclei of the regions from which charge is removed, as well as on the positioning of the regions in which it is accumulated. Consider, for example, the DDM for H_2 for $R = 6.0$ and 8.0 au (Fig. 2-4). While charge is accumulated in the binding regions for these separations, yielding attractive forces on the nuclei, one finds that V has *increased* relative to the separated atoms. In this stage of the formation of the H_2 molecule, the rather diffuse distribution of charge accumulated in the binding region is removed from regions localized in the immediate vicinity of each nucleus. That is, one has a clear situation in which $\Delta\rho(r) < 0$ in regions where the PE is lower than it is in the single region where $\Delta\rho(r) > 0$. Thus the observation that $\Delta V > 0$ is understandable. However, these same redistributions of charge *decrease* the kinetic energy (KE) of H_2 relative to the separated atoms (*13*) and thus E is also decreased even though $\Delta V > 0$.

At the equilibrium separation R_e of a bound molecular system the virial theorem imposes additional restraints, namely, $\Delta V < 0$ and $\Delta T > 0$. *Thus the charge redistributions shown in the BDMs (which refer to R_e for a bound system), must necessarily denote changes in $\rho(r)$ which decrease the PE of the system and simultaneously increase its KE relative to the separated atoms.*

What one cannot do, as some have done, is to isolate a *portion* of the region where $\Delta\rho(r) > 0$ and argue that this particular part of the charge increase does not contribute to the lowering of PE. For example, Feinberg and Ruedenberg

---

Fig. 2-11. Contour maps of the bond densities for the ground states of the homonuclear diatomics $Li_2 \rightarrow F_2$. The same scale of length applies to all maps. The dotted lines (shown in full for N_2) separate the binding from the antibinding regions. The Li and N atomic densities are spherical; those for B and C contain one and two p_π electrons, respectively; and those for O and F contain one p_σ electron with the remaining p electrons being averaged over the π orbitals. (Reproduced from Refs. 6 and 7, courtesy the American Institute of Physics.)

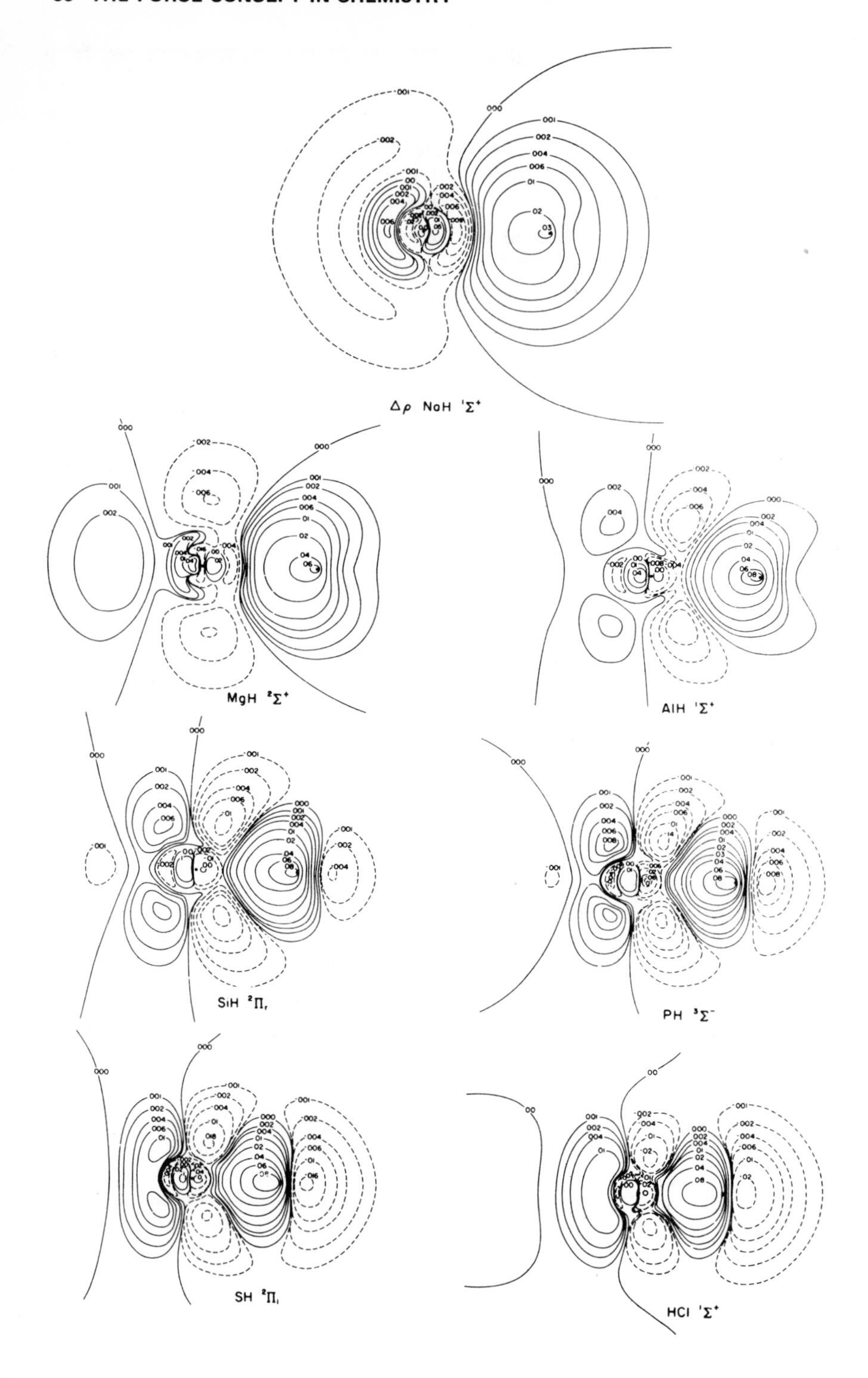
Δρ NaH ¹Σ⁺
MgH ²Σ⁺
AlH ¹Σ⁺
SiH ²Πr
PH ³Σ⁻
SH ²Πi
HCl ¹Σ⁺

(*48*) have argued that the charge increase in the "bond or overlap region" (a region not clearly defined but apparently being a region centrally located in the binding region) does not lead to a PE lowering, and the decrease in V is instead obtained from the orbital contraction around the nuclei. This observation was based on the BDM for H_2^+, which is similar to that for H_2. The region in which $\Delta\rho(r) > 0$ is everywhere of lower potential than the region where $\Delta\rho(r) < 0$. The argument of Feinberg and Ruedenberg presumes that one has a priori knowledge that the charge accumulated in one particular part of the region in which $\Delta\rho(r) > 0$ came from a region of still lower potential, say from the immediate vicinity of each nucleus–and that this step was then followed by a transfer of charge to the regions of the nuclei, since $\Delta\rho(r) > 0$ there as well. One can equally argue that the charge increase in the "bond region" came from regions of higher potential far removed from the nuclei! Such discussions are, of course, meaningless. All one knows is that the total charge redistribution shown in the BDMs (i.e., for a bound system at R_e) depicts a change in $\rho(r)$ which lowers the PE of the molecule.

While it is the changed nuclear potential of the molecule compared to that of the separated atoms which is partly responsible for the redistribution of charge density, the charge redistributions themselves do not always have the property of being described by or being the result of a simple contraction toward the nuclei. The H-F theorem demands that $\Delta\rho(r) > 0$ in some portion of the binding region for a bound molecule at R_e, but does not exclude the possibility of $\Delta\rho(r) > 0$ in the antibinding region. It does, however, limit the extent of charge accumulation in the antibinding regions relative to that accumulated in the binding region by the restraint of zero forces.

All of this is discussed in greater detail in Section 2-6. We introduce the discussion here since the relationship between the patterns of charge increase and decrease in a BDM and the ultimate lowering of the PE of the system is one which must be kept in mind during the present characterization of the BDMs. For H and He [in a bound system such as HeH^+ (*37*)], there is a single region of charge increase which encompasses the nucleus and extends primarily into the binding region. When H is in combination with an element from the first or second columns of the periodic table, this single region of charge increase is the only important characteristic, as the region of charge decrease is localized in the vicinity of its bonded partner, which in these systems transfers charge to the proton (see, e.g., $\Delta\rho(r)$ for NaH or MgH, Fig. 2-12). When H is in combination with itself or with an atom of approximately equal or greater electro-

---

Fig. 2-12. Contour maps for the bond densities of the ground states of the diatomic hydrides NaH → HCl at their experimental internuclear separations. The $M_L = 0$ component atomic state was used for the atomic density of Na, Mg, Al, P, and Cl, and an equal mixture of the $M_L = \pm 1$ component states was used for Si and S. (Reproduced from Ref. 33, courtesy the American Institute of Physics.)

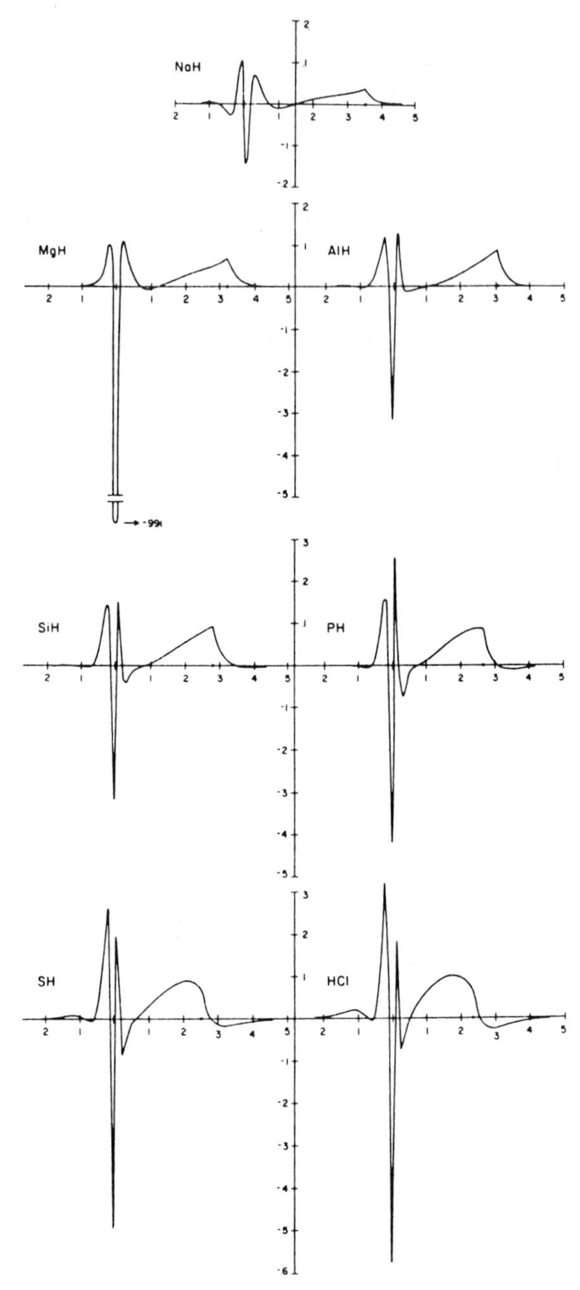

Fig. 2-13. Profiles of the bond density maps for NaH → HCl along the internuclear axis. The density difference and the distances along the internuclear axis are in au.

negativity, an area of charge removal is found which is always concentrated in the antibinding region of the proton, e.g., $\Delta\rho(r)$ for H_2 at $R_e = 1.4$ au in Fig. 2-4 or $\Delta\rho(r)$ for HCl → SiH in Fig. 2-12. ($\Delta\rho(r)$ for the region of H in AlH is transitional, as is that for BeH in the second-row series.) In summary, the charge rearrangement in the vicinity of a proton upon bond formation corresponds to a charge increase which is confined primarily to its binding region and a charge decrease which, when present, is confined primarily to its antibinding region. This dipolar pattern of charge rearrangement, from the antibinding region to the binding region, is the simplest manner in which the electrostatic requirement of zero forces on the nuclei can be met.

The elements Li and Na from family I (and perhaps the remaining members) are transitional in their behavior with regard to charge rearrangement on bond formation. In formation of a covalent species, such as Li_2, they resemble H_2 in that the valence charge density is decreased in the antibinding region and increased in the binding region (Fig. 2-11). The total molecular densities indicate that the decrease in charge density for Li or Na in their antibinding region corresponds to an almost total removal of the loosely bound valence density from this region. While this is anticipated in, say, LiF (Fig. 2-14) it is almost as pronounced in Li_2 or LiBe. When Li is in combination with any element other than itself, the charge increase is more (as in LiF) or less (as in LiB) localized in the neighborhood of the second nucleus. The behavior of the valence density is, as for H, a transfer from the antibinding to what is for the Li nucleus a binding region. The elements of the first family, however, possess a core, a *K* or $1s^2$ core in Li and a *KL* or $1s^2 2s^2 2p^6$ core in Na, and it is the polarization of this core density which makes their BDMs and the details of bond formation more complex than those for hydrogen.

In all ground-state molecular charge distributions in which Li is simply bonded in a linear field, the core is polarized in a direction counter to the direction of valence charge transfer; i.e., in the spatial region associated with the core, charge density is decreased in the binding region and increased in the antibinding region. This may be simply referred to as a "back polarization" and is exemplified in the BDMs for Li_2 (Fig. 2-11) and LiF (Fig. 2-14).[6] The core of Na is more complex, and three successive polarizations are observed (*33*) but the dominant polarizations (the innermost and outermost) are again, as for Li, counter to the direction of valence charge transfer (see, e.g., NaH in Fig. 2-12). The forces exerted on the Li and Na nuclei by their core densities are antibinding, in agreement with the observed back polarization of their core densities. The BDMs for Li are best described as resulting from two opposing dipolar polarizations.

BDMs in the regions of Li and Na are clearly transitional in their characteristics between those found for H and those for the elements Be → Ne and Mg → Ar. The BDMs for $B_2 \rightarrow F_2$ shown in Fig. 2-11 and for MgH → HCl in Fig. 2-12 exhibit the quadrupolar type of polarization which is typical of these latter

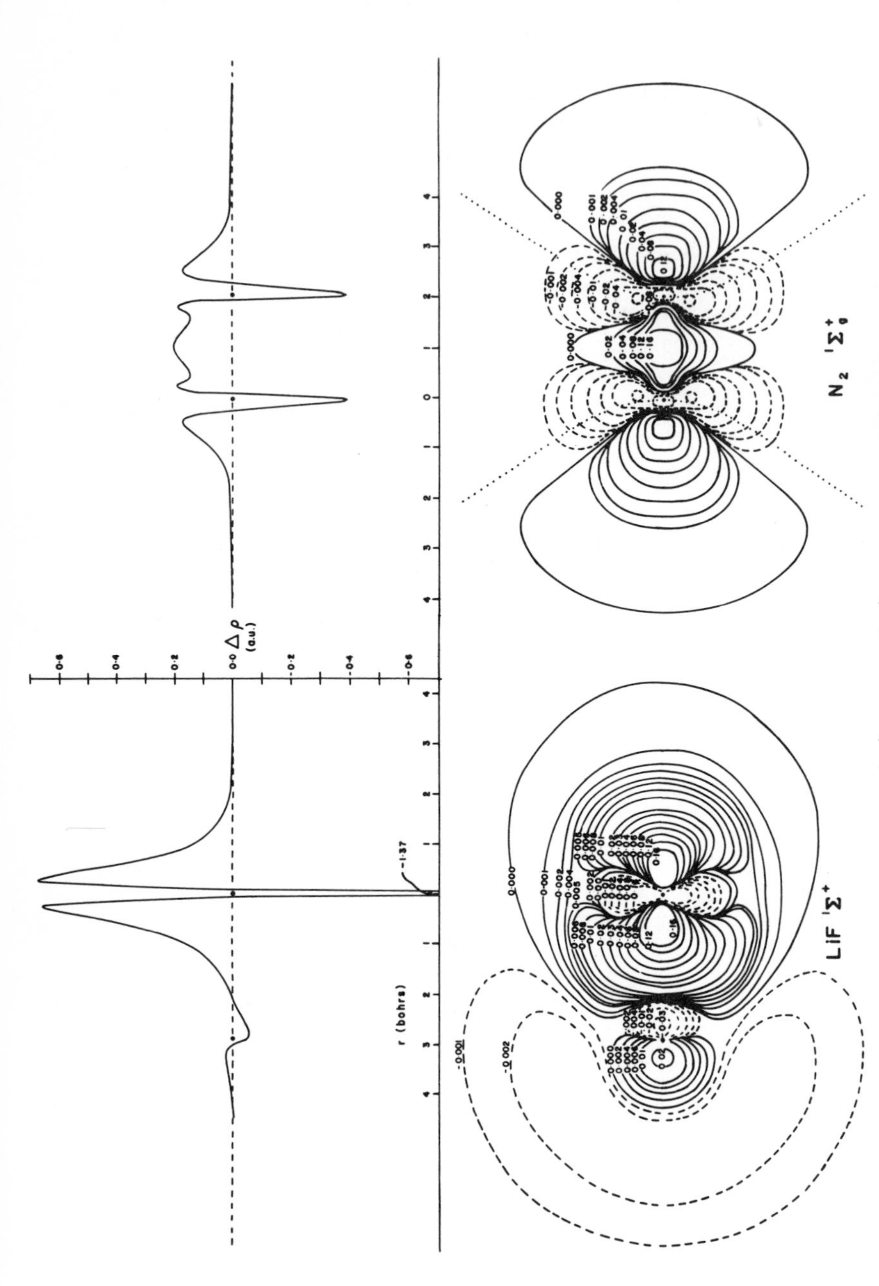

Fig. 2-14. Bond density maps for the ground states of $N_2(^1\Sigma_g^+)$ and $LiF(^1\Sigma^+)$ and their profiles along the internuclear axis. These maps contrast the two possible extremes of the manner in which the original atomic charge densities may be redistributed to obtain a bound molecular state in electrostatic equilibrium (all contours in au). The contour values of the density difference maps increase (solid contours) or decrease (dashed contours) from the zero contour line (first solid contour adjacent to a dashed contour) in steps of 2×10^n, 4×10^n, and 8×10^n, with n beginning at -3 and increasing in steps of unity to -1.

elements. There is, as for Li_2, a separate increase of charge in the binding region of each of these nuclei, one which excludes the nucleus itself. There is, as in Li_2, a separate increase in charge density in the antibinding region of each nucleus. However, in contrast to the situation with Li or Na, these increases do not arise from a back polarization of the core density, but represent diffuse accumulations of valence charge density which extend far into the antibinding region.[7] Finally, the regions of charge removal are unlike those for Li, Na, or H, and instead correspond to toruslike regions perpendicular to the bond axis and located either partially or totally in the binding region. A visual integration of $\Delta\rho(r)$ for N_2 indicates a net transfer of charge from the binding to the antibinding regions. In the hydrides AH, with A = Al → Cl, the toruslike region of charge removal from the vicinity of the A nucleus is essentially totally confined to the binding region of the molecule. The same is true of O_2 and F_2. An inspection of the $\Delta\rho(r)$ map for A_2 and AH for both second- and third-row elements indicates that the region of charge removal shifts more into the binding region as the nuclear charge is increased across a given row of the periodic table. Another important characteristic of the BDMs for these elements is the presence of a region of charge removal at or adjacent to the nuclear position (e.g., the $\Delta\rho(r)$ profiles of N_2 and LiF in Fig. 2-14).

From a study of H_2^+, Feinberg and Ruedenberg (*48*) have argued that the most important characteristic of this system is what they describe as a *contraction* of the orbitals and hence of the charge density toward the vicinity of each nucleus. This feature, they argue, is of paramount importance to bonded systems. The BDM for H_2^+ is similar in all characteristics to that shown in Fig. 2-4 for H_2 at $R = 1.4$ au. Theirs is a possible description of the BDMs for H_2 and H_2^+, one which stresses the contraction of the charge toward each nucleus at the expense of ignoring the dominant increase in charge across the whole of the binding region (their description does describe well the DDMs for the unbound system He_2). This description is not applicable to systems beyond H_2. In the A_2 systems (Fig. 2-11) there is a separate region of charge increase, which excludes the nuclei, *centrally* located in the binding region. For these systems, $\Delta\rho(r)$ attains its maximum value at the bond midpoint and does not have the appearance of being contracted toward the nuclei, as it does in H_2 where there is a single, continuous region of charge increase, $\Delta\rho(r)$ being a minimum at the bond midpoint and a maximum at the nuclei.

While verbal descriptions of the BDMs are not important, the interpretations of the charge rearrangements they depict are. As will be demonstrated below, quantitative measures of $\rho_m(r)$ and $\Delta\rho(r)$ can be made to determine whether or not charge density is simply contracted toward the nuclei in the formation of a molecule. This is found to be the case for H and He but not, in general, for other atoms. From this survey of BDMs and the cataloging of some of their principal features one might anticipate that H_2^+ and H_2 do not provide a universal model for the formation of even covalent bonds. As will be illustrated below, this is indeed the case. What one does find is that the details of different bonding mechanisms correlate with the patterns of charge rearrangement depicted in the BDMs.

2-3-5. Characterization of Ionic and Covalent Binding

The concept of the ionic character of a bond is rooted in the language of valence bond theory, the percentage of ionic character being related to the relative weighting of the wavefunction for the ionic structure to that for the covalent structure. Relating this concept to the parameters in an approximate wavefunction makes it difficult to give an exact mathematical or physical interpretation to it. The inadequacies of past definitions of partial ionic character have been detailed by Shull (*82*). The original purpose in defining ionic and covalent character and the closely related concept of electronegativity was to obtain, by empirical methods, crude estimates of how the valence electrons were distributed or shared in a molecule. In this sense the use of the words *ionic* and *covalent* was predictive in nature. Since the electron density can now be calculated theoretically with some precision, it might be argued that the concept of relative ionic-covalent character of a bond could be discarded. The physical property of interest is, after all, the electron density and not the wavefunction. However, the terms ionic and covalent are still useful in a descriptive sense when they are defined in terms of the charge density and its dependent properties. In what follows the terms *ionic* and *covalent* are defined in terms of the density and the forces which it exerts on the nuclei.

Charge is transferred to the regions of charge increase in the BDMs in order to obtain a state of electrostatic equilibrium and a chemical bond. Thus, it is natural to characterize the bond according to the location relative to the nuclei of the charge increase which binds the nuclei. The BDMs for homonuclear diatomic molecules exhibit an increase in charge density symmetrically placed between the nuclei in the binding region. It is the force exerted by this shared density which binds the nuclei in these molecules as it represents the increase in the density relative to a distribution which does not place sufficient density in the binding region to balance the force of nuclear repulsion. Thus the definition of a covalent bond is one for which the bond density map exhibits a density increase in the binding region which is shared equally by both nuclei.

The nature of the charge increase in the binding region found in the BDM for LiF (Fig. 2-14) is distinct from that found for the homonuclear diatomic molecules. In LiF the density increase which binds the nuclei is localized on F. The $\Delta\rho$ map for LiF exhibits the characteristics of ionic binding as defined in the electrostatic approach (*4*). These characteristics are: (i) a transfer of charge from one atom to the other, the charge increase being localized on one atom, as indicated by the fact that the positive contours are approximately centered on one of the nuclei and the region of increase is bounded by a zero contour which encompasses only a single nucleus; (ii) a polarization of the density increase localized on the anion and of the density remaining on the cation (the core density) in a direction counter to the direction of the charge transfer. This latter characteristic is a direct consequence of the extreme localization of the valence density on a single nucleus. It is clear that the transfer of charge to a region which is localized near F and excludes the Li nucleus will lead to the creation of a net negative electric field at the Li nucleus. Furthermore, if the localized charge was symmetrically placed with respect to the F nucleus, this nucleus would experience a net positive electric field originating from the partially descreened Li nucleus. Thus, to achieve electrostatic equilibrium in the presence of such complete charge transfer, the density distribution localized on the F must be polarized towards the Li. Such a polarization exerts a force on the anionic nucleus which counterbalances the net force of repulsion due to the positive electric field. Similarly, the density in the immediate vicinity of the Li nucleus must be polarized away from F to counterbalance the net force of attraction exerted by the density transferred to F. These polarizations are evident in the BDM for LiF,[8] but are even more evident in a $\Delta\rho(r)$ map constructed by the subtraction of the superposed densities of the Li^+ and F^- ions from the molecular charge distribution (*4*).

BDMs and their profiles along the internuclear axis for N_2 and LiF (Fig. 2-14) illustrate the two possible extremes of the manner in which the original atomic densities may be redistributed to obtain a chemical bond. These models represent the definitions of the limiting cases of covalent and ionic binding. These same definitions have been used to classify the binding in intermediate cases, as illustrated in the studies of the diatomic hydrides and AB systems mentioned above.

2-4. ORBITAL ANALYSIS OF FORCES IN MOLECULES

Any general interpretive approach to chemical binding cannot depend critically on a particular form of the molecular wavefunction, i.e., the conclusions must remain unchanged when the wavefunction is subjected to an arbitrary unitary transformation. The values of the physically measurable properties of a system are invariant to such a transformation of the wavefunction. The charge density, being a measurable property, enjoys this invariance and can be used to develop a completely general interpretive theory of chemistry. This is an ultimate goal and progress is being made in this direction (see, e.g., Ref. 97). Meanwhile, models of electronic structure based on MO theory (*64*) have played, and continue to play, a role in interpretative theories of chemistry. We review here the classification of orbitals as binding or antibinding in terms of the forces which the individual orbital densities exert on the nuclei. This classification is, of course, suggested by and complementary to the definitions of *bonding* and *antibonding* introduced by Mulliken (*61, 65–67*) to describe in a succinct manner the energetic consequences of the occupation or change in occupation of a given orbital in a molecular system.

2-4-1. Binding and Antibinding Orbital Densities

To attempt an understanding of binding in a molecule, it is useful to relate the change in the molecular density relative to that of the separated atoms to the corresponding change in the forces which the electron density exerts on the nuclei. Since the total force exerted on the nuclei is zero both for the separated atoms and for the molecule at its equilibrium separation, such an analysis requires a partitioning of the total force into components which reflect changes in the charge distribution. The decomposition of the total electronic force into orbital contributions meets this requirement. First, the decomposition is straightforward, as the total density is simply the sum of the orbital densities ρ_i,

$$\rho = \sum_i \rho_i,$$

and hence the electronic force may correspondingly be equated to a sum of orbital contributions. Second, in the Hartree-Fock model each orbital density correlates with some set of AOs centered on the nuclei of the separated atoms, thereby allowing the total change in the charge density and their forces to be related to a sum of changes in the individual orbital contributions. One expresses the total force on nucleus A in a diatomic molecule AB as

$$F_A(R) = (Z_A/R^2)\left[Z_B - \sum_i f_i^A(R)\right], \tag{2-7}$$

where the $f_i^A(R)$ are defined in terms of the orbital densities ρ_i as

$$f_i^A(R) = R^2 \int \rho_i (\cos\theta_{\nu A}/r_{\nu A}^2)\, d\tau. \tag{2-8}$$

Since $f_i^A(R)$ is an electric field at nucleus A multiplied by R^2, it has the dimension of charge. Each $f_i(R)$ is equal to the number of electronic charges which when placed at the position of nucleus B, will exert the same force on nucleus A as does the *i*th MO density. The reason for defining the charge equivalent of the force exerted by each orbital density is as follows: For very large values of R, the wavefunction for AB will reduce to a simple product of separately antisymmetrized atomic functions, and correspondingly the total density will reduce to a sum of atomic densities. Each orbital density will similarly be expressible as a sum of atomic densities on A and B in a manner determined by the correlation properties of the orbital with the states of the separated atoms. For large values of R, the contribution to f_i^{A} from the atomic density on A will be zero. In this same limiting situation, $\cos\theta_{\nu\mathrm{A}} \rightarrow 1, r_{\nu\mathrm{A}} \rightarrow R$ and

$$f_i^{\mathrm{A}}(R\rightarrow\infty) = \int \rho_i(\mathrm{B})\,d\tau = N_i(\mathrm{B}). \tag{2-9}$$

Thus, for the separated atoms, f_i^{A} reduces to the number of electrons $N_i(\mathrm{B})$, equal to 0, 1, or 2, which correlate with atom B from the ith MO. For a molecular state which dissociates into neutral atoms, one has

$$\sum_i f_i^{\mathrm{A}}(\infty) = N(\mathrm{B}) = Z_{\mathrm{B}} \tag{2-10}$$

and the force on A vanishes because the nuclear charge of B is screened by an equal number of electrons, as expressed in (2-7). In the case of a homonuclear diatomic molecule, to be used as an illustration below, each AO correlates with an identical singly occupied AO on A and on B, and the limiting value of each f_i is unity.

During the approach of two atoms which form a bound molecular state the sum of the f_i^{A} values will exceed the limiting value $N(\mathrm{B}) = Z_{\mathrm{B}}$ corresponding to the existence of an attractive force drawing the nuclei together in accordance with the general behavior of the force curve for the formation of a bound state (Fig. 2-4). The attractive force vanishes at $R = R_e$. Thus, in the equilibrium state of the molecule one again finds, from (2-7), that the charge equivalents of the orbital forces must sum to Z_{B}, as they do for the separated atoms. However, the f_i values in the molecular case may be greater than, equal to, or less than unity depending on the details of the charge redistribution accompanying molecule formation. An f_i value greater than unity means that relative to the separated atoms the ith MO density is built up in the region between the nuclei and therefore exerts an attractive force in excess of the repulsion due to one nuclear charge, i.e., electrons in the ith MO shield more than one nuclear charge. Correspondingly, any f_i value less than unity implies that relative to the separated atoms the charge density associated with $|\phi_i(r)^2|$ is removed from between the nuclei and is thus less effective in shielding the nuclear repulsion of one positive charge, i.e., electrons in the ith MO shield less than one nuclear charge.

These observations, together with the association of MOs with distinct separated-atom AOs, suggest that the f_i values for the molecule can be used to classify the MOs as binding, antibinding, or nonbinding. This classification is based on the scale set by the separated-atom case (*2*, *6*, *7*): (i) $f_i > 1$, *binding MO*; (ii) $f_i < 1$, *antibinding MO*; (iii) $f_i \sim 1$, *nonbinding MO*. If the value of f_i remains close to unity at R_e the density is termed nonbinding, as it plays the same role in the molecule as in the separated atoms, that of screening one unit of nuclear charge on B from nucleus A. Of the two electronic charges in such an MO, only one contributes to the field at nucleus A (or B). Inner-shell MOs possess f_i values close to unity, indicating that the density is indeed tightly bound to each nucleus in close to spherical distributions. An f_i value less than unity usually indicates that the density has been redistributed to regions behind each nucleus, with the result that the nuclei are descreened in molecule formation. The transfer of density to the antibinding region may be so great as to give a negative value to f_i. Such an MO is strongly antibinding, since it no longer even screens a single nuclear charge on B from A ($f_i = 0$), but it actually exerts a repulsive force on the nuclei which pulls one nucleus away from the other.

In summary, if $f_i < 1$, the quantity $1 - f_i$ determines the net positive electric field at the two nuclei and hence the net force of repulsion acting in the molecule due to the *i*th MO. If $f_i > 1$, then the quantity $1 - f_i$ is a quantitative measure of the net negative electric field and of the binding provided by this MO density over and above the screening of one nuclear charge. For the ground state or singly ionized state of a homonuclear diatomic molecule, the total force on either nucleus, and working expression, is written as

$$F_A = (Z/R^2) \sum_{i \subset AB} (\delta_{ii} - f_i), \tag{2-11}$$

where the sum is understood to be over each of the occupied orbitals, the $\pm\lambda$ components of a doubly degenerate orbital being counted separately.

Table 2-1 lists the f_i value for the homonuclear diatomic molecules formed from first-row atoms (*6*, *7*). These values are calculated from the Hartree-Fock ground-state wavefunctions at the experimental value for R_e except for Be_2, which does not form a stable molecule. The forces for Be_2 (Table 2-1) are for a distance of 3.5 bohr. With the exception of Be_2, the sum of the f_i values is close to Z in each case, as required for electrostatic equilibrium.[9] For Be_2 the sum is less than Z, a result which correctly predicts the ground state of this molecule to be repulsive.

One can obtain a quantitative measure of the binding or antibinding nature of each MO by comparing its f_i value with unity. With the exception of the $1\sigma_g$ and $1\sigma_u$ orbitals, the binding or antibinding character for a given MO is the same for all molecules in the series. The $1\sigma_g$ and $1\sigma_u$ MOs are slightly binding for N_2, O_2, and F_2 and essentially nonbinding for Be_2, B_2, and C_2. In Li_2, however, the density in the $1\sigma_u^2$ or $1\sigma_g^2$ orbitals is considerably altered

Table 2-1. Orbital Forces in First-Row Homonuclear Diatomic Molecules.

Ground state	Molecule	$f(1\sigma_g)$	$f(1\sigma_u)$	$f(2\sigma_g)$	$f(2\sigma_u)$	$f(1\pi_u)$	$f(3\sigma_g)$	$f(1\pi_g)$	$\Sigma_i f_i$	Net force (au)
$^1\Sigma_g^+$	Li_2	0.706	0.658	1.591					2.955	0.005
$^1\Sigma_g^+$	Be_2[a]	1.051	1.028	2.003	−0.399				3.683	0.103
$^3\Sigma_g^-$	B_2	0.979	0.971	2.305	−0.492	1.188			4.951	0.027
$^1\Sigma_g^+$	C_2	0.969	0.954	2.250	−0.436	1.125[b]			5.987	0.015
$^1\Sigma_g^+$	N_2	1.160	1.085	2.682	−0.463	1.216[b]	0.150		7.046	−0.075
$^3\Sigma_g^-$	O_2	1.232	1.138	2.934	−0.518	1.302[b]	0.174	0.426	7.990	0.016
$^1\Sigma_g^+$	F_2	1.243	1.123	2.447	−0.168	1.232[b]	0.516	0.656[b]	8.937	0.080

[a]The $^1\Sigma_g^+$ state of Be_2 is a repulsive one. These results refer to an internuclear distance of 3.5 bohr.
[b]All of the f_i values are quoted for double occupation of the orbitals for comparative purposes. The values marked by *b* are to be doubled to obtain the total electronic force, as they refer to filled pi orbitals.

(in molecule formation) relative to the larger molecules. The f_i values for the $1\sigma_g$ and $1\sigma_u$ MOs for Li_2 are significantly less than unity and the overall disposition of the 1σ density results in a force of repulsion equivalent to placing approximately three-tenths of a positive charge on each nucleus. The $2\sigma_g$ orbitals are uniformly strongly binding. In the N_2 and O_2 molecules the $2\sigma_g$ density exerts an attractive force on the nuclei which is almost three times greater than the simple screening or nonbinding value of unity. The $f(2\sigma_u)$ values are negative in every case. Thus, the $2\sigma_u$ orbital is strongly antibinding in the absolute sense that $\rho(2\sigma_u)$ alone exerts a repulsive force on the nuclei which not only does not counter the partial nuclear repulsion but tends to aid the latter in pushing the nuclei apart.

The $2\sigma_g$ and $2\sigma_u$ densities give only a small net binding for the N_2, O_2, and F_2 molecules, since the sum of $f(2\sigma_g)$ and $f(2\sigma_u)$ (2.22, 2.42, and 2.28, respectively) is only slightly greater than 2.0 for these molecules. However, this corresponding sum for Be_2 is considerably less than 2.0, and for B_2 and C_2 is slightly less than 2.0. These results indicate that in Be_2, B_2, and C_2 the combined $2\sigma_g$ and $2\sigma_u$ density is antibinding, as it does not screen an equivalent number of nuclear charges. From these observations it is clear that the Be_2 molecule with the configuration $1\sigma_g^2 1\sigma_u^2 2\sigma_g^2 2\sigma_u^2$, $^1\Sigma_g^+$ will be unstable as the nuclei in the molecule will be imperfectly screened and experience a net force of repulsion.

The $1\pi_u$ orbital is only weakly binding even in the N_2 and O_2 molecules. In fact, the $1\sigma_g$ density, which is usually considered to be nonbonding, exerts as large a binding force on the nuclei as does the $1\pi_u$ density in the case of F_2, and is only slightly less effective in this regard for O_2 and N_2. A surprising feature of Table 2-1 is that the $3\sigma_g$ orbital which correlates with the $2p_\sigma$ AOs in the separated atoms is classified as antibinding rather than binding. While the density in this orbital does exert an attractive force on the nuclei, the magnitude of this force is considerably less than that required to screen one unit of nuclear charge. In the N_2 and O_2 molecules the magnitude of the force exerted by the $3\sigma_g$ density is such that the value of $1 - f_i$ is equivalent to the force obtained from the presence of eight-tenths of a positive charge on each nucleus. The $1\pi_g$ orbital is slightly more antibinding than the $1\pi_u$ is binding. The complete filling of the $1\pi_u$ and $1\pi_g$ orbitals in F_2 leads to a small net force of repulsion, as the electric field exerted by the eight electrons in these π orbitals no longer balances the field due to four positive charges on each nucleus.

It is useful to have a more detailed picture of the relationship which exists between the exact disposition of the orbital densities and their binding or antibinding character. This information is easily obtained from the present wavefunctions as the basis sets employed in the expansions are composed of Slater-type functions centered on each of the nuclei. Thus, each MO density may be broken down into atomic and overlap populations and a separate contribution of each population to the f_i value may be determined. Therefore, each f_i can be written as

$$f_i = R^2(f_i^{(AA)} + f_i^{(BB)} + f_i^{(AB)}).$$

Considering again the forces exerted on nucleus A in the diatomic molecule AB, we make the following definitions.

Atomic force (from $f_i^{(AA)}$). The force exerted on nucleus A by the atomic charge population on A. If this density is undeformed, then it exerts no net force on the A nucleus as it then possesses a center of symmetry. However, any polarization of the atomic distribution, as described by $s - p$ or $p - d$ hybridization, results in a force on the A nucleus in the same direction as that of the polarization.

Overlap force (from $f_i^{(AB)}$). The force exerted on either nucleus by the overlap density. The positive overlap of orbitals centered on A and B results in the transfer of charge density (the overlap population) to the region between the two nuclei, and the overlap force provides a quantitative measure of the effectiveness of this transferred density in binding the two nuclei together.

Screening force (from $f_i^{(BB)}$). The force exerted on nucleus A by the atomic density centered entirely on nucleus B. It is a measure of the electronic shielding of nucleus B from nucleus A by electrons on nucleus B. The screening force provides the sole contribution to the f_i values for large values of R.

The conceptual advantage of interpreting a chemical bond through a consideration of the electrostatic forces exerted on the nuclei lies in the fact that these forces are directly determined by the one-electron density. This density exists in real space and thus its pictures may be used to obtain a physical insight into the forces acting on the nuclei. With this aspect in mind we have reproduced in Fig. 2-15 the orbital density diagrams for the O_2 molecule. The caption under each diagram gives the force exerted by each orbital density and its breakdown into atomic, overlap, and screening contributions as described above.

The $1\sigma_g$ and $1\sigma_u$ diagrams reiterate that the density in these orbitals is contained in inner-shell, atomlike orbitals centered on each nucleus and should be essentially nonbinding, i.e., a real "core." The screening contributions of unity for these orbitals bear this out and indicate that each MO places sufficient density on each nucleus to screen a single unit of nuclear charge. The near-zero overlap contributions indicate again that no significant amount of charge has been transferred from either atom in the formation of these "core" orbitals. The force analysis does show, however, that the $1s$-like $1\sigma_g$ or $1\sigma_u$ density on each atom is polarized toward the other nucleus. It is this polarization of the 1σ density rather than an overlap force which accounts for the slight binding char-

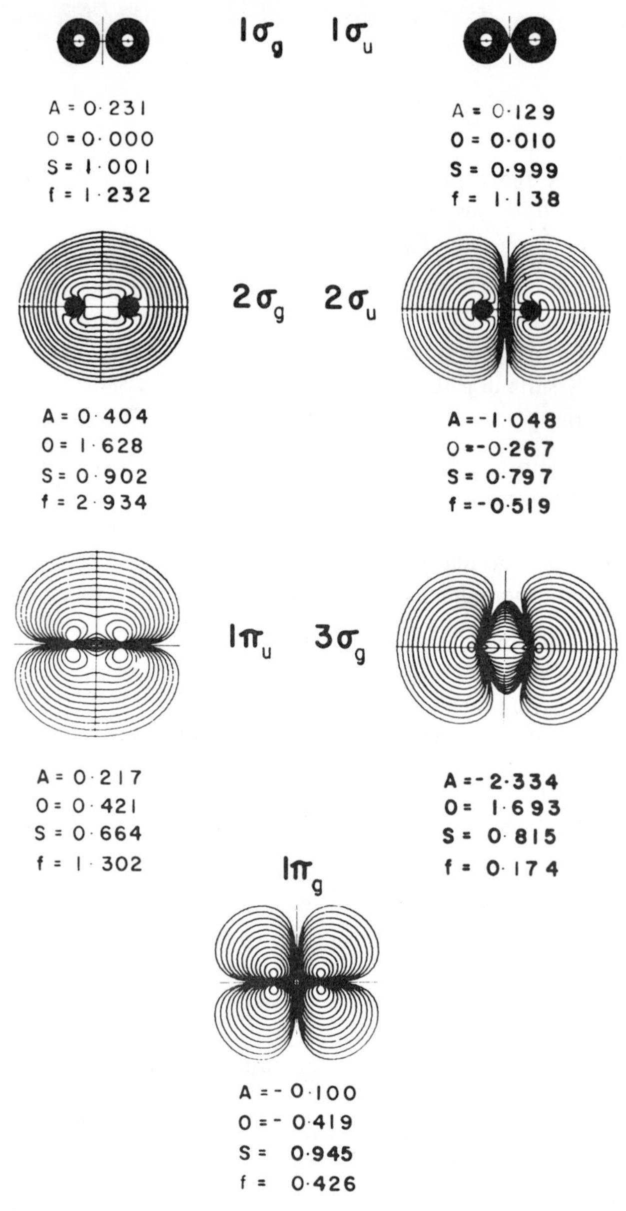

Fig. 2-15. Contour maps of the Hartree–Fock molecular orbital charge densities for $O_2(X^3\Sigma_g^-)$ for the equilibrium separation of 2.282 au. The numerical values are the charge equivalents of the electric fields exerted at an oxygen nucleus by the atomic (A), overlap (O), and screening (S) populations for each orbital density.

acter possessed by these two MOs. The same description applies to the $1\sigma_g$ and $1\sigma_u$ orbitals of F_2 and N_2. The slight antibinding noted for these orbitals in C_2, B_2, and Li_2 molecules is caused by a reversal of the direction of the polarization of inner-shell atomlike MOs.

The strong binding force exerted by the $2\sigma_g$ density may be interpreted as the result of two effects: the accumulation of charge in the overlap region and an inward polarization of the remaining atomic density on each nucleus. The transfer of charge to the binding region is reflected in the screening value being less than unity. This disposition of the $2\sigma_g$ density is characteristic for all the molecules except C_2 and N_2, for which the atomic force contribution is small and negative.

The large amount of density accumulated in the antibinding region by the $2\sigma_u$ orbital may be interpreted as the result of both an overlap and a polarization effect. The negative sign obtained for the overlap contribution to the force indicates that the charge which is transferred in the formation of this orbital is placed predominantly in the regions behind each nucleus. This transfer of charge results in a considerable reduction of the screening contribution and thus in an unshielding of the nuclei relative to the separated atoms. In addition, the atomic densities are strongly polarized and exert a large atomic force drawing the nuclei apart. A similar description of the $2\sigma_u$ density applies to the other molecules in this series. The resultant features of $2\sigma_g$ and $2\sigma_u$ densities are net attractive force from the overlap density (the $2\sigma_g$ orbital placing more density in this region that that removed by the $2\sigma_u$ orbital) and a net antibinding force from the atomic densities (the back polarization of the $2\sigma_u$ orbital outweighing the inward polarization of the $2\sigma_g$). These two resultant properties dominate the $\Delta\rho$ maps for B_2 and C_2, with the result that their bonds can be considered sigma bonds rather than pi bonds. In B_2, for example, the binding force exerted by the overlap density in the $2\sigma_g$ orbital is three times larger than that exerted by the $1\pi_u$ overlap density.

The $1\pi_u$ density for O_2 is typical of the π bond found in this series of molecules. There is a considerable descreening of the nuclei relative to the separated atoms (the average value of the screening contribution for the series being 0.5), indicating that a substantial amount of charge is transferred to the overlap region. However, as symmetry requires the presence of a nodal surface along the internuclear axis, the overlap is not particularly effective in binding the nuclei. In fact, for this series of molecules the sum of the overlap and screening contributions for the $1\pi_u$ orbitals is unity, the nonbinding value. Thus, the result of unshielding the nuclei and transferring the charge density to the overlap region does not in itself, in the case of a pi bond, result in any significant net binding. This is totally unlike the case of a sigma bond, for which the sum of the overlap and screening values is considerably greater than unity. Since the overlap force is canceled by the accompanying unshielding of the nuclei, the small net binding force exerted by the $1\pi_u$ density may be ascribed to the polarization of the π atomic densities.

The surprising fact that the $3\sigma_g$ orbital is classified as antibinding can be explained from the sign of its atomic contribution. The charge density transferred into the binding region as the result of the overlap of $2p_\sigma$ AOs is seen to be very contracted along the internuclear axis where it exerts a correspondingly large binding force on the nuclei. Indeed for O_2 and F_2 (where the overlap contribution is 2.07) the $3\sigma_g$ overlap contribution is larger than even that of the $2\sigma_g$ orbital. It is the very contracted nature of the $3\sigma_g$ overlap density which accounts for the greater stability of N_2 compared to C_2, even though twice the amount of charge is transferred to the binding region in the formation of C_2 than in the formation of N_2. However, the atomic densities in the $3\sigma_g$ orbital are very strongly back polarized, resulting in an atomic force term which more than negates the large positive overlap contribution to the force.

The appearance of the density lobes pictured for the $1\pi_g$ orbital is the result of a transfer of overlap density to the antibinding region and of a direct polarization of the atomic densities. In both O_2 and F_2 the $1\pi_g$ orbital is slightly more antibinding than the $1\pi_u$ orbital is binding.

Similar analyses of the binding and antibinding character of orbitals in the AH diatomics have also been made (*6, 7, 33*). The formation of H_2 (*8*) and that of Li_2 (*35, 92*) have been interpreted in terms of the variation in the f_i values as functions of R. Quantities analogous to the atomic, overlap, and screening contributions defined as above in terms of the AO basis set and summed over all orbitals have been used by Nakatsuji to obtain an electrostatic model for the discussion of molecular interactions and molecular geometry (*68, 70, 98–100*). Deb and co-workers (*43–46*) have also employed the concept of atomic and overlap forces in their discussion of molecular geometry.[10]

2-4-2. A Comparison of Binding and Bonding

The definition of binding ($f_i > 1$), nonbinding ($f_i = 1$), and antibinding ($f_i < 1$) based on a consideration of orbital forces is in general agreement with the usual definitions of bonding, nonbonding, and antibonding. The latter set of definitions is theoretically associated with the nodal characteristics and energies of MOs. The bonding or antibonding character of MOs is thus defined by the absence or presence, respectively, of a nodal surface between the nuclei, and the degree of bonding or antibonding by the state of promotion of the MO in the correlation scheme between the separated and united atoms (*65, 66*). The $2\sigma_g$ and $1\pi_u$ orbitals are both binding and bonding while the $2\sigma_u$ and $1\pi_g$ orbitals are both antibinding and antibonding. The only qualitative discrepancy between the two sets of definitions concerns the $3\sigma_g$ orbital, which is classed as bonding in terms of orbital energies and antibinding in terms of the orbital forces. The $3\sigma_g$ orbital is classified as weakly bonding because of its incipient promotion to a $3s$ orbital in the united atom. Correspondingly, it is not a strongly antibinding orbital, since it does not exert a force pulling the nuclei away from one another (a negative f_i value), but rather an attractive force which is less than that required to balance the force of nuclear repulsion due to one positive charge on each nucleus. A change in the definition of binding and antibinding to relate only the sign of the force ($f_i \lessgtr 0$), corresponding to the orbital forces drawing the nuclei together or pushing them apart, is not as satisfactory as the definition which has been proposed above. In the present example, the $3\sigma_g$ orbital would become weakly binding but the $1\pi_g$ orbital would also have to be classified as binding. Furthermore, restricting the terms antibinding to cases where $f_i < 0$ would destroy one of the most useful features of the definition of bonding or binding, that of predicting molecular stability. A molecular state will be an unstable one as long as the sum of the f_i values is less than Z, indicating that the nuclear repulsive force is not balanced by the attractive force of the electrons, as for example in Be_2. It is only at relatively short interatomic distances that antibonding orbitals obtain f_i values that are negative. The f_i values of antibonding orbitals are, however, always less than 1 (except perhaps for very large distances), and thus equal occupation of bonding and antibonding orbitals will lead to an unstable molecular state as the sum of the f_i values will be less than Z. This can result even though none of the f_i values are negative.

The classification of an orbital density as binding, antibinding, or nonbinding in terms of the f_i values is a static one, in the sense that is is based on the contribution of an orbital density to the total force of the static system. This may or may not correctly reflect the change in the system's properties when an electron is ionized or excited from a given orbital. This latter view, which could be called a dynamic one, was suggested by Mulliken (*67*) as a basis for the classification of the bonding properties of a given orbital. Thus, a given orbital could be classified as antibonding or bonding depending on whether the dissociation energy increases or decreases, or the equilibrium bond length decreases or increases, upon excitation or ionization of an electron from the orbital.

To recover this type of predictive ability using the force criterion it has been proposed (*54*, *93*) that the binding or antibinding character of an orbital be related to the derivative of its energy. Tal and Katriel (*93*) show that

$$\partial \epsilon_i / \partial R = -\partial F / \partial n_i,$$

i.e., this derivative is a direct measure of the change in the total force on a nucleus with respect to a change in the occupation number of a particular orbital. They point out that the Hartree-Fock approximation satisfies the H-F theorem in two different ways. (i) From the Fock eigenvalue equation

$$\hat{F}\phi_i = \epsilon_i \phi_i, \quad \text{where} \quad \hat{F} = \hat{h} + \hat{G}$$

it follows that

$$\partial\epsilon_i/\partial R = \langle\phi_i|d\hat{F}/dR|\phi_i\rangle = (Z_A/R^2)\, f_i + \langle\phi_i|d\hat{G}/dR|\phi_i\rangle,$$

which is the H-F theorem with respect to the orbital energy. (ii) The Hartree-Fock wavefunction also satisfies the usual form of the theorem, i.e., with respect to the total energy and the exact Hamiltonian,

$$dE_{HF}/dR = \langle\psi_{HF}|d\hat{H}/dR|\psi_{HF}\rangle = (Z_A/R^2)\sum_i f_i.$$

The orbital force in the first of the above two definitions is more closely related to Mulliken's criteria, stressing the effect of a single-electron ionization on the binding characteristics. The f_i's have the advantage of partitioning the total binding force into a sum of additive one-electron contributions. The difference between the two definitions is a consequence of the fact that $E_{HF} \neq \Sigma_i n_i \epsilon_i$, as was also pointed out by Goscinski and Palma (*54*). These latter authors also allow for and discuss the effects of electronic relaxation on the derivatives of the orbital energies. Qualitatively, the two definitions of binding agree rather well, except with regard to the core orbitals, which the dynamic analysis finds to be antibinding in homonuclear diatomics.

2-5. CORRELATION OF ELECTROSTATIC BINDING WITH DISSOCIATION ENERGIES

One cannot obtain a correlation between the dissociation energy D_e and the attractive force binding the nuclei by identifying the latter with the total electronic force (increasing as Z^2). In the total electronic force at R_e most of the density continues to simply screen nuclear charge as in the separated atoms, and in no way reflects the change in density or forces, and hence in energy, due to molecule formation.

The pertinent changes in $\rho(r)$ are reflected in the atomic, overlap, and screening contributions to the forces (*6, 7*). In a covalent bond the large overlap population or shared density between the two nuclei results in a descreening of *both* of them. The resulting net repulsive force between the nuclei is counterbalanced by the attractive force exerted by the overlap density (this is the binding force in a covalent bond). For separated atoms the total screening contribution equals Z, the nuclear charge, and the atomic and overlap contributions are zero. Data for first-row homonuclear diatomics (Table 2-2) reveal that the nuclei are indeed unshielded in molecule formation, as the screening contribution is less than Z in every case. The difference between Z and total shielding (Table 2-2) is a measure of the amount of nuclear charge (in the molecule) which is no

Table 2-2. Total Atomic, Overlap, and Screening Contributions to the Forces in Stable First-Row Homonuclear Diatomic Molecules.

	Atomic contribution, $\Sigma_i f_i^{(AA)}$	Overlap contribution, $\Sigma_i f_i^{(AB)}$	Screening contribution, $\Sigma_i f_i^{(BB)}$	Z minus screening, $Z - \Sigma_i f_i^{(BB)}$	D_e (eV)
Li_2	−0.563	0.927	2.591	0.409	1.106
B_2	−0.644	1.708	3.887	1.113	2.884
C_2	−0.735	2.198	4.523	1.477	6.251
N_2	−1.943	3.853	5.136	1.864	9.909
O_2	−2.284	3.486	6.788	1.212	5.181
F_2	−1.949	2.505	8.381	0.619	1.647

longer screened by atomic densities. Nuclear descreening occurs because of distortion of atomic densities and charge migration to the overlap region. There is, in fact, a correlation between the amount of descreening and the magnitude of the overlap force, both increasing to a maximum at N_2 and then decreasing through O_2 to F_2; this parallels the variation in the amount and disposition of the shared density increase exhibited in the BDMs. Likewise, the atomic contributions provide a quantitative measure of the antibinding effect exerted by the density increase in the lone-pair or nonbonded region. In every case the overlap contribution exceeds the deficit created by nuclear descreening and the excess attractive force of the former is balanced by the negative atomic force.

Thus, in a covalent bond the sum of atomic and overlap forces provides a measure of the net attractive force binding each unscreened nuclear charge by the electron density which has been redistributed in forming the molecule. Since $\Sigma_i(f_i^{(AA)} + f_i^{(AB)})/R_e^2$ is the force per unscreened nuclear charge on A, it is a measure of the net attractive electric field on A and correlates linearly (*6, 7*) with D_e (Fig. 2-16). The only molecule which is seriously out of line is Li_2, whose bond involves primarily the overlap of *s* AOs and not *p* AOs as in other molecules of Fig. 2-16 (there may well be different slopes for different bond types). For the molecular ions O_2^+, O_2^-, and N_2^+ the sum $\Sigma_i(f_i^{(AA)} + f_i^{(AB)})$ is corrected by adding 0.5 for negative ions and subtracting 0.5 for positive ions to take account of the fact that there is initially present, between the nuclei, an attractive force for negative ions and a repulsive force for positive ions. One can of course argue that there is no a priori reason why the graph (Fig. 2-16) should be linear and that the breakdown of the total electronic force into atomic, overlap, and screening contributions is not unique. Nevertheless, the existence of such a linear relationship suggests that the bond strength is related to, if not determined by, the magnitude of the net electrostatic field at the nuclei due to the charge redistribution accompanying molecule formation.

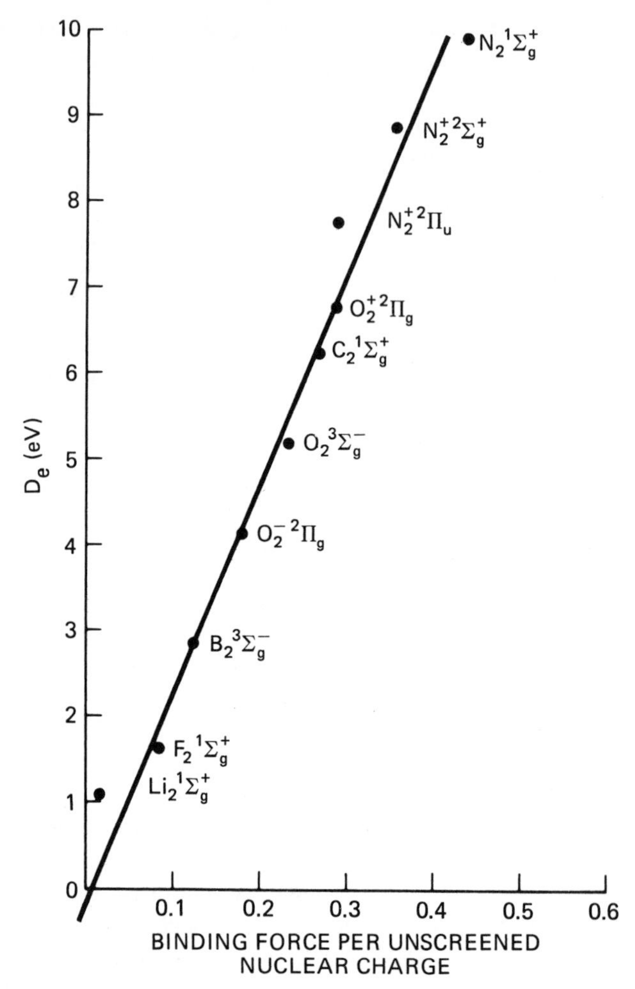

Fig. 2-16. A plot of the binding force per unscreened nuclear charge (the quantity $\Sigma_i(f_i^{(AA)} + f_i^{(AB)})/R_e^2 = (Z - \Sigma_i f_i^{(BB)})/R_e^2$) versus the molecular dissociation energy. The linearity of the plot illustrates the existence of a relationship between the net binding electric field exerted on the nuclei by the charge density and the bond strength.

We shall now discuss a similar but more detailed study by Hirshfeld and Rzotkiewicz (*56*) of the correlation between a net binding force and dissociation energies for AH and A_2 systems. The basis for their model is the resolution of the molecular charge density $\rho_m(r)$ into a sum of (spherical) atomic densities $\rho_a(r)$, each referred to its nuclear position as origin, and the density difference $\Delta\rho(r)$. The atomic density $\rho_a(r)$ represents an intermediate "promolecule" state whose ultimate distortion to yield $\rho_m(r)$ is described by the charge migration $\Delta\rho(r)$. Correspondingly, the electrostatic field E_A at a nucleus A in a molecule AB may be equated to the field in the promolecule plus the field resulting from the charge migration $\Delta\rho(r)$.

$$E_A = E_{p,A} + E_{\Delta,A}. \tag{2-12}$$

The field $E_{p,A}$ must be repulsive (Section 2-3) and equal to

$$E_{p,A} = q_B/R_e^2,$$

where q_B is the charge of atom B lying beyond a sphere centered on B of radius R_e. The charge q_B is simply a measure of the unscreened nuclear charge on B, for if at R_e nucleus A has penetrated the charge density on atom B such that a sphere of radius R_e centered on B contains N_B electrons, then $q_B = Z_B - N_B$ and $E_{p,A}$ is

$$E_{p,A} = (Z_B - N_B)/R_e^2. \tag{2-13}$$

$E_{p,A}$ is called the penetration field; it is clearly another measure of nuclear descreening. N_B is set equal to the number of charges on B which would exert the same field at A as would a spherical atomic population[11] on B of radius R_e.

At R_e the net field E_A must vanish and hence

$$E_{\Delta,A} = -E_{p,A}, \tag{2-14}$$

i.e., the charge migration acts to bind the nuclei together against the repulsive fields acting from mutual atomic penetration at R_e. This is another attempt to dissect the vanishing H-F field into repulsive and attractive parts, with the penetration field corresponding to the repulsive field resulting from nuclear descreening and the migration field to the net attractive field exerted by the overlap and atomic contributions. Thus, the magnitudes of the penetration (or migration) fields correlate postively with dissociation energies. $E_{\Delta,A}$ is characteristic of atom A and not of its ligand B. Thus the $E_{\Delta,H}$ fall on one curve, while the $E_{\Delta,A}$, whether for A_2 or AH, fall on another. The binding behavior of H, however, appears to be distinct from that of the other elements.

The total migration field is decomposed into contributions from the core, σ and π densities as determined by the orbital description of total density of the molecule, and separated atoms. The contribution to the migration field dominates binding in all molecules with occupied π orbitals. This result is obtained when one first sums the core and σ valence contributions to the migration field, contributions which are consistently opposed to one another.

Hirshfeld and Rzotkiewicz provide a very useful and detailed radial analysis of the contributions of $\Delta\rho(r)$ to the migration field. A radial distribution $\epsilon_A(\tau)$ is defined as

$$\epsilon_A(\tau) = -2\pi \int_0^\tau dr \int_0^\pi \Delta\rho \cos\theta \sin\theta \, d\theta, \tag{2-15}$$

which specifies the contribution to $E_{\Delta,A}$ of the difference density lying within a sphere of radius τ centered at nucleus A. Values of τ at which $\epsilon_A(\tau)$ has negative slope identify spherical shells about nucleus A in which forward polarization of the electronic charge favors binding of that nucleus. Similarly, regions of positive slope imply backward polarization whose electrostatic effect is antibinding.

The simplest behavior is found when the origin is placed at a proton. The radial migration field $\epsilon_H(\tau)$ has almost uniform negative slope out to $\tau \sim R$, reflecting a general forward polarization of $\rho(r)$ at all distances. This observation is in total accord with the generalized description given previously of the dominant forward charge polarization near a proton found in the BDMs. Also in accord with the conclusions made from the BDMs is the marked contrast in the pattern for $\epsilon_A(\tau)$ when A is a nucleus from Li $\rightarrow$ F. In these cases a complex pattern of alternating slopes is found that clearly reflects the competing influences of the several components, core, σ valence, and π valence densities.

The same authors also calculate the total classical electrostatic energies and forces of the promolecule. This procedure, however, goes beyond any well-defined quantum mechanical procedure. An interested reader may wish to refer to the results so obtained. We reproduce the summary (*56*) of this electrostatic analysis of the charge migration field generated by the density difference $\Delta\rho(r)$. (a) The contribution $E_{\Delta,A}$ of the charge migration $\Delta\rho$ to the H–F field at a nucleus correlates positively with molecular binding energy. (b) The slope of this correlation is much steeper in hydrogen than in the first-row atoms. (c) The σ valence density around a first-row nucleus is polarized oppositely in its inner and outer regions, the reversal occurring sharply at the radius of the atomic 2*s* node. The outer portion of this density makes the larger contribution to $E_{\Delta,A}$. This contribution varies regularly with atomic number, from binding in Li to strongly antibinding in F. (d) The core of a first-row atom is significantly polarized, in the same sense as is the inner portion of the valence density, because of exchange interaction between the interpenetrating charge distributions. Consequently, the net electrostatic effect of the orbitals–core plus valence–is near zero. (e) A first-row atom can bind strongly only if it has occupied π orbitals. These lack radial nodes and are readily polarized forward, contributing greatly to the electrostatic binding even in molecules having no formal π bonds. This π binding in the first-row is the proper counterpart of the familiar σ binding in hydrogen. (f) *Charge contraction toward the atomic* *nuclei, regarded by some as an essential feature of covalent binding, is characteristic of the hydrogen atom but plays almost no role in the first-row atoms.*

The opposing polarization of the core density relative to the direction of transfer of valence density has been noted previously for LiF and diatomic hydrides (*6, 7*). The BDMs for NaH $\rightarrow$ HCl (Figs. 2-12, 2-13) indicate a rather complex polarization of the core densities in these molecules. It would be sensible to speak of the changes in $\Delta\rho(r)$ near the A nucleus as arising from the very slight polarization of the neonlike *KL* shell ($1s^2 2s^2 2p^6$) relative to the free-atom situation.

Analysis of individual f_{iA} contributions for the core orbital densities indicates two opposing core polarizations. Thus, in NaH and MgH the $1\sigma^2$, $2\sigma^2$, and $1\pi^4$ densities exert a repulsive field tending to pull A from H, but the $3\sigma^2$ density exerts an attractive field; for AlH, SiH, PH, SH, and HCl the roles of these two types of densities are reversed. The spherical ($1\sigma^2 \sim 1s^2$, $2\sigma^2 \sim 2s^2$) and perpendicular ($1\pi^4 \sim 2p\pi^4$) core polarizations dominate the axial ($3\sigma^2 \sim 2p\sigma^2$) core polarization in every case.

In summary, the transfer of valence density from the Na^+-like core to the proton (Fig. 2-12) induces an opposing polarization in the $2s^2$ and $2p\pi^4$ densities (namely, $\overleftarrow{A}$—H) which is countered by a deeper axial attractive polarization of the $2p\sigma^2$ density (namely, $\overrightarrow{A}$—H) which, in turn, is opposed by a smaller back polarization of the $1s^2$ density ($\overleftarrow{A}$—H) in the neighborhood of the Na nucleus. As the direction, extent, and loci of the charge transfer comprising the area in which $\Delta\rho(r) > 0$ in the binding region change, the core response changes, as shown by these three polarizations. In MgH the polarizations are reduced in value from those in NaH and are more evenly balanced. Through the sequence AlH $\rightarrow$ HCl, increasing charge is transferred from H to A with an increasing axial contribution and reversal of the *direction* of polarization. Thus, in HCl the outermost core polarization (of the $2\sigma^2$ and $1\pi^4$ densities) is directed towards the proton in response to the transfer of valence density from H to behind Cl, the axial polarization is directed away from H and the inner $1\sigma^2$ polarization is directed toward H.

While the charge equivalent of the net field due to the "A nucleus + core" equals the number of valence electrons on A for both *K*-shell and *KL*-shell cores, the field strength itself is reduced in second-row hydrides because of the increased size of the *KL*-shell core relative to molecular size as compared to that of the K-shell core. This reduction in the effective field is largely responsible for the differences between binding of the proton in first- and second-row hydrides (*33*).

2-6. FROM THE THEORY OF BINDING TO THEORIES OF BONDING

Since force is only *one* of many constraints on the molecular density $\rho(r)$ such that the latter reproduces the observed values for various properties of the system, one cannot argue that $\rho(r)$ is distributed so as to yield a state of electrostatic equilibrium. The fundamental constraint which determines the properties of a quantum mechanical system may be stated as an extremum principle, namely, that the appropriate energy integral be minimized with respect to all arbitrary variations in an antisymmetrized wavefunction. The calculus of variations yields Schrödinger's equation as an equivalent statement of this extremum principle. Thus, *charge is distributed in a molecule so as to minimize the energy*. Although at present we do not have a complete and exact expression relating $\rho(r)$ directly to the energy E, the well-known Hohenberg-Kohn theorem (*58*) states that the ground-state energy is a unique functional of $\rho(r)$. There is also growing evidence (see, e.g., Ref. 97) that a variational procedure (*63*) exists for the energy in terms of the density functional (Section 2-7). The charge density of a system uniquely determines the external PE operator and thus the Hamiltonian. Therefore the KE and wavefunction are also uniquely determined. Thus, it is meaningful to relate the observable E and its potential and kinetic components to the measurable $\rho(r)$, and thereby obtain a totally physical model of the system.

2-6-1. The Virial Theorem, the Electrostatic Theorem, and Chemical Bonding

The H-F and virial theorems are regarded as "two of the most powerful theorems applicable to molecules and solids" (*84*). However, the interpretations of chemical binding resulting from these theorems have been strongly criticized (*48*). We present both viewpoints here by first showing how the two theorems combine to yield relationships between forces exerted by the charge density and the changes in kinetic (T) and potential (V) energies which accompany chemical bonding.

The virial theorem for diatomics is (*85*)

$$2T = -V - R(dE/dR). \tag{2-16}$$

At $R = R_e$ or ∞, T equals $-\frac{1}{2}V$. Since at R_e, $F(R)$ is zero and the binding energy $\Delta E(=E_m - E_a) < 0$, we have $\Delta V < 0$, $\Delta T > 0$ and $2\Delta T = -\Delta V$.[12] Thus, the virial theorem ascribes molecular stability to a decrease in potential energy, a change which necessarily demands that the KE increase. Since, next to nuclear positions, the internuclear region has the lowest potential, such an interpretation appears to be consistent with change accumulation in the binding region.[13] These two conclusions, i.e., (a) that molecular stability arises from a decrease in V in spite of an increase in T and (b) that this lowering in V is due to charge accumulation in the binding region, have been criticized (*48*, *79*). Before discussing these criticisms we must fully consider the electrostatic and virial approaches to bonding.

One cannot ascribe molecular stability to one particular component of total energy. The decrease in total energy must be considered, and thus the KE and PE have complementary roles. By combining the virial and electrostatic theorems one can obtain a set of constraints on ΔT and ΔV for any given ΔE in terms of the forces exerted on the nuclei.

Figure 2-4 shows a typical force curve for a stable molecule. One can write

$$\Delta E(R') = E(R') - E(\infty) = \int_{\infty}^{R'} F(R)\, dR, \tag{2-17}$$

the last term indicating the area defined by the force curve and the R axis. Thus, the binding energy $\Delta E(R_e)$ is determined by the forces and their increments with R, which are related to corresponding changes in $\rho(r)$. If during the approach of the atoms $\rho(r)$ is continuously redistributed so as to exert large binding forces on the nuclei for intermediate values of R, the resulting molecular state will be tightly bound. This is possible only if $\rho(r)$ in the binding region increases continuously as R decreases. From (2-16), $T < \frac{1}{2}|V|$ or $\Delta T < \frac{1}{2}|\Delta V|$ over the range of R in which the net force is binding, and $\Delta T = \frac{1}{2}|\Delta V|$ at R_e. The formation of an unbound state is characterized by $\Delta T > \frac{1}{2}|\Delta V|$, indicating that the forces on nuclei are repulsive over the whole range of R (except for certain large R where the van der Waals minimum occurs). Thus, forces on nuclei, whether they are binding or antibinding, place restraints on T and ΔT relative to V and ΔV.

The above restraint on ΔT in molecule formation is determined just by the relative sign (i.e., attractive or repulsive) of the force. Knowledge of the complete force curve and its derivative at R places additional general constraints on dT/dR and dV/dR, and on ΔT relative to ΔV for bound molecules. Subtracting T from each side of (2-16) and differentiating with respect to R yields

$$dT(R)/dR = 2F(R) + R\, dF(R)/dR, \tag{2-18}$$

where $-dE(R)/dR$ has been set equal to $F(R)$. The addition of $2V$ to each side of (2-16) and differentiation with respect to R yields a corresponding expression for $dV(R)/dR$,

$$dV(R)/dR = -3F(R) - R\, dF(R)/dR. \tag{2-19}$$

The points at which $F(R)$ equals zero and its minimum value, enable one to define ranges for values of R in which the signs of $F(R)$ and its derivative are uniquely determined (Table 2-3). The force curve is the derivative of the energy curve $E(R)$ and the inflexion point in $E(R)$ corresponds to the point at which

$F(R)$ is a minimum, a point labeled R_i. At this point the binding forces exerted on the nuclei by the charge density attain their maximum magnitude.

In range I, where $\infty > R > R_i$, $F(R)$ and $dF(R)/dR$ have opposite signs, and the signs of dT/dR and dV/dR are not uniquely determined (Table 2-3). However, for large R, dT/dR and dV/dR ought to be determined by the sign of $dF(R)/dR$ [see (2-18) and (2-19)]. This was found for H_2 (*13*), using beyond-Hartree-Fock wavefunctions (*41*). The data (Table 2-4) indicate that KE initially *decreases* ($dT/dR > 0$) during the approach of two H atoms, and this continues up to $R_i \sim 2$ au (Fig. 2-4). Correspondingly, PE increases initially ($dV/dR < 0$), with, e.g., $\Delta V = +0.0569$ au at $R = 4$ au. The DDM for H_2 (Fig. 2-3) indicates that at $R < R_i$ the charge buildup in the binding region requires charge removal from the cusplike confinement in the region of each nucleus. This is the primary cause of the initial decrease in KE and increase in PE of the system as the two atoms approach each other, leading, respectively, to a negative and a positive contribution to ΔE. However, since the charge removed from the nuclei is transferred to the binding region, the resultant forces on the nuclei are (always) *attractive*, and hence $\Delta E < 0$.[14] Thus, the electrostatic and virial theorems provide a detailed physical description and understanding of the long-range attraction between atoms in terms of the charge redistribution which accompanies their approach (see below).

At $R = R_i$, where the binding force attains its highest magnitude, the signs of $dT/dR (<0)$ and $dV/dR (>0)$ are uniquely determined (Table 2-3). Thus, the decrease in energy for $R >> R_i$ can result from a decrease in T and in spite of an increase in V. However, for $R \leqslant R_i$, up to R_e, T and V must, respectively, begin to increase and decrease. For example, as R in H_2 decreases from infinity, initially $dV/dR < 0$ and $\Delta V > 0$. However, at $R = R_i = 2$ au, $\Delta V = -0.0625$ au and hence dV/dR changes in sign before $R = R_i$. Thus, from the present general analysis, the charge rearrangements in H_2 (Fig. 2-3) for $R \leqslant 2$ au lead to a decrease in the PE of the system and to an increase in the KE. The

Table 2-3. General Behavior of dT/dR and dV/dR in the Formation of a Bound Molecular State.

	Range of value of R	$F(R)$	$dF(R)/dR$	$dV(R)/dR$	$dT(R)/dR$
I	$\infty > R > R_i$	<0	>0	$\lesseqgtr 0$ <0 for $R >> R_i$,	$\lesseqgtr 0$ >0 for $R >> R_i$
	$R = R_i$	<0	0	>0	<0
II	$R_i > R > R_e$	<0	<0	>0	<0
	$R = R_e$	0	<0	>0	<0
III	$R < R_e$	>0	<0	>0	<0
IV	$R << R_e$	>0	<0	<0	>0

Table 2-4. Values of $T_{\parallel}$ and $T_{\perp}$ for H_2 and He_2 as a Function of R.

R (a.u.)	T (a.u.)	$T_{\parallel}$ (a.u.)	$T_{\perp}$ (a.u.)	$(T_{\perp} - T_{\parallel})/T$ (a.u.)
		H_2		
1.0	1.4332	0.4032	1.0299	0.4373
1.2	1.2775	0.3465	0.9310	0.4575
1.4	1.1569	0.3043	0.8527	0.4740
2.0	0.9293	0.2343	0.6949	0.4956
4.0	0.9289	0.2885	0.6404	0.3789
8.0	0.9868	0.3288	0.6580	0.3336
∞	1.0000	0.3333+	0.6666+	0.3333+
		He_2		
1.0	6.6190	2.4137	4.2043	0.2707
2.0	6.0803	2.1594	3.9209	0.2897
4.0	5.7381	1.9160	3.8221	0.3322
6.0	5.7249	1.9084	3.8165	0.3333
∞	5.7233	1.9078	3.8156	0.3333+

contributions from $F(R)\,dR$ to the energy integral in the range $R_i \geqslant R \geqslant R_e$ must dominate over those in the range $\infty > R > R_i$, since the total changes in T and V determining $\Delta E(R_e)$ must be such that $\Delta T > 0$ and $\Delta V < 0$.

The signs of dT/dR and dV/dR at $R = R_e$ will persist for some $R < R_e$, since initially $F(R)$ is small and dF/dR is large. However, in this range $F(R)$ rapidly increases (tending to infinity) while dF/dR remains finite. Thus eventually, for some $R < R_e$ the signs of dT/dR and dV/dR will again change, and V and T will respectively increase and decrease with decreasing R.

We now summarize the requirements for the interpretation of a bound molecular state in terms of the electrostatic theorem. A bound molecular state is obtained if the electrostatic forces on the nuclei are predominantly attractive over the internuclear range from $R = \infty$ to R_e.[15] If the attractive forces in this interval are larger in magnitude, the resulting molecule is more stable. To obtain attractive forces, $\rho_m(r)$ in the binding region must continuously increase during the approach of the two nuclei, to yield electron-nuclear forces which exceed and eventually balance the ever increasing nuclear repulsion. For any R, $\rho_m(r)$ must exceed $\rho_a(r)$ obtained by superimposing atomic densities within some portion of the binding region (i.e., $\Delta\rho(r) > 0$). This is a necessary but not sufficient condition to ensure attractive forces. Thus, as R decreases $\rho_m(r)$ must increase faster than $\rho_a(r)$ within some portion of the binding region.

In view of the criticisms of the electrostatic approach (see below), the interpretation of the incremental contributions $F(R)\,dR$ to the total binding energy

$\Delta E(R_e)$ require a careful analysis. Berlin (*31*) first drew attention to the result that since the electrostatic force is determined by just the change in the potential energy operator $\hat{V}$ averaged over the wavefunction,

$$F(R) = -dE/dR = -\langle \psi, (\partial \hat{V}/\partial R)\, \psi \rangle = -(dV/dR)_1, \tag{2-20}$$

the change in the *average* KE must be equal and opposite to that part of the change in the *average* PE which is determined by the change in the wavefunction. That is, since

$$\langle \partial\psi/\partial R, (\hat{T} + \hat{V})\, \psi \rangle + \langle \psi (\hat{T} + \hat{V}), \partial\psi/\partial R \rangle = 0 \tag{2-21}$$

then

$$dT/dR + (dV/dR)_2 = 0. \tag{2-22}$$

The total change in the average PE is given by

$$dV/dR = (dV/dR)_1 + (dV/dR)_2,$$

where $(dV/dR)_1$ is the H-F force term. By virtue of the cancellation in (2-22) one obtains the interpretation of a change in E being determined by an electrostatic force acting through some distance

$$dE = -F(R)\, dR. \tag{2-23}$$

One may thus view the infinitesimal change dE as being the work done in moving the nuclear charges a distance dR in a field determined by a purely electrostatic potential (ESP); in turn, the binding energy may be viewed as the sum of such infinitesimal contributions to the work required in moving the nuclei from $R = \infty$ to $R = R_e$. This is the basis of the electrostatic approach and the source of its conceptual simplicity.

In accordance with this view of the electrostatic theorem (*87*), let us write the ESP $\Phi(\mathbb{R}|R'_\alpha)$ generated at a point R'_α by the electronic distribution and the remaining nuclei, for some particular configuration of the nuclei denoted by $\mathbb{R}$, as

$$\Phi(\mathbb{R}|R'_\alpha) = -\int \frac{\rho(\mathbb{R}|r)}{|r - R'_\alpha|}\, dr + \sum_{\beta(\neq\alpha)} \frac{Z_\beta}{|R_\beta - R'_\alpha|}. \tag{2-24}$$

The force exerted on nucleus α is given by the gradient of this potential at the point $R'_\alpha = R_\alpha$,

$$F_\alpha(\mathbb{R}) = -Z_\alpha [\nabla'_\alpha \Phi(\mathbb{R}|R'_\alpha)]_{R'_\alpha = R_\alpha}. \tag{2-25}$$

The difference in energy ΔE between two configurations of the nuclei, $\mathbb{R}_1$ and $\mathbb{R}_2$, is equal to the electrostatic work done in moving nucleus α from its position $(R_{1\alpha}, \mathbb{R}_1)$ in the configuration $\mathbb{R}_1$ to its position $(R_{2\alpha}, \mathbb{R}_2)$ in the configuration $\mathbb{R}_2$ through the gradient of the potential $\Phi(\mathbb{R}|R_\alpha)$ which is itself a function of the nuclear coordinates $\mathbb{R}$,

$$\Delta E = -Z_\alpha \int_{(R_{1\alpha}, \mathbb{R})}^{(R_{2\alpha}, \mathbb{R})} [\nabla'_\alpha \Phi(\mathbb{R}|R'_\alpha)]_{R'_\alpha = R_\alpha} \, dR_\alpha. \tag{2-26}$$

Early efforts (*1*, *59*) to calculate molecular binding energies by the electrostatic method were motivated by the great simplicity of this approach. But the method yields only approximate results, as one must model $\rho(r)$ as a function of R if (2-26) is to be used in an a priori way to calculate ΔE. However, the conceptual link between ΔE and the ESP as determined by the charge distribution is the important result provided by the electrostatic view of chemical binding.

The electrostatic interpretation of chemical bonding on the basis of the cancellation in (2-22) has been criticized (*48*, *78*) by saying that the H–F theorem "does not even make statements regarding the properties of dV/dR and dT/dR, not to speak of the undifferentiated quantities E, T, and V." It is true that the H–F force at a single value of R yields no information besides the value of $F(R)$, and such a single bit of information is obviously insufficient to account for a chemical bond. However, knowledge of a single value of $E(R)$ without the knowledge of its value at $R = \infty$ does not indicate whether or not bonding has occurred, or yield values of T and V either. The above criticism is given in a paper (*48*) which considers bond formation in H_2^+ from the separated atoms in terms of the *calculated* changes in E, T, and V over the entire range of R. In a discussion of chemical bonding, one is concerned with the change in energy relative to the separated atoms, and the variation with R of the physical quantities involved seems to be an obvious and necessary choice. Granting the same degree of knowledge in the electrostatic approach, i.e., a knowledge of $F(R)$ as a function of R, then, as demonstrated above, one can make statements regarding the signs and magnitudes of dV/dR and dT/dR, and hence about the values of V and T, and what is most important, *one can relate these changes in a direct physical manner to the changes in the charge density via the quantities $F(R)$ and $dF(R)/dR$.* The same paper (*48*) states that "the strong negative values of $-dV/dR$ upon approach to equilibrium arise *not* from the Hellmann–Feynman contribution [i.e., not from $(dV/dR)_1$] but from $(dV/dR)_2$." First, the quantity $(dV/dR)_2$ may be related to the H–F force and its derivative, for it is simply the negative of dT/dR as given in (2-22). Second, while $F(R)$ does vanish at the point R_e, the *electronic* contribution to the gradient of the ESP does not. As demonstrated in Section 2-5, there is a correlation between the binding energy ΔE and the magnitude of a *net* binding electrostatic field exerted on the nuclei at R_e. This provides a static interpretation of binding in terms of the forces generated by the net reorganization of the charge density found at one value of R, namely, R_e. Alternatively, one may use (2-26) to relate ΔE to the sum of the increments $F(R)\,dR$ from $R = \infty$ to R_e. In this approach, the more negative the values attained by the electrostatic force over the complete range of R values, the more negative the value of ΔE and the stronger the binding.

In either view, static or as a function of R, the binding energy ΔE is determined by $-(dV/dR)_1$, the electrostatic force, and its electronic contribution. Alternative interpretations of ΔE, while possible, cannot yield a conclusion at variance with the electrostatic result, since (2-26) is, after all, a statement of quantum mechanics.

Ruedenberg (*48*) has criticized statements which relate the lowering in PE to the accumulation of charge in the bond region:

> Since the charge in the binding region generates attractive forces and since charge does, in fact, accumulate in the bond region, it has been speculated that the potential-energy lowering, observed for a stable molecule *in the neighborhood of the equilibrium distance*, might arise from this charge accumulation in this region. This conjecture cannot be upheld in view of the detailed quantitative analysis of the energy contributions involved in the variational process, as summarized in the preceding sections. For example, we have seen that, for internuclear distances larger than 4.0 au, incipient binding occurs due to a drop in kinetic energy and in spite of an increase in potential energy, and that these two energetic effects are associated with accumulation of charge in the bond.

In answer to these criticisms we must first stress that there is an electrostatic interpretation of molecular stability based on the properties of the ESP and that this interpretation is distinct from other interpretations, say one involving changes in KE and PE. *The views are complementary, but they must yield identical conclusions in the interpretation of phenomena they discuss in common.* First, in the electrostatic approach charge *must* be increasingly concentrated in the binding region to cause a continuing decrease in the *electronic contribution* to the ESP as R is decreased, for it is the gradient of this potential which determines the force which when integrated, yields ΔE. If the molecule is to be bound the forces must be attractive–hence charge density must be increased in the binding region. To deny this is to deny the validity of (2-26), which is simply an integrated form of the H-F theorem. To follow the argument through, since $\Delta E < 0$ demands an increase in $\rho(r)$ in the binding region and since *at the equilibrium separation* $\Delta V < 0$, the same *redistribution of charge* which yielded the binding forces must also describe the decrease in the total PE. The BDMs shown earlier indicate that $\rho_m(r)$ is increased in regions other than the binding region relative to $\rho_a(r)$ at R_e. This observation, while important and discussed more fully below, does not invalidate the extremely useful and simple conclusion of the electrostatic interpretation of bonding, namely, that if $\Delta E(R_e) < 0$, the forces must be attractive and hence $\rho_m(r)$ must increase in the binding region relative to $\rho_a(r)$.

Ruedenberg (*78*, *79*), and Feinberg and Ruedenberg (*48*) discuss the formation of a chemical bond as occurring through a number of imagined, discrete stages. In particular, in the discussion of H_2^+ (*48*), each successive stage is described by a wavefunction of increasing accuracy (as judged by their increasingly better variational estimates of the average energy). The argument which these authors then present to disprove that charge accumulation in the "bond region" leads to a lowering of PE runs as follows. The values of ΔT and ΔV for H_2^+ predicted by this series of wavefunctions are compared. The crudest of these wavefunctions is simply the linear combination of 1*s* hydrogen orbitals. This function not only fails to satisfy the virial theorem at any finite value of R, *it totally violates the principal statement of the virial theorem.* Thus, at R_e this function predicts $\Delta T < 0$ and $\Delta V > 0$, and hence binding is obtained as a result of a lowering in T!

The $\Delta\rho(r)$ map obtained from this function is similar in its characteristics to those obtained from the proper wavefunctions for H_2^+ and H_2 at large values of R, i.e., $\Delta\rho(r) > 0$ in the center of the binding region and $\Delta\rho(r) < 0$ in the region of each nucleus. As previously discussed, such a redistribution corresponds to a migration of charge from regions of low to high potential, thereby yielding $\Delta V > 0$ and $\Delta T < 0$. This is the correct description of the system for large values of R, but a physically impossible description of the system in the neighborhood of the equilibrium separation. However, since this function does give a $\Delta\rho(r)$ map which is positive over a portion of the binding region, the authors conclude that accumulation of charge in the "bond" region does not lead to a lowering in the PE. This conclusion, based on an impossible model of the equilibrium system, cannot be generalized to describe the real physical situation of the system in the neighborhood of R_e. If it could, one would in effect be using a result obtained from a wavefunction which does *not* obey the virial or H-F theorems to disprove a conclusion based on these theorems.

The next wavefunction they consider is a scaled version of the first. A scaling of the electron coordinates ensures the satisfaction of the virial theorem, and indeed this function predicts $\Delta V < 0$ and $\Delta T > 0$ at R_e. The effect of scaling is to increase the exponents in the 1*s* orbital functions over the separated-atom value of unity. This results in a contraction of the orbital and the density, and to an even larger accumulation of charge in the binding region. Hence their conclusion that the lowering of PE is a result of orbital contraction.

In an earlier criticism of the virial theorem (*78*), Ruedenberg stated that it was essential that the KE and PE of a molecule be compared not with the same quantities for the separated atoms (which yields $\Delta V < 0$ and $\Delta T > 0$) but with those of atoms in special "promotion states." The fictitious promotion state corresponds to one in which the valence electrons contract toward the nucleus. Such a "contractive promotion" has two obvious effects on the energy of the system; the PE is decreased below and the KE increased above their final equilibrium values. The formation of a molecule from atoms in promoted states thus results in $\Delta V > 0$ and $\Delta T < 0$, at variance with the virial theorem. The "promotion state," aside from being imprecisely defined, is again one which does *not* (even according to Ruedenberg) satisfy the virial theorem.

In a study of bonding or binding which is restricted to the use of charge distributions derived from wavefunctions which *do* satisfy those theorems of quantum mechanics which are subsequently used in the interpretation of the properties of the charge distributions, the subjective role of the observer should be severely limited. The observations made in such a study will remain true; personal bias is then restricted to an ordering of the importance of the observations and to their possible interpretation in terms of models.

As will be demonstrated in the next section, a study of such theoretically acceptable charge distributions indicates that the KE does indeed play a complex role in the formation of a bond. It is possible to demonstrate (*13*) the existence of spatial regions in which the KE is decreased relative to the separated-atom values as determined by the correct molecular density and the correct separated-atom densities.

The further statement (*48*) that for large internuclear separations $\Delta T < 0$ and $\Delta V > 0$ is irrelevant to their original statement, which was concerned with the effect of charge accumulation in the bond region in the neighborhood of R_e. The virial and H-F theorems dictate that in this region V must be decreasing with decreasing R and ΔV must be less than zero, as discussed above. As for the observations that $\Delta T < 0$ and $\Delta V > 0$ for large values of R, no inconsistencies arise when the electrostatic approach is used to interpret this result. Indeed, as discussed previously, by focusing attention on the changes in $\rho(r)$ for large R these changes in T and V are both predicted and rationalized. The $\Delta\rho(r)$ maps for H_2 at $R = 8$ and 6 au represent a distribution of charge which lowers the KE, increases the PE, and simultaneously, since $\Delta\rho(r) > 0$ in the binding region, decreases the gradient of the ESP.

Central to Ruedenberg's interpretation of bonding is the concept of a contraction of the orbitals toward the nuclear centers. The point is emphasized to the extent that the charge redistribution shown in the BDM for H_2^+ is interpreted not primarily as representing an accumulation of charge in the "bond region" but rather as illustrating the contraction of charge around the nuclei. While admitting that the two effects are not clearly separable, Feinberg and Ruedenberg (*48*) make the following statement immediately after the above criticisms of the PE lowering being the result of charge accumulation in the bond:

> We have seen furthermore that the drop in potential energy upon approach to equilibrium is the direct result of the orbital contraction described by the increase in the orbital exponents and that it occurs because on each side of the symmetry plane which bisects the internuclear axis, electronic charge from the outlying regions moves into closer vicinity of the nucleus on that side.

The statement is made with reference to a BDM for H_2^+ which is similar in all its characteristics to that given here for H_2 (Fig. 2-3). This shift in emphasis in the interpretation of BDMs raises a number of questions. First, the DDMs for He_2 (Fig. 2-5) represent an extreme example of orbital contraction and a consequent contraction of charge into "regions centered around the nuclei." Yet this system is not bound; $\Delta\rho(r) < 0$ over most of the binding region.[16] Second, as discussed in Section 2-3, BDMs for H_2^+ and H_2 are atypical. BDMs for molecules containing atoms other than H or He, in general show a great reduction in charge density at the positions of the nuclei and/or in their immediate vicinity and do not exhibit a contraction of charge toward the nuclei. Instead, for other homonuclear diatomics there is a central increase in the binding region, reaching a maximum at the bond midpoint, away from the nuclei and separate increases in the antibinding regions. The analysis in the next subsection demonstrates that these latter accumulations of density are less contracted than in the separated atoms, as they lead to a local lowering in the KE.

Further criticisms of the electrostatic approach (*48*) are based on an artificially restricted use of the H-F theorem. The final comment of Feinberg and Ruedenberg, that "Berlin's force analysis in terms of a binding and antibinding region is of extremely broad validity and *does not differentiate between situations of quite diverse bonding character*" is a reflection of their belief that H_2^+ is the prototype of the chemical bond. But it is precisely through the study of BDMs and their spatial properties relative to the electrostatically defined binding and antibinding regions that one is able to (a) determine the existence of different bond types and bonding mechanisms, and (b) classify them. Thus, one finds that molecules which exhibit similar characteristics in the redistribution of charge, as illustrated in their DDMs, exhibit similar characteristics in other properties which provide further insight into the mechanism of bonding. For example, the behavior of the KE on bonding is the same for H_2^+ and H_2. Just as the $\Delta\rho$ maps for He_2 represent a reversal of the principal features found in the BDMs for H_2^+ and H_2, so the KE of He_2 is found to change in a manner opposite to that for H_2^+ and H_2. Similarly, the causes of the KE changes in systems such as N_2, whose BDMs exhibit dominant quadrupolar rather than simple dipolar charge redistributions, are again similar to one another, but different from that found for H_2.

Thus, the principal features of the charge redistribution depicted in BDMs serve to categorize the binding mechanism in the electrostatic approach and the bonding mechanism in terms of changes in the PE and KE. This is illustrated in the following subsection in a study which relates the KE of a molecule to the topographical features of the charge density.

2-6-2. Kinetic Energies of Molecular Charge Distributions and Molecular Stability

In order to clarify the role of KE in molecule formation and to relate the former to the charge distribution in a system, we define (*13*) a KE density function $G(r)$ such that

$$\int G(r)\, dr = T, \tag{2-27}$$

where $G(r)\, dr$ is the contribution from the volume element dr to the total kinetic energy T of the system. Taking the natural orbital expansion of the first-order density matrix,[17]

$$\rho(r|r') = \sum_i \lambda_i \phi_i^*(r)\, \phi_i(r') = \sum_i \rho_i(r|r')$$

we have

$$G(r) = \tfrac{1}{2} \sum_i \lambda_i \nabla\phi_i^*(r) \cdot \nabla\phi_i(r) = \tfrac{1}{2}\, [\nabla \cdot \nabla' \rho(r|r')]_{r=r'} \tag{2-28}$$

$$= \frac{1}{8} \sum_i \frac{\nabla\rho_i(r) \cdot \nabla\rho_i(r)}{\rho_i(r)}. \tag{2-29}$$

The operator in (2-29) is the quantum analog of the classical KE, $p^2/2m$:

$$\frac{p^2}{2m} \Rightarrow \frac{1}{2m}(-i\hbar\nabla)^* \cdot (-i\hbar\nabla') = \frac{\hbar^2}{2m} \nabla \cdot \nabla'. \tag{2-30}$$

The density $G(r)$ is everywhere positive and finite, with discontinuities at the nuclear positions due to cusp conditions on the wavefunction (*16*). Although this definition of a KE density is not unique (*13*), it has the desirable properties of being positive and finite, and the operator form in (2-30) appears to be the most fundamental definition of the KE operator (*88*).

$G(r)$ allows one to determine the spatial contributions to the KE of a system and to relate them to the spatial distribution of the charge density $\rho(r)$. For example, what is the consequence of charge accumulation in the binding region to the KE of a system. We review here the results for H_2 and He_2 (*13*) as well as those for other diatomics which exhibit more complex charge redistribution on binding than does H_2.

In his pioneering studies on molecular momentum distributions, Coulson (*39, 40*) observed that in H_2 the mean component of the velocity along the bond direction is decreased while that perpendicular to the bond direction is increased. This is strikingly displayed by $G(r)$, which also locates the spatial regions in which the momentum attains minimum or maximum magnitudes. The differing behavior of $\nabla\rho(r)$ parallel and perpendicular to the bond axis as revealed in $G(r)$ are so pronounced for H_2 that it is reflected in the *average* values of the parallel and perpendicular contributions to the KE,

$$T_{\|} = \frac{1}{2}\left\langle \frac{\partial\psi}{\partial z}\frac{\partial\psi}{\partial z} \right\rangle = -\frac{1}{2}\left\langle \psi \left| \frac{\partial^2}{\partial z^2} \right| \psi \right\rangle$$

$$T_{\perp} = \frac{1}{2}\left(\left\langle \frac{\partial\psi}{\partial x}\frac{\partial\psi}{\partial x} \right\rangle + \left\langle \frac{\partial\psi}{\partial y}\frac{\partial\psi}{\partial y} \right\rangle\right) = -\frac{1}{2}\left\langle \psi \left| \frac{\partial^2}{\partial x^2} + \frac{\partial^2}{\partial y^2} \right| \psi \right\rangle.$$

For separated spherical atoms, $T_{\|} = \frac{1}{2}T_{\perp}$ and hence $(T_{\perp} - T_{\|})/T = \frac{1}{3}$. The change in the value of $(T_{\perp} - T_{\|})/T$ from its atomic value of $\frac{1}{3}$ measures the extent to which molecule formation differentiates between parallel and perpendicular components of the KE. This yields a pertinent summary of the role of KE in the bonding process (*13*).

Figure 2-17 depicts contour and profile plots of $G(r)$, $\rho(r)$, $\Delta G(r)$, $\Delta\rho(r)$ for H_2 at R_e. The integral $\int \Delta G(r)\,dr$ yields ΔT, or minus the total binding energy of the molecule if $R = R_e$. Aside from the binding region, $G(r)$ contours are similar to those for $\rho(r)$. In the binding region, $G(r)$ attains ridgelike maximum values on either side of the internuclear axis, decreasing to a deep minimum in a region around the bond midpoint. In the natural orbital expansion for H_2 (*41*), $\rho_{1\sigma_g}$ accounts for 98% of (total) $\rho(r)$, and the properties of $G(r)$ are determined almost entirely by $\nabla\rho_{1\sigma_g}$. Since $\nabla\rho_{1\sigma_g}$ (or any $\nabla\rho_{\sigma_g}$), both parallel and perpendicular to the internuclear axis, vanish at the bond midpoint, the central minimum in the $G(r)$ map is understandable. That $G(r) = 0.0165$ au at the bond midpoint is because of the small contribution from $\rho_{1\sigma_u}$. From (2-28) a u orbital will contribute to $G(r)$ at the bond midpoint, since its node introduces a nonzero slope for $\phi_i(r)$ along the bond axis in the nodal plane. Thus, the bond-

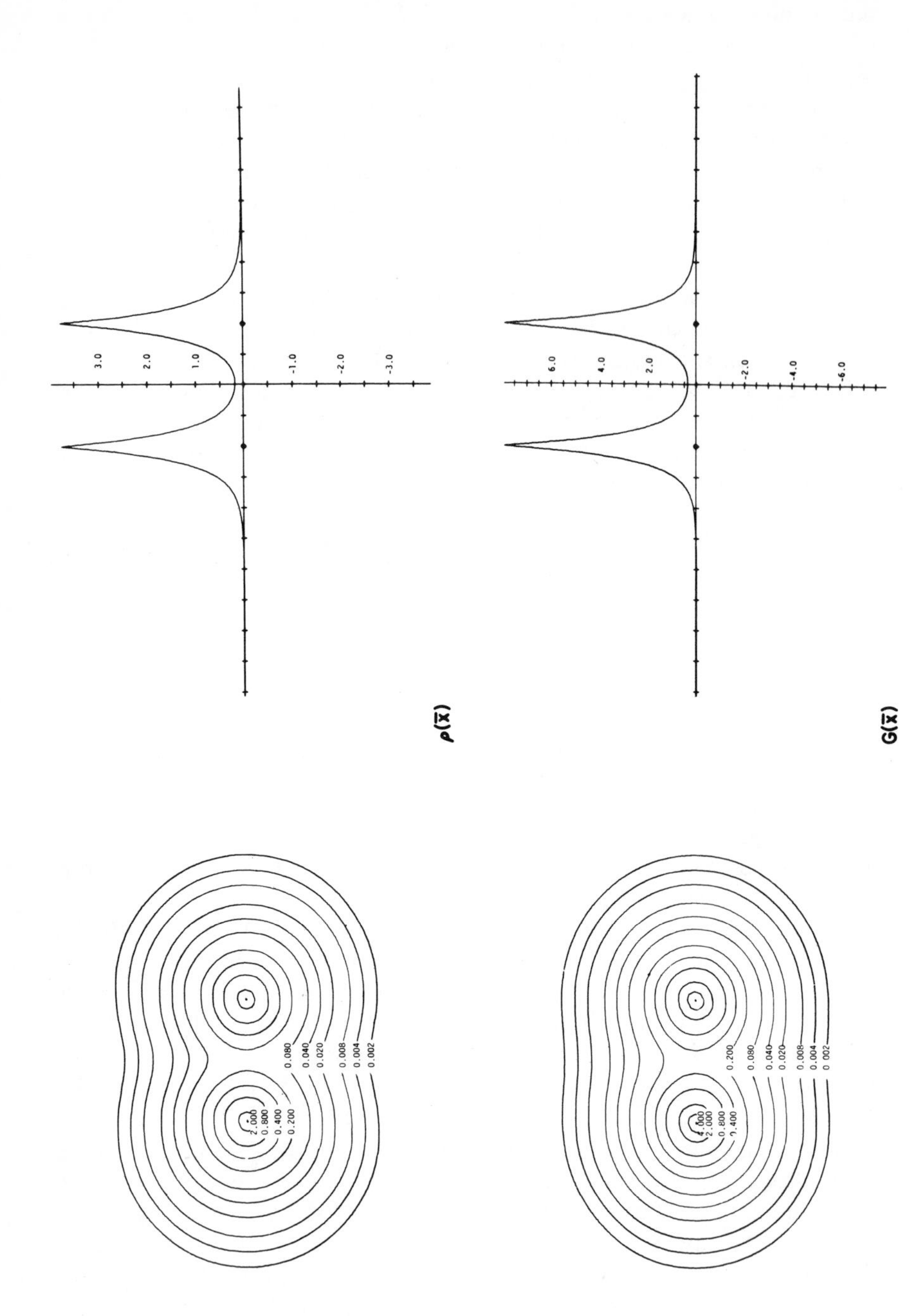

3.0
2.0
1.0
-1.0
-2.0
-3.0
ρ(x̄)
6.0
4.0
2.0
-2.0
-4.0
-6.0
G(x̄)
2.000
0.800
0.400
0.200
0.080
0.040
0.020
0.008
0.004
0.002
4.000
2.000
0.800
0.400
0.200
0.080
0.040
0.020
0.008
0.004
0.002

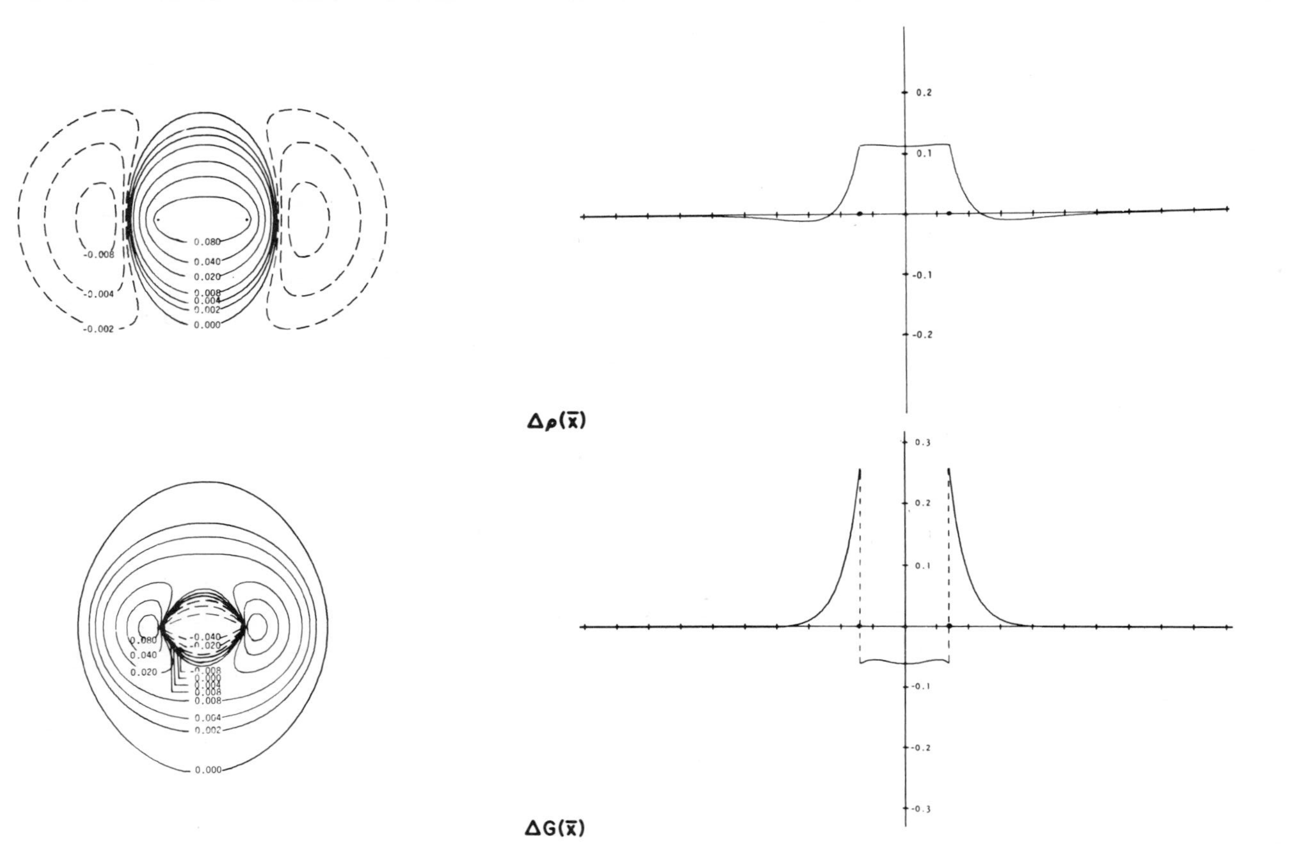

Fig. 2-17. Contour maps of the molecular charge distribution $\rho(r)$ and profile, of the kinetic energy density $G(r)$, and of their corresponding difference functions referenced to the atomic states for $H_2(X^1\Sigma_g^+)$ at $R_e = 1.4$ au. The scale on the internuclear axis of the profile diagrams is 0.5 au. All distributions are given in atomic units.

ing (g symmetry) or antibonding (u symmetry) character of an orbital (or, in general, the presence or absence of a node between the nuclei) has a *pronounced* effect on $G(r)$ in the binding region and on this region's contribution to the KE.

Thus, only for one-electron systems where $\rho(r) = \rho_1(r)$ is KE precisely determined by $\nabla\rho(r)$. The KE will be approximately determined by $\nabla\rho(r)$ for two-electron systems where the ground-state $\rho(r)$ is dominated by a single natural orbital, i.e., $\rho(r) \sim \rho_1(r)$, as in H_2 at $R = R_e$. In particular, $G(r)$ in H_2 may be related to the parallel and perpendicular components of $\nabla\rho(r)$.

The charge accumulated in the binding region of H_2 yields low values for the parallel component of $\nabla\rho(r)$, leading to a decrease in the contribution of $T_{\parallel}$ to T. Since the perpendicular gradients of $\rho(r)$ are zero along the internuclear axis (except at nuclear positions where they are undefined), the profile of $G(r)$ along this axis isolates the effects of the parallel gradients of $\rho(r)$ on the parallel component of the KE density (small in the binding region). The discontinuous decrease in $G(r)$ on the binding side of each nucleus (Fig. 2-17) dramatizes the great reduction in the contributions of parallel gradients of $\rho(r)$ along the whole binding region to the total KE. Because of the contracted nature of $\rho(r)$ perpendicular to the bond axis in the binding region, the perpendicular components are large compared to the parallel component. In the antibinding regions, where $\rho(r)$ is contracted toward the nuclei in both parallel and perpendicular directions, both contributions to $G(r)$ are large.

This differing behavior of the components of the KE density is reflected in the average values of $T_{\parallel}$ and $T_{\perp}$. At $R = R_e = 1.4$ au (Table 2-4), $T_{\parallel} < \frac{1}{2}T_{\perp}$ and $T_{\parallel}(R_e) < T_{\parallel}(\infty)$, while $T_{\perp}(R_e) > T_{\perp}(\infty)$. The considerable differentiation between $T_{\parallel}$ and $T_{\perp}$ caused by the differing parallel and perpendicular gradients of $\rho(r)$ is reflected in the deformation index $(T_{\perp} - T_{\parallel})/T$, which changes from 0.3333 (spherical atoms) to 0.4740 (molecule). The decrease in T for large R (e.g., $R = 8$ au in Table 2-4) is due to almost equal decreases in $T_{\parallel}$ and $T_{\perp}$, since the deformation index remains almost unchanged from its value at $R = \infty$. These observations are consistent with a decrease in $\rho(r)$ in the region of each nucleus, a change that decreases $\nabla\rho(r)$ in both parallel and perpendicular directions. Interestingly, $(T_{\perp} - T_{\parallel})/T$ reaches its highest value near $R = 2$ au, as is true of the electrostatic force. At this R the positive values of $\Delta\rho(r)$ extend across the entire binding region, but not into the antibinding region. Thus, $T_{\parallel}$ should and does attain its minimum value.

The difference distribution $\Delta G(r)$ reinforces the above observations. Note in particular the dramatic decrease in the KE contributions from the charge density accumulated in the binding region relative to the separated-atom values.

From general quantum mechanical principles, the contraction of charge density into regions of low potential is expected to increase KE. This increase is restricted (a) by the virial theorem, namely, $\Delta T = \frac{1}{2}|\Delta V|$ at $R = R_e$, and (b) by the electrostatic requirement, namely, $\Delta T < \frac{1}{2}|\Delta V|$ at $R > R_e$ for binding forces on nuclei. Thus, one is led to consider how in the course of molecule formation charge density must be distributed in space such that (a) and (b) are satisfied. From the previous discussion it is clear that for the H_2 molecule study of the KE density provides the answer to this question.

The charge accumulated in the binding region of H_2 is distributed in such a way as to keep the KE increase to a minimum. This increase, necessary for a decrease in PE (virial theorem), is restricted primarily to perpendicular contributions to $G(r)$. The relaxation of the parallel gradient of $\rho(r)$ in the binding region greatly reduces contributions to $G(r)$ and hence to T, at the same time decreasing V. The contraction of $\rho(r)$ toward the nuclei in the antibinding regions and perpendicular to the axis in the binding region also lowers V, but only at the cost of an increase in T.[18]

Consider now the unstable He_2 system (Fig. 2-18). Although $\rho(r)$ is very peaked at the nuclei, its value at the "bond" midpoint (0.164 au) is much less than that for H_2 (0.268 au). A comparison of the $\Delta\rho(r)$ maps for H_2 and He_2 indicates that apart from an increase in $\rho(r)$ in the nuclear regions in both cases, the charge distributions exhibit opposite behavior with respect to where charge is removed and where it is accumulated. The large $\Delta\rho(r)$ deficit in the binding region of H_2 is a direct result of the antisymmetry requirement (Pauli exclusion principle). From electrostatic analysis, a charge distribution like the $\Delta\rho(r)$ in He_2 is inherently incapable of balancing the forces of nuclear repulsion.

The differing topography of $\rho(r)$ for H_2 and He_2 is reflected in their $G(r)$, particularly in the binding region. The contours of $G(r)$ and $\rho(r)$ in He_2 are similar in shape, indicating that the equidensity contours represent lines of almost constant value for the classical-like contribution to the KE. However, although $\rho(r)$ in the He_2 internuclear region is greatly reduced compared to H_2, $G(r)$ for He_2 in this region is much greater than that for H_2, the values at the bond midpoints being 0.3344 and 0.0165 au, respectively. Due to the ρ_{σ_u} component in He_2, $G(r)$ does not exhibit a minimum at the bond midpoint as it does in H_2. Instead, both parallel and perpendicular contributions to $G(r)$ and hence to T are large in He_2: at the discontinuity in $G(r)$ (too small to be shown in Fig. 2-18), the parallel gradient on the bonded side of a nucleus exceeds in magnitude that on the nonbonded side. Just the opposite behavior is found in the $G(r)$ profile for H_2. Further, both $T_{\parallel}$ and $T_{\perp}$ are increased above their atomic values (Table 2-4), $T_{\parallel}$ increasing more than $T_{\perp}$. Thus, the ratio $(T_{\perp} - T_{\parallel})/T$, instead of increasing from the limiting atomic value of $\frac{1}{3}$ as in H_2, decreases to 0.290. This decrease indicates a greater overall tightening of the charge density parallel to the bond axis, rather than perpendicular to it as in the case of H_2.

The $\Delta G(r)$ map for He_2 strikingly illustrates that the contributions to the KE from molecular charge density are everywhere greater than they are for the separated atoms. In spite of greatly reduced values of $\rho(r)$ in the binding re-

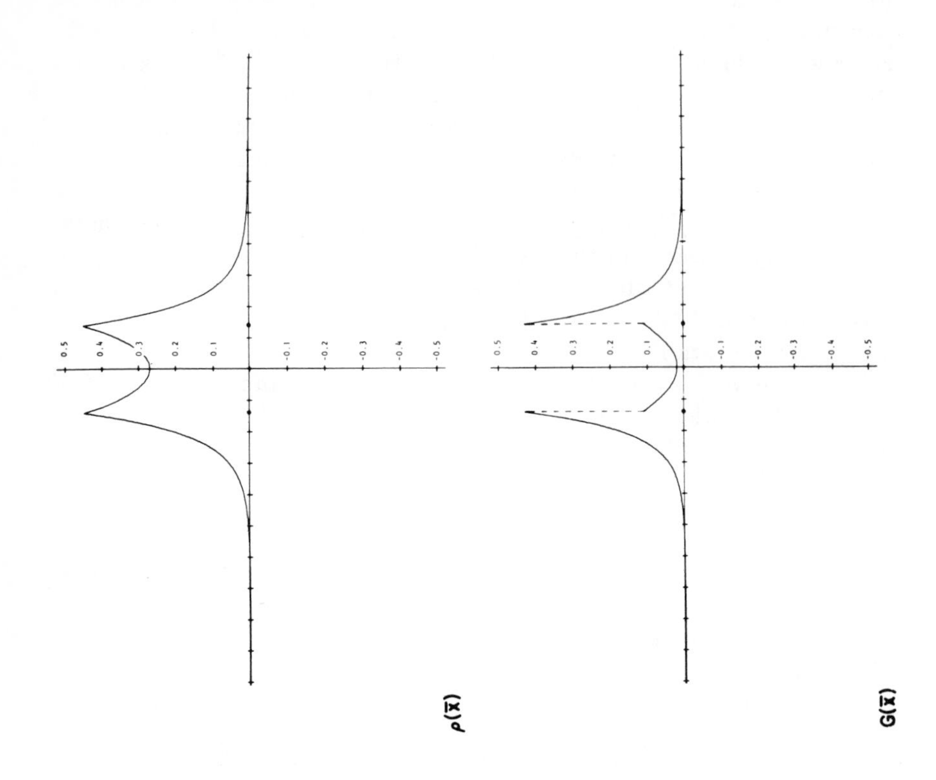

$\rho(\bar{x})$
$G(\bar{x})$

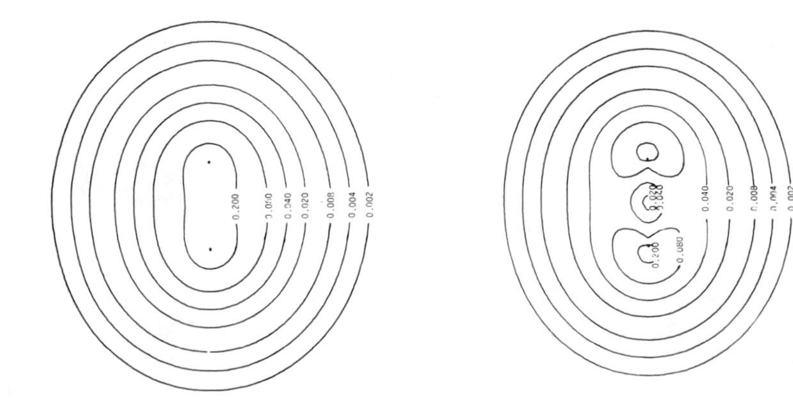

0.200
0.080
0.040
0.020
0.008
0.004
0.002

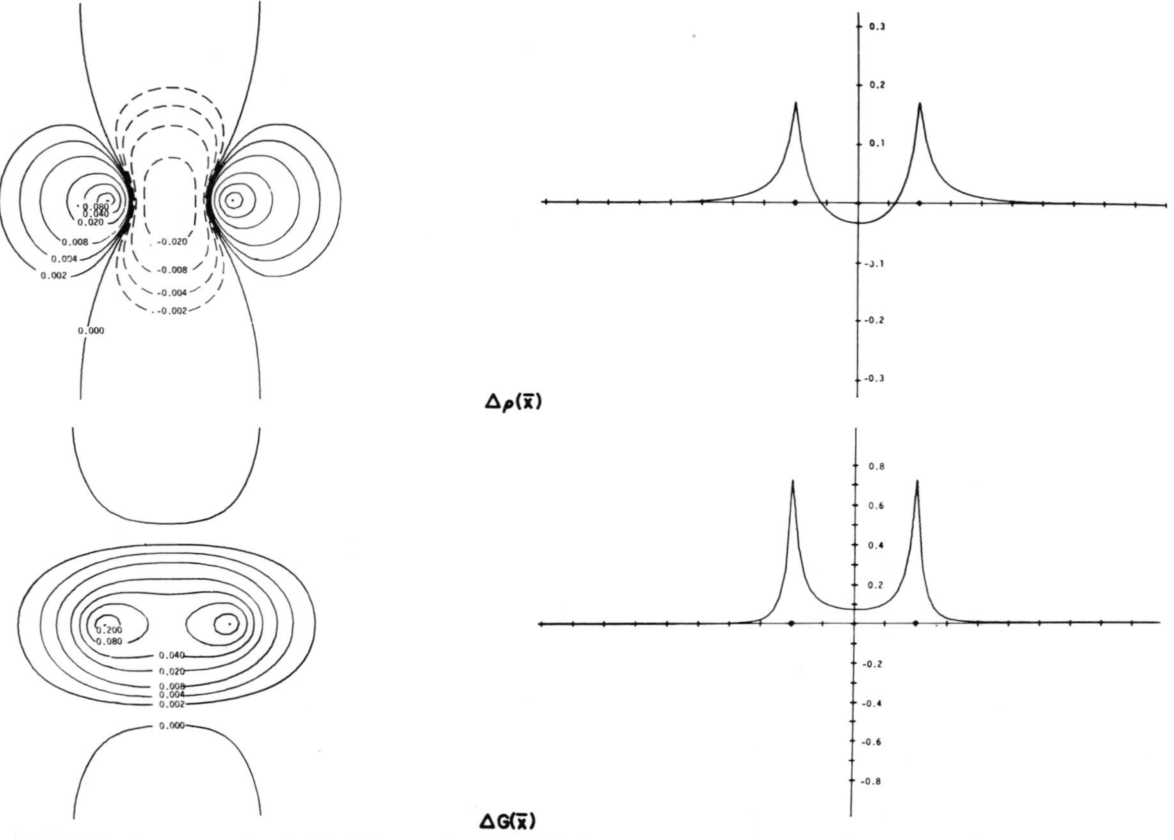

Fig. 2-18. Contour maps of $\rho(r)$, $G(r)$, $\Delta\rho(r)$, and $\Delta G(r)$ for He_2 at $R = 2.0$ au. All quantities are in atomic units.

gion, contributions from the binding region exceed those from the antibinding regions. The contrast in the behavior of $T_{\parallel}$ in H_2 and He_2 is quite evident from the contrasting properties of their $\Delta G(r)$ maps. Because of the softening of the parallel gradient of $\rho(r)$ in the H_2 binding region, $\Delta G(r)$ is negative in this region and attains its minimum value along the bond axis. In contrast, $\Delta G(r)$ for He_2 exhibits a maximum along the bond axis.

In contrast to H_2, He_2 *shows no regions of space in which charge accumulation lowers the PE and simultaneously yields low values for the KE.* Instead, electronic PE is decreased only at the expense of an increase in KE in all regions of space, with the result that $\Delta T > \frac{1}{2}|\Delta V|$, repulsive forces dominate, and He_2 is not bound.

Next we consider KE distributions for molecules containing nonhydrogenic atoms, i.e., systems whose $\Delta\rho(r)$ maps exhibit quadrupolar rather than dipolar polarizations around each nucleus. For such systems, unlike H_2, one can no longer approximately relate $G(r)$ to gradients in $\rho(r)$, since because of the increased number of electrons (2-29) is really determined by a sum of orbital density gradients and not by the gradient of the sum, i.e., of $\rho(r)$. Using Hartree-Fock wavefunctions we will limit our discussion to the equilibrium separation and to only the major features of $G(r)$ and $\Delta G(r)$. It is to be hoped that these features may persist beyond the Hartree-Fock approximation.[19]

In terms of (2-29), one can understand why N_2, unlike H_2, has a $G(r)$ which does not show a pronounced decrease near the bond midpoint (Fig. 2-19). The ground-state electronic configuration of N_2 is $1\sigma_g^2 1\sigma_u^2 2\sigma_g^2 2\sigma_u^2 3\sigma_g^2 1\pi_u^4, {}^1\Sigma_g^+$. The $2\sigma_u$ orbital contributes substantially to $G(r)$ near the bond midpoint (like $1\sigma_u$ in He_2) because of its nodal surface perpendicular to the bond axis, while the $1\pi_u$ orbital contributes to $G(r)$ along the whole of internuclear axis which lies on the nodal planes. Thus, the softening of $\nabla\rho(r)$ in the binding region does not yield small values of the parallel components of $G(r)$. The contributions from the σ_g densities are zero at the bond midpoint and small in its immediate vicinity. However, the σ_u and π_u contributions in this region are large and $G(r)$ exceeds $\rho(r)$ along the internuclear axis in the binding region. This is opposite to the case of H_2, where $G(r)$ is primarily determined by a σ_g orbital.

The $\Delta G(r)$ distribution in N_2 has positive values over most of the binding region, with a pronounced drop at the bond midpoint. The regions in which $\Delta G(r)$ is most negative are localized around each nucleus, but are most predominant in the antibinding regions. Fig. 2-19 indicates that the region of charge removal in the $\Delta\rho(r)$ map corresponds to a region of decreased KE. Since, other things being equal, charge depletion in a given region decreases the KE contribution from that region, one is more interested in regions where (i) $\Delta\rho(r) < 0$ and $\Delta G(r) > 0$, e.g., the He_2 binding region, or (ii) $\Delta\rho(r) > 0$ and $\Delta G(r) < 0$, e.g., the H_2 binding region.

Regions of type (ii) occur in the antibinding regions of N_2. Thus, *a major portion of the charge increase along the axis in the antibinding regions, a characteristic feature of nonhydrogenic atoms, lowers the KE of the molecule relative to separated atoms.* Notice that, unlike the situation in H_2, this charge accumulation cannot be interpreted as a contraction towards the nuclei since then $\nabla\rho_i$ would increase relative to separated atoms for each ρ_i and $\Delta G(r) > 0$. Instead, since $\Delta G(r) < 0$, the charge accumulation in the antibinding regions softens each individual $\nabla\rho_i$. Note that the values of $\Delta G(r)$ which exceed zero in the antibinding regions are extremely small. Furthermore, in N_2 the parallel and perpendicular components of KE are both increased and decreased throughout space by essentially equal amounts, causing the value of $(T_\perp - T_\parallel)/T = 0.3331$ to remain almost unchanged from the separated-spherical-atoms value 0.3333. Thus, compared to H_2, KE changes in a rather different way in the formation of N_2.

The ground state of Li_2 has the configuration $1\sigma_g^2 1\sigma_u^2 2\sigma_g^2, {}^1\Sigma_g^+$. Again, the diffuse density increase in the binding region makes small positive contributions to $\Delta G(r)$, unlike the case of H_2. The Li core polarizations, very dominant in the BDM, contribute the most to $\Delta G(r)$. For each core, the region of charge removal on the binding side and of charge increase on the antibinding side yield, respectively, positive and negative values for $\Delta G(r)$. Also, $(T_\perp - T_\parallel)/T$ is 0.3334, almost unchanged from the separated-atoms value.

$\Delta G(r)$ maps referenced to the same separated-atom states as in BDMs have been obtained for B_2, C_2, O_2, and F_2 (*73*). The $B_2({}^3\Sigma_g^-)$ and $C_2({}^1\Sigma_g^+)$ maps are similar in all major respects to that of $N_2({}^1\Sigma_g^+)$ while the $O_2({}^3\Sigma_g^-)$ and $F_2({}^1\Sigma_g^+)$ maps exhibit areas in which $\Delta G(r) < 0$ in the antibinding regions with the same characteristic shape as for N_2. In addition, O_2 has a region around the bond midpoint with $\Delta G(r) < 0$ while a similar region for F_2 extends almost throughout the binding region. In both O_2 and F_2, the decrease in $\Delta G(r)$ in the antibinding regions is greater than that in the binding region.[20]

That the spatial positionings of these KE decreases are characteristic of their atoms and their associated charge densities is substantiated by the $\Delta G(r)$ maps of, e.g., $LiH(X^1\Sigma^+)$ and $NH(X^3\Sigma^-)$, both referenced to spherical atoms (Fig. 2-20). $\Delta G(r)$ near the Li nucleus in LiH is similar in form to that in Li_2, i.e., in the antibinding region $\Delta G(r) < 0$ where $\Delta\rho(r) > 0$ and in the binding region $\Delta G(r) > 0$ where $\Delta\rho(r) < 0$, the charge distributions describing Li core polarization. In contrast, $\Delta G(r)$ near the proton is similar in form to that in H_2, i.e., $\Delta G(r) < 0$ in the binding region and $\Delta G(r) > 0$ in the antibinding region. Similarly, in NH the KE decrease is localized around the N nucleus and in its antibinding region, where $\Delta\rho(r) > 0$, in a manner observed for N_2. Near the proton, $\Delta G(r)$ again shows the characteristic pattern for this atom. The manner in which $G(r)$ and $\Delta G(r)$ are directly determined by gradients of $\rho(r)$ in the re-

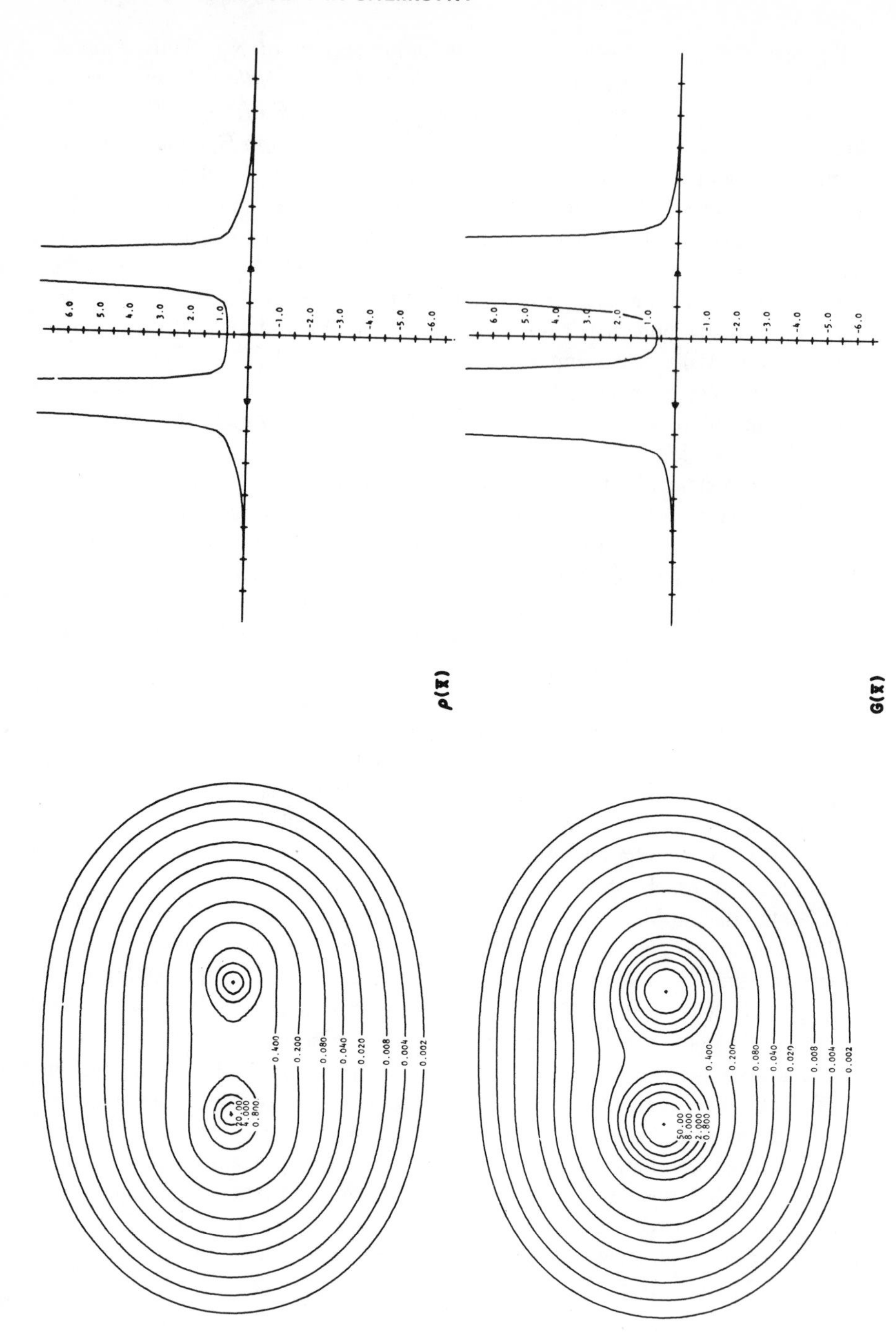

ρ(X)
G(X)
6.0
5.0
4.0
3.0
2.0
1.0
-1.0
-2.0
-3.0
-4.0
-5.0
-6.0
20.00
4.000
0.800
0.400
0.200
0.080
0.040
0.020
0.008
0.004
0.002
50.00
8.000
2.000
0.800

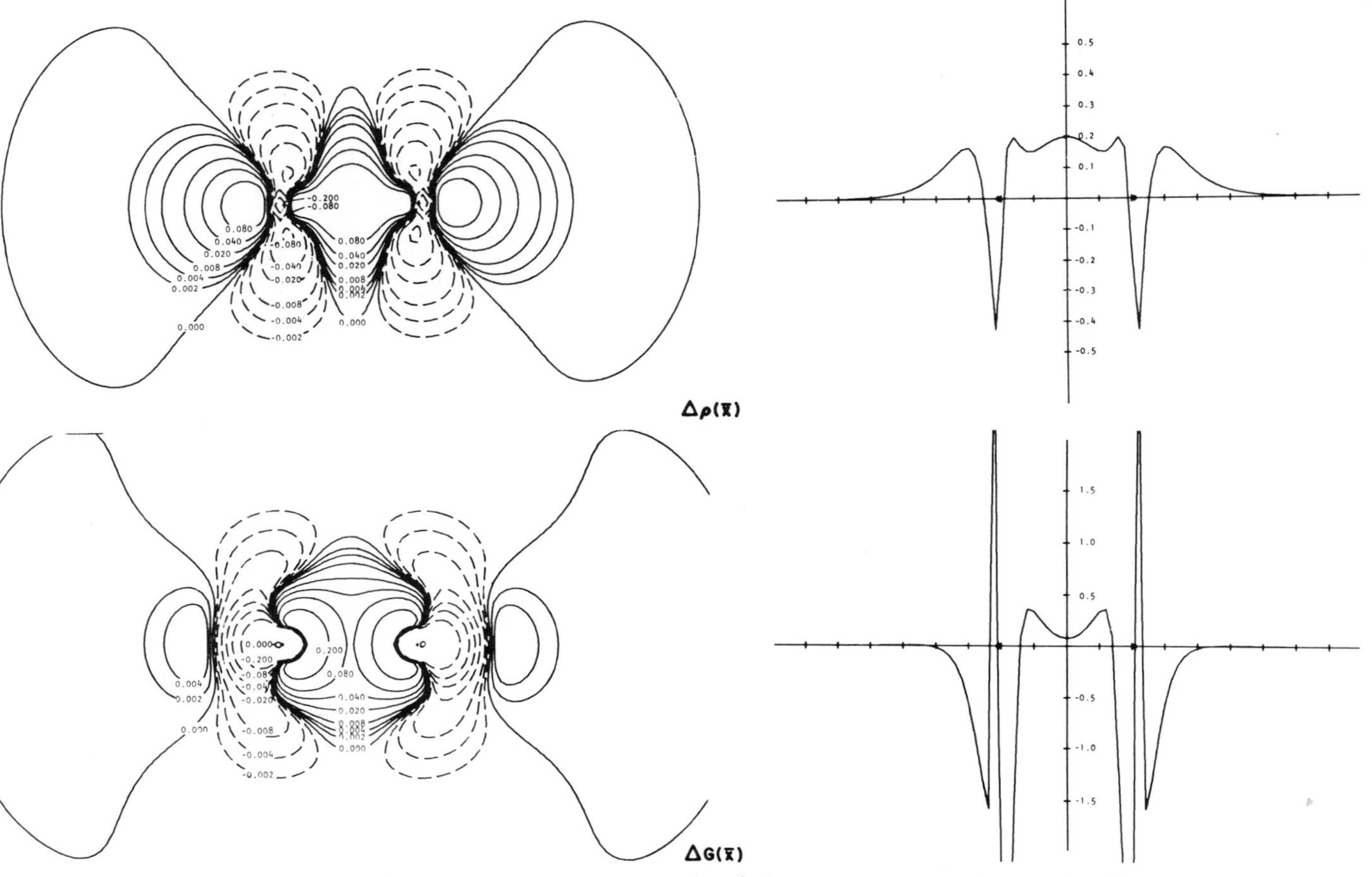

Fig. 2-19. Contour maps for $\rho(r)$, $G(r)$, $\Delta\rho(r)$, and $\Delta G(r)$ for $N_2(X^1\Sigma_g^+)$ at $R_e = 2.068$ au. All quantities are in atomic units.

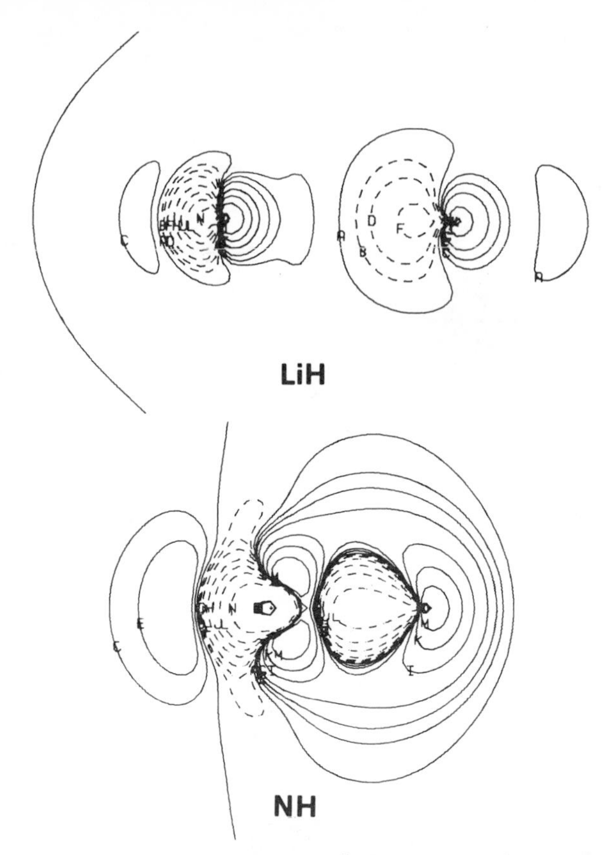

Fig. 2-20. Contour maps of $\Delta G(r)$ for $LiH(X^1\Sigma^+)$ and $NH(X^3\Sigma^-)$ at their respective equilibrium separations of 3.015 and 1.961 au. Both maps are referenced to the appropriate spherical atom contributions. The contours increase (solid contours) or decrease (dashed contours) from the zero contour (first solid contour adjacent to dashed contour) in the same order of values as the $\Delta G(r)$ plot for N_2 in Fig. 2-19.

gion of a proton illustrates that its physical space, in some set of natural orbitals, is described largely by a single orbital function. The NH molecule exemplifies the two principal mechanisms whereby charge density may be accumulated in regions of low potential (low relative to the principal regions of charge removal in the BDM) and simultaneously lead to a local decrease in KE.

Thus, from all the bound molecular states so far investigated (A_2 and AH, $A = H \rightarrow F$) one indeed finds regions of space, localized in regions of low potential, in which the KE assumes low values. These values decrease to an amount determined by the corresponding atomic contributions. Similarly, for unbound He_2 and Be_2, $G(r)$ indicates that the KE is essentially everywhere increased relative to separated atoms *(73)*.

Other workers (*78, 79, 95, 96*) have devised orbital partitionings of the binding energy which place particular emphasis on changes in KE quantities. Ruedenberg's approach (*78, 79*) is based on a partitioning of the molecular $\rho(r)$ and $\rho(r, r')$. $\rho(r)$ is written as a sum of a quasi-classical and an interference density:

$$\rho(r) = \rho_{qc}(r) + \rho_i(r).$$

$\rho_{qc}(r)$ is equal to the sum of the atomic densities as expressed in terms of the same AOs used in the expansion of the final molecular wavefunction, but with different coefficients such that $\rho_{qc}(r)$ sums to the total number of electrons in the system. $\rho_i(r)$ is then defined as the remainder of the total density and integrates to zero.[21]

In the one-electron H_2^+ system this partitioning of $\rho(r)$ and $\rho(r, r')$ is sufficient to yield a partitioning of the total energy. Ruedenberg uses the Finkelstein-Horowitz wavefunction for H_2^+, so that $\rho_{qc}(r)$ is equal to $\frac{1}{2}(1s_a'^2 + 1s_b'^2)$, where the prime denotes a variationally determined exponent different from the value unity. For discussing binding in this scheme one first determines the energy changes involved in changing the initial states of the separated atoms into the imaginary system of separated atoms as described by $\rho_{qc}(r)$. Since any change in the description of the isolated atoms without allowing interaction between them must increase the energy of the system, this contribution to the energy change is called a promotion energy. For H_2^+ the promoted state corresponds to one in which the density is more contracted around each nucleus than in the separated atoms, since the exponent in the $1s'$ functions increases from its value of unity as R is decreased. Such an "orbital contraction" raises the KE and decreases the PE of the promoted state relative to the energies of the separated atoms, i.e., with the subscript p denoting promoted-state quantities, $\Delta V_p \ll 0$ and $\Delta T_p > 0$. The atoms in the promoted states are then allowed to interact to form the molecule, a process described by the density change $\rho_i(r)$. The promoted state *necessarily* violates the virial theorem and for the case of H_2^+ does so in such a way as to greatly increase T and greatly decrease V from their virially allowed values ($2T = -V$) of the separated atoms. Thus it is not surprising to find that for H_2^+ $\rho_i(r)$ yields the energy changes (subscripted by i) $\Delta T_i < 0$ and $\Delta V_i > 0$.

In an earlier discussion (*78*), Ruedenberg placed great emphasis on the large negative value found for ΔT_i, its magnitude being approximately twice ΔE, even though the magnitude of the decrease in ΔV_p is greater. Since $\Delta V_p \ll 0$ one necessarily finds ΔV_i to be greater than zero. After the observation that one component ($T_\parallel$) of the real KE undergoes a dramatic decrease on bonding (*13, 48*), a decrease which dominates the real energy changes for large values of R, Ruedenberg (*79*) then states that ΔT_i and ΔV_i determine the binding at large R, while ΔT_p and ΔV_p dominate the behavior around R_e. All the above final physical observations are, of course, obtainable from the analysis of the real energy changes ΔV, $\Delta T_{\|}$, and $\Delta T_{\perp}$ in conjunction with the real density change $\Delta\rho(r)$.

It is not clear that the important features of the analysis involving the fictitious promotion state (i.e., $\Delta V_p \ll 0$ and $\Delta T_i < 0$) will apply to systems other than H and He. For example, if one compares the exponents for the AOs used in the expansion of the Hartree-Fock atomic wavefunction for nitrogen (*27*) with the exponents of the AOs used in the Hartree-Fock description of N_2 (*32*), one finds that each and every exponent has decreased in value in the molecular description. This comparison is a very fair one, as the exponents were separately and carefully optimized in both the atomic and molecular calculations and the atomic set (suitably augmented with polarizing functions) was the starting set for the initial iteration in determining the molecular basis set; e.g., both the atomic and molecular basis sets contain 3*s* functions, whose exponents are, respectively, 7.334 and 7.040. Thus Ruedenberg's promoted state for N_2 as described by ρ_{qc} would in this case correspond to one in which the atomic densities have expanded rather than contracted, a result which is not at variance with the BDM for N_2. The virial theorem for the promoted state would be violated in a direction opposite to that found for H_2^+, in that $\Delta V_p > 0$ and $\Delta T_p < 0$. Consequently, the energy changes associated with ρ_i would be reversed as well, with $\Delta T_i > 0$ and $\Delta V_i < 0$. The "interference kinetic energy" rather than being the crucial quantity for bonding, would in the case of N_2 act to oppose bonding and decrease ΔE.[22]

Wilson and Goddard (*95, 96*) appeal to the original idea of Ruedenberg (*78*), partially discarded by him later (*79*) as follows: "it is a decrease in the kinetic energy which is responsible for chemical bonding." The KE referred to by these authors is not the total KE but a quantity determined by an exchange-type density which must be defined in terms of the set of MOs from a *G*1 type of wavefunction (*52*). The exchange density ρ^x of Wilson and Goddard is that component of $\rho(r)$ or $\rho(r, r')$ which, when expanded in terms of *G*1 orbitals, contains the cross products of different *G*1 orbitals; the *G*1 orbitals do not form an orthogonal set. The exchange density, like Ruedenberg's interference density, integrates to zero over all space. They find that if the energy is expressed as

$$E = T^x + \omega,$$

where T^x is a KE as determined by ρ^x, and ω is, as they state, "everything else," then the change in T^x is negative for the bound systems H_2, LiH, and BH, and positive for the unbound systems HeH, H_2H, He_2, HeH_2, and $(H_2)_2$. In addition $|\Delta T^x| > |\Delta\omega|$, and for H_2 ΔT^x is of the same order of magnitude as the binding energy.

To determine why T^x is negative for H_2, the $G1$ wavefunction constructed from H 1s orbitals for H_2 (which is just the usual valence bond function in this case) is considered in detail. The KE density t^x which yields T^x upon integration is expressed in terms of the H 1s orbitals on the two centers. The quantity t^x is expressed in terms of dot products of orbital gradients. In doing so they find a term $\nabla\phi_a \cdot \nabla\phi_b$, where ϕ_a and ϕ_b are H 1s orbitals on the two centers. For H_2 this product is negative in the binding region, for here the orbitals on a and b have gradients of opposing sign. This negative contribution to T^x is called the contragradience, as opposed to the rest of T^x, which is called the non-contragradient contribution. A plot of t^x along the internuclear axis for H_2 shows a function which is negative across the whole of the binding region and positive outside of it–a plot which is remarkably similar to the $\Delta G(r)$ plot for H_2 (Fig. 2-17). The behavior of the KE along the internuclear axis is determined entirely by $T_\parallel$. As Wilson and Goddard point out, T^x does parallel the behavior of $T_\parallel$. The same contragradient behavior may be isolated from $G(r)$. If one expresses $G(r)$ in terms of the AO basis set rather than in terms of the MOs themselves, one obtains the cross term $[S/(1 + S^2)]\ \nabla\phi_a \cdot \nabla\phi_b$ for H_2 using the same valence bond wavefunction. To ascribe the low values of $T_\parallel$ and the negative contributions to $\Delta G(r)$ in the binding region to a decrease in the parallel gradient of the MO or its density or to the contragradient contribution $\nabla\phi_a \cdot \nabla\phi_b$ are all equivalent explanations. Similarly, for an unbound molecule such as He_2, one obtains the terms $[S/(1 + S^2) - S/(1 - S^2)]\ \nabla\phi_a \cdot \nabla\phi_b$ in both the orbital expression for $G(r)$ and in the T^x function, an expression which yields a net positive contribution to the total KE density across the binding region (Fig. 2-18). In the orbital expression for $G(r)$ in He_2, this effect is attributed to the opposing contributions of $\nabla\rho_{\sigma_g}$ and $\nabla\rho_{\sigma_u}$ in the binding region.

Wilson and Goddard (*96*) propose that bond formation be viewed as follows: (1) freeze the ($G1$) orbitals of the separated atoms; (2) bring the atoms to their positions in the molecule (this leads to T^x–frozen, as discussed above in H_2–and to about the correct binding energy); and (3) allow the orbitals to relax to their SCF forms.

The important physical observation regarding the behavior of KE on bonding is that there are local regions of space in which the KE density attains low values. In H_2 and H_2^+ with densities determined primarily by the overlap of s orbitals, the low values of $G(r)$ are most pronounced in its parallel component along the internuclear axis in the binding region. To observe this behavior of KE one does not have to employ $G1$ orbitals and wavefunctions, nor is it necessary to define an exchange KE referenced to an imaginary "frozen" state of the atoms.

The thesis originally developed from the studies of KE behavior in H_2 (*13*) and H_2^+ (*48*), is that the charge accumulation in the binding region (or in the "bond" region of Ruedenberg) leads to a local decrease in the KE of the molecule relative to the separated atoms and, because of this phenomenon, the increase in KE obtained as a result of the contraction or accumulation of charge in the nuclear potential field could be tolerated within the bounds determined by the virial theorem, namely, $\Delta T = \frac{1}{2}|\Delta V|$. Is this local lowering of the KE in the binding region an essential or necessary requirement for the formation of a chemical bond? Not when it is stated in a manner which restricts the possible decrease in the KE to the internuclear region and relates it to a softening of the parallel gradients in $\rho(r)$ resulting from the charge accumulation in the binding (or bond) region. This is a property unique to hydrogen (and perhaps helium in a bound molecular state such as HeH^+). We have seen that the more general quadrupolar-type polarizations, or dipolar core polarizations, as exhibited by BDMs for atoms other than hydrogen, are characterized by decreases in KE contributions which are localized primarily in the antibinding regions and the immediate vicinity of the nucleus. Thus the separate accumulation of charge density in the antibinding region, a feature characteristic of all BDMs except for hydrogen, plays the same role in contributing to the stability of the molecule via a local lowering of the KE as does the charge accumulated in the binding region of hydrogen.

These observations suggest that localized decreases in KE may be a necessary requirement for the formation of a bound molecular state, but they do not prove it. The only necessary conditions which can be rigorously imposed on the kinetic energy change in the formation of a bound state are the virial restraints, namely, $\Delta T > 0$ and $\Delta T = \frac{1}{2}|\Delta V|$. The above observations regarding local decreases in KE are rationalizations after the fact; they are not proven requirements for the formation of a bound molecular state.

2-7. THE ROLE OF THE CHARGE DENSITY IN CHEMISTRY

2-7-1. The Variational Principle for a Subspace

Chemical observations made on a system are determined by the morphology of the system's charge distribution and by its evolution with time. This may be demonstrated by first determining the universal properties of molecular charge distributions and then establishing the existence of a mapping of the experimentally determined concepts of chemistry onto this set of universal properties. One distinguishes between those primary concepts which are necessary for the understanding of chemistry and the models which have evolved to rationalize these concepts. One hopes to discover the concepts intact in the properties of the charge density; the models may survive totally, partially, or not at all.

The primary concepts, those without which there would be no correlation and no prediction of the observations of descriptive chemistry, are: (i) the existence of atoms in molecules; (ii) the ability to identify an atom (or a functional grouping of atoms) in a molecule by its characteristic properties; and (iii) the concept of bonding–that molecular stability may be understood by assuming the existence of particularly strong interactions (bonded interactions) between atoms within the molecule. In addition, to understand the saturation of bonding, valency, and ultimately the geometry of molecules, one must in general assume that certain pairs of atoms in a molecule are bonded to one another while other pairs are not.

On the basis of extensive studies of the properties of molecular charge distributions (*16*, *21*, *22*, *101*), important universal properties of the molecular charge density $\rho(r)$ have been determined. Such a study is essentially a study of the topological properties of the charge density. It has been found (*22*, *80*) that the universal properties may be characterized in terms of the vector field of the charge density, $\nabla\rho(r)$. The properties of this field, and hence the principal characteristics of a charge distribution, are totally determined by the number and character of its critical points, points at which the field vanishes. The trajectories of $\nabla\rho(r)$, all of which terminate at particular critical points, define the atoms in a molecule. Those which originate at the same critical points isolate the number, locations, and directions of those lines throughout a charge distribution that link particular pairs of atoms along which the charge density is a maximum. This network of lines faithfully reproduces the features outlined above as being essential to the concept of bonding. In this specific sense these particular trajectories of $\nabla\rho(r)$, called bond paths, define the number, locations, and directions of chemical bonds in a molecule. The atoms so defined are also observed to behave in a manner consistent with the chemical concept of an atom in a molecule. In particular, such topologically defined atoms are by their very definition the most transferable pieces of a molecule in real space, i.e., at the charge density level, and further, their degree of transferability is found to coincide with chemical expectations.

Of equal importance to this density-based approach to chemistry is the finding that the same topological property of the charge density which defines the boundary of an atom in a molecule also defines the boundary of a subspace in a total system whose variational properties exhibit maximum correspondence with the expressions of all-space quantum mechanics (*25*, *88*, *89*). From quantum mechanics one obtains both an explanation and a prediction of the existence of additivity schemes in chemistry, and what is more important, of the experimental observation that atomic fragments or polyatomic functional groups may exhibit characteristic sets of properties which vary between relatively narrow limits.

Collard and Hall (*37a*) have demonstrated the utility of orthogonal trajectories, i.e., paths traced out by the gradient vectors of a scalar field and their associated critical points, in the analysis of scalar functions of several variables. Their analysis is used to classify the general topological properties of charge distributions of molecules in their equilibrium geometries. These definitions of basic chemical concepts apply not only to charge distributions of molecules in equilibrium geometries, but also to a dynamic system. Collard and Hall point out that the analysis of the discontinuous change in the topological characteristics of a molecular distribution resulting from the continuous change in its nuclear coordinates is given by the catastrophe theory of René Thom (*94*). The extension of the theory to the dynamic case provides precise meanings to the concepts of the making and breaking of chemical bonds, for these are the catastrophes of chemical change.

We are only on the threshold of understanding and exploiting the properties of molecular distributions: witness the recent identification of the chemical potential of density functional theory with the concept of electronegativity (*72*). The single most important feature of a theory of chemistry stated in terms of the charge density is the possibility of basing it solely on the observed properties of $\rho(r)$ and on their governing relationships as derived from quantum mechanics. Future studies will uncover other properties and discover new relationships, but the observations so far obtained will stand, as will the quantum relationships so far derived (see also Ref. 55).

The universal properties of the charge density $\rho(r; X)$, where X denotes a point in nuclear configuration space, may be characterized in terms of its associated vector field $\nabla\rho(r; X)$. The properties of this field, and hence the principal characteristics of a charge distribution, are totally determined by the number and character of its critical points, points at which this field vanishes (*22*, *26*, *80*). A particular type of critical point defines a surface and a unique axis perpendicular to this surface. The axis defines a line through the charge distribution along which $\rho(r)$ is a maximum—the bond path. The set of surfaces generated by these critical points contained within a molecular distribution partitions the space of the molecule into a collection of chemically identifiable atomlike regions. The network of bond paths defined by the same set of critical points coincides with the network obtained by linking together those pairs of atoms which are considered to be bonded to one another on the basis of chemical considerations. For the diborane molecule (Fig. 2-21), the particular critical points are the saddle points in $\rho(r)$ in the bridging plane. Shown in heavy lines are the pair of gradient paths which originate at each of these critical points and terminate at each neighboring nucleus, and the pair of gradient paths which terminate at each of these saddle points. In three dimensions the collection of all gradient paths which terminate at one of these saddle points defines a surface—the atomic surface separating the regions of the neighboring atoms linked by a bond path, as discussed previously.

The partitioning surfaces (Fig. 2-22) are unique because of the topological properties of this particular type of critical point. Their unique character may

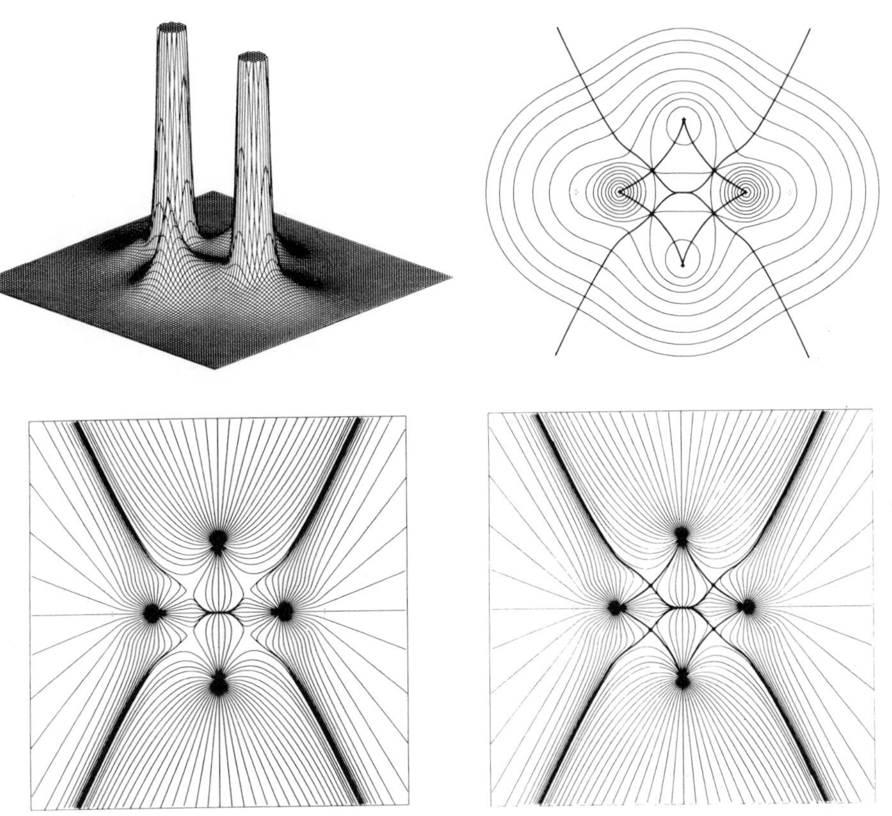

Fig. 2-21. Representations of $\rho(r)$ and of the paths of $\nabla\rho(r)$ in the bridging plane of diborane. There are five saddle points in this plane–four bridging saddle points and the ring saddle point. In this plane the ring saddle point appears as a minimum in $\rho(r)$. Two maps of the gradient paths are shown, one omitting and the other including the gradient paths associated with the bond saddle points. Transfer of these particular gradient paths to the contour map of $\rho(r)$ yields a definition of the boundaries between the hydrogen and boron fragments in this plane and of the bond paths which link these fragments. One extra contour of $\rho(r)$ is indicated in this diagram. Its value, 0.1195 au, is the value of $\rho(r)$ at a bridging saddle point. The value of $\rho(r)$ at the ring saddle point is 0.1059 au.

be expressed in another way: they are the only closed continuous surfaces $S(\boldsymbol{r})$ within a molecular charge distribution which satisfy the boundary condition of zero flux,

$$\nabla\rho(\boldsymbol{r}) \cdot \boldsymbol{n}(\boldsymbol{r}) = 0 \text{ and } \boldsymbol{r} \in S(\boldsymbol{r}). \tag{2-31}$$

The $\boldsymbol{n}(\boldsymbol{r})$ is normal to the surface at $\boldsymbol{r}$. It is this surface condition which provides the link with quantum mechanics, for one finds that subspaces of a total

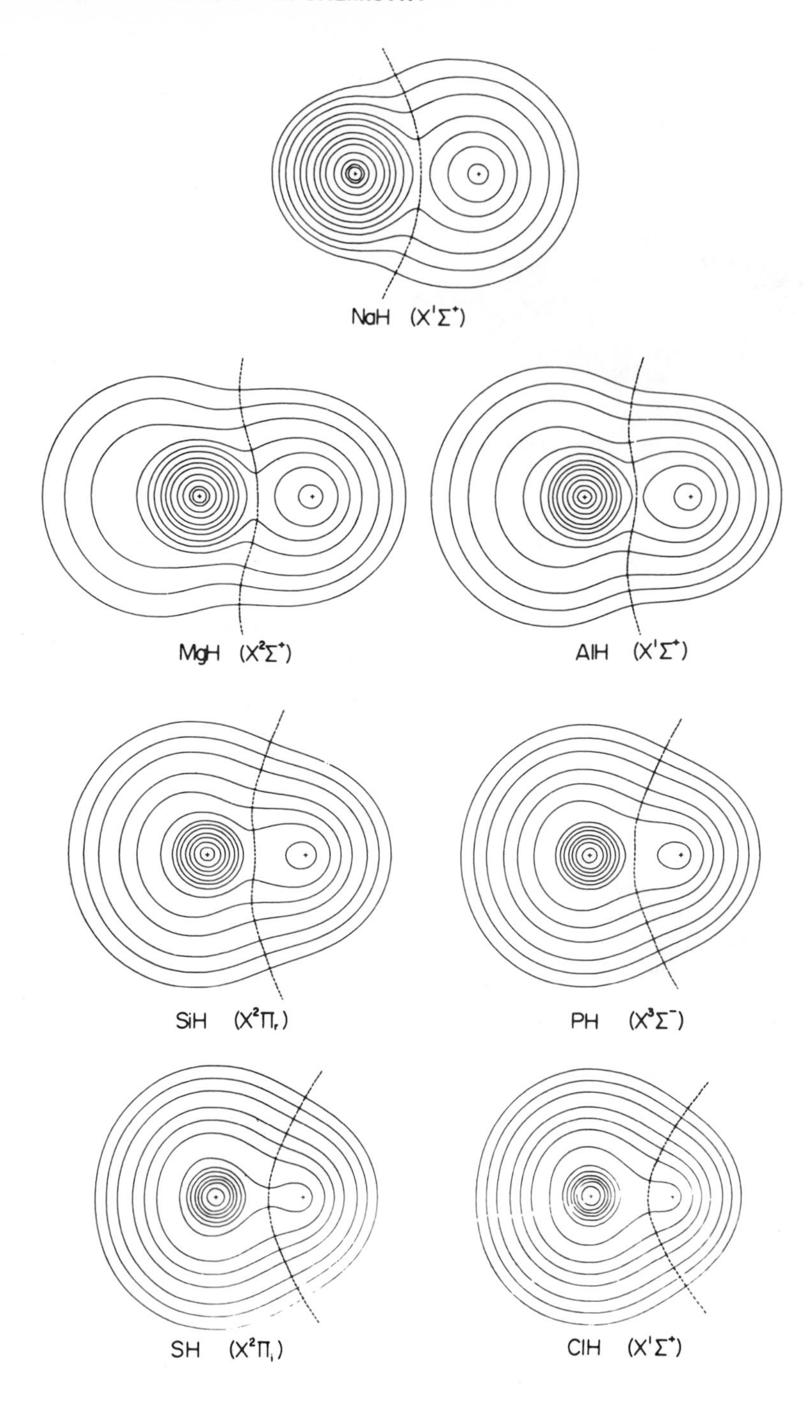
NaH $(X^1\Sigma^+)$
MgH $(X^2\Sigma^+)$
AlH $(X^1\Sigma^+)$
SiH $(X^2\Pi_r)$
PH $(X^3\Sigma^-)$
SH $(X^2\Pi_i)$
ClH $(X^1\Sigma^+)$

system emcompassed by boundaries satisfying (2-31) exhibit variational properties in maximum correspondence with the expressions of all-space quantum mechanics (*25*, *88*, *89*). It has been proposed that the usual variational expressions of (all-space) quantum mechanics be replaced by the more general forms

$$\delta E(\psi, \Omega) = \tfrac{1}{2}\{\langle[\hat{H}, \hat{A}]\rangle_\Omega + \text{c.c.}\}/\langle\psi, \psi\rangle_\Omega \tag{2-32}$$

for a stationary state,

$$\delta L(\psi, \Omega) = -\tfrac{1}{2}\{\langle[\hat{H}, \hat{A}]\rangle_\Omega + \text{c.c.}\}$$

for a time-dependent system, and a corresponding expression for the change in the *action* of a region of space, for the most general case, which involves kinematic and dynamic changes. The energy functional $E(\psi, \Omega)$ is defined as

$$E(\psi, \Omega) = \int_\Omega dr_1 \int dr_2 \cdots \int dr_N \left(\tfrac{1}{2} \sum_i \nabla_i \psi^* \cdot \nabla \psi + \hat{V}\psi^*\psi\right) \Big/ \langle\psi, \psi\rangle_\Omega . \tag{2-33}$$

$L(\psi, \Omega)$ is the corresponding quantum Lagrangian integral for a time-dependent system, $\hat{H}$ is the Hamiltonian operator, and $\hat{A}$ is the generator of an infinitesimal unitary transformation. The above expressions apply to any region Ω of real space encompassed by a boundary satisfying the zero-flux surface condition in (2-31). This includes the total (isolated) system with boundaries at infinity (which is but one special case) and, because of the fundamental topological properties of $\rho(r)$, it includes an atom in a molecule, a molecule in a system of interacting molecules, or an ion in a solid–in brief, those structural subunits which are recognized as having physical significance in the interpretation of experimental results. (This physical boundary leads to the definition of a molecular shape). The boundary defined in the usual formulations of quantum mechanics is the one at infinity, where the boundary condition on the variation of ψ is satisfied. These approaches are necessarily restricted in their applications to a total, isolated system. One may ask whether it is possible to obtain from quantum mechanics the definition of another, more general physical boundary. In so doing one obtains the generalized statements of the variational

---

Fig. 2-22. Contour plots of the ground-state total molecular charge distributions of the diatomic hydrides AH, A = Na → Cl, showing the virial partitioning surfaces which define the (A) and (H) fragments. The populations of the (H) fragments in this series of molecules are, respectively, beginning with NaH: 1.810, 1.796, 1.825, 1.795, 1.579, 1.094, and 0.759. (Reproduced from Ref. 20, courtesy the National Research Council of Canada.)

principle given above. For example, consider the intermediate time-dependent kinematic expression in (2-32). Satisfaction of this result for all generators of infinitesimal unitary transformations ensures that the evolution of the *total* system is described by Schrödinger's time-dependent equation. It is obtained for the case $\Omega \equiv$ all space, through the removal of variational constraints on ψ. This expression is then found to remain unchanged when the additional variational constraint that

$$\int_{t_1}^{t_2} \delta \left\{ \int_{\Omega} \nabla^2 \rho(r)\, dV \right\} dt = 0 \qquad (2\text{-}34)$$

is imposed on the problem. The additional constraint in (2-34) is satisfied if ψ generates a density such that the zero-flux surface condition and its variation is satisfied. Thus, for a total system described by Schrödinger's equation of motion, the average properties of any region of this total system bounded by a zero-flux surface are determined by the same combined variational-hypervirial relationship (2-32).

As a result of the quantum description of an atomic fragment and its properties, any property of a molecule may be partitioned into a sum of atomic contributions, i.e., all the properties of an atomic fragment are defined and their average values may be determined. The single most important concept in the definition of the properties of a subspace is that they are all determined by corresponding one-electron operators. Thus, one may define for every property a corresponding density distribution in real space, and obtain the subspace average value of the property by integration over a volume of space out to a surface of zero flux.

The generalized variational principle which applies to any system bounded by a surface of zero flux in $\nabla\rho(r)$ yields a variational statement of the hypervirial theorem for stationary or time-dependent systems. A particular but fundamental application of this result is to obtain a variational derivation of the subspace (or atom in a molecule) virial theorem. The subspace virial theorem in turn leads to a unique spatial partitioning of the total energy of a molecule into a sum of atomic contributions by yielding a quantum prescription for the definition of a one-electron energy density in a many-electron system.

2-7-2. Properties of Atoms in Molecules (Virial Fragments)

We deal here with only two properties of an atomic fragment, namely, its electron population and energy. The population of a fragment Ω is obtained by integrating $\rho(r)$ over the fragment subspace.

$$N(\Omega) = \int_{\Omega} \rho(r)\, dr. \tag{2-35}$$

The populations and net charges $C(\Omega)$, with $C(\Omega) = Z_{\Omega} - N(\Omega)$, have been obtained for a large number of atomic fragments in both polyatomic and diatomic systems; these agree well with previous chemical expectations (*17, 20, 22*).[23]

Setting the operator $\hat{A}$ equal to $-r_1 \cdot \nabla_1$ in the generalized variational integral, (2-32) yields a variational derivation of the subspace virial theorem,

$$2T(\Omega) = -V(\Omega), \tag{2-36}$$

where $T(\Omega)$ is the KE of the subspace (Ω),

$$T(\Omega) = \int_{\Omega} \tfrac{1}{2}[\nabla \cdot \nabla' \rho(r, r')]_{r=r'}\, dr, \tag{2-37}$$

and $V(\Omega)$, equal to the PE of the fragment, is given by the virial of all the forces exerted on the fragment,

$$V(\Omega) = \int_{\Omega} r_1 \cdot \nabla_1 \left(\sum_{\alpha} Z_{\alpha}/r_{\alpha 1} \right) \rho(r_1)\, dr_1 + \int_{\Omega} dr_1 \int dr_2\, \Gamma^{(2)}(r_1, r_2)/r_{12}. \tag{2-38}$$

The first integral on the right-hand side of (2-38) yields the attractive interaction of all the nuclei in the system with the charge density in (Ω) and the contribution of the total nuclear–nuclear repulsive interactions to the PE of (Ω). The second term is the contribution of the electron–electron repulsive energy to (Ω). These quantities are discussed more fully elsewhere (*21, 22*). We point out here how the definition of a fragment PE as the virial of the forces exerted on the charge density in (Ω) yields, via the H–F theorem, a unique spatial partitioning of the nuclear repulsive PE and hence to a spatial partitioning of the total energy of the system.

The application of the virial operator to the electron–nuclear PE operator in (2-38) yields the following expression for the contribution of this term to the electronic energy of a fragment.

$$\int_{\Omega} dr_1\, \rho(r_1) \left\{ -\sum_{\alpha} Z_{\alpha} [(|r_1 - R_{\alpha}|)^{-1} + R_{\alpha} \cdot \nabla_{\alpha} (|r_1 - R_{\alpha}|)^{-1}] \right\}.$$

The first term in the sum averages to the electron–nuclear attractive PE of (Ω), while the second yields the sum of the virials of the H–F forces which the charge density in (Ω) exerts on each of the nuclei in the system, the nuclear virial $V_n(\Omega)$,

$$V_n(\Omega) = -\sum_{\alpha} R_\alpha \cdot F_\alpha^e(\Omega). \tag{2-39}$$

The nuclear virial is equal to the virials of the nuclear repulsive forces and of any external forces acting on the system. The nuclear virial is in turn equal to the nuclear–nuclear repulsive potential, and thus for a system at electrostatic equilibrium

$$\sum_{\Omega} V_n(\Omega) = \sum_{\alpha<\beta} Z_\alpha Z_\beta / R_{\alpha\beta}.$$

Because the forces F_α^e are determined by $\rho(r)$, the term $V_n(\Omega)$ leads directly to a partitioning of the nuclear repulsive potential by determining what fraction of the force exerted on each nucleus α is exerted by the charge density in the fragment (Ω). If the nuclei are not in their equilibrium positions, then $V_n(\Omega)$ includes a contribution to the energy of the fragment arising from the action of the external forces.

Subtraction of $T(\Omega)$ from both sides of (2-36) yields the subspace analog of the all-space statement of the virial theorem which relates T and E, namely,

$$-T(\Omega) = E(\Omega) = T(\Omega) + V(\Omega). \tag{2-40}$$

The fragment energy has the additive property

$$\sum_{\Omega} E(\Omega) = E,$$

which is a consequence of the related property

$$E(\Omega) = E \text{ for } (\Omega) = \text{all space}.$$

Thus, as a consequence of the application of the variational principle to a particular class of subspace, the subspace itself being defined by the variationally derived boundary condition of zero flux (2-31), one obtains a rigorous and nontrivial (independent of any model) partitioning of the total energy of a molecular system.

The definition of a subspace energy in terms of a single-particle virial operator suggests a general reformulation of the Hamiltonian operator of a many-electron system. This alternative formulation of the Hamiltonian demonstrates that the energy of a molecular system at its equilibrium geometry is rigorously expressible as a sum of one-electron contributions (*88*).

Briefly, the definition of a virial sharing operator $\hat{P}_i$ for the ith particle in a system as

$$\hat{P}_i = -r_i \cdot \nabla_i$$

allows one to define a single-particle PE within a system of interacting particles. That is, the action $\hat{P}_i$ on the total PE overator $\hat{V}$ projects out that share of the PE operator (or PE in a classical system) which belongs to the ith particle. Since

$$\sum_i \hat{P}_i \hat{V} = \hat{V}, \tag{2-41}$$

one may partition the total Hamiltonian of a many-particle system into a sum of single-particle contributions,

$$\hat{H} = \sum_i \hat{H}_i = \sum_i (-\tfrac{1}{2}\nabla_i^2 + \hat{P}_i \hat{V}), \tag{2-42}$$

and the total energy may in turn be expressed as a sum of single-particle energies

$$\langle \hat{H} \rangle = E = \sum_i E_i = \sum_i \langle \hat{H} \rangle_i. \tag{2-43}$$

The use of (2-42) leads immediately to the subspace energy given in (2-40),

$$E(\Omega) = \sum_i \langle \hat{H}_i \rangle_\Omega = N\langle \hat{H}_1 \rangle_\Omega = T(\Omega) + V(\Omega), \tag{2-44}$$

with the PE of (Ω) defined in terms of the average PE of a single electron as determined by the virial sharing operator. The interesting point of this formulation of the energy in terms of an average over a single-particle Hamiltonian is that while E over all space may be expressed in this manner, the energy of a subspace can be defined and expressed *only* in terms of the single-particle Hamiltonian. Thus, the formulation of $\hat{H}$ given in (2-42) is more general than the usual expression, in that it enables one to determine the energy of a quantum subspace as well as the energy of the total system.

2-7-3. Chemical Significance of Virial Fragments

A cornerstone of chemistry is the observation that fragments or groups of fragments in different molecular systems, or ions in various crystal environments, can have characteristic sets of properties which vary between relatively narrow limits.[24] This retention of properties is, in many instances, so close as to give rise to additivity schemes for properties, including the energy. We illustrate how the concept of an atomic fragment accounts for these observations and how their property of satisfying the virial theorem allows one to state the physical condition which must be obeyed if a fragment is to possess an identical energy and population in two different systems.

Assuming both systems in which the fragment (Ω) occurs are in electrostatic equilibrium, then because of the fragment virial theorem, the requirement that $E(\Omega)$ be the same in both systems requires that $T(\Omega)$ and hence *the virial* $V(\Omega)$ *of all the forces exerted on* (Ω) *be identical in the two systems.* A more detailed statement of this restraint is possible. The total virial of a fragment may be equated to the virial of the forces which originate within the fragment, the inner virial $V^{(i)}(\Omega)$, plus the virial of the forces exerted on the fragment by the nuclei and charge density outside the fragment, the outer virial $V^{(o)}(\Omega)$ (*21*, *22*). Thus

$$E(\Omega) = -T(\Omega) = \tfrac{1}{2}(V^{(i)}(\Omega) + V^{(o)}(\Omega)).$$

The interesting point is that $T(\Omega)$ and the inner virial $V^{(i)}(\Omega)$ are determined solely by the distribution of charge and quantum properties within the fragment (Ω). Requiring the fragment to be identical in all respects in both systems will result in the inner virial $V^{(i)}(\Omega)$ being the same in both systems. Thus, *for a given fragment to possess identical properties in two different systems, the virial of all the external forces exerted on the fragment must be identical in the two systems.* It is important to note that while the various contributions to the external forces exerted on a fragment (electron-nuclear, electron-electron, nuclear-nuclear) must necessarily be different in different systems, the requirement for the retention of properties of the fragment is only that the *sum* of the virials of all external forces remain the same.

Thus an atomic fragment, as a consequence of its ability to act as a single unit, responding only to the *total* change in the external forces exerted on it, accounts for the observation that fragments in molecular systems can have characteristic sets of properties which vary between relatively narrow limits.

An example of this criterion for constancy in the properties of a fragment is afforded by the (H) fragment in BeH and BeH_2 (*18*). The charge density $\rho(r)$ is very similar for the (H) fragments in both systems, and their populations and energies differ by only $0.007e^-$ and 6 kcal/mole, respectively. Using Hartree-Fock wavefunctions one finds that the virial of the forces exerted on (H) by (Be) in BeH, the outer virial $V^{(o)}(H)$, is -0.7174 au while the virial of the forces exerted on (H) by the (BeH) fragment in BeH_2 is -0.7196 au, a difference of only ~2 kcal/mole. This near constancy in $V^{(o)}(H)$ is found in spite of the large changes which occur in the individual contributions to this quantity when the adjacent fragment is changed from (Be) to (BeH).

It has been observed that when an atom is defined by a surface of zero flux, the constancy in its kinetic and total energy is paralleled by a corresponding constancy in the distribution of electronic charge within the fragment (*16*, *18*, *21*). In particular, it is found that $\rho(r)$ and $G(r)$ are the same throughout a given atom in different systems to the extent that the virial of all the forces exerted on each element of $\rho(r)$ remains unchanged. The extent to which the properties of an atom remain unchanged in different systems is, therefore, determined by the extent to which $\rho(r)$ remains unchanged. Based on the above observations, if $\rho(r)$ of an atom remains unchanged on transfer, so does its total virial $V(\Omega)$ and, since the atom obeys the virial theorem, $T(\Omega)$ and $E(\Omega)$ are conserved as well. Thus, it is assumed that $\rho(r)$ *is the fundamental carrier of information in a system.*

An atomic fragment, therefore, maximizes the possible transfer of properties between systems because the zero-flux surface—by the nature of its definition in terms of $\rho(r)$—maximizes the extent to which the charge distribution of the fragment is transferable between systems. That is, any other choice of surface will either include part of the neighboring fragment, which may change radically on transfer, or will omit a portion of the charge density which changes little on transfer. The same zero-flux condition defines a fragment which is variationally defined and obeys the virial theorem, has its own set of definable properties, and behaves maximally as an isolated system, acting as a single unit in response to changes in the external forces exerted on it.

The concept of an atomic fragment demonstrates that certain well-defined fragments may have transferable properties if certain realizable constraints are satisfied. These fragments are *not* bondlike in nature. Thus the physical basis for additivity has its origin in a constancy of the charge distribution and properties of atomlike fragments which are defined in real space. Such atomic fragments represent the smallest unit of transferable information between systems.

2-7-4. Virial Fragments and Interpretive Theories

The quantum topological partitioning as discussed here provides an unbiased and model-independent decomposition of all molecular properties into a sum of fragment contributions. This method has been used to analyze molecular binding energies and the energy changes of chemical reactions (*17*, *19*, *22*). The binding energy ΔE, for example, may be equated to a sum of energy changes for each atom → atomic fragment in the system,

$$\Delta E = \sum_{\Omega} \Delta E(\Omega) = -\sum_{\Omega} \Delta T(\Omega).$$

The change in the KE and the changes in the individual contributions to the virial (the PE) over the spatial region of each fragment are thus determined.

One may employ atomic populations and forces to illustrate a previously proposed electrostatic model of ionic binding (7). The application of the model (Section 2-3-5) assumes a partitioning of $\rho(r)$ and the forces it exerts on the nuclei into cationic and anionic contributions. This was previously accomplished in terms of the BDM and an orbital breakdown of the forces, both steps involving arbitrary assumptions.

The present procedure removes these obstacles by providing a quantum mechanical definition of a subspace and its properties. For an ionic system, the charge density is so distributed that the zero-flux boundary condition yields fragments which correspond with one's intuitive concept of an ionic fragment. Consider the illustrative examples of LiH($X^1\Sigma^+$) and LiF($X^1\Sigma^+$) (Fig. 2-23 and Table 2-5). The net charges are close to the values of ±1 obtained by the transfer of one electron. In an ideal ionic system with the transfer of one electron to form A^+B^-, the (B) fragment will, if its charge distribution approaches full spherical symmetry, exert a force on the A nucleus equivalent to the presence of one excess negative charge on (B), i.e., $Z_B - F_A(B)(R^2/Z_A) = -1$, where $F_A(B)$ is the force exerted on nucleus A by the charge density of (B). Similarly, the (A) fragment should exert a force on the B nucleus equivalent to the presence of one excess positive charge on (A), i.e., $Z_A - F_B(A)(R^2/Z_B) = +1$.

The excess $0.94e^-$ on (F) in LiF exerts a force at Li equivalent to $-0.76e^-$. Similarly, the excess $0.91e^-$ on (H) exerts a field at Li equivalent to $-0.62e^-$. Thus the charge distributions of the anionic fragments, because of their diffuse

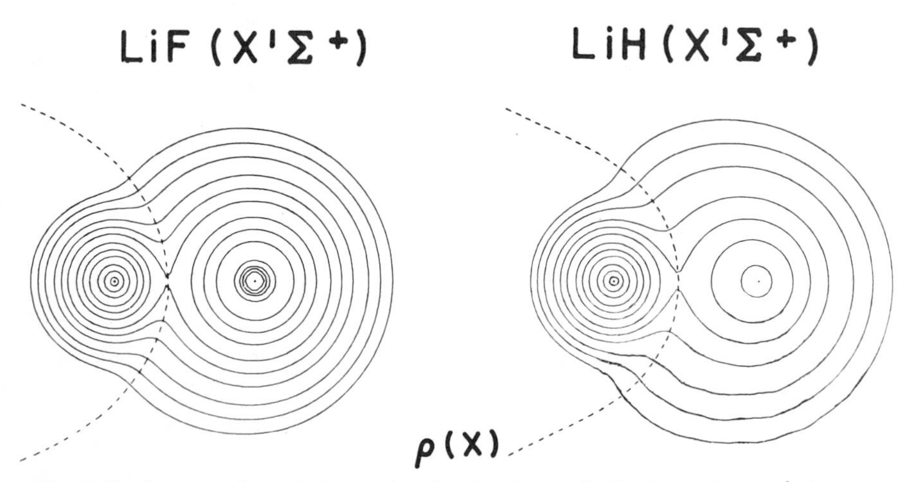

Fig. 2-23. Contour plots of the total molecular charge distributions of LiF($X^1\Sigma^+$) and LiH($X^1\Sigma^+$), showing the virial partitioning surfaces which define the (Li), (F), and (H) fragments.

Table 2-5. Virial Fragment Populations and Forces in LiH and LiF.

AB (R_e)	$C(A) = Z_A - N(A)$	$C(B) = Z_B - N(B)$	Charge equivalent of force exerted on: Li nucleus by (F) or (H)	F or H nucleus by (Li)
LiF ($X^1\Sigma^+$) (2.955)	+0.937	−0.937	−0.763	+0.959
LiH ($X^1\Sigma^+$) (3.015)	+0.911	−0.911	−0.615	+0.918

and very polarized nature (Fig. 2-23), approach but do not attain the distribution of an idealized negative ion in terms of the forces they exert on the cationic nucleus. It is apparent from Fig. 2-23 that the cationic distribution, that for (Li), is very compact and close to being spherical in both molecules. Consequently, one finds a close correspondence between the actual net charge on the (Li) fragment and the number of charges which, when placed in a spherical distribution on Li, would exert the same force on the anionic nucleus (Table 2-5). These forces which one fragment exerts on its neighboring nucleus are, of course, balanced by a force resulting from the polarization of the charge density of the neighboring fragment. For example, the (Li) charge density is polarized away from (F) and from (H) to an extent sufficient to create an atomic force equal and opposite to the net attractive force exerted on the Li nucleus by the (F) and (H) fragments. Similarly, the charge distributions of (F) and (H) are inwardly polarized to exert forces balancing the net repulsive forces exerted by the (Li) fragment.

The (Li) fragment in LiH and LiF provides a simple example of a fragment whose charge distribution and properties remain nearly constant in spite of a change in the identity of its neighboring fragment from (F) to (H). The (Li) fragments are very similar in their charge distributions (Fig. 2-23) and electron populations (Table 2-5). The virial definition of a fragment energy yields (for Hartree-Fock wavefunctions) $E(\text{Li}) = -T(\text{Li}) = -7.354$ au in LiF and -7.368 au in LiH. The similarity in the energies of the (Li) fragments (they differ by only ~9 kcal/mole) reflects the similarity in their charge distributions. The Hartree-Fock energies of a Li atom and a Li^+ ion are -7.433 au and -7.236 au, respectively. Thus the energies of the (Li) fragments in LiH and LiF are readily rationalized as being those of polarized Li^+ ion in an external (attractive) field. The definition of a fragment energy in terms of the virial of the forces exerted on it always yields values in agreement with chemical expectations. It should be recalled that the definition of a fragment energy involves a partitioning of the nuclear repulsive energies in a nonarbitrary way and, in some cases, with unexpected results. For example in LiH, because of the pronounced back polarization of the (Li) fragment density, the sign of the nuclear virial V_n(Li) is negative, implying that the nuclear–nuclear repulsive potential contributes to an energy lowering of the (Li) fragment. One first observes that for (Li) in LiH, V_n(Li) is required to be less than zero to satisfy the fragment virial theorem. Secondly, the definition of a virial sharing operator which yields the single-particle PE operator in a many-particle system (*88*) illustrates that the term $\Sigma_{\alpha<\beta} Z_\alpha Z_\beta / R_{\alpha\beta}$ is part of the molecular electronic energy, the energy of the (motionless) nuclei at the equilibrium configuration in the Born–Oppenheimer approximation equaling zero. The contribution of the term $\Sigma_{\alpha<\beta} Z_\alpha Z_\beta / R_{\alpha\beta}$ to the electronic energy of a fragment (Ω) is determined by the forces which the charge density in (Ω) exerts on the nuclei. These forces may be net attractive or repulsive and hence $V_n(\Omega)$ may be greater or less than zero.

It is interesting to observe that the accumulation of charge density in the antibinding regions of nuclei other than the proton, a feature of the BDMs illustrated and discussed previously, will lead to a lowering of the energy of a fragment as a consequence of its negative contributions to $V_n(\Omega)$.

2-8. SUMMARY

This chapter presents a discussion and classification of chemical binding based on the properties of molecular electronic charge distributions and, in particular, on the forces which they exert on the nuclei. Chemical binding is discussed in terms of the density difference distributions, which picture the redistribution of charge that results from the formation of a molecule. The forces exerted on the nuclei in a molecule are related to changes in the charge distribution depicted in the density difference maps, and a quantitative discussion of the manner in which electrostatic equilibrium is attained to give a stable molecule is given in terms of the forces. The concepts of bonding and antibonding as applied to orbitals are compared with binding and antibinding, the latter two concepts being defined in terms of the forces exerted on the nuclei by MO densities.

Definitions of ionic and covalent binding based on the density difference distributions and the manner in which they determine the corresponding states of electrostatic equilibrium are presented. A partitioning of the total electronic force in accordance with the density difference maps demonstrates that in covalent binding the nuclei are bound by the density which is shared between them, while in ionic binding the nuclei are bound by the density which is localized about a single nucleus.

A discussion is given of the correlation between the strength of the electric field binding the nuclei, which results from the accumulation of charge density in the binding region, with molecular binding energies. By combining the results of the virial and electrostatic theorems it is shown that general constraints may be placed on the signs and relative magnitudes of the kinetic- and potential-energy changes over the complete range of internuclear separations involved in the formation of a molecule. These constraints are stated in terms of the behavior of the forces and their derivatives with respect to the internuclear separation, and hence they may be directly related to the charge density and the manner in which it changes as the atoms approach one another. A detailed discussion of the changes in kinetic energy which accompany molecule formation is given. The important physical observation regarding the behavior of the kinetic energy on bonding is the existence of local regions of space in which it attains particularly low values, low in absolute terms or relative to the isolated atom values. The positionings of these regions relative to the nuclei in different systems may be predicted from and interpreted in terms of the different possible charge rearrangements depicted in the density difference maps. Thus the behavior of the kinetic energy in the region of hydrogen or helium is very different from that found in the region of, say, carbon, nitrogen, or oxygen. The changes in the charge density on bonding and the associated changes in the forces and kinetic and potential energies are relatively simple for hydrogen and helium and are readily rationalized. The same changes incurred on bonding for atoms other than hydrogen or helium, while tabulated here, are more complex and have not been well rationalized in terms of any physical model.

The chapter concludes with a review of some new quantum mechanical results which indicate the possibility of the electron charge density playing a more fundamental role in future predictive and interpretive theories of chemistry.

Notes

1. We do not exclude the possibility of the existence of a small potential-energy barrier for large values of R. In this situation, the integral of $F(R)\,dR$ over the range in which $F(R)$ is repulsive must be smaller in magnitude than the integral of $F(R)\,dR$ where $F(R)$ is binding.

2. The difference between the observed X-ray structure factors and those calculated from the spherical-atom scattering factors using neutron positional and temperature factors yields in a Fourier display the contribution of the distortions of the free-atom charge distributions to the charge distribution of the system, i.e., a display of the $\Delta\rho(r)$ distribution. Experimentalists label the density difference function obtained in this manner as $\Delta\rho(r)_{x-n}$. The experimental result corresponds to an average over the thermal motions of the nuclei.

3. The forces on some nuclei in a polyatomic system may be zero as a result of a balance of the nuclear repulsions. For example, in an AX_3 system of D_{3h} symmetry, the force on the A nucleus will be zero for all internuclear separations which maintain D_{3h} symmetry.

4. Thus, the attractive part of the potential energy (PE) curve is due to electron–nuclear attraction arising from the electron density accumulated in the binding region, and the repulsive part of the curve is due to nuclear–nuclear repulsion and electron–nuclear attraction arising from the electron density accumulated in the antibinding region.

5. The necessary approximations for data interpretation are given in Ref. 62.

6. This observation remains true however one defines the "core" distribution: (a) by identifying it with the correspondingly localized MO, e.g., $1\sigma_g$ and $1\sigma_u$ in Li_2 or 2σ in LiF; (b) by identifying it with the total number and pair densities which are localized on the Li nucleus (*23, 42*).

7. The cores of N, O, and F and of Al → Cl are in any event polarized into the binding region.

8. The $\Delta\rho$ map for LiF shows that while the outer, more diffuse increase in charge density on F is polarized away from Li, the charge increase close to the F nucleus is polarized toward Li. Since the force is determined by averaging over the coordinate r_F^{-2}, the direction of the force and electric field is determined by the polarization of the inner density. This is substantiated by a calculation of the force exerted on the F nucleus:

9. The net force would be still smaller if the calculations had employed wavefunctions at the Hartree–Fock values for R. For example, the net force on the N nuclei at the Hartree–Fock minimum for N_2 is 0.024 au.

10. The models of Nakatsuji and Deb are discussed at length in Chapter 3.

11. Bader et al. (*6, 7*) set $N_B(\equiv \Sigma_i f_i^{(BB)})$ equal to the number of electronic charges on B which exerts the same field on A as does an actual atomic population on B.

12. Since $2T_a = -V_a$, (2-16) becomes $2\Delta T = -\Delta V - R(dE/dR)$.

13. This interpretation is totally consistent with the correlation between D_e and net attractive fields on the nuclei (Section 2-5).

14. The differing numerical coefficients of $F(R)$ in (2-18) and (2-19) indicate that dV/dR changes sign at a value of R larger than that at which dT/dR does. Thus, V begins to decrease with a further decrease in R, at some $R > R_i$, before T beings to increase.

15. One may have a situation in which the forces are repulsive for large R and attractive for smaller R. The forces on the nuclei are then zero at infinity, at some intermediate point R', and at R_e. The resulting molecular state will be bound (relative to separated atoms) if the integral of the forces between $R = \infty$ and R' is of smaller magnitude than the corresponding integral evaluated between the limits R' and R_e.

16. At the positions of the nuclei in He_2, $\Delta\rho(r) = 0.2$ au for $R = 1.5$ au, while for H_2, at $R = 1.4$ au, $\Delta\rho(r) \sim 0.12$ au at the positions of the nuclei.

17. $\{\phi_i(r)\}$ may also represent a set of SCF orbitals when the wavefunction is single-determinantal. Then λ_i equals unity for each occupied spin orbital.

18. Some time after the appearance of the work discussed above (*13*), Feinberg and Ruedenberg (*48*) reported calculations of $T_\parallel$ and $T_\perp$ for the H_2^+ molecule ion as functions of R. The behavior of T, $T_\parallel$, and $T_\perp$ in the formation of

H_2^+ is similar to that reported above for H_2. They claim, however, to have a different interpretation of the results. We quote here their summary of their interpretation.

> The extension of the system from one atom to two atoms weakens the "kinetic energy pressure" because the bond parallel component is partially annihilated. As a result the "nuclear solution" can be more effective in drawing the electron cloud deeper in the potential wells near the nuclei than is possible in the atomic one-center system.
>
> It is in the behavior of the kinetic-energy integral, in particular $T_{\parallel}$, that the molecule differs essentially from the free atom. At large distances, this leads to incipient binding due to kinetic-energy lowering. Near the equilibrium distance it leads to binding by lowering the kinetic resistance against the nuclear suction. This results in a closer attachment of the electronic cloud to the nuclei with a concomitant lowering of the potential energy and increase in the kinetic energy.

19. The principal features of $G(r)$ and $\Delta G(r)$ at R_e and the conclusions based on them for H_2 will not change if the single-determinantal wavefunction for the configuration $1\sigma_g^2$ had been used in place of the extended CI wavefunction.

20. The Hartree–Fock binding energies (in eV) for N_2, O_2, and F_2, together with experimental values in parentheses, are, respectively, 5.179 (9.902), 1.283 (5.213), and -1.37 (1.647), with the wavefunctions employed here. This indicates that the results given above for O_2 and F_2 must be treated with great caution, particularly for F_2, for which $\int \Delta G(r)\, dr$ will yield (incorrectly) a negative value for ΔT.

21. $\rho_i(r)$ is not the same as $\Delta\rho(r)$, since the latter is defined with respect to *separated* atomic densities. Thus, while $\rho_a(r)$ and $\Delta\rho(r)$ are independent of the orbital composition of the wavefunctions, or of the form of the wavefunction, $\rho_i(r)$ is tied to an orbital expansion and its explicit form is determined by the particular basis set employed in the expansion.

22. It is not clear how one defines ρ_{qc} in the case of a large extended basis set. Does one include the polarizing functions in the atomic descriptions of the promoted state? This would not appear to be possible, as only the σ and π components of the $3d$ and $4f$ orbitals are included in the molecular set and the complete sets are of course required for the description of an S state. Furthermore, the AO coefficients are normalized for the MOs and not for the atomic case. Some prescription must be devised to redetermine and renormalize the coefficients for the determination of ρ_{qc}.

23. For example, for LiF$(X^1\Sigma^+)$ $C(\text{F}) = -C(\text{Li}) = -0.940e^-$; for $BF_3(X^1A_1')$ $C(\text{B}) = +2.592e^-$, $C(\text{F}) = -0.864e^-$; for $BH_3(X^1A_1')$ $C(\text{B}) = +2.137e^-$, $C(\text{H}) = -0.712e^-$.

24. It has been recently pointed out (*43, 44*; Section 3-4-4c) that "fragment" shape is another property which is transferable from one molecule to another.

References

1. Bader, R. F. W., 1960, *Can. J. Chem.*, **38**, 2117.
2. —— and Jones, G. A., 1961, *Can. J. Chem.*, **39**, 1253.
3. ——, 1964, *J. Am. Chem. Soc.*, **86**, 5070.
4. —— and Henneker, W. H., 1965, *J. Am. Chem. Soc.*, **87**, 3063.
5. —— and Preston, H. J. T., 1966, *Can. J. Chem.*, **43**, 1131.
6. ——, Henneker, W. H., and Cade, P. E., 1967, *J. Chem. Phys.*, **46**, 3341.
7. ——, Keaveny, I., and Cade, P. E., 1967, *J. Chem. Phys.*, **47**, 3381.
8. —— and Chandra, A. K., 1968, *Can. J. Chem.*, **46**, 953.
9. —— and Bandrauk, A. D., 1968, *J. Chem. Phys.*, **49**, 1666.
10. —— and Bandrauk, A. D., 1968, *J. Chem. Phys.*, **49**, 1653.
11. —— and Ginsburg, J. L., 1969, *Can. J. Chem.*, **47**, 3061.
12. ——, 1970, *An Introduction to the Electronic Structure of Atoms and Molecules* (Clarke, Irwin and Co. Ltd., Toronto).
13. —— and Preston, H. J. T., 1969, *Int. J. Quantum Chem.*, **3**, 327.
14. ——, Keaveny, I., and Runtz, G., 1969, *Can. J. Chem.*, **47**, 2308.
15. —— and Gangi, R. A., 1971, *J. Am. Chem. Soc.*, **93**, 1831; *J. Chem. Phys.*, **55**, 5369.
16. —— and Beddall, P. M., 1972, *J. Chem. Phys.*, **56**, 3320.
17. —— and Beddall, P. M., 1973, *J. Am. Chem. Soc.*, **95**, 305.
18. ——, Beddall, P. M., and Peslak, J., 1973, *J. Chem. Phys.*, **58**, 557.
19. ——, Duke, A. J., and Messer, R. R., 1973, *J. Am. Chem. Soc.*, **95**, 7715.
20. —— and Messer, R. R., 1974, *Can. J. Chem.*, **52**, 2268.
21. ——, 1975, *Acc. Chem. Res.*, **8**, 34.
22. —— and Runtz, G. R., 1975, *Mol. Phys.*, **30**, 117, 129.
23. —— and Stephens, M. E., 1975, *J. Am. Chem. Soc.*, **97**, 7391.
24. ——, 1975, *Physical Chemistry*, series two, vol. 1 (International Review of Science), eds. A. D. Buckingham and C. A. Coulson (Butterworths, London).
25. ——, Srebrenik, S., and Nguyen-Dang, T. T., 1978, *J. Chem. Phys.*, **68**, 3680.
26. ——, Anderson, S. G., and Duke, A. J., 1979, *J. Am. Chem. Soc.*, **101**, 1389.
27. Bagus, P. S. and Gilbert, T. L., 1967, as reported by McLean, A. D. and Yoshimine, M., in *Tables of Linear Molecule Wave Functions*, supplement to *IBM J. Research*, November 1967.
28. Balfour, W. J. and Douglas, A. E., 1970, *Can. J. Phys.*, **48**, 901.
29. Bartell, L. S. and Brockway, L. O., 1953, *Phys. Rev.*, **90**, 833.
30. Becker, P., Coppens, P., and Ross, F. K., 1973, *J. Am. Chem. Soc.*, **95**, 7604.
31. Berlin, T., 1951, *J. Chem. Phys.*, **19**, 208.
32. Cade, P. E., Sales, K. D., and Wahl, A. C., 1966, *J. Chem. Phys.*, **44**, 1973.
33. Cade, P. E., Bader, R. F. W., Henneker, W. H., and Keaveny, I., 1969, *J. Chem. Phys.*, **50**, 5313.
34. —— and Pelletier, J., 1971, *J. Chem. Phys.*, **54**, 3517.
35. Chandra, A. K. and Sundar, R., 1971, *Mol. Phys.*, **22**, 369.

36. ——, 1972, *Chem. Phys. Lett.*, **14**, 577.
37. —— and Sebastian, K. L., 1976, *Mol. Phys.*, **31**, 1489.
37a. Collard, K. and Hall, G. G., 1977, *Int. J. Quantum Chem.*, **12**, 623.
38. Coppens, P., 1967, *Science*, **158**, 1577.
39. Coulson, C. A., 1941, *Proc. Cambridge Phil. Soc.*, **37**, 55, 74.
40. Coulson, C. A. and Duncanson, W. E., 1941, *Proc. Cambridge Phil. Soc.*, **37**, 67, 406.
41. Das, G. and Wahl, A. C., 1966, *J. Chem. Phys.*, **44**, 87.
42. Daudel, R., Bader, R. F. W., and Stephens, M. E., 1974, *Can. J. Chem.*, **52**, 1310.
43. Deb, B. M., 1974, *J. Am. Chem. Soc.*, **96**, 2030; 1975, **97**, 1988.
44. ——, 1975, *J. Chem. Educ.*, **52**, 314.
45. ——, Bose, S. K., and Sen, P. N., 1976, *Indian J. Pure Appl. Phys.*, **14**, 444.
46. ——, Sen, P. N., and Bose, S. K., 1974, *J. Am. Chem. Soc.*, **96**, 2044.
47. Dedieu, A. and Veillard, A., 1972, *J. Am. Chem. Soc.*, **94**, 6730.
48. Feinberg, M. J. and Ruedenberg, K., 1971, *J. Chem. Phys.*, **54**, 1495; **55**, 5804.
49. Feynman, R. P., 1939, *Phys. Rev.*, **56**, 340.
50. Fink, M., Gregory, D., and Moore, P. G., 1976, *Phys. Rev. Lett.*, **37**, 15.
51. Gilbert, T. L. and Wahl, A. C., 1967, *J. Chem. Phys.*, **47**, 3425.
52. Goddard, W. A., 1967, *Phys. Rev.*, **157**, 81.
53. Gordon, R. G. and Kim, Y. S., 1972, *J. Chem. Phys.*, **56**, 3122.
54. Goscinski, O. and Palma, A., 1977, *Chem. Phys. Lett.*, **47**, 322.
55. ——, 1976, *Quantum Science*, eds. J. L. Calais, O. Goscinski, J. Linderberg, and Y. Öhrn (Plenum Publishing Company, New York).
56. Hirshfeld, F. L. and Rzotkiewicz, S., 1974, *Mol. Phys.*, **27**, 1319.
57. Hirschfelder, J. O. and Eliason, M. A., 1967, *J. Chem. Phys.*, **47**, 1164.
58. Hohenberg, P. and Kohn, W., 1964, *Phys. Rev.*, **136**, B864.
59. Hurley, A. C., 1954, *Proc. Roy. Soc.* (London), **A226**, 170, 179, 193.
60. ——, 1956, *Proc. Roy. Soc.* (London), **A235**, 224.
61. ——, 1966, *Quantum Theory of Atoms, Molecules and the Solid State*, ed. P.-O. Löwdin (Academic Press, New York), p. 571.
62. Kohl, D. A. and Bartell, L. S., 1969, *J. Chem. Phys.*, **51**, 2891, 2896.
63. Kohn, W. and Sham, L. J., 1965, *Phys. Rev.*, **137**, A1697.
64. Mulliken, R. S., 1928, *Phys. Rev.*, **32**, 186, 761.
65. ——, 1932, *Rev. Mod. Phys.*, **4**, 1.
66. ——, 1939, *Phys. Rev.*, **56**, 778.
67. ——, 1966, *Quantum Theory of Atoms, Molecules and the Solid State*, ed. P.-O. Löwdin (Academic Press, New York), p. 231.
68. Nakatsuji, H., 1973, *J. Am. Chem. Soc.*, **95**, 345, 354, 2084.
69. —— and Koga, T., 1974, *J. Am. Chem. Soc.*, **96**, 6000.
70. ——, 1974, *J. Am. Chem. Soc.*, **96**, 24, 30.
71. —— and Parr, R. G., 1975, *J. Chem. Phys.*, **63**, 1112.
72. Parr, R. G., Donnelly, R. A., Levy, M., and Palke, W. E., 1978, *J. Chem. Phys.*, **68**, 3801.
73. Preston, H. J. T., 1969, Dissertation, McMaster University.

74. Ransil, B. J. and Sinai, J. J., 1967, *J. Chem. Phys.*, **46**, 4050.
75. Roux, M., Besnainou, S., and Daudel, R., 1956, *J. Chim. Phys.*, **54**, 218.
76. ——, 1958, *J. Chim. Phys.*, **55**, 754.
77. ——, Cornille, M., and Burnelle, L., 1962, *J. Chem. Phys.*, **37**, 933.
78. Ruedenberg, K., 1962, *Rev. Mod. Phys.*, **34**, 326.
79. ——, 1975, *Localization and Delocalization in Quantum Chemistry*, vol. I, eds. O. Chalvey, R. Daudel, S. Diner, and J. P. Malrieu (D. Reidel, Dordrecht, Holland), p. 223.
80. Runtz, G. R., Bader, R. F. W., and Messer, R. R., 1977, *Can. J. Chem.*, **55**, 3040.
81. Schwendeman, R. H., 1966, *J. Chem. Phys.*, **44**, 556.
82. Shull, H., 1962, *J. Appl. Phys. Suppl.*, **33**, 290.
83. Slater, J. C., and Johnson, K. H., 1972, *Phys. Rev.*, **B5**, 844.
84. ——, 1972, *J. Chem. Phys.*, **57**, 2389.
85. ——, 1933, *J. Chem. Phys.*, **1**, 687.
86. Smith, J. R., Ying, S. C., and Kohn, W., 1973, *Phys. Rev. Lett.*, **30**, 610.
87. Srebrenik, S., Weinstein, H., and Pauncz, R., 1974, *J. Chem. Phys.*, **61**, 5050.
88. —— and Bader, R. F. W., 1975, *J. Chem. Phys.*, **63**, 3945.
89. ——, Bader, R. F. W., and Nguyen-Dang, T. T., 1978, *J. Chem. Phys.*, **68**, 3667.
90. Stevens, R. M. and Lipscomb, W. N., 1964, *J. Chem. Phys.*, **41**, 184, 3710.
91. ——, Switkes, E., Laws, E. A., and Lipscomb, W. N., 1971, *J. Am. Chem. Soc.*, **93**, 2603.
92. Sundar, R. and Chandra, A. K., 1974, *Indian J. Chem.*, **12**, 145.
93. Tal, Y. and Katriel, J., 1977, *Theoret. Chim. Acta.* (Berlin), **46**, 173.
94. Thom, René, 1975, *Structural Stability and Morphogenesis*, American edition (W. A. Benjamin Inc., Reading, Mass.).
95. Wilson, C. W. and Goddard, W. A., 1970, *Chem. Phys. Lett.*, **5**, 45.
96. ——, 1972, *Theoret. Chim. Acta.* (Berlin), **26**, 195, 211.

Supplementary References

97. Bamzai, A. S. and Deb, B. M., 1981, *The Role of Single-Particle Density in Chemistry, Rev. Mod. Phys.*, forthcoming.
98. Koga, T., Nakatsuji, H., and Yonezawa, T., 1978, *J. Am. Chem. Soc.*, **100**, 7522.
99. Nakatsuji, H., Kanayama, S., Harada, S., and Yonezawa, T., 1978, *J. Am. Chem. Soc.*, **100**, 7528.
100. Nakatsuji, H., Koga, T., Kondo, K., and Yonezawa, T., 1978, *J. Am. Chem. Soc.*, **100**, 1029.
101. Smith, V. H., 1977, *Phys. Scrip.*, **15**, 147.
102. Tal, Y., Bader, R. F. W., and Erkku, J., 1980, *Phys. Rev.*, **A21**, 1.
103. Theophilou, A., 1979, *J. Phys. C: Sol. St. Phys.*, **12**, 5419.

3
Force Models for Molecular Geometry

H. Nakatsuji and T. Koga
Department of Hydrocarbon Chemistry
Faculty of Engineering
Kyoto University
Kyoto, Japan

Contents

3-1. INTRODUCTION

In order to explain the great variety and regularity in structural chemistry, general principles and conceptual models are of fundamental importance, since they provide a unified viewpoint for interrelating diverse chemical phenomena

(*142*, *225*). Until recently, most models for molecular geometry were constructed on energetic grounds. In this chapter our main purpose is to explain the force concept as applied to molecular geometry by fully utilizing the physical simplicity and visuality of the electrostatic Hellmann-Feynman (H-F) theorem (*101*, *133*)

$$F_A = Z_A \int r_{A1}/r_{A1}^3 \rho(r_1)\, dr_1 - Z_A \sum_{B \neq A} Z_B R_{AB}/R_{AB}^3, \qquad (3\text{-}1)$$

where $\boldsymbol{F}_A$ is the force acting on nucleus A of charge Z_A, $\rho(r_1)$ the 3-D electron density, $\boldsymbol{r}_{A1}$ and $\boldsymbol{R}_{AB}$ position vectors from nucleus A to electron 1 and nucleus B, respectively (see chapters 1 and 2).

Among earlier workers, Mulliken (*204*) and Walsh (*294*) considered molecular shapes, drawing angular correlation diagrams[1] in which the energies of simple "empirical" MOs were plotted against the valence angle of a molecule. This gave rise to the celebrated *Walsh rules*, which provided a beautiful correlation between molecular shapes and the number of valence electrons.[2] In spite of minor exceptions like highly ionic molecules, the Mulliken-Walsh MO model has had much influence on later developments in structural chemistry (*134*) and theoretical chemistry (*142*). About the same time, Sidgwick and Powell (*274*), Gillespie and Nyholm (*121*) and Gillespie (*118-120*) developed the valence-shell electron-pair repulsion (VSEPR) model for molecular shapes based on Pauli repulsions between electron pairs. The comparatively recent second-order Jahn-Teller (SOJT) theory for molecular shapes is based on second-order perturbation theory (*13*, *14*, *28*, *48*, *232*, *233*, *235*). These energetic models are reviewed briefly in Section 3-2.

More recently, the H-F theorem has been applied to molecular geometries with great success, e.g., the relation between electron density and molecular shapes of simple hydrides (*20-22*), the origin (*127*, *277*) of the internal rotation barrier in ethane (see Chapter 4), the FOJT effect (*56*, *65*, *71*), and a reinterpretation of the Walsh diagram (*66*). Subsequently, Nakatsuji and co-workers (*207-211*, *215*, *216*, *219-221*), and Deb and co-workers (*76-79*, *82*) have proposed versatile force models for molecular geometry which are very successful in explaining the variety and regularity in structural chemistry. Nakatsuji's electrostatic force (ESF) theory is applicable to both isolated and interacting molecules, including long-range forces (*167*, *215*) and chemical reactivity. These two force models will be discussed in detail in Sections 3-3 and 3-4. In Section 3-5 we study the dynamic behavior of electron density during nuclear rearrangement processes which occur in molecular geometry, molecular vibrations and chemical reactions, using both H-F and integral H-F theorems (*166*, *227*; see also Chapter 4). Next we will study a border area between molecular geometry and chemical reaction, namely, the change in the geometries of reactant molecules during a chemical reaction.

3-1-1. Comparison of the Energy and Force Viewpoints in Their Exact Limits

In the energetic viewpoint, a stable molecular geometry corresponds to the minimum in the potential surface which is calculated by

$$E = \langle \psi, H\psi \rangle$$

$$= \int \left[-\tfrac{1}{2} \Delta \rho(r_1' | r_1) \right]_{r_1' = r_1} dr_1 - \sum_{A} \int (Z_A / r_{A1}) \, \rho(r_1) \, dr_1$$

$$+ \int (1/r_{12}) \, \Gamma(r_1 r_2 | r_1 r_2) \, dr_1 \, dr_2 + \sum_{A<B} Z_A Z_B / R_{AB}, \qquad (3\text{-}2)$$

where ψ is the exact wavefunction, H the Born–Oppenheimer (*39*) Hamiltonian, and ρ and Γ are the first- and second-order reduced density matrices (*180, 181*; see Chapter 8), respectively. In contrast to the force expression (3-1), the energy expression (3-2) is much more cumbersome, since (i) one simultaneously considers four energy terms which are usually competitive with one another; (ii) one requires both first- and second-order density matrices in contrast to just the electron density in (3-1); (iii) (3-1) permits a classical electrostatic interpretation *even in the exact limit*.

The H–F force has a severe drawback in that it is more sensitive than energy to inaccuracies of the (approximate) wavefunctions used. This has prevented quantification from this approach (however, see Refs. 220, 221, and 221a), except for some simple systems (*16*, *140*, *215*). Although the so-called energy gradient (see Chapter 9) is as accurate as the energy and has been employed to study potential surfaces (*103*, *115*, *169*, *188*, *196*, *248*), it does not have a simple meaning like the H–F force.

A wavefunction which satisfies the H–F theorem is called "floating" (*67*, *147*, *148*) or "stable" (*129*). When we expand the wavefunction in terms of an incomplete basis set (e.g., LCAO-MO), it becomes unstable (nonfloating) if the basis AOs are fixed on the constituent nuclei, since then electronic coordinates are not treated as free from nuclear coordinates. Floating wavefunctions have been calculated for H_2 (*273*), H_2O, NH_3, CH_3^+, and NH_3^+ (*220, 221*). The floating spherical Gaussian orbital (FSGO) wavefunctions (*106*) also satisfy the H–F theorem. Recently, a new method of obtaining a stable wavefunction was given[3] and shown to be useful in the force theoretic approach (*221*a).

As seen in Section 3-2, since the energetic models of molecular geometry start from more or less drastic approximations to (3-2), their underlying concepts themselves are approximate. On the other hand, the force models (especially that in Sections 3-3 and 3-5) start from the exact equation (3-1) and, although certain approximations are introduced for practical reasons, the concept itself is rigorous and preserved in both approximate and exact treatments.

3-2. ENERGETIC MODELS FOR MOLECULAR SHAPE

Although one can nowadays perform ab initio geometry calculations for molecules of moderate size, in good agreement with experiment, such calculations by themselves are usually ineffective in explaining the various trends in structural chemistry (*26*). Since "to calculate a molecule is not to understand a molecule" (*228*), the need for molecular geometry models is as great as ever. We review here briefly the more important energetic models so that we can compare them with the force models described later.

3-2-1. Walsh's Correlation Diagram

Figure 3-1 gives an example of Walsh's correlation diagrams (*294*) for AH_2 molecules. It plots the MO "binding" energies for valence electrons against the HAH angle. The MO energy variations can be easily rationalized (Fig. 3-2). Thus, in the π_u-b_1 correlation, the p_{zA} AO (MO) having a node on the AH_2 plane is not affected much by bending in the xy plane. In the π_u-a_1 (lone-pair) correlation, the p_{yA} AO at linear shape is mixed by bending with the s_A AO of lower energy and the two hydrogenic AOs can overlap in phase with the bottom lobe of p_{yA} AO (*123*) and with each other. Therefore, bending rapidly stabilizes the a_1 (lone-pair) MO. In the σ_u-b_2 correlation, bending pulls two s_H AOs

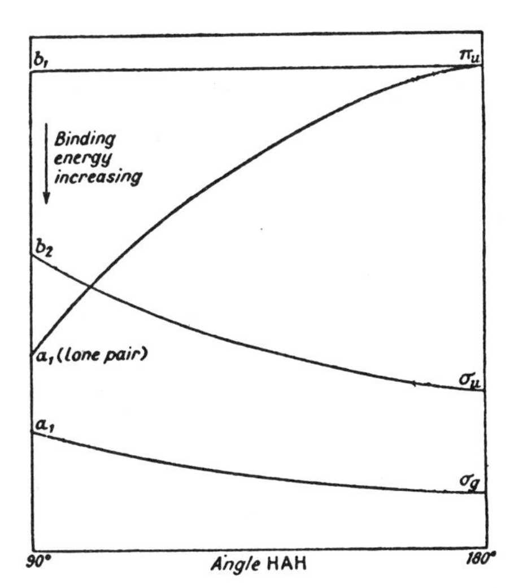

Fig. 3-1. Orbital correlation diagram for AH_2 molecules. (Reproduced from Ref. 294, courtesy the Chemical Society, London.)

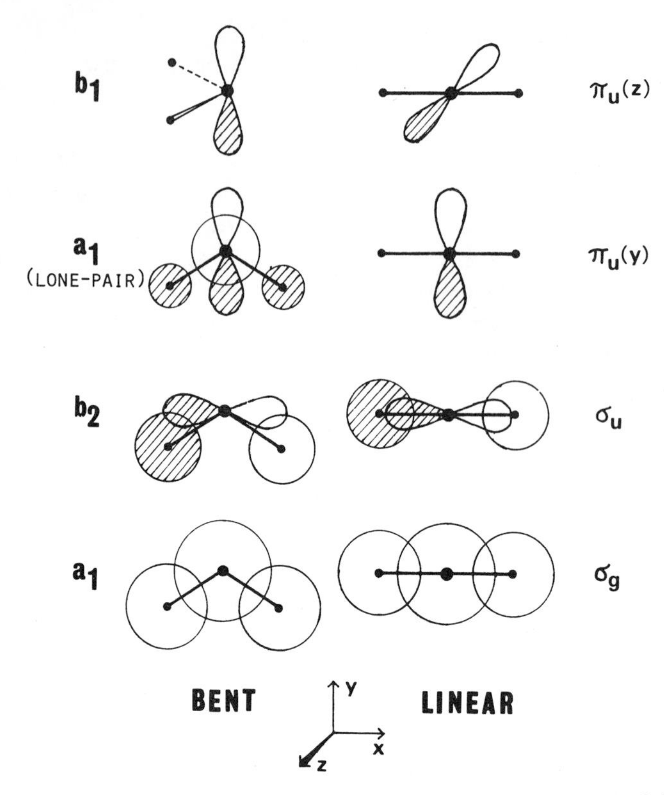

Fig. 3-2. Schematic MOs of linear and bent AH_2 molecules (see also Fig. 3-16).

out of overlap with the lobes of p_{xA} AO and destabilizes the b_2 MO as compared to the σ_u MO (*123*). The slope of the σ_g-a_1 curve in Fig. 3-1 is of the wrong sign; the a_1 MO will be lower in energy than the σ_g MO, mainly because of the overlap among two s_H and p_{yA} AO's (*4, 66, 123, 240*; Fig. 3-16).

With this correction in Fig. 3-1, the shapes of AH_2 molecules (see Table 3-3) can be explained as follows: For molecules with 4 valence electrons, the lowest two MOs are occupied and therefore the shape is linear. For molecules with 5-8 valence electrons, the steep π_u-a_1 (lone pair) leads to a bent structure. With 2 valence electrons, the *corrected* Walsh diagram rightly predicts a bent structure, e.g., LiH_2^+ and H_3^+ (*245, 249*). Further, when an electron jumps from the a_1 (lone-pair) to the b_1 MO, the HAH angle increases, e.g., in NH_2 and PH_2.

The Walsh rules (Table 3-1) generally agree well with experimental data (*134*), though interesting exceptions exist, e.g., highly ionic molecules like alkaline earth dihalides with heavy metal-light halogen combinations (Section 3-3-7d)

Table 3-1. Walsh Rules Linking the Shapes of Ground-State Molecules with the Numbers of Valence Electrons.[a]

Molecule	Shape (Number of valence electrons)
AH_2	linear (4), bent (5–8)
AH_3	planar ($\leqslant$6), pyramidal (7–8)
HAB	linear ($\leqslant$10), bent (10–14), linear (16)
AB_2, ABC	linear ($\leqslant$16), bent (17–20), linear or nearly linear (22)
H_2AB	planar (12), nonplanar (13–14)
AB_3	planar ($\leqslant$24), pyramidal (25–26)[b], planar or nearly planar (28)
HAAH	linear (10), bent but planar (12), bent and nonplanar (14)

[a]Reprinted from Ref. 294.
[b]The BAB angle will be less for 26-electron molecules than for 25-electron molecules.

and the sensitive problem of the shapes of 7-valence electron AH_3 molecules like CH_3, NH_3^+, and BH_3^- (Section 3-3-7a). The Walsh diagram predicts the former type to be linear and the latter type to be nonplanar. While molecules like SiH_3, GeH_3, and SnH_3 (*151*, *152*, *197*), having heavier central atoms, and CH_2F, CHF_2, and CF_3, having electronegative substituents (*100*), are pyramidal, the radical CH_3 is planar or almost planar, in that the potential barrier to inversion is very shallow (experimental: *12*, *59*, *91*, *98*, *99*, *134*, *195*, *285*; theoretical: *51*, *87*, *194*, *198*, *282*, *306*). BH_3^- and NH_3^+ are experimentally reported as planar (*58*, *285*, *286*). The ESF theory (Section 3-3) satisfactorily explains these trends.

In the Walsh model the total molecular energy is supposed to be expressible as the sum of "effective" MO energies. However, in spite of many attempts, the reasonings for such a summation based on more refined quantum theories are not completely satisfactory, because of electron-electron and nuclear-nuclear repulsion terms (*4*, *5*, *34–37*, *44*, *46*, *66*, *69*, *240*, *242*, *256*, *280*, *303*, *304*).

Extensions and modifications of the Walsh diagram have been suggested (*45*, *113*, *122–124*). Hayes (*131*) considered outer *d* AOs for alkaline earth dihalides (see also Ref. 126). Other workers (*266*, *287–289*) viewed the Walsh diagram in ways that are related to the VSEPR model.

3-2-2. Valence-Shell Electron-Pair Repulsion (VSEPR) Model

The VSEPR model, extensively developed by Gillespie (*118–120*), regards Pauli repulsions among bonding and nonbonding electron pairs in the valence shell of the central atom as the most important factor in determining stereochemistry. For example, when the central atom has four equivalent bonding pairs the interpair repulsion will lead to the tetrahedral arrangement, e.g., CH_4. The main postulates concerning such electron-pair repulsions are as follows: (1) since nonbonding electron pairs occupy larger and less confined orbitals (they are influenced by one nucleus, in contrast to *two* nuclei for a bonding electron pair), they repel adjacent electron pairs more strongly than bonding electron pairs. (2) The repulsions due to bonding electron pairs decrease with increasing electronegativity of the ligand. (3) Multiple-bond orbitals repel other orbitals more strongly then single-bond orbitals, the order being triple bond > double bond > single bond.

Consider the shapes of AX_2 molecules. For molecules having no nonbonding electron pair, the repulsion between two bonding pairs makes the molecules linear (see Tables 3-3 to 3-5). Molecules with one and two nonbonding pairs are bent, with $\angle XAX < 120°$, because of rule (1) above. With three nonbonding pairs the shape becomes linear again, e.g., ICl_2^-, XeF_2, etc.

The VSEPR model is intuitive and quite successful. However, it has certain limitations (*86*). It is restricted only to ground-state molecules, neglects interaction between ligands, and is not applicable to the internal rotation problem. In fact, the model's underlying concept itself may be incorrect (see below). It is indeed interesting that the Walsh model (which does not explicitly consider electron repulsion) and the VSEPR model (which considers *only* electron-pair spin repulsion) both give similar shape predictions that agree with experiment (*27*; see also Ref. 29).

The underlying concept of the VSEPR model, namely, that Pauli repulsions between electron pairs determine molecular shapes, has been questioned (*22*, *33*). The Pauli repulsion was defined as the effect of antisymmetrization for the overlapping orbital bases including two electrons of antiparallel spins in each. In the simplest case of two orbitals, e.g., the He–He short-range interaction, antisymmetrization leads to strong repulsion (exclusion shell, see Ref. 258). However, in the more general case, the effect leads to incorrect molecular shape, contrary to rule (1) above. Second, the effect can be quite arbitrary, depending on the choice of basis orbitals. Further, if one identifies Pauli repulsion as the "Fermi hole" between parallel spins, this Fermi hole is actually quite local and does not have the expected effect on either the average interelectronic distance or repulsion energy, contrary to the familiar explanation of Hund's rule (*1*, *72*, *73*, *130*, *163*, *164*, *170*, *173*, *193*). Thus, the Pauli repulsion appears to be a misleading fiction (*33*). Nevertheless, question remains: why is the VSEPR model so successful? For a possible answer, see Section 3-3-9 and Ref. 265a.

3-2-3. Second-Order Jahn–Teller (SOJT) Theory

When a molecule is distorted by a small nuclear displacement Q_i along the *i*th normal symmetry coordinate, its energy is given by second-order perturbation theory as (*13, 14, 48*)

$$E(Q_i) = E_0 + Q_i \langle \psi_0 | \partial H / \partial Q_i | \psi_0 \rangle + \tfrac{1}{2} Q_i^2 \langle \psi_0 | \partial^2 H / \partial Q_i^2 | \psi_0 \rangle$$

$$+ Q_i^2 \sum_{k \neq 0}^{\infty} |\langle \psi_0 | \partial H / \partial Q_i | \psi_k \rangle|^2 / E_0 - E_k), \quad (3\text{-}3)$$

where ψ_0 and ψ_k are the ground and kth excited states of the undistorted molecule ($Q_i = 0$), E_0 and E_k the corresponding energies, and H the Hamiltonian; the derivatives are taken at $Q_i = 0$. We are interested in only such Q_i's that change molecular shapes, and hence are *not* totally symmetric.

Now, while H and $\partial^2 H / \partial Q_i^2$ belong to the totally symmetric representation, $\partial H / \partial Q_i$ has the same symmetry as Q_i. Since the ground electronic states ψ_0 of most molecules are nondegenerate and totally symmetric, the integral in the term linear in Q_i in (3-3) vanishes for such molecules. This first-order term is nonvanishing only for molecules in degenerate electronic states (e.g., benzene cation and anion) and causes the well-known Jahn–Teller distortion (*92, 153*; Section 3-4-3). The term involving $\partial^2 H / \partial Q_i^2$ in (3-3) is clearly positive and will cause destabilization, indicating that the electron density at $Q_i = 0$ operates to resist the nuclear displacement Q_i. The last term, called relaxability (*261*), is always negative and causes stabilization, representing the effect of electron reorganization induced by nuclear displacement. In this infinite sum, *which includes continuum states*, only those ψ_k for which the direct-product representation of ψ_0 and ψ_k contains the representation of Q_i will contribute to relaxability. Writing the second-order terms as $\frac{1}{2} Q_i^2 (f_{00} - f_{0k})$, where $f_{00} - f_{0k}$ is an approximate force constant for Q_i, one obtains *three* situations, assuming all first-order distortions to have already occurred: (i) $f_{00} > f_{0k}$, i.e., the original configuration is stable; (ii) $f_{00} < f_{0k}$, i.e., the original configuration changes spontaneously along the normal mode Q_i; (iii) $f_{00} \simeq f_{0k}$, i.e., the two structures are readily interconvertible, being separated by a low energy barrier.

Using simple MO theory, let us approximate the ground and excited states by the electronic configurations:

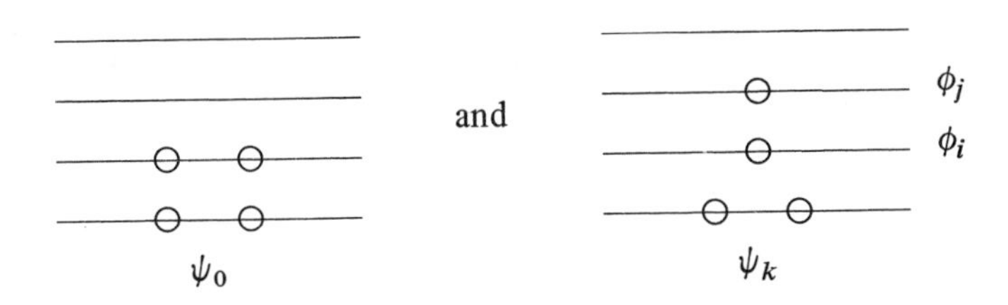

The symmetry rules for molecular structure that decide whether the original configuration will relax spontaneously along the normal mode Q_i are as follows (*28, 232-237, 261*): (1) Replace the infinite sum in f_{0k} by only the first term, corresponding to HOMO-LUMO transition, $\phi_i \rightarrow \phi_j$. The energy gap for this one-electron excitation must be less than 4 eV. (2) The direct-product representation of ϕ_i (HOMO) and ϕ_j (LUMO) must contain the representation of Q_i. (3) The transition density $\psi_0 \psi_k$, which in this approximation is proportional to $\phi_i \phi_j$, must be concentrated in the regions of nuclear displacement Q_i.

As an illustrative example, consider the stability of linear AH_2 molecules. The MO order in increasing energy is $(1\sigma_g)(1\sigma_u)(1\pi_u)(2\sigma_g)(2\sigma_u)$. For molecules having 4 valence electrons the HOMO-LUMO direct product is $\sigma_u \times \pi_u = \pi_g$, which is not the mode that converts linear into bent form; hence the linear form is stable. For molecules with 5-8 valence electrons the direct product is $\pi_u \times \sigma_g = \pi_u$, which is the converting mode. Since the $1\pi_u$-$2\sigma_g$ energy gap may not exceed 4 eV, the molecules will bend spontaneously. A 10-valence-electron molecule (e.g., NeH_2) will be linear, since $\sigma_g \times \sigma_u = \sigma_u$. However, the procedure is not applicable to the stability of the bent form, since the normal mode which converts bent into linear form is totally symmetric. The SOJT conclusions generally agree with known results, including Walsh rules (Tables 3-1 and 3-3).

Since the SOJT model is largely based on symmetry considerations, some difficulties may arise with less symmetric molecules. Also, to know the MO energy sequence is in some cases a delicate problem. The model is inapplicable to internal rotation about a single bond (*233, 234*), but can be extended to excited states (*235*).

The basic equation (3-3) of SOJT theory is quite general if Q_i is small. It can be applied to a wide range of nuclear rearrangement processes, e.g., potential interaction constants (*13*), chemical reactivity (*14, 234, 236, 237, 262*), vibrationally induced stabilization of vertically excited states (*83*), etc. We will again refer to the SOJT approach in Section 3-5-3.

3-2-4. Other Recent Models

3-2-4a. The Model Due to Takahata, Schnuelle, and Parr. Assuming certain empirical force laws for the Pauli repulsion in VSEPR model, several authors have tried to make semiquantitative predictions for molecular geometries and force constants (*40, 53, 117, 229, 266, 271, 272, 287, 289, 290*). For example, Searcy (*271, 272*) assigned empirical "electrostatic repulsion numbers" to various bonding and lone-electron pairs, and minimized the resulting effective point-charge repulsion energy with respect to bond angle. The calculated molecular shapes were in reasonable agreement with experiment.

The model proposed by Takahata, Schnuelle, and Parr contains four assumptions, the first three being related to a localized description of bonding and lone-pair electrons, while the last gives a rule for calculating stable geometries (*266*).

(1) Simple VB structures are adequate to describe the bonding. (2) Structures with closed shells on terminal atoms are preferred.[4] Between such alternative structures, the one with all atoms neutral is preferred. (3) The electrons left after procedure 2 constitute lone pairs on the central atom. The lone pairs are described by crystal-field theory, with the terminal closed-shell atoms as field-generating ligands. For example, consider AX_2 molecules in linear and bent forms,

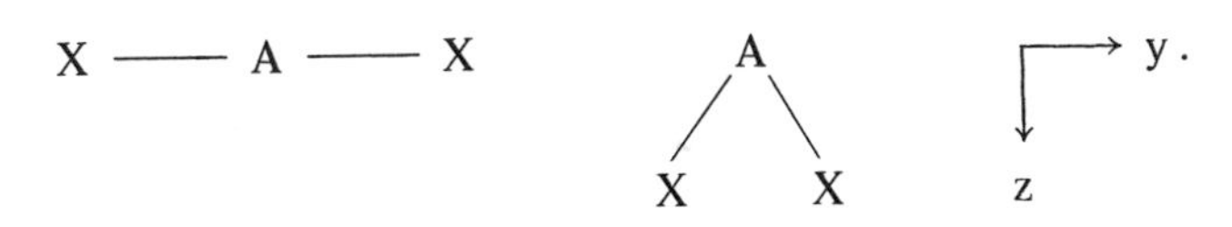

Using crystal-field theory the energy levels of central-atom AOs may be depicted as follows:

Linear	Bent
p_y —	— p_y
	— sp_z^+
p_x, p_z — —	— p_x
s —	— sp_z^-

When 1 or 2 electrons are left on the central atom, they occupy the lowest sp_z^- lone-pair orbital, leading to a bent molecule. When 3 or 4 electrons are left, the sp_z^- and p_x AOs are occupied; since the latter are insensitive to the bending mode, little change will occur in the apex angle. With 5 electrons, the apex angle increases since the sp_z^+ AO is now occupied, whereas with 6 electrons the doubly occupied s, p_x, and p_z AOs make the linear configuration more stable. Such conclusions generally agree with known shapes (Tables 3-3 and 3-5). (4) Using assumptions 1-3, and by assigning integral charges on terminal atoms and on the centroid of the Slater sp lone-pair orbital, apex angles and force constants can be calculated by minimizing the classical point-charge repulsion potential, in satisfactory agreement with experimental results.

3-2-4b. Liebman's Fragmentation Model. Liebman (*176*) divides a molecule into two fragments and uses the number of valence electrons in each fragment to qualitatively predict molecular shapes. The model was also applied to excited states (*177*), inversion barriers, and singlet–triplet energy differences (*178*).

As an illustration, the shapes of ABC molecules can be predicted as follows. The triatomic molecule is divided into an atomic and a diatomic fragment. The former is blocked if it has either a closed-shell electronic configuration or one electron less; otherwise it is porous. The diatomic fragment may belong to one of four classes: (i) σ-rich species, like 4-valence-electron $BH(\sigma^4)$, 10-valence-electron $CO(\sigma^6\pi^4)$, and their isoelectronic analogs (*30, 31, 62, 175*). (ii) π-rich species like 8-valence-electron $HF(\sigma^4\pi^4)$, 14-valence-electron $F_2(\sigma^6\pi^8)$, and their isoelectronic analogs. (iii) H_2 and its isoelectronic analogs, alkali-metal hydrides and dimers. (iv) He_2, Ne_2, and their isoelectronic analogs.

Now divide the ABC molecule into AB^{q+} and C^{r-} such that AB^{q+} is either σ-rich or π-rich and C^{r-} is either blocked or porous; q and r are zero or integers (±). The shape of ABC molecule is given by the following rule:

	C^{r-}	
AB^{q+}	blocked	porous
σ-rich	bent	linear
π-rich	linear	bent

This rule is quite successful (see Tables 3-3 and 3-5). Thus, CO_2 can be divided into either σ-rich CO plus porous O or π-rich CO^{2+} plus blocked O^{2-} fragments; both possibilities yield a linear molecule. Note, however, that for C_3 two such fragmentations ($C_2^{2-} + C^{2+}$, $C_2 + C$) give contradictory predictions (linear and bent, respectively).

Liebman (*176*) explained the relations of this model to previous energetic models. For example, the connection with VSEPR model is as follows: σ-rich + porous → no lone pair (on central atom); σ-rich + blocked → one lone pair; π-rich + porous → two lone pairs; π-rich + blocked → three lone pairs. This indicates that the two models make identical predictions.

As mentioned before, the predictions from the models of Liebman and Takahata and co-workers generally agree with known results. However, the physical principles underlying both models and the reasons for their success seem to be unclear. Although both models may be related to the VSEPR concept, the latter itself has some problems (see Sections 3-2-2 and 3-3-9).

3-3. ELECTROSTATIC FORCE THEORY APPLIED TO MOLECULAR SHAPE

3-3-1. Introduction

In this section we describe the ESF theory of Nakatsuji and co-workers as applied to molecular shapes, although the theory is applicable to a wide range of phenomena such as molecular structures, chemical reactions, and long-range

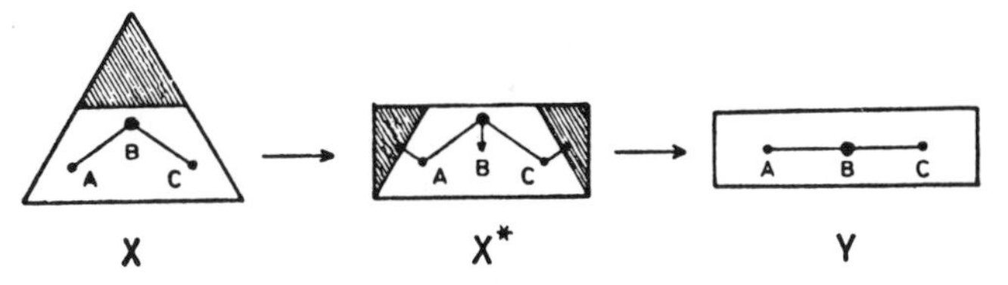

Fig. 3-3. Illustration of forces and electron clouds. (Reproduced from Ref. 207, courtesy the American Chemical Society.)

forces (see Section 7-3). We decompose the H-F force into three pictorial forces (related to chemical concepts such as bond and lone pair), and examine molecular structure in ground and excited states as a stable balance of these separate forces. This provides another explanation of Walsh rules (Table 3-1). Various effects of changes in concerned atoms and their neighboring substituents are systematized and a simple rule for molecular shape is devised. The origin of the internal rotation barrier is also clarified.

It will be instructive to start from the following imaginary example: Consider a stable bent molecule X whose nuclei A, B, C are embedded in the triangular electron cloud of Fig. 3-3. If one transfers (e.g., by electron excitation) the shaded portion of the electron cloud of X to that of X* (e.g., as in the Franck-Condon state), the nuclei A and C in X* receive attractive forces from the shaded portions of electron cloud, while the nucleus B receives a downward (recoil) force due to the break of electrostatic equilibrium in X. These forces induce nuclear movements (arrows in Fig. 3-3) until they die away in structure Y. This example illustrates how the H-F force determines molecular geometry and why the bent molecule X (ground state) becomes linear in the excited state Y.

3-3-2. Partitioning of the Hellmann-Feynman Force

In order to obtain an insight into the relation between force and electron cloud, it is convenient to expand the electron density $\rho(r_1)$ by a sufficiently large AO basis set $\{\chi_r\}$ as

$$\rho(r_1) = \sum_{r,s} P_{rs}\chi_r(r_1)\,\chi_s(r_1), \qquad (3\text{-}4)$$

where P_{rs} is the (generalized) bond order between χ_r and χ_s. Then the H-F theorem (3-1) is rewritten as

$$F_A = Z_A\left\{\sum_{r,s} P_{rs}\langle\chi_r|f_A|\chi_s\rangle - \sum_{B(\neq A)} Z_B R_{AB}/R_{AB}^3\right\}, \qquad (3\text{-}5)$$

where $f_A = r_A/r_A^3$ is the force operator per unit nuclear charge.

For simplicity, we assume that our basis $\{\chi_r\}$ has its center on respective nuclei and its angular part is expressed by the spherical harmonics. We designate the center of basis by subscript, like χ_{rA}, and classify the force integrals in (3-5) into: one-center type $\langle \chi_{rA} | f_A | \chi_{sA} \rangle$; two-center types $\langle \chi_{rA} | f_A | \chi_{sB} \rangle$ (exchange integral) and $\langle \chi_{rB} | f_A | \chi_{sB} \rangle$ (shielding integral); and three-center type $\langle \chi_{rB} | f_A | \chi_{sC} \rangle$. Within the one-center type, the diagonal one ($\chi_{rA} = \chi_{sA}$) vanishes identically from symmetry, since f_A is a vector operator. The off-diagonal integrals which survive are of types $\langle \boldsymbol{s}_A | f_A | \boldsymbol{p}_A \rangle$, $\langle \boldsymbol{p}_A | f_A | \boldsymbol{d}_A \rangle$, etc., where $\boldsymbol{s}, \boldsymbol{p}, \boldsymbol{d}, \ldots$ are sets of $s, p, d, \ldots$ orbitals. These one-center integrals are important when the electron density in atomic region is polarized in the direction of f_A. For the two-center exchange integrals, we introduce a net-exchange-force integral defined by

$$\langle \chi_{rA} | (f_A)_0 | \chi_{sB} \rangle \equiv \langle \chi_{rA} | f_A | \chi_{sB} \rangle - \tfrac{1}{2} S_{rAsB} \langle \chi_{sB} | f_A | \chi_{sB} \rangle, \tag{3-6}$$

where S_{rAsB} is the overlap integral between χ_{rA} and χ_{sB}. The net-exchange integral represents the net attractive (or repulsive) force due to the accumulation (or depletion) of electron density in the overlap region between χ_{rA} and χ_{sB}. This is seen as follows: Like the electrons of atom B, the electrons in the overlap region also shield the nuclear charge Z_B. This effect is given by $\frac{1}{2} S_{rAsB} \langle \chi_{sB} | f_A | \chi_{sB} \rangle$, which is just the Mulliken (*205*) approximation of the first term. The net-exchange integral is obtained by subtracting this shielding effect from the exchange integral. Further, considering Ruedenberg's (*255*) interference partitioning of electron density into quasi-classical part and quantum interference part, the net-exchange integral represents just the quantum interference effect [see (3-34)].

We now partition the H-F force equation (3-5) into three parts:[5] The *atomic dipole (AD) force* coming from one-center integrals, the *exchange (EC) force* coming from net-exchange force integrals, and the rest called *extended gross charge (EGC) force*, namely,

$$\boldsymbol{F}_A = \boldsymbol{F}_A^{AD} + \boldsymbol{F}_A^{EC} + \boldsymbol{F}_A^{EGC}, \tag{3-7}$$

where

$$\boldsymbol{F}_A^{AD} = 2Z_A \sum_{r \geqslant s}^{A} \sum^{A} P_{rAsA} \langle \chi_{rA} | f_A | \chi_{sA} \rangle \tag{3-8}$$

$$\boldsymbol{F}_A^{EC} = \sum_{B(\neq A)} \boldsymbol{F}_A^{EC}(AB) \tag{3-9a}$$

$$\boldsymbol{F}_A^{EC}(AB) = 2Z_A \sum_{r}^{A} \sum_{s}^{B} P_{rAsB} \langle \chi_{rA} | (f_A)_0 | \chi_{sB} \rangle \tag{3-9b}$$

$$F_{\text{A}}^{\text{EGC}} = \sum_{\text{B}(\neq \text{A})} F_{\text{A}}^{\text{EGC}}(\text{AB}) \tag{3-10a}$$

$$\begin{aligned} F_{\text{A}}^{\text{EGC}}(\text{AB}) = & Z_{\text{A}} \sum_r^{\text{B}} \sum_s^{\text{B}} P_{r\text{B}s\text{B}} \langle \chi_{r\text{B}} | f_{\text{A}} | \chi_{s\text{B}} \rangle \\ & + Z_{\text{A}} \sum_r^{\text{A}} \sum_s^{\text{B}} P_{r\text{A}s\text{B}} S_{r\text{A}s\text{B}} \langle \chi_{s\text{B}} | f_{\text{A}} | \chi_{s\text{B}} \rangle \\ & + Z_{\text{A}} \sum_{\text{C} \neq \text{A},\text{B}} \sum_r^{\text{B}} \sum_s^{\text{C}} P_{r\text{B}s\text{C}} \langle \chi_{r\text{B}} | f_{\text{A}} | \chi_{s\text{C}} \rangle \\ & - Z_{\text{A}} Z_{\text{B}} R_{\text{AB}} / R_{\text{AB}}^3. \end{aligned} \tag{3-10b}$$

Here Σ_r^{A} implies summation over the AO bases of atom A. The first three terms in (3-10b) represent the shielding of the bare nuclear repulsion given by the last term.

The physical interpretations of the AD, EC, and EGC forces defined above are as follows: The AD force represents the attraction on nucleus A by the centroid (weight factor $1/r_a^3$) of the polarized atomic density (e.g., lone pair), as illustrated by

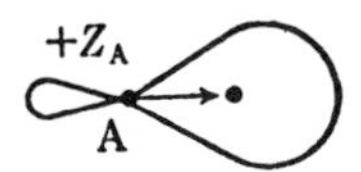

Such polarizations occur typically because of s–p and p–d AO mixing. The EC force represents the attraction (or repulsion) on nucleus A by the electron density accumulated (or depleted) in the overlap region between A and B, as illustrated by

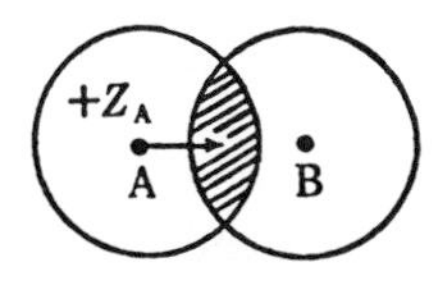

The EGC force represents the interaction between nuclei A and B, the latter shielded by the remainder of electron cloud.

Although the above definitions of AD, EC, and EGC forces are rigorous and include no integral approximation, the EGC force may be further interpreted by invoking the following integral approximations: (1) A point-charge approximation[6] for the shielding integral

$$\langle \chi_{sB} | r_A / r_A^3 | \chi_{sB} \rangle \simeq R_{AB} / R_{AB}^3, \tag{3-11}$$

and (2) the Mulliken approximation for the smaller integrals $\langle \chi_{rB} | f_A | \chi_{sC} \rangle$ (B, C $\neq$ A and r $\neq s$),

$$\langle \chi_{rB} | f_A | \chi_{sC} \rangle \simeq \tfrac{1}{2} S_{rBsC} \{ \langle \chi_{rB} | f_A | \chi_{rB} \rangle + \langle \chi_{sC} | f_A | \chi_{sC} \rangle \}. \tag{3-12}$$

Then, (3-10) is approximated as

$$F_A^{EGC}(AB) \simeq F_A^{GC}(AB), \tag{3-13a}$$

$$F_A^{GC}(AB) = Z_A \delta_B R_{AB} / R_{AB}^3, \tag{3-13b}$$

where $\delta_B = Z_B - N_B$ is the gross charge (GC) on atom B and $N_B = \Sigma_r^B \Sigma_s^{all} P_{rs} S_{rs}$ is gross atomic population (*205*). The GC force in (3-13b) represents the electrostatic interaction between Z_A and δ_B, as illustrated by

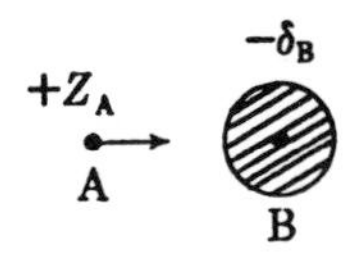

Since the shielding of nuclear charge is usually incomplete,[6a] nuclear–nuclear repulsion is more stressed in the EGC force than in the GC force. Further, since the approximations (3-11) and (3-12) might be inadequate (see also Refs. 52 and 74) we will not use them in force calculations of actual systems.

3-3-3. Construction of a Model for Molecular Shape

We first examine how the AD, EC, and EGC forces operate in a real molecule, e.g., NH_3 [Fig. 3-4(a)]. The AD force (especially important for lone-pair electrons[7]) pulls the N nucleus upwards and operates to make the molecule nonplanar.[8] But, the vector sum of three EC forces due to N—H bond electrons pulls N downwards, operating to make the molecule planar. The EGC or GC forces on N due to three H's and those between the H's are repulsive but small (see below), and the balance of all these forces make the NH_3 molecule pyramidal. However, for CH_3^+, which has no lone pair, the three EC forces make the molecule D_{3h}. Generally, molecules with no lone pair acquire maximum symmetry in which the EC forces balance (e.g., CH_4).

For molecular shape, the respective ratios of AD, EC, and EGC (or GC) forces are important. Comparing these forces on A in A—A and A—H pairs

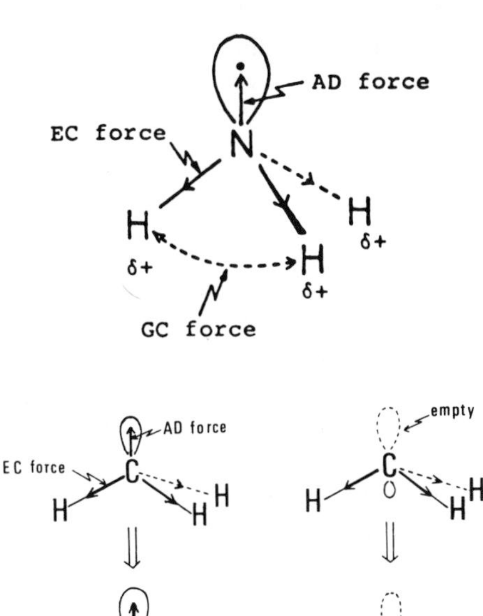

Fig. 3-4 (a). Important forces acting in (i) NH_3, (ii) CH_3^-, and (iii) CH_3^+. (Reproduced from Ref. 207, courtesy the American Chemical Society.)

(A is not hydrogen), Nakatsuji (*207*) came to the following conclusions: (1) The AD/EC ratio is larger than unity and is approximately constant for atoms belonging to the same period.[9] (2) The EC force increases with increasing bond multiplicity. For a single bond, EC forces on A in A—A and A—H bonds are approximately equal. One obtains the result

$$\text{AD force} > \text{EC force (triple bond} > \text{double bond} > \text{single bond)} > \text{EGC force.} \tag{3-14}$$

This sequence is consistent with the fact that in the force operator $\mathbf{r}_{A1}/r_{A1}^3$ the nearer electron 1 is to nucleus A the larger is the operator. Thus, the AD and EC forces on nucleus A will mainly determine the shapes of neutral molecules. For ions or highly unshielded atoms, the EGC force may be appreciable.[10,11]

Some examples of rule (3-14) follow: In NH_3 (Fig. 3-4) the AD (>EC) force makes the HNH angle smaller $(107.8°)$[12] than the tetrahedral value $(109.47°)$. In isobutene and ethylene, ∠C—C—C (111.5°, 117.6°) is smaller than ∠C—C=C

(124.3°, 121.2°), since the EC force along C=C bond is larger than that along C—C and C—H bonds.

3-3-4. Various Influences on the AD and EC Forces

The most important reason for the great variety of molecular shapes is that the valence-electron distribution near A varies from molecule to molecule and state to state. Since the AD and EC forces are dominant factors for molecular shape, it is worthwhile to consider the effects on these forces due to changes in A and the neighboring substituent B, in order to apply the theory to a wide range of molecules. For the AD force, we consider only *s–p* mixing, since with, e.g., H_2S the *p–d* mixing was found to be small.

3-3-4a. Central-Atom Effects on AD and EC Forces. If one replaces the central atom A with heavier atoms in the same group of the periodic table, then the extent of *s–p* (or *p–d*) mixing increases if the energy difference between valence *s* and *p* AOs decreases. This energy difference may be measured by the difference in Mulliken's orbital electronegativity (Fig. 3-4(b); *137, 203, 207*). From Fig. 3-4(b) one then concludes that *within the same group, the extent of s–p mixing and hence the AD/EC ratio will increase with increasing atomic number.*[13] Further, because of the large differences between the first- and second-row elements of groups IV to VII, *there are frequent sharp decreases in valence angles for these elements.*

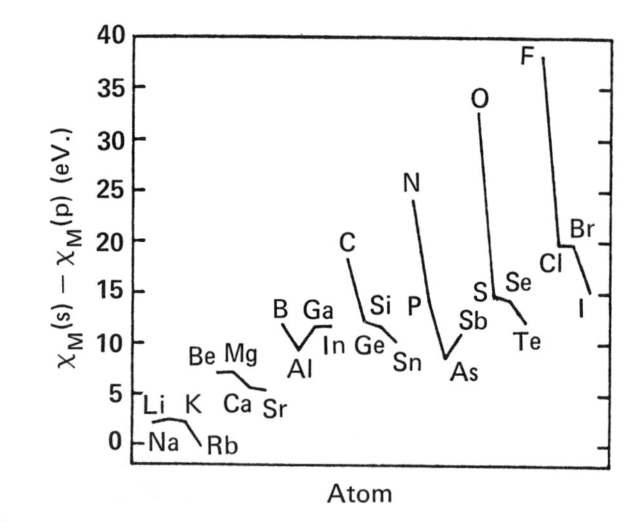

Fig. 3-4(b). Differences in the Mulliken orbital electronegativity between valence *s* and *p* AOs for the elements from the first to fourth rows in the periodic table. (Reproduced from Ref. 207, courtesy the American Chemical Society.)

Consider now the EC force. Since the inner-shell (core) radius increases with atomic number within the same group (see, e.g., Ref. 231), the bond electrons are kept away from the nucleus, and *the EC force is reduced.* Here, too, a sharp difference is found between first- and second-row elements.

The above central-atom effects on AD and EC forces are cooperative, making the molecule with a heavier central atom more bent or pyramidal; for example,

Group V: NH_3(107.8°), PH_3(93.3°), AsH_3(92°)
Group VI: H_2O(105.2°), H_2S(92.2°), H_2Se(91°), H_2Te(90.25°).

The sharp decrease in valence angle from NH_3 to PH_3 and from H_2O to H_2S is as expected. We find the same trend in Table 3-3 even for excited states.

3-3-4b. Central-Symmetry Effect on the AD Force. This is the effect of local symmetry of the electron cloud near A. For a distribution which is spherically symmetric around A, or symmetric along the x axis (e.g., when s and p_x AOs are completely filled), no AD force is generated.[14] Examples of this central-symmetry effect can be seen with FHF^-, $ClHCl^-$, $BrHBr^-$ (*95-97, 149*), $LiHLi^-$, $BeHBe^+$ (*154, 247*), $HHHe^+$, He_3^+, $LiHeH^+$ (*245*), ICl_2^-, I_3^-, $IBrCl^-$, Br_3^- (*283, 284*), XeF_2 (*250, 291*). The first eight molecules have H or He as central atom and, since the AD force does not arise due to large $1s$-$2p$ promotion energy, they are linear. The latter molecules are also linear since here the s, p_x, and p_y AOs are completely filled and no important AD force arises even if the molecules are bent from linearity. The same is true for ClF_3, BrF_3, and the planar Jahn-Teller systems LiH_3^+, HeH_3^+ (*245*). In the former two molecules (anchor shape), $\angle FClF$, $\angle FBrF < 90°$, since the in-plane AD force attracts the central atom more than do the EC forces along the bonds.

3-3-4c. Inductive Substituent Effects on AD and EC Forces. Consider the effect of substituent B in the A—B fragment. Since the AD force on A arises mainly from s-p mixing, we look for a σ-inductive effect and a π-inductive effect. For the latter, a π-donating (or attracting) substituent B increases (or decreases) the coefficient of $p_{\pi A}$ AO and the AD force on A. The less direct σ-inductive effect is just the reverse of the π-effect (*207*). The π-effect is more effective than the σ-effect, although the two are usually cooperative[15] (see below).

Consider the effect on the EC force at a fixed A—B distance. If B is more electronegative than A, B draws electron density in the A—B bond region toward itself and thus the EC force on A decreases. Hence, at fixed A—B distance, the more electronegative the substituent B, the smaller is the EC force on A.

As an example, consider the out-of-plane angles (and also the corresponding force constants) in fluoromethyl radicals (*100*). The systematic increase, CH_3 (0°), CH_2F(<5°), CHF_2(~12.5°) CF_3(~17.8°) is explained as follows: Since F is π donating and σ attracting, it increases the AD force on C; and since F is more electronegative than H the EC force on C diminishes. The two effects cooperate to make the fluorosubstituted molecule nonplanar. The planarity of the CH_3 radical is due to a sensitive balance of AD and EC forces (see Section 3-3-6). These considerations also explain why ESR experiments reveal the series CCl_3, CCl_2F, $CClF_2$, CF_3 to be increasingly nonplanar (*60*).

3-3-4d. Overlap Effects on AD and EC Forces. Orbital overlap is quite important for both molecular structure and chemical reaction. The overlap effect usually causes larger changes in electron density than the inductive effect. For a homopolar A—B fragment, the bonding and antibonding MOs are

$$\phi_b = (\chi_{rA} + \chi_{sB})/(2 + 2S)^{1/2} \tag{3-15a}$$

$$\phi_a = (\chi_{rA} - \chi_{sB})/(2 - 2S)^{1/2}, \tag{3-15b}$$

where S is the overlap integral between AOs χ_{rA} and χ_{sB}. Equations (3-15) indicate that in the bonding interaction the electrons in each AO flow into the overlap region and, in the antibonding interaction, the electrons in the overlap region are transferred back into each atomic region because of the presence of a node.

Table 3-2 summarizes the overlap effects on AD and EC forces. $D(\chi_{rA})$ denotes the electron density in the χ_{rA} AO. Since an increase in the coefficients of s_A and p_A AOs facilitates the formation of atomic dipole, the change in $D(\chi_{rA})$ parallels that in the AD force.[16] In the bonding interaction, the AD force decreases, since $D(\chi_{rA})$ decreases from 1 to $1/(1+S)$ and the EC force increases by $2I_{EC}/(1+S)$. Both effects *cooperate* to make the molecule linear or planar. The antibonding overlap effect is just the reverse and facilitates bending. When both χ_{rA} and χ_{sB} are doubly occupied, the antibonding effect surpasses the bonding effect and again facilitates bending. Generalization of Table 3-2 to more than two interacting AO systems is easy. There, we have to consider the nonbonding interaction (recall the second π MO of allyl radical) where the AD and EC overlap effects are approximately zero (see Section 3-3-7c).

Clearly, the overlap effect[17] becomes appreciable when (1) the AOs χ_{rA} and χ_{sB} overlap significantly, (2) their energy levels are close, and (3) electrons in both AOs are not tightly bound.

The fact that EC force increases with bond multiplicity is a manifestation of overlap effect. Compare the shapes [Fig. 3-4(c)] of acetylene and triplet $CH_2(^3B_1)$. In linear form, the $p_{\pi C}$ AO in CH_2 does not have overlap, but that

Table 3-2. Overlap Effect on Molecular Geometry and Chemical Reaction.

Interaction	$D(p_{\pi A})$ AD force	EC force[a,b]	Corresponding phenomena: Molecular shape	Corresponding phenomena: Chemical reaction	
Bonding interaction[c]	$\frac{1}{1+S}$	$\frac{2}{1+S} I_{EC}$	linear or planar	driving force	Conservation of molecular orbital symmetry
Antibonding interaction[d]	$\frac{1}{1-S}$	$-\frac{2}{1-S} I_{EC}$	bent or pyramidal	repulsive force	
Interaction between fully occupied AOs	$\frac{2}{1-S^2}$	$-\frac{4S}{1-S^2} I_{EC}$	bent or pyramidal	repulsive force	(i) Non-least motion reaction path (ii) Exchange repulsion

[a] $I_{EC} = \langle \chi_A | (f)_0 | \chi_{sB} \rangle$. The prefactor of I_{EC} is the bond order between χ_{rA} and χ_{sB}.
[b] Plus and minus signs correspond to the attractive and repulsive forces between A and B.
[c] The values correspond to the electronic configuration $(\phi_b)^2 (\phi_a)^0$.
[d] The values correspond to the electronic configuration $(\phi_b)^0 (\phi_a)^2$.
SOURCE: Ref. 209. Courtesy the American Chemical Society.

in acetylene strongly overlaps with the adjacent $p_{\pi C}$ AO. Since $D(p_{\pi C})$ is 1.0 for CH_2 but ~0.75 for acetylene, the AD force on C produced by small bending is stronger in CH_2. Second, because of overlap the EC force on C is stronger in acetylene. Thus, while CH_2 is bent (*135*) acetylene is linear.

Overlap effect is also important for molecules in which the central atom p AO overlaps with lower lying vacant AOs of adjacent atoms (e.g., $p\pi$-$d\pi$ conjugation).[18] Consider the change in valence angle in the ammonia derivatives, NF_3(102.5°), NH_3(107.8°), $N(SiH_3)_3$(120°; Ref. 132). The change from NH_3 to NF_3 is due to inductive-substituent effects on AD and EC forces. The larger change from NH_3 to $N(SiH_3)_3$ is mainly due to overlap conjugation between $p_{\pi N}$ and $d_{\pi Si}$ AOs, which reduces the AD force on N and strengthens the EC force due to the multiple-bond character of the N—Si bond.[19] The less important inductive-substituent effect on the EC force also facilitates this change. Similar examples are seen in the difference in the COC angles of aliphatic (108–110°) and aromatic (120–125°) ethers (*116*), planarity (*125*) around N in $N_2(SiH_3)_4$, and in the water derivatives OF_2(103.8°), OH_2(105.2°), $O(SiH_3)_2$(144.1°; Ref. 6), $O[TiCl_2(C_5H_5)]_2$(180°; Ref. 61). The larger bond angle in OCl_2(110.8°) compared with H_2O indicates that the overlap effect is more important than the inductive effect. However, the two effects are cooperative in alkali metal derivatives of H_2O (Section 3-3-7b). When the central atoms N and O are replaced by heavier atoms such strong p_π-d_π interaction does not seem to occur (*116*).

The overlap effect on the EC force is quite important for choosing chemical reaction pathways. The attractive EC force due to bonding overlap between χ_{rA} and χ_{sB} (A and B are reaction sites) provides the driving force of the reaction (*207, 216*), but the antibonding overlap and overlap between fully occupied AOs cause repulsive forces. When this knowledge is combined with HOMO-LUMO interactions (*109–112*) one obtains the well-known Woodward-Hoffmann

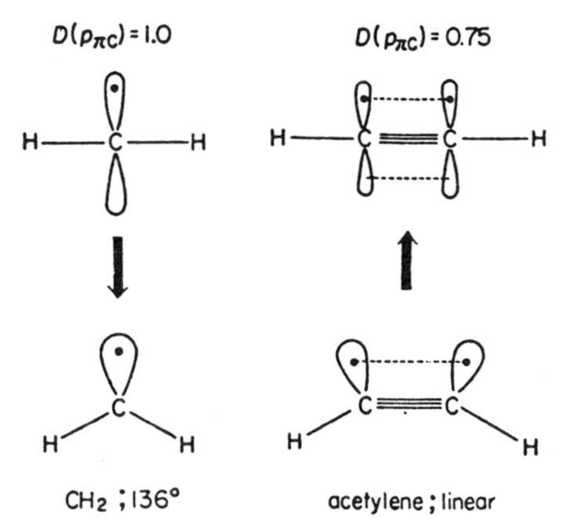

Fig. 3-4 (c). Structures of triplet carbene and acetylene as examples of overlap effect.

rule (*108*, *302*). The overlap stabilization proposed by Fukui (*107*) is an energetic expression of the overlap effect on the EC force. The repulsive EC force between fully occupied AOs expresses short-range exchange repulsion and corresponds to the so-called exclusion shell[20] (*258*, *260*). For example, the non-least motion reaction path studied by Hoffmann et al. (*143*) is determined by a combination of the attractive and repulsive EC forces.

3-3-5. Changes in Force Due to Changes in Electronic State

We will employ MO theory in dealing with changes in electronic state such as electron excitation, ionization, and attachment (recall Fig. 3-3). Let $\{\phi_i\}$ be a set of MOs with occupation numbers $\{m_i\}$. The electron density $\rho(r_1)$ is

$$\rho(r_1) = \sum_i m_i \phi_i^*(r_1)\, \phi_i(r_1) = \sum_i m_i \rho_i(r_1). \tag{3-16}$$

Denoting the *i*th MO contribution to the force as f_{Ai}

$$f_{Ai} \equiv \int f_A(r_1)\, \rho_i(r_1)\, dr_1, \tag{3-17}$$

we obtain the H–F theorem in the form

$$F_A = Z_A \sum_i m_i f_{Ai} - Z_A \sum_{B(\neq A)} Z_B R_{AB}/R_{AB}^3. \tag{3-18}$$

If we neglect the changes in MOs following the change in electronic state, $\alpha \rightarrow \beta$ (vertical state), the change in the force acting on A is

$$\Delta F_A^{\alpha \rightarrow \beta} = Z_A \sum_i (m_i^\beta - m_i^\alpha) f_{Ai}. \tag{3-19}$$

For example, for excitations between states with the same electronic configurations,

$$\Delta F_A^{\alpha \rightarrow \beta} = 0, \tag{3-20}$$

which implies that the differences in the geometries, force constants, etc. of the two states are quite small. Hurley (*148*) confirmed this.

For the excitation from *i*th to *j*th MO,

$$\Delta F_A^{i \rightarrow j} = Z_A (f_{Aj} - f_{Ai}); \tag{3-21}$$

for ionization from ith MO,

$$\Delta F_{\mathrm{A}}^{i \to \infty} = -Z_{\mathrm{A}} f_{\mathrm{A}i}; \tag{3-22}$$

and for electron attachment to ith MO,

$$\Delta F_{\mathrm{A}}^{\infty \to i} = Z_{\mathrm{A}} f_{\mathrm{A}i}. \tag{3-23}$$

For cations (R^+), radicals ($R\cdot$), and anions (R^-) having electronic configurations $(\phi_i)^0$, $(\phi_i)^1$, and $(\phi_i)^2$, the force at radical geometry satisfies

$$\Delta F_{\mathrm{A}}(\mathrm{R}\cdot \to \mathrm{R}^+) = -\Delta F_{\mathrm{A}}(\mathrm{R}\cdot \to \mathrm{R}^-), \tag{3-24}$$

implying that the force-dependent properties of R· should always be intermediate between those of R^+ and R^- (*207*).

Consider now the changes in the AD, EC, and EGC forces. We rewrite (3-21) as

$$\Delta F_{\mathrm{A}}^{i \to j} = Z_{\mathrm{A}} \int f_{\mathrm{A}}(r_1)\, \Delta\rho(r_1)\, dr_1, \tag{3-25}$$

where $\Delta\rho(r_1) = \rho_j(r_1) - \rho_i(r_1)$ is the change in electron density due to excitation. The change in the atomic region changes the AD force, while that in the bond region changes the EC force. As an example, consider electronic transition from the lone-pair orbital to the Rydberg *ns* or *np* orbital in NH_3. Since the electron densities in these Rydberg orbitals are almost symmetric with respect to the N nucleus, they exert no force on N and hence the geometry in these excited states should be planar (*134*), like NH_3^+. (For more detailed studies on such geometry changes, see Refs. 38, 49, 68, 155, 214, and 241).

3-3-6. Illustrative Applications of the ESF Model and a Simple Rule for Molecular Shape

Consider the roles of AD and EC forces when the molecules CH_3^-, CH_3^+, and CH_3 are slightly bent from planar shape.[21] With CH_3^- (Fig. 3-4), an atomic dipole rapidly forms on the C atom and pulls it upwards while the vector sum of three EC forces pulls it downwards. In view of relation (3-14), the molecule continues to bend until the two opposing forces on C balance. Thus, CH_3^- is pyramidal (experimental: *47*; theoretical: *43, 87, 88, 162, 282*). For CH_3^+, the upward atomic dipole is not generated, since the corresponding orbital is empty, and the three EC forces balance each other by restoring the molecule into planar shape (D_{3h}). Similarly, isoelectronic BH_3 and BeH_3^- are planar D_{3h}

(*240*). According to (3-24), CH_3 is intermediate between CH_3^- and CH_3^+, and its shape is the result of a sensitive balance between AD, EC, and EGC forces. It is planar (*134*) like its isoelectronic analogs NH_3^+ (*58*) and BH_3^- (*286*). The qualitative arguments given here were justified by ab initio force calculations with floating wavefunctions for NH_3, CH_3^+, and NH_3^+ which satisfy the H–F theorem (*221*). Further, the out-of-plane bending force constant should be larger for CH_3^+ than for CH_3.

On the basis of such examples as those given above, we shall now formulate a simple and quite general rule that links molecular shape with $D(p_{\pi A})$, A being the central atom. As $D(p_{\pi A})$ we take the value when the molecule is set in *planar or linear* form.

The relation of $D(p_{\pi A})$ with planarity (or linearity) and bending force constant is given by the simple rule in Fig. 3-5. AXY and AXYZ molecules differ mainly in the number of EC forces, two for the former and three for the latter, e.g., CH_2 (bent) and CH_3 (planar) although $D(p_{\pi C})$ is unity for both molecules.

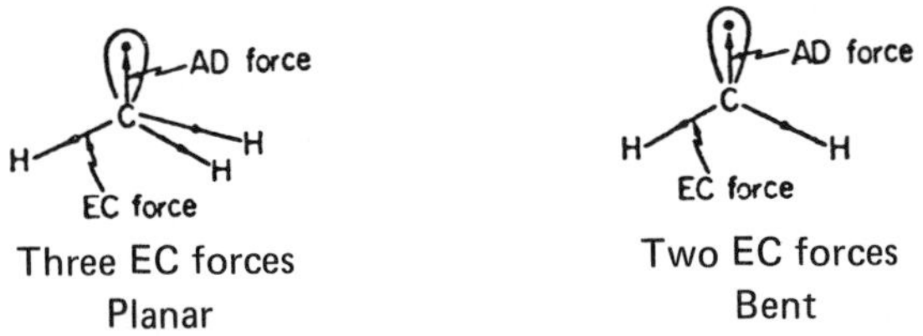

Three EC forces
Planar

Two EC forces
Bent

The critical values of $D(p_{\pi A})$ in Fig. 3-5, separating shaded from dark regions, are chosen from this example. The generality of this rule for AXY and AXYZ molecules arises from the fact that the inductive substituent effects on the AD and EC forces are generally cooperative, as are the overlap effects on the AD and EC forces[22] (Sections 3-3-4c, d). The quantity $D(p_{\pi A})$ is taken as an indicator common to these effects. *Figure 3-5 is quite useful in predicting molecular shapes in ground and excited states*, e.g., the vast number of planar aromatic hydrocarbons for which $D(p_{\pi C})$ is always less than unity. This is also reminiscent of Walsh rules, although $D(p_{\pi A})$ is likely to be more useful for shape predictions than the number of valence electrons, since the former is a better index for local electron distribution. In fact, this rule applies to excited states as well as ground states.

It should be noted, however, that the simple rule in Fig. 3-5 has certain limitations. (i) It is affected by the central-atom and central-symmetry effects, the

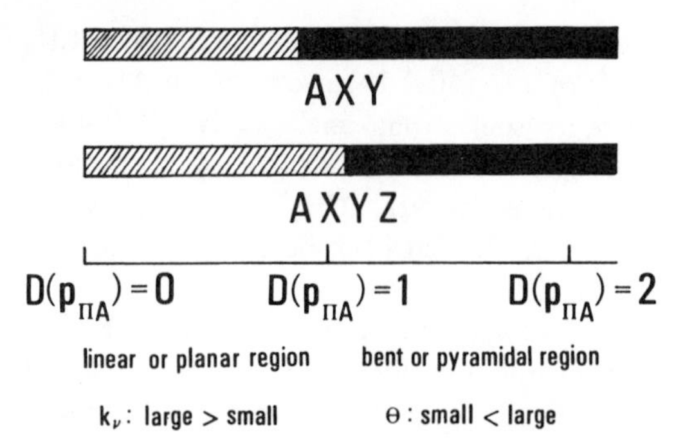

Fig. 3-5. A simple rule for the shapes of AXY and AXYZ molecules where A is the central atom and X, Y, Z are the substituents. $D(p_{\pi A})$, k_ν, and θ are, respectively, the electron density of the $p_{\pi A}$ AO in planar or linear structure, the out-of-plane bending force constant, and the out-of-plane angle. (Reproduced from Ref. 207, courtesy the American Chemical Society.)

former being especially important near the critical values of $D(p_{\pi A})$. (ii) It implicitly assumes near constancy in the nature of s_A AOs for various AX_n molecules. Since s_A electrons are more tightly bound than p_A ones, this assumption is usually justified, except when the number of valence electrons around A increases to fill the high-lying s_A-σ_X antibonding orbital (Fig. 3-6 below). Here, both s_A and $p_{\pi A}$ AOs are completely filled and, because of the central-symmetry effect, the molecule becomes linear or planar (e.g., XeF_2, ClF_3) even though $D(p_{\pi A}) \geqslant 2$.

3-3-7. Applications of the ESF Model to Molecular Shapes

3-3-7a. AH_2 and AH_3 Molecules. Based on the previous discussions, the reader will now be able to rationalize the shapes of those molecules in ground and excited states (Tables 3-3 and 3-4). $D(p_{\pi A})$ is the occupation number of the $n(\bar{\pi})$ orbital. Being symmetric about nucleus A, the electron densities in π and Rydberg orbitals do not exert force on A. When an electron jumps from the $n(\bar{\pi})$ to the π orbital, the apex angle increases substantially, e.g., $^*NH_2$ and $^*PH_2$ (Table 3-3). But when an electron jumps from π to Rydberg orbital, there is little change, e.g., $^{\ddagger}H_2O$. The central-atom effect is observed for both ground and excited states; a sharp change occurs between first- and second-row elements. Limiting ourselves to the ground state, $D(n(\bar{\pi})) = 0, 1, 2$ corresponds to 4, 5, 6-8 valence electrons, respectively. Thus, the ESF model reproduces the Walsh rules (Table 3-1). This is also true for more complex molecules.

For the 2-valence-electron bent molecules H_3^+, Li_3^+, LiH_2^+ (*245, 249*), a direct application of the above model seems inadequate. For example, with LiH_2^+ it may be wrong to imagine a bond between Li and H, since the resulting EC force on Li will give a linear molecule. Rather, the electron cloud in these systems will

Table 3-3. Shapes of AH_2 Molecules.

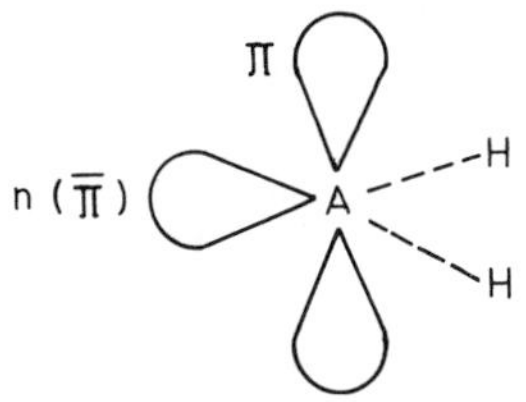

	Occupation no. of orbital[a]			
Shape	$n(\bar{\pi})$	π	Rydberg	AH_2[b]
linear	0	0 or 1	0 or 1	BeH_2, BH_2^+, ‡BH_2, ‡AlH_2
bent	1	0	0	BH_2(131°), AlH_2(119°)
bent	1	1	0	CH_2($^3B_1$, 136°), CH_2($^1B_1$, ~140°), NH_2^+[$^3B_1$, 143.2°][c]
bent	1	2	0	*NH_2(144°), *PH_2(123.1°)
bent	2	0	0	CH_2($^1A_1$, 102.4°), BH_2^-[102°], NH_2^+[110.5°][c], SiH_2(–)
bent	2	1	0 or 1	NH_2(103.4°), PH_2(91.5°), ‡OH_2(106.9°)
bent	2	2	0	OH_2(105.2°), SH_2(92.2°), SeH_2(91°), TeH_2(90.25°), NH_2^-(104°)

[a]See the diagram at the top of the table.
[b]Values in ( ) and [ ] mean the apex angles obtained experimentally or from ab initio calculations, respectively. The asterisk and double dagger mean, respectively, the $n \rightarrow \pi^*$ and Rydberg excited states.
[c]M. M. Heaton and R. Cowdry, *J. Chem. Phys.*, **62**, 3002 (1975).
SOURCE: Ref. 208. Reprinted courtesy the American Chemical Society.

Table 3-4. Shapes of AH_3 Molecules.

Shape	Occupation no. of π orbital	AH_3[a]
planar	0	HeH_3^+, LiH_3^+, BeH_3, BeH_3^-, BH_3, CH_3^+, ‡CH_3, SiH_3^+[b]
planar	1	CH_3, BH_3^-, NH_3^+, ‡NH_3, ‡PH_3
pyramidal	1	SiH_3(113.5°), GeH_3(115°), SnH_3(117°)
pyramidal	2	CH_3^-(–), NH_3(107.8°), PH_3(93.3°), AsH_3(92°), OH_3^+(117°), SiH_3^-[b]

[a]The double dagger means that the molecule is in the Rydberg excited state.
[b]B. Wirsam, *Chem. Phys. Lett.*, **18**, 578 (1973).
SOURCE: Ref. 208. Reprinted courtesy the American Chemical Society.

shield the three nuclei (or cores) more effectively if it accumulates within the nuclear triangle.

For AH_3 molecules (Table 3-4), almost all features have been discussed previously, including the sensitive problem of 7-valence-electron molecules. Since $D(p_{\pi A})$ is related to the number of valence electrons, the Walsh rules are again reproduced.

3-3-7b. Alkali Metal Derivatives of H_2O. The linear geometries of Li_2O (*42, 44, 128, 297*), LiOH (*44, 128*), KOH, RbOH and CsOH (*2, 172, 174, 186*) are quite interesting. Because of the highly ionic character of, e.g., Li_2O and LiOH, ab initio (*44*) and extended Hückel (*5*) angular correlation diagrams and the SOJT model (Section 3-2-3) were unsuccessful for these molecules.

The ESF model explains these linear shapes by noting that here all the substituent effects on the AD, EC, and EGC forces cooperate to increase the apex angle considerably. Since the highly electropositive alkali metal has vacant p valence AOs, it is a σ-donating and π-overlapping (like p_π-d_π overlap, Section 3-3-4) substituent and decreases the AD force on oxygen. Also, since the metal-oxygen bond electrons are highly polarized toward oxygen, the EC force on O increases considerably. The EGC force between terminal atoms is also repulsive. The linearity of Li_2O is also explained from the central-symmetry effect, since its electronic structure is almost $Li^+O^{2-}Li^+$(*4, 5*).

3-3-7c. XAY Molecules. The relevant MOs of these molecules (Fig. 3-6), in ascending energy order, are: $\bar{\pi}_u(0.37)$, $\sigma_u(0.0)$, $\bar{\pi}_g(0.0)$, $\bar{\pi}_u^*(0.77)$, $\sigma_s^*(0.0)$, the numbers indicating contribution to $D(p_{\bar{\pi}A})$ as calculated for the allyl radical, and the asterisk indicating antibonding orbitals. $\bar{\pi}_g$ is a nonbonding MO. Consider bending in the $\bar{\pi}$-plane. The π_u, π_g, and π_u^* MOs, perpendicular to the $\bar{\pi}$-plane, do not contribute to $D(p_{\bar{\pi}A})$ but contribute to the EC force through overlap effect (bond multiplicity). The σ_s^* MO is s_A-σ_X and s_A-σ_Y antibonding, and is closely related with central-symmetry effect.

From Fig. 3-5, we obtain the following predictions parallel to the Walsh rules:

1. When the $\bar{\pi}_u^*$ MO is empty (<16 valence electrons in ground state), $D(p_{\bar{\pi}A}) < 1$, at most $2 \times 0.37 = 0.74$. Hence, the molecule will be linear (Table 3-5).
2. When $\bar{\pi}_u^*$ is singly occupied (17 valence electrons in ground state), $D(p_{\bar{\pi}A}) > 1$, e.g., $0.74 + 0.77 = 1.51$, and the molecule is bent (experimentally, 122–136°; Table 3-5).
3. When $\bar{\pi}_u^*$ is doubly occupied (18–20 valence electrons in ground state), $D(p_{\bar{\pi}A})$ increases further, e.g., $0.74 + 2 \times 0.77 = 2.3$ and the molecule becomes more bent (experimentally, 100–126°).
4. When the σ_s^* MO is filled (22 valence electrons in ground state), the s_A, $p_{\bar{\pi}A}$ and $p_{\pi A}$ AOs are completely filled and the central-symmetry effect makes the molecule linear.

These conclusions are also applicable to excited states (Table 3-5).

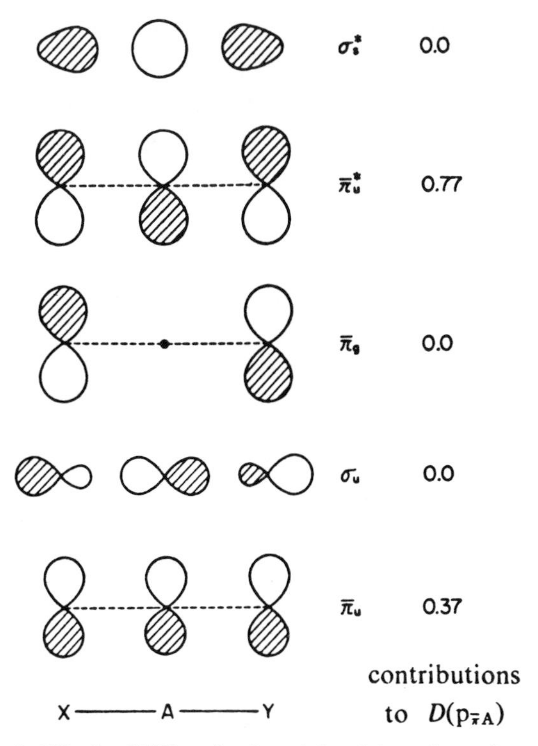

Fig. 3-6. Schematic MOs for XAY molecules. A hatch is made to show the bonding and antibonding character of the MOs. The right-hand number shows the contribution of each MO to $D(p_{\overline{\pi}A})$ calculated from the allyl system. (Adapted from Ref. 208, courtesy the American Chemical Society.)

3-3-7d. Alkaline Earth Dihalides. These 16-valence-electron XAY molecules are very interesting, since their shapes depend upon the metal-halogen combinations (Table 3-6), whereas according to Walsh rules they should all be linear. Actually, the light metal-heavy halogen combinations are linear, while the heavy metal-light halogen combinations are bent (*41*). If one modifies the Walsh diagram by considering the effect of normally unoccupied *d* AOs of the metal atom, then these trends can be explained (*64, 126, 131*).

The ESF model explains these trends as follows: Since the metal has vacant *p* valence AOs and low-lying vacant *d* AOs,[23] the halogen *p* electrons will flow into these vacant AOs by both overlap and π-donating inductive effects. Since these effects clearly increase from I to F and from Be to Ba, the AD force on metal increases in this order (the *p*-*d* mixing may also be important here[23]). Due to the increase in metal-halogen electronegativity difference the EC force increases in the same order (σ-inductive effect). Since both AD and EC forces are cooperative, this explains the trends in Table 3-6.

Table 3-5. Shapes of XAY Molecules.

Geometry and electronic configuration[a]	XAY[b]
linear $(\pi_u)^2(\sigma_u)^0$	OLi_2
linear $(\pi_u)^4(\sigma_u)^m(\pi_g)^n$ $m \leqslant 2$ $n \leqslant 4$	$C_3(2, 0), C_3^*(1, 1)$; $CCSi(2, 0)$, $CCSi^*(1, 1)$; $CCN(2, 1)$, $CCN^*(1, 2)$; $CNC(2, 1)$, $CNC^*(1, 2)$; $NCN(2, 2)$, $NCN^*(1, 3)$; $BO_2(2, 3)$, $BO_2^*(1, 4)$; $NCO(2, 3)$, $NCO^*(1, 4)$; $CO_2^+(2, 3)$, $CO_2^{+*}(1, 4)$; $OCS^+(2, 3)$, $OCS^{+*}(1, 4)$; $CS_2^+(2, 3)$, $CS_2^{+*}(1, 4)$; $N_3(2, 3)$, $N_3^*(1, 4)$; $NNO^+(2, 3)$, $NNO^{+*}(1, 4)$; $NNO(2, 3)$
linear $(\pi_g)^4(\pi_u^*)^0$	CO_2, CS_2, NCO^-, NCS^-, N_3^-, OCS, NNO, NO_2^+
bent $(\pi_g)^3(\pi_u^*)^1$; $(\pi_g)^4(\pi_u^*)^1$	$CO_2^*(122°)$, $CS_2^*(135.8°)$, $NCO^{-*}(–)$, $NCS^{-*}(–)$, $N_3^{-*}(–)$, $NO_2(4, 1, 134.1°)$
bent $(\pi_g)^m(\pi_u^*)^n$ $3 \leqslant m \leqslant 4$ $2 \leqslant n \leqslant 4$	$NO_2^*(3, 2, 121°)$[c]; $CF_2(4, 2, 104°)$, $CF_2^*(3, 3, –)$; $CCl_2(4, 2, –)$; $SiF_2(4, 2, 101°)$, $SiF_2^*(3, 3, –)$; $NO_2^-(4, 2, 115.4°)$, $NO_2^{-*}(3, 3, –)$; ONF(4, 2, 110°), $ONF^*(3, 3, –)$; $SO_2(4, 2, 119.5°)$, $SO_2^*(3, 3, 126.1°)$; SSO(4, 2, 118°), $SSO^*(3, 3, –)$; ONCl(4, 2, 116°), ONBr(4, 2, 117°), $O_3(4, 2, 116.8°)$, $NF_2(4, 3, 104.2°)$, $ClO_2(4, 3, 117.6°)$, $OF_2(4, 4, 103.8°)$, $OCl_2(4, 4, 110.8°)$, $SCl_2(4, 4, 101°)$, $TeBr_2(4, 4, \sim 98°)$, $ClO_2^-(4, 4, 110.5°)$
linear $(\pi_u^*)^4(\sigma_s^*)^2$	$IBrCl^-$, Br_3^-, ICl_2^-, $ClIBr^-$, IBr_2^-, I_3^-, XeF_2

[a]The MO density distributions are illustrated in Fig. 3-6. Both π and $\bar{\pi}$ MOs are written as π.
[b]Values in parentheses are m, n, and experimental apex angles. The asterisk means that the molecule is in the excited state.
[c]The structure of the Rydberg excited state of NO_2, $(\pi_g)^4(\pi_u^*)^0(3p\sigma)^1$ is linear.
SOURCE: Ref. 208. Reprinted courtesy the American Chemical Society.

3-3-8. Force along Internal Rotation and the Shapes of X_m ABY$_n$ Molecules

Since the shapes of X_mAB and ABY$_n$ fragments are readily understood from previous discussions, we are now concerned with a new problem, namely, internal rotation about the AB bond (single, double, or triple). By extension of these two treatments, the shapes of general $AX_k(BY_l)_mCZ_n$ molecules can be explained.

Table 3-6. Geometries of Alkaline Earth Dihalides.[a,b]

Central metal	Halogen F (4.0)	Cl (3.0)	Br (2.8)	I (2.5)
Be (1.5)	*l*	*l*	*l*	*l*
Mg (1.2)	*l*[c] or *b* (158°)[d]	*l*	*l*	*l*
Ca (1.0)	*b* (140°)[d]	*l*	*l*	*l*
Sr (1.0)	*b* (108°)[d]	*b* (120°)[e]	*l*	*l*
Ba (0.9)	*b* (~100°)[d]	*b* (~100°)[e]	*b*	*b*

[a] *l*, linear; *b*, bent. Values in parentheses are the Pauling electronegativities (*231*).
[b] L. Wharton, R. A. Berg, and W. Klemperer, *J. Chem. Phys.*, 39, 2023 (1963); Refs. 41, 42.
[c] M. Kaufman, J. Muenter, and W. Klemperer, *J. Chem. Phys.*, 47, 3365 (1967).
[d] V. Calder, D. E. Mann, K. S. Seshadri, M. Allavena, and D. White, *J. Chem. Phys.*, 51, 2093 (1969).
[e] D. W. White, G. V. Calder, S. Hemple, and D. E. Mann, *J. Chem. Phys.*, 59, 6645 (1973).

3-3-8a. Orbital Following and Preceding. In this section we shall see that the dominant origin of internal rotation barrier arises from an interesting behavior of electron density during rotation, namely, orbital following (*179*) and preceding. The ESF concept explained here is quite general and provides a new key to understanding nuclear rearrangement processes such as molecular structure, molecular vibration, and chemical reaction.

Consider the internal rotation barrier of ethylene (Fig. 3-7). Energetically, the

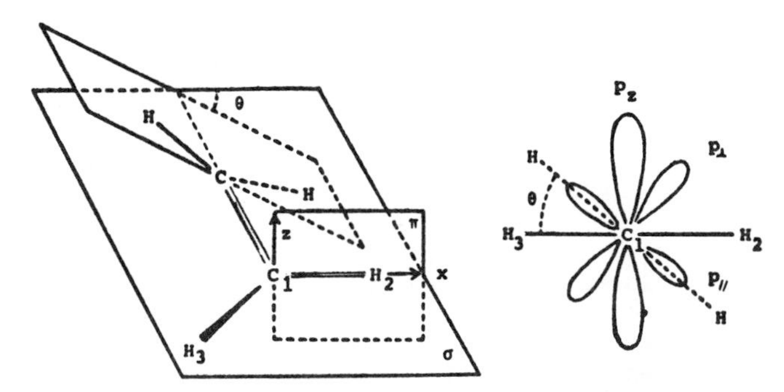

Fig. 3-7. The ethylene molecule twisted by the angle θ. The right-hand side shows the Newman-type diagram. Front-side and back-side CH_2 fragments are shown by solid and dashed lines, respectively. The p_z AO on the front-side CH_2 is decomposed into $p_\perp$ and $p_\parallel$ AOs, where the notations $\perp$ and $\parallel$ refer to the back-side CH_2 plane. (Reproduced from Ref. 209, courtesy the American Chemical Society.)

planar form is stable because the bonding interaction between two $p_{\pi C}$ AOs is maximum at planarity (*161,201,202,204*). The ESF explanation considers the behavior of electron density in the C_1—H_2 bond region (Fig. 3-7): We switch off the interaction between two CH_2 fragments, rotate the back-side CH_2 by angle $-\theta$, and then resolve the p_z AO of the front-side CH_2 into $p_\perp$ and $p_\parallel$ components on the π-plane. When we switch on the interaction, electrons in $p_\perp$ AO flow into the C—C bond region owing to the bonding overlap with the p_π AO of back-side CH_2, but electrons in the $p_\parallel$ AO suffer little change, since it interacts with back-side C—H bond orbital which tightly binds the electrons (Section 3-3-4d). Consequently, the electron density along the C_1—H_2 bond changes (see Ref. 209 for detailed contour plot at $\theta = 45°$, using the extended Hückel method).[24] Indeed, the C—H bond electron cloud follows *incompletely* (*orbital following*) the rotation of the C—H axis in the θ direction (Fig. 3-8). This

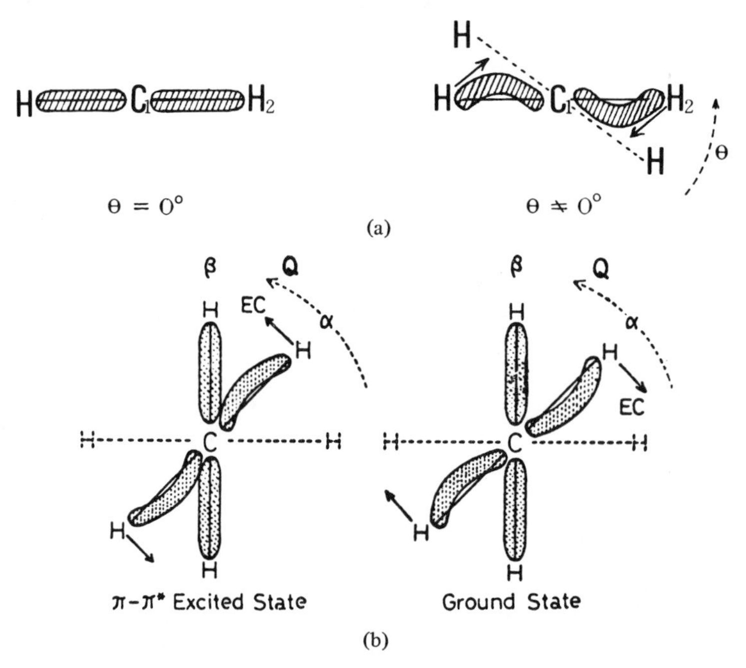

Fig. 3-8. General features of the C—H bond electrons during internal rotation of the ground and π–π^* excited states of ethylene. (a) Incomplete following of the C—H bond orbital in the ground state. In (b) the dashed C—H bonds belong to the back-side CH_2 group. Q is the coordinate of internal notation from planar to bisected configuration, α corresponds to the configuration of the intermediate rotational angle, and β to the bisected configuration. The arrows starting from protons at α show the EC forces acting on protons due to the preceding (π–π^* excited state) and incomplete following (ground state) of the C—H bond electron clouds. (Reproduced from Ref. 210, courtesy the American Chemical Society.)

causes an EC force (arrow in Fig. 3-8) which does not lie on the C—H axis and resists internal rotation. Extended Hückel calculations (*209*) for the $(\pi)^2(\pi^*)^0$ electronic configuration of ethylene reveal that the EC(H_2—C_1) force component (-0.0286 au) is the dominant contribution to $F_z(H_2)$, the total rotational force component (-0.0262 au).[25] Thus, ESF theory concludes that orbital following and the resulting resisting force are responsible for the planarity of ethylene (valid also for a wide range of double-bonded hydrocarbons).

On the other hand, there are situations where the C—H bond electron cloud *precedes* the rotation of C—H axis (*orbital preceding*). Consider the difference density map (Fig. 3-9) for the π-π^* excited state of ethylene, $(\pi)^1(\pi^*)^1$. The antibonding (π^*) overlap effect on the $p_\perp$ AO (Fig. 3-7) exceeds the bonding (π) overlap effect (Table 3-2) and causes orbital preceding. The resulting EC force on the proton further accelerates the rotational motion (Fig. 3-8) and makes the molecule stable in the bisected form, $\theta = 90°$. Extended Hückel calculations (*209*) again show that the EC(H—C) component is the major contribution

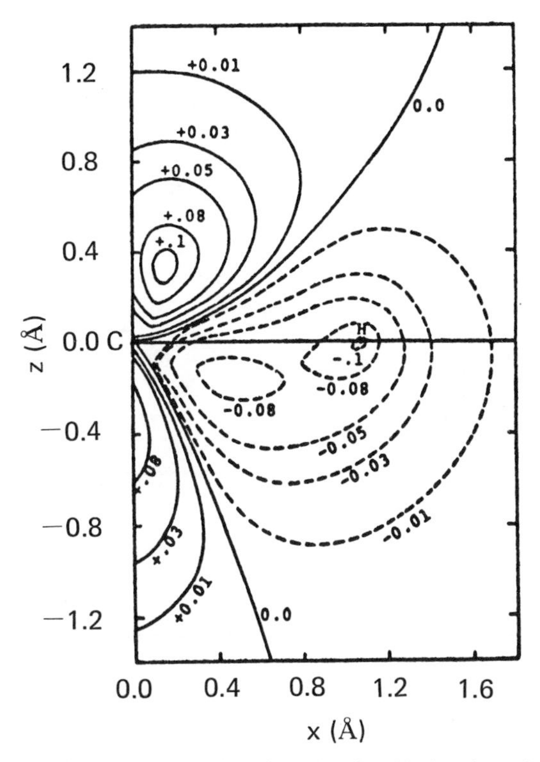

Fig. 3-9. The change in electron density along the C_1—H_2 bond on the π-plane induced by twisting the π-π^* excited state of ethylene through $\theta = 45°$. The electronic configuration is $(\pi)^1(\pi^*)^1$. (Reproduced from Ref. 209, courtesy the American Chemical Society.)

(0.0144 au) to the total accelerating force (0.0178 au). The same is true for the $(\pi)^0(\pi^*)^0$ configuration. Indeed, the π-π^* excited state of ethylene (*187, 298*), BH_2BH_2 (Frost, quoted in Ref. 233), and B_2F_4 (*114, 230*) all have the bisected form, $\theta = 90°$.

When both the p_π AOs of the two rotors are doubly occupied, $(\pi)^2(\pi^*)^2$, the overlap effect becomes similar to $(\pi)^1(\pi^*)^1$, although the extent of orbital preceding is now smaller (*209*). In the $(\pi)^2(\pi^*)^2$ case, both $D(p_{\pi A})$ and $D(p_{\pi B})$ are greater than 2 and both H_2AB and ABH_2 fragments become bent. Since the A—B bond here is regarded as a single bond, we shall discuss these structures in Section 3-3-8b.

It has been confirmed that for all ethylenic configurations depicted in Fig. 3-10, the $EC(H_2—C_1)$ force due to orbital following (negative region) and preceding (positive region) is the dominant factor in $F_z(H_2)$. Orbital following also occurs for the configuration $(\pi)^2(\pi^*)^1$. Interestingly, for the cation configuration $(\pi)^1(\pi^*)^0$ orbital preceding occurs when $0° \leqslant \theta \leqslant 45°$, but orbital

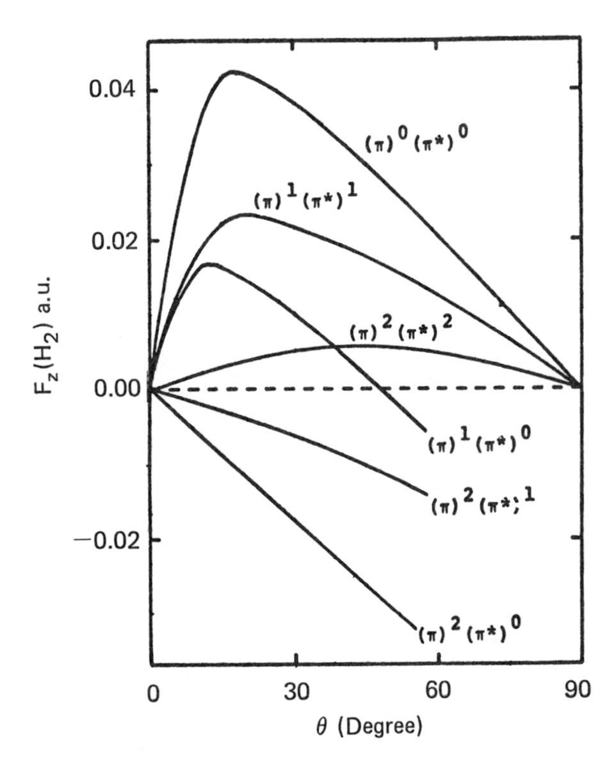

Fig. 3-10. Curves of the rotational force, $F_z(H_2)$ versus the twist angle θ, for various electronic configurations arising from ethylenic MOs calculated by the extended Hückel method. Since π and π^* MOs become degenerate at $\theta = 90°$, the plot is given only in the region $0° \leqslant \theta \leqslant 45°$ for the configurations $(\pi)^1(\pi^*)^0$, $(\pi)^2(\pi^*)^0$, and $(\pi)^2(\pi^*)^1$. (Reproduced from Ref. 209, courtesy the American Chemical Society.)

following occurs when $\theta > 45°$, suggesting the stable conformation to be near $\theta = 45°$. This agrees with other theoretical (*161, 206*) and experimental studies (*190–192*). The latter studies reported that for the first Rydberg excited state of ethylene $\theta \simeq 25°$, and suggested a similar possibility for the ethylene cation.

In summary, *orbital following and preceding are the origin of the rotational barrier and arise from the difference in interactions (overlap effects) of the* $p_\perp$ *and* $p_\parallel$ *AOs with the electron cloud of the rotor of another side.* If the bonding overlap effect on $p_\perp$ AO is larger than that on $p_\parallel$ AO, orbital following occurs; in the reverse case, orbital preceding occurs. The B—Y bond orbital of the other rotor does not affect the $p_\parallel$ AO significantly, since they are quite different in energy.

3-3-8b. Rotation about a Single Bond. The sensitive problem of the origin of the barrier to internal rotation about a single bond has been the subject of extensive theoretical studies (see Ref. 3 and Chapter 4). ESF theory gives a successful intuitive account of this problem.

Consider the following three factors for internal rotation about the A—B single bond in $H_m ABH_n$ molecules: (a) orbital following and preceding, (b) interaction of the concerned proton H_1 on one side with the B—H_2 bond on another side, and (c) interaction of the concerned proton H_1 on one side with the lone-pair electrons on another side. Factor a has already been discussed in Section 3-3-8a. For factor b, illustrated below, the electron cloud of the

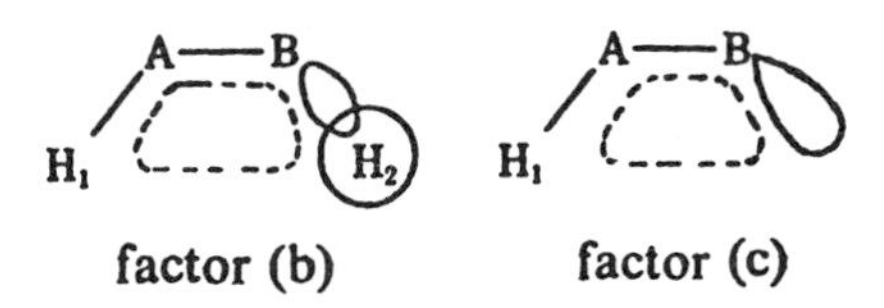

B—H_2 bond region and H_2 atomic region (enclosed by solid lines) attracts the H_1 proton, but this attraction is offset by the bare nuclear repulsion between H_1 and H_2. The force on H_1 due to the electron cloud shown by dotted lines will usually be repulsive, since the bond order between a nonbonded pair is usually negative (negative EC force). In the case of ethane (*216*) *the total effect of factor b seems repulsive.* For factor c, the interaction between the H_1 proton and the lone-pair electron cloud at B (shown by the solid ellipse) is attractive, but that due to the lone-pair electron cloud shown by dotted lines may be either attractive or repulsive. However, since the electron density in the solid ellipse is greater than that in the dotted region, the former attractive force will predominate and *the total effect of factor c will be attractive.* Note that in such arguments for factors b and c we have assumed complete following. From previous discussions, we expect that *for rotation about a single bond factor a will be most important, while factors b and c are of secondary importance.*

If the A—B bond has partial double-bond character, the molecule will be planar because of orbital following, e.g., butadiene (see Ref. 182 for numerous examples[26]). Molecules having no π bonds, e.g., BH_2BH_2 and BF_2BF_2, are also coplanar because of orbital following (Fig. 3-10). Factor a in fact includes the so-called hyperconjugation effect (*269, 301*).

In contrast, when both A and B have lone pairs, orbital preceding becomes important. Since two lone pairs on either side lie in similar energy levels, they interact through the overlap effect (interaction between fully occupied AOs in Table 3-2). However, since $D(p_{\pi A})$ and $D(p_{\pi B})$ exceed 2 for these molecules, the bending of H_mAB and ABH_n fragments will also have to be considered. The top left-hand side of Fig. 3-11 shows the bisected form of NH_2NH_2 due to orbital preceding in the $(\pi)^2(\pi^*)^2$ configuration (factor a). Each NH_2 fragment bends from this form, since $D(p_{\pi N}) \simeq 2$, in agreement with experimental geometry. The same is true for NH_2OH (bottom left-hand side of Fig. 3-11), for which

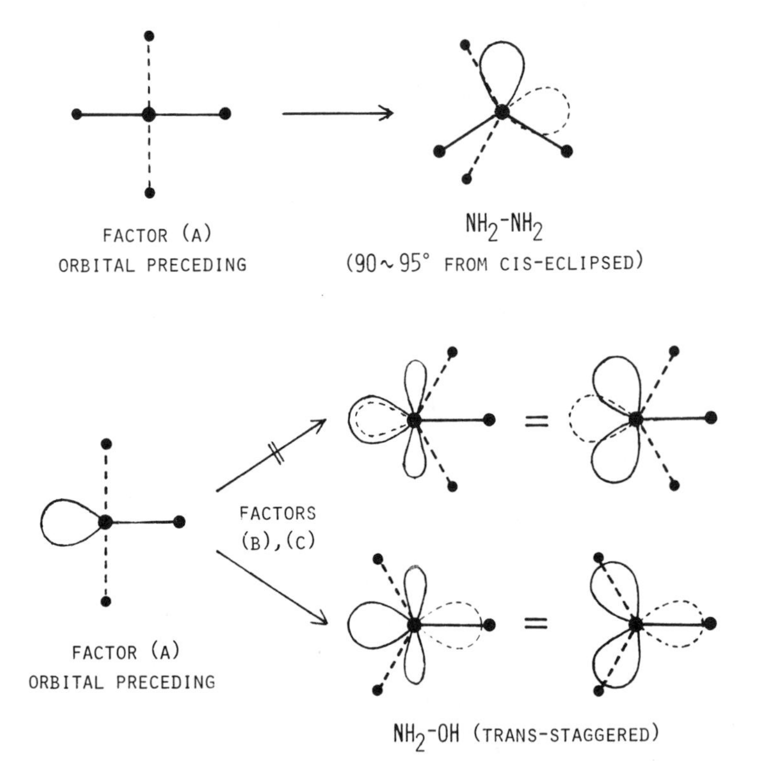

Fig. 3-11. Explanations for the structures of NH_2—NH_2 and NH_2—OH. The solid lines show the front-side NH_2 rotor (for NH_2—NH_2) or the OH rotor (for NH_2—OH) while the dotted lines show the back-side NH_2 rotors. The larger closed curves denote lone-pair orbitals; the equality sign implies equivalence.

one N—H bond is replaced by the σ lone pair of oxygen. For NH_2OH, the NH_2 fragment (dotted line) in the bisected form can bend in two different ways. But factors b and c favor the lower *trans* staggered form (in agreement with experiment) in which the O—H and N—H bonds are as far apart as possible (factor b) and eclipse with adjacent lone-pair orbitals (factor c). For HOOH and its derivatives the situation is simpler. Replacing one N—H bond of each rotor (see Fig. 3-7) with a σ lone-pair orbital, we predict from factor a that the relative orientation of the two rotors will be $\sim 90°$. The experimental data (*209*) are HOOH (111.5°, 119.8°), HSSH (90.6°), FOOF (87.6°), FSSF (87.9°), ClSSCl (82.5°), BrSSBr (83.5°). Note that factors b and c cooperate for such geometries, since two OH bonds repel each other (factor b) and they eclipse the adjacent π lone pair which is nearer than the σ lone pair. These factors lower the *trans* barrier (1.10 kcal/mole for H_2O_2) compared with the *cis* barrier (7.03 kcal/mole for H_2O_2). *In summary, when both A and B have lone pairs, we first write down the bisected form (factor a) and then bend each fragment considering factors b and c.*

When only one of A and B has a lone pair (or lone pairs), or neither A nor B has a lone pair, no significant orbital preceding occurs, since the A—H bond electrons are more tightly bound than lone-pair electrons. One may then use factors b and c. Lowe (*182*) summarized these molecules as ethanelike, $(X_1X_2X_3)A—B(Y_1Y_2Y_3)$ where Y_1 and/or Y_2 may be lone pairs, having staggered conformation; e.g., for ethane, factor b leads to a staggered form. Also, the staggered forms of molecules like CH_3NH_2, CH_3OH, having a lone pair (or lone pairs) on one side, may reflect factor b, since factor c will lead to the eclipsed form.

Actual force calculations (*216*) show that even in ethane factor a predominates for the staggered form. Throughout rotation, the EC(H—C) force arising from incomplete following of the C—H bond orbital predominates. The EC (other) force and EGC force, reflecting factor b, are minor repulsive forces. Both the EC(H—C) and EC (other) forces[27] may be interpreted as arising from the repulsive exclusion of the two approaching C—H bond orbitals,

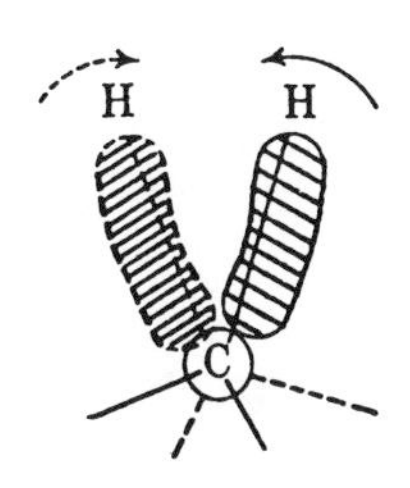

which is just the overlap effect between these orbitals. This is similar to the origin of the repulsive force between two He atoms at the overlap region.

3-3-8c. Geometries of H_2ABH Molecules. As another example of the previous discussions, consider the shapes of H_2AB and ABH fragments, and their relative orientations:

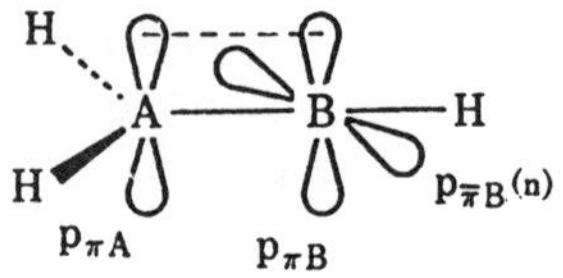

The $p_{\pi A}$ and $p_{\pi B}$ AOs form π and π^* MOs, and $p_{\bar{\pi} B}$ forms a nonbonding (n) orbital, with the energy order $\pi < n < \pi^*$. The shape of the H_2AB fragment can be predicted as follows (Fig. 3-5): When the π^* MO is (i) empty, $D(p_{\pi A}) < 1$, H_2AB is planar; (ii) singly occupied, $D(p_{\pi A}) > 1$, H_2AB is pyramidal;[28] (iii) doubly occupied, H_2AB is more pyramidal. For the ABH fragment, if bending occurs in the π-plane, the prediction is similar to H_2AB. If bending occurs in the $\bar{\pi}$-plane, then when (a) $p_{\bar{\pi} B}$ is empty, ABH is linear; (b) $p_{\bar{\pi} B}$ is singly or doubly occupied, $D(p_{\bar{\pi} B}) = 1$ or 2, ABH is bent.

Now consider the relative orientation of H_2AB and ABH fragments. With both π and π^* MOs empty, orbital preceding occurs, and the H_2A and ABH planes bisect each other. No experimental geometries are known for such excited states, $(\pi)^0 (n)^{1 \sim 2} (\pi^*)^0$. With π MO doubly occupied and π^* MO empty or singly occupied, orbital following occurs (Fig. 3-10) and the shape becomes planar [for $(\pi^*)^0$] or 90° from the *cis* staggered form [for $(\pi^*)^1$]. With both π and π^* doubly occupied, orbital preceding occurs (factor a) and factors b and c lead to a *trans* staggered form (Fig. 3-11). Rotational force calculations on vinyl (see Ref. 209) confirm these conclusions.

The above discussions can be combined to make predictions about the shapes of H_2ABH molecules (Table 3-7), in agreement with known shapes. For H_2COH, ESF theory predicts the relative orientation to be 90° from *cis* staggered form; no experimental evidence is known as yet.

3-3-9. Comparison with the VSEPR Concept

A possible answer for the considerable success of the VSEPR model (Section 3-2-2) may be found by comparing the rule for the relative magnitude of Pauli repulsions,

$$\text{lone pair} > \text{bond pair (triple} > \text{double} > \text{single)},$$

with the relative importance, as in (3-14), of AD and EC forces,

$$\text{AD force} > \text{EC force (triple} > \text{double} > \text{single)}.$$

Table 3-7. Shapes of H_2ABH Molecules.

planar C_{2v} planar C_s nonplanar

Shape[a]	No. of valence electrons	Configuration			H_2ABH
		π	$n_{\bar{\pi}B}$	π^*	
planar C_{2v}	10	2	0	0	H_2CCH^+
planar C_s	11	2	1	0	H_2CCH
planar C_s	12	2	2	0	H_2CNH, H_2COH^+
nonplanar	13	2	2	1	H_2COH
nonplanar	14	2	2	2	H_2NOH[b]

[a]See representations above.
[b]*Trans* staggered configuration.
SOURCE: Ref. 209. Reprinted courtesy the American Chemical Society.

When applied to known molecules, e.g., NH_3, and isobutene (Section 3-3-3), they lead to same predictions in spite of essential differences in their basic ideas. The ESF concepts (e.g., AD, EC, and EGC forces) are well-defined and easily verifiable concepts (*221*), while the VSEPR concept is not clearly defined and justified (Section 3-2-2). Hence, it should be noted that this parallelism in predicted results, showing the success of VSEPR model, does not necessarily imply justification of its underlying concept.

3-3-10. Applications of ESF Theory

In comparison with other models of molecular structure, the ESF model has a remarkably wide applicability, e.g., to ground and excited states and the problem of the internal rotation barrier. It has been successfully applied so far to AH_2, AH_3, HAX, XHY, H_2AX, XAY, HAAH, HABX, H_2ABH, and H_2AAH_2 molecules and their various substituents (*207–209*). The underlying well-defined concepts, whose physical meanings are always very clear, have been verified by actually calculating AD, EC, and EGC forces from ab initio floating wavefunctions that satisfy the H–F theorem (*168*a, *220*, *221*, *221*a). Indeed, the ESF model has been very successful in explaining and predicting the regularity and variety in structural chemistry. It has recently been applied to predictions of geometries of molecules in strong electric fields (*221*b).

ESF theory is also applicable to various phenomena, e.g., chemical reactions (*16*, *168*a, *216*, *219*) and long-range forces (*215*; Section 7-2). Such versatility of the theory, due essentially to the simplicity and exactness of the H-F theorem, would be quite useful for a unified understanding of these phenomena. For example, chemical reactions (molecular interactions) and molecular structure are closely related, since most chemical reactions proceed with changing the geometry of reactants, e.g., in a biological system. Sometimes a single EC force between reaction sites is a common origin of the driving force of the reaction and the change in the geometry of reactants (*207*). In some cases, such geometrical changes cause forces (e.g., the AD force) which play an essential role in driving the reaction. An example, the dimerization of two methyl radicals (*216*), is briefly described in Section 3-5-6 (see also Section 6-7).

3-4. A SIMPLE MECHANICAL FORCE MODEL FOR MOLECULAR GEOMETRY

3-4-1. Introduction

In this section we discuss a simple force model for molecular geometry proposed by Deb and co-workers (*76–79, 82*). This model considers basically the MO *force* correlation diagram (fcd) in which the individual MO contributions to the H-F transverse force on the terminal nuclei are plotted against the bending angle (cf., the Walsh diagram, Section 3-2-1). The essential features of the fcd are depicted in a shape diagram which arranges the MOs according to their roles in planar (or linear)-bending correlation. The shape diagram was constructed intuitively by considering the qualitative MO densities and the model was greatly simplified by introducing the HOMO postulate (*76*) which assumes that the behavior of the HOMO mainly governs gross equilibrium geometry. The validity of the postulate was established by applying the model to many types of molecules, e.g., AH_2, AH_3, AH_4, AH_5, AB_2, HAB, ABC, AB_3, AB_4, AB_5, HAAH, BAAB, H_2AB, HAB_2, B_2AC, etc.

In order to provide the necessary background, we shall first discuss two earlier H-F force treatments of molecular geometry: (i) a reinterpretation study of the Walsh diagram (*66*) and (ii) the first-order Jahn–Teller distortion in VCl_4 (*65*).

3-4-2. Reinterpretation of the Walsh Diagram

Coulson and Deb (*66*) gave the following reasoning for the Walsh diagram. In the Hartree–Fock MO approximation, the H-F force on nucleus A is given by (3-18), where $f_{Ai} = \langle \phi_i | f_A | \phi_i \rangle$ is the MO contribution to the total electronic force which is a simple sum of all MO contributions. By integrating f_{Ai} over the displacement of nucleus A,[29] we obtain the change in electronic energy (relative to the lower limit of integration) as a simple sum of MO contributions. This quantity is identical with that calculated directly from the Hartree–Fock method, if *exact* Hartree–Fock orbitals are used throughout displacement. Thus, in principle the electronic energy is expressible as a simple sum of MO contributions, although this is not true with the conventional Hartree–Fock energy formula (*252*) because of electron–electron repulsion terms. This summation provides a clue to the Walsh diagram, although this reasoning excludes the nuclear–nuclear repulsion energy of (3-18).

The above argument can be employed to construct Walsh-type diagrams. Consider an AH_2 molecule with apex angle α and A—H length λ. If the molecule is bent by symmetric transverse motions of the two hydrogens, with λ and the A atom fixed, then the total energy $E(\alpha)$, relative to the linear form ($\alpha_0 = 180°$), is

$$E(\alpha) = 2 \sum_{i}^{occ} w_e^i(\alpha) + \frac{1}{2}\left(\operatorname{cosec}\frac{\alpha}{2} - 1\right). \tag{3-26}$$

The two terms in (3-26) represent changes in electronic energy and nuclear repulsion energy, respectively, and

$$w_e^i(\alpha) = \lambda \int f_\perp^i \, d\alpha \tag{3-27}$$

represents the work done by *i*th MO contribution to the transverse force, $f_\perp^i$, acting on a proton during bending from 180° to α. Coulson and Deb (*66*; see also *64*) calculated $f_\perp^i$ for H_2O from the MOs of Ellison and Shull (*90*) and obtained $w_e^i(\alpha)$ by numerical integration. The resulting $w_e^i(\alpha)$ correlation diagram resembles very closely the Walsh–Allen (*240*) and the corrected Walsh diagrams[30, 31] (Fig. 3-12). Note that the corrected Walsh diagram has a rising $2a_1$ curve, instead of a falling one as obtained earlier due to certain unrealistic assumptions of Walsh (Section 3-2-1; *136*, *165*, *263–265*). Thus, the above treatment gives a successful explanation of the Walsh and Walsh–Allen diagrams for H_2O, although it omits nuclear repulsion terms and $w_e^i(\alpha)$ is not always a good approximation to the difference in orbital ionization potentials between the bent (α) and linear molecules. Similar conclusions are also valid for NH_3.

Through a similar procedure, the internal rotation[32] in H_2O_2 was studied by calculating the transverse rotational force on a proton, using the wavefunctions of Kaldor and Shavitt (*160*). The resulting w_e^i correlation diagram (Fig. 3-13), relative to the *trans* planar form, resembles the corresponding Walsh–Allen diagram, even in slopes, although the unoccupied MO curves and the orders of the occupied MO curves do not match in the two diagrams. From zero net force, the equilibrium dihedral angle is ~65°, whereas the corresponding minimum-energy value is 120°. It is not surprising that both values are wrong (the experimental value is 111.5°), since the *trans* barrier of H_2O_2 is very difficult to reproduce (*102*, *226*, *238*).

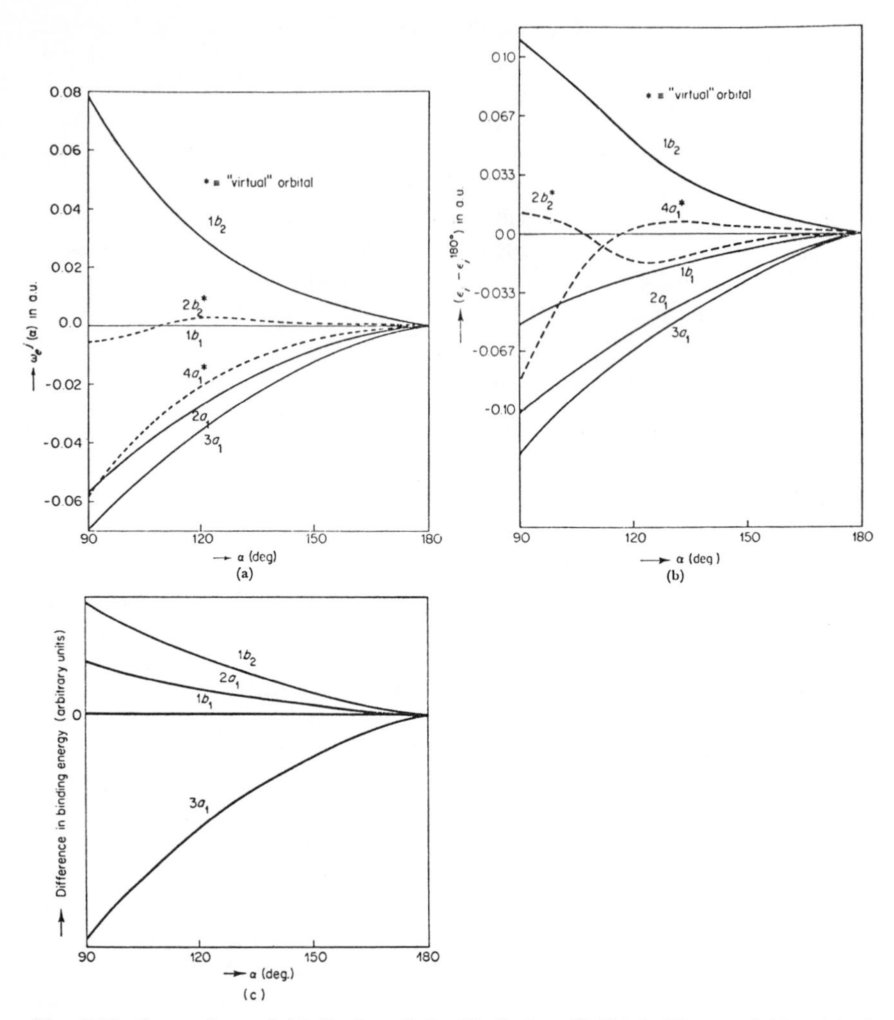

Fig. 3-12. Comparison of (a) Coulson–Deb, (b) "reduced" Walsh–Allen, and (c) original Walsh diagrams for H_2O. ϵ_i^α refers to the ith MO energy at the valence angle α. (Reproduced from Ref. 66, courtesy John Wiley & Sons, Inc.)

In Section 3-4-4 we shall consider the fcd's for these molecules which plot $f_\perp^i$ against an angle and are just the derivatives of Coulson–Deb diagrams.

3-4-3. First-Order Jahn–Teller Effect

In Section 3-2-3 we have seen that (3-3) gives the energy of a molecule, under a small displacement Q_i, according to second-order perturbation theory. The

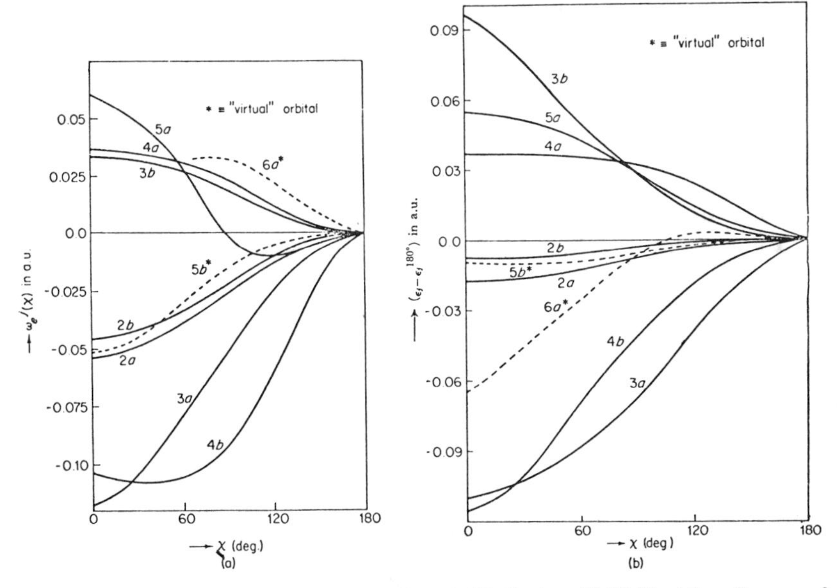

Fig. 3-13. Comparison of (a) Coulson–Deb and (b) "reduced" Walsh–Allen diagrams for H_2O_2. χ is the dihedral angle between two HOO planes (χ = 180°, *trans*-planar). (Reproduced from Ref. 66, courtesy John Wiley & Sons, Inc.)

significance of the second-order terms for molecular structure has already been discussed. We now discuss the role of the first-order term which survives only when the molecular wavefunction ψ_0 belongs to a degenerate energy level. The nonlinear[33] polyatomic molecule then tends to distort to a lower symmetry (first-order Jahn–Teller effect; Refs. 92, 153).

The H–F force formulation (*56*) of the static[34] JT effect is more advantageous than the energetic treatment in its conceptual simplicity and requires fewer assumptions for deriving the JT theorem. Only the Born–Oppenheimer approximation is assumed; there is no need to assume explicit Q_i dependence of the Hamiltonian, or invoke perturbation theory. The formulation has been applied to JT distortions in CH_4^+, CF_4^+, excited states of NH_3^+, NH_3 (*71*), and VCl_4 (*65*). We discuss the VCl_4 example below.

The electronic ground state (doubly degenerate) of tetrahedral VCl_4 may be written as $\ldots(a_1)^2(t_2)^6(t_1)^6(e)^1$, ^2E (*24*). One therefore expects that the static JT effect will distort the molecule from its idealized tetrahedral shape. We assume (i) that the electronic structure of VCl_4 differs from that of the hypothetical VCl_4^+, a closed-shell tetrahedral molecule, only in the presence of one electron in the nonbonding *e* MO (the frozen MO assumption; Refs. 55, 56); and (ii) that the doubly degenerate *e* MOs responsible for the distortion are composed of either the $3d_{z^2}$ or $3d_{x^2-y^2}$ AO of the central vanadium atom.[35] Then the H-F force acting on a Cl nucleus in tetrahedral VCl_4 is given by

$$F_{Cl}(VCl_4) = F_{Cl}(VCl_4^+) + Z_{Cl}\int \phi^*\phi(r_{Cl}/r_{Cl}^3)\,d\tau, \tag{3-28}$$

where Z_{Cl} is the effective nuclear charge of Cl (assumed to be unity) and ϕ is either $3d_{z^2}$ or $3d_{x^2-y^2}$ AO. Putting $F_{Cl}(VCl_4^+)$ as zero for tetrahedral VCl_4^+, we obtain the JT force tending to cause deformation as

$$F_{Cl}(VCl_4) = Z_{Cl}\int \phi^*\phi(r_{Cl}/r_{Cl}^3)\,d\tau. \tag{3-29}$$

Using the Urey-Bradley (*293*) model, we rewrite (3-29) as

$$(K_1 + 4K_3)\Delta r = Z_{Cl}\int \phi^*\phi(z_{Cl}/r_{Cl}^3)\,d\tau \tag{3-30a}$$

$$K_2 p = Z_{Cl}\int \phi^*\phi(x_{Cl}/r_{Cl}^3)\,d\tau, \tag{3-30b}$$

where z_{Cl} and x_{Cl} are, respectively, along and perpendicular to a V—Cl bond; K_1, K_2, K_3 are Urey-Bradley force constants; Δr is the decrease in V—Cl bond length; and p is the transverse displacement of a Cl nucleus.

Interestingly, the following conclusions can be made (*75*) with the help of Fig. 3-14, even without any numerical calculations. (a) The attractive force on

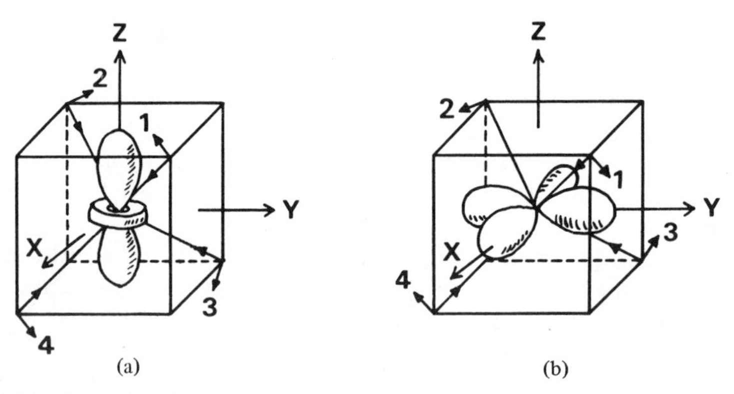

Fig. 3-14. The static Jahn-Teller effect in VCl_4, using (a) metal d_{z^2} and (b) metal $d_{x^2-y^2}$ AOs. The arrows on ligand atoms indicate the directions of distortion. Arrows 1,2 and 3,4 lie in the planes defined by the metal z-axis and the relevant V—Cl bonds. (Reproduced from Ref. 75, courtesy the American Institute of Physics.)

Cl due to d_{z^2} or $d_{x^2-y^2}$ density causes a totally symmetric decrease in V—Cl bond lengths; this does not reduce the tetrahedral symmetry. (b) Due to the symmetry of the d_{z^2} and $d_{x^2-y^2}$ AOs, the JT distortion [p in (3-30b)] occurs in the plane defined by the metal z-axis (S_4 or C_2) and the relevant V—Cl bond. (c) If the e electron occupies the d_{z^2} AO, the cube will be elongated along the metal z-axis (i.e., positive p), and if the $d_{x^2-y^2}$ AO is occupied, the cube will be flattened (i.e., negative p). These two distortions are distinguishable, resulting in the two distorted forms having different stabilities. There are three equivalent distortions of each type. (d) The two e orbitals lead to opposite and distinguishable distortions; the e electron is more likely to occupy (after deformation) either d_{z^2} or $d_{x^2-y^2}$ orbital rather than a linear combination of them, since this leads to greater bending distortion energy which is an upper bound for the barrier to interconversion among the three equivalent distorted forms via a tetrahedral form.

By using two sets of spectroscopic values for K_1, K_2, and K_3, and both Slater and SCF AOs for two vanadium configurations (*295*, *296*), Coulson and Deb (*65*) made detailed calculations for Δr, p, and the corresponding stabilization energies. The bending distortion energies for flattened and elongated tetrahedra were slightly different, implying that both forms are equally probable for VCl_4. This conclusion is consistent with low-temperature ESR results of Johannesen et al. (*156*), who reported a mixture of 58% flattened and 42% elongated forms.

3-4-4. A Simple Mechanical Model for Molecular Geometry

3-4-4a. Construction of the Model. According to the basic ideas outlined at the beginning of this section, we now explain the necessary steps in constructing the model. (1) qualitative construction of symmetry-adapted LCAO-MOs using valence s and p AOs as basis functions: the symmetry-group orbitals are combined in bonding, nonbonding, and antibonding ways; (2) arranging the MOs in energy order according to their bonding, nonbonding, and antibonding character, and by their symmetry properties; (3) assigning the role of each MO in, e.g., linear–bent correlation by using the qualitative picture of the H–F theorem to decide whether a given MO density causes a bending or linearizing force on a terminal nucleus [for example, in Fig. 3-15 if the MO accumulates more density inside the AB_2 triangle, it will cause a positive (bending) transverse force on nuclei B_1 and B_2, and vice versa; all these considerations are summarized in the shape diagram, which is a simplified version of calculated fcd[36]]; (4) Aufbau feeding of electrons into the MOs and then using the HOMO postulate to determine gross equilibrium molecular geometry. If the role of HOMO is to bend (or linearize) the molecule, the equilibrium shape will be bent (or linear). In case the HOMO is insensitive to a valence angle, the angular behavior of the

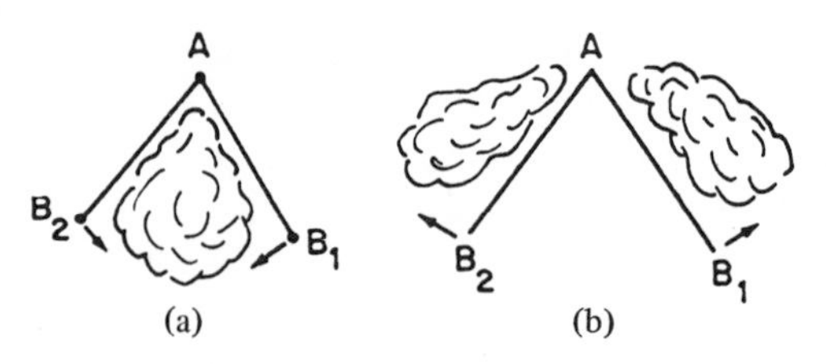

Fig. 3-15. Electron clouds, transverse forces, and terminal nuclear motions leading to bent or linear AB_2 molecules. The atom A and the A—B bond length are kept fixed during such motions. (Reproduced from Ref. 78, courtesy the American Chemical Society.)

next lower MO, if sensitive, will determine the shape. If this MO is insensitive too, then the next lower MO is to be examined, and so on. Cases where it is difficult to determine a unique HOMO occur when (a) the electronic states are orbitally degenerate (FOJT, Section 3-4-3), and (b) the HOMO and the MO next to it have opposite influences on a valence angle and cross each other in energy (accidental degeneracy; Ref. 76). In case b the next influence of these MOs is to be considered.[37] We shall now explain the stepwise applications of the model to AH_2, AH_3, AB_2, and HAAH molecules. The reader may compare these results with those from energetic models (Section 3-2) and the ESF model (Section 3-3).

3-4-4b. Illustrative Applications of the Model. *1.* AH_2 *Molecules.* From the three valence-group orbitals s_a, p_{za} (along the symmetry axis), and $h_1 + h_2$ belonging to the A_1 representation of C_{2v} (*76*) we can construct three a_1 MOs. The $1a_1$ MO is fully bonding (in-phase overlap) between all atom pairs (Fig. 3-16). The $2a_1$ MO is approximately nonbonding (lone pair), and is almost localized on the s_a and p_{za} AOs. If this MO is occupied, the variational principle will lead to a small bonding character for the p_{za}, $h_1 + h_2$ pair, since it is most effective to stabilize the MO. The $3a_1$ MO is strongly antibonding between A and H atoms. Similarly, from the two group orbitals p_{ya} (in the molecular plane) and $h_1 - h_2$ belonging to the B_2 representation, the bonding ($1b_2$) and antibonding ($2b_2$) MOs are constructed (Fig. 3-16). The nonbonding $1b_1$ MO is formed by p_{xa} AO alone (perpendicular to molecular plane).

The energy order of MOs is closely related to the number of nodes and the nature of AO interactions. Thus, Deb (*76*) summarized the order: bonding < feeble bonding (e.g., lone pair) < nonbonding < feeble antibonding < antibonding; and the suborders: more symmetric < less symmetric, π-antibonding < σ-antibonding. Using these orders, the AH_2 valence MOs may be arranged as $1a_1 < 1b_2 < 2a_1 < 1b_1 < 3a_1 < 2b_2$.

Consider now the transverse MO forces acting on terminal nuclei. Since each MO (assumed real) is given by, $\phi_i = \Sigma_p C_{ip} \chi_p$, where χ_p is an AO, the MO density becomes $\phi_i^2 = \Sigma_p C_{ip}^2 \chi_p^2 + \Sigma_{p \neq q} C_{ip} C_{iq} \chi_p \chi_q$, the first term giving rise to

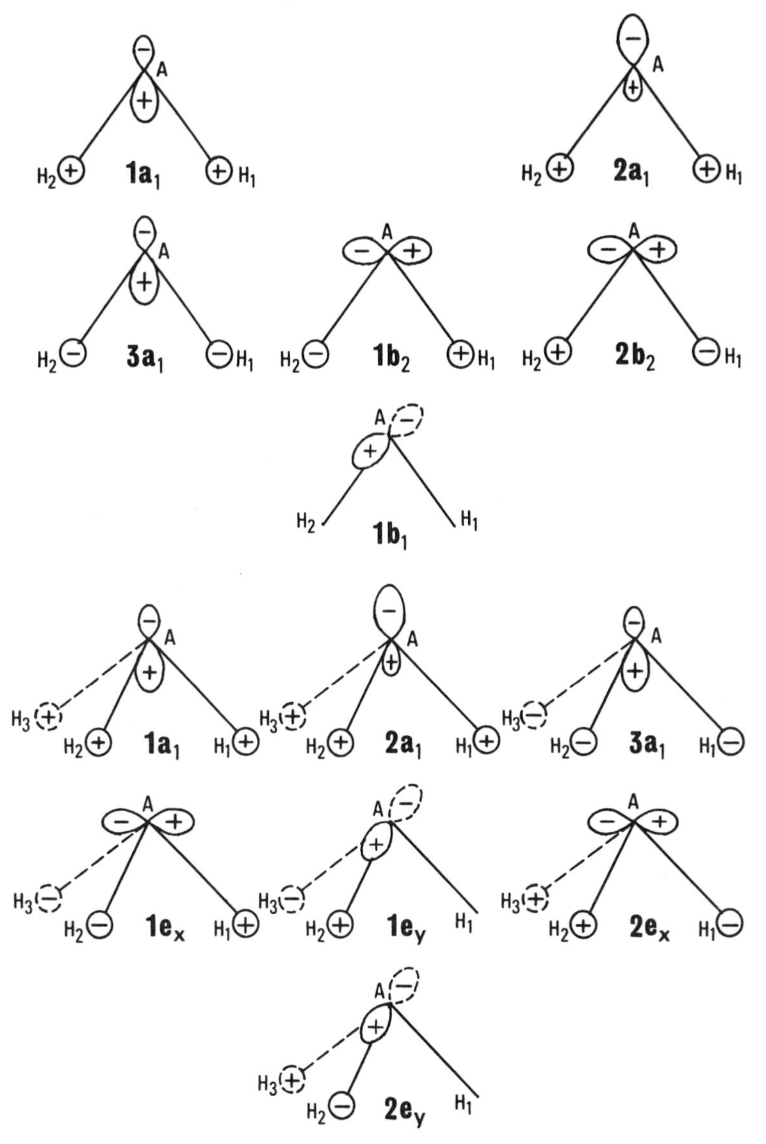

Fig. 3-16. Schematic MOs for AH_2 (top) and AH_3 (bottom) molecules. (Reproduced from Ref. 76, courtesy the American Chemical Society.)

atomic force and the second to overlap force. In case of the $1a_1$ MO (Fig. 3-16), both atomic and overlap densities are larger inside the triangle than outside and therefore this MO density exerts a positive transverse force on the terminal protons (see calculated fcd in Fig. 3-17). For the $2a_1$ MO, the atomic density on A

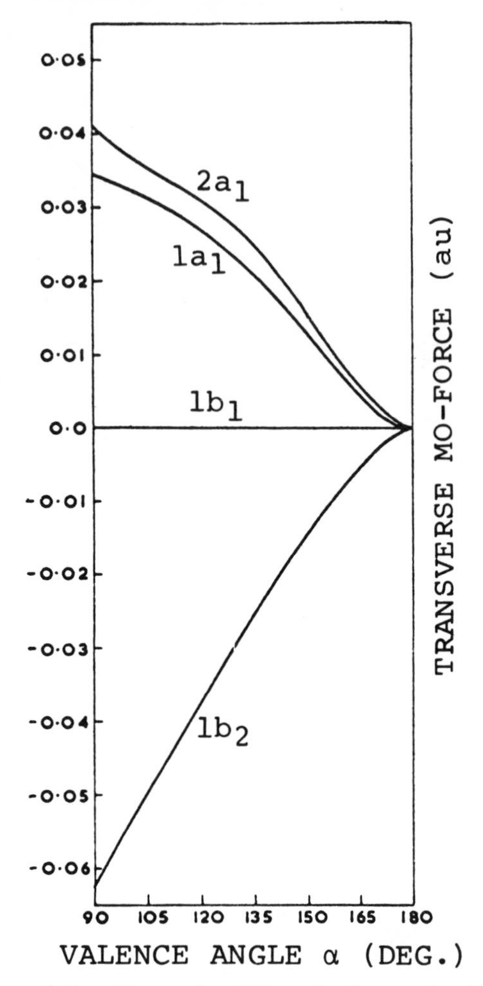

Fig. 3-17. Force correlation diagram for AH_2 molecules, constructed with the data for H_2O molecule. (Reproduced from Ref. 76, courtesy the American Chemical Society.)

pulls up the nucleus A (AD force in ESF theory, Section 3-3-2), but here we do not consider this effect because we are considering the transverse forces on terminal protons. This atomic density on A is larger outside the triangle than inside, and causes a negative transverse force on the protons. But the other atomic and overlap densities lead to positive forces and the net transverse force due to the $2a_1$ MO will be positive (Fig. 3-17). For the $1b_2$ MO, both atomic and overlap densities concentrate more charge outside, and therefore it gives a negative transverse force. The $1b_1$ MO does not exert any transverse force on the protons because of symmetry. These conclusions are summarized in the shape diagram, Fig. 3-18. The electrons are filled in the circles, according to the given energy order, and then the HOMO postulate is applied.

For ground-state molecules having 1, 2 and 5, 6 valence electrons the HOMOs are $1a_1$ and $2a_1$. Therefore, such molecules will be bent. For molecules with 3, 4 valence electrons the HOMO is $1b_2$, and so the shape will be linear. For

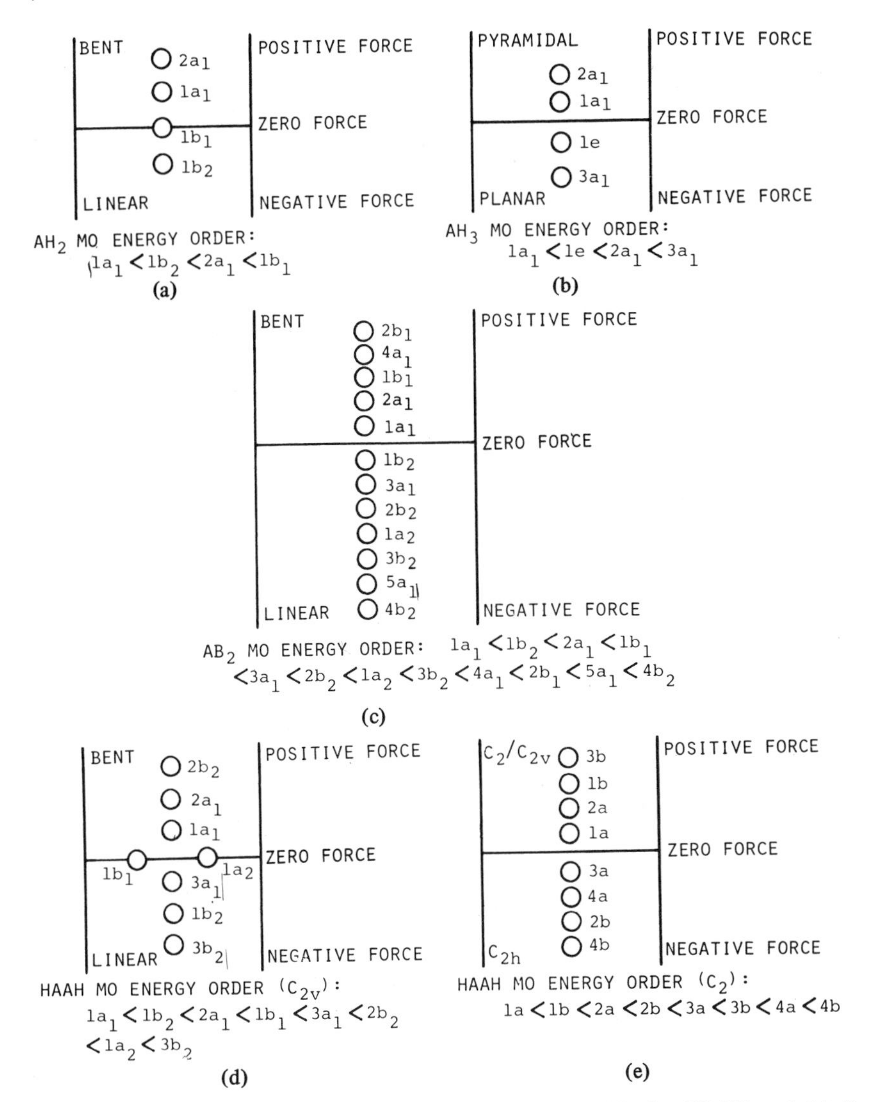

Fig. 3-18. Shape diagrams for (a) AH_2, (b) AH_3, (c) AB_2, (d) C_{2v} HAAH, and (e) C_2 HAAH molecules. (Reproduced from Refs. 76, 82, courtesy the American Chemical Society.)

Table 3-8. Geometry Predictions for Four Molecular Classes.

Molecule class	No. of valence electrons	Predicted ground-state geometry	Examples	Predicted excited-state geometry	Examples	Apparent exceptions
AH_2	1, 2	bent	H_3^+, LiH_2^+, HeH_2^{2+}	linear (HOMO $1b_2$)		
	3, 4	linear	BeH_2, HeH_2^+, BH_2^+	bent (HOMO $2a_1$)	BeH_2	BeH_2^+
	5–8	bent	CH_2, NH_2, OH_2, BH_2^-, BH_2, AlH_2, NH_2^+, NH_2^-	bent (HOMO $1b_1$) or linear (HOMO $1b_1$)	CH_2, NH_2 (both bent), BH_2 (linear)	
AH_3	1, 2	pyramidal		planar (HOMO $1e$)		
	3–6	planar	LiH_3^+, HeH_3^+, GaH_3, BeH_3, CH_3^+, BeH_3^-, BH_3	pyramidal (HOMO $2a_1$)		
	7, 8	pyramidal	CH_3 (?), NH_3, OH_3^+, CH_3^-, SbH_3, NH_3^+ (?), SiH_3	planar (HOMO $4a_1$)	NH_3	CH_3, NH_3^+
	9, 10	planar				
AB_2	1, 2	bent	Li_3^+	linear (HOMO $1b_2$)		Li_2H^+
	3, 4	linear	He_2H^+, $BeLi_2$, Li_2H^-, Be_2H^+	bent (HOMO $2a_1$)		
	5–8	bent		linear (HOMO $3a_1$)		He_3^+, Li_2O

	9–16	linear	HF_2^-, CO_2, CO_2^+, CS_2, N_3^-, BeF_2, $HgCl_2$, $CuCl_2^-$, $Ag(NH_3)_2^+$, UO_2^{2+}, UO_2^+, MoO_2^{2+}, PuO_2^{2+}, C_3, NC_2, CN_2, $C(CH_2)_2$, $C(CO)_2$, BO_2, NO_2^+	bent (HOMO $4a_1$ or $2b_1$)	NO_2^+, CO_2, CS_2	BaF_2, $BaCl_2$, $BaBr_2$, BaI_2, SrF_2, $SrCl_2$, CaF_2
	17–20	bent	NO_2^-, NO_2, ClO_2^-, ClO_2, SO_2, SCl_2, O_3, OCl_2, $SnCl_2$, PbI_2, $(CH_2)_3$, $InCl_2$, $Se(SCN)_2$, S_3^{2-}, $O(CH_3)_2$, CF_2, NF_2, ICl_2^+	linear (HOMO $5a_1$ or $4b_2$)		
	21–24	linear	XeF_2, I_3^-, ICl_2^-, F_3^{2-}, Cl_3			
HAAH	1, 2	bent (C_{2v}/C_2)				
	3, 4	linear				
	5, 6	bent (C_{2v}/C_2)				
	7, 8	bent (C_{2h})				
	9, 10	linear	C_2H_2, $N_2H_2^{2+}$	bent (HOMO $2b_2$)		
	11, 12	bent (C_{2v}/C_{2h})	N_2H_2, $(CH)_2H_2$			
	13, 14	bent (C_2)	H_2O_2	linear (HOMO $3b_2$)		
	15, 16	linear				

SOURCE: Refs. 76, 82. Reprinted courtesy the American Chemical Society.

molecules having 7, 8 valence electrons the HOMO ($1b_1$) is insensitive to valence angle, so the next lower MO ($2a_1$) determines the shape (bent). These predictions (Table 3-8) agree with known results except for BeH_2^+.[38] Apart from individual predictions, the model can also predict many trends. For example, since CH_2 ($^1A_1$, 104°), has one more electron in the $2a_1$ MO than BH_2 (131°), it should have a smaller bond angle. In the two series NH_2^+, NH_2 (103.3°), NH_2^- (104°), and BH_2^- (102°), CH_2 ($^1A_1$, 104°), NH_2 (103.3°), H_2O (105.2°), the apex angle should be more or less the same, since the molecules differ only in the occupancy of the indifferent $1b_1$ MO. The decrease in apex angle in the series H_2O (105.2°), H_2S (92.2°), H_2Se (91°), H_2Te (89.5°) may be explained as follows. From O to Te the valence *s* and *p* AOs become more diffuse (i.e., greater mean radii) and the tail of the lone-pair ($2a_1$ MO in Fig. 3-16) *sp* hybrid on A can overlap more effectively with the hydrogen AOs. Therefore, bending is facilitated more in, e.g., H_2S than in H_2O. For the excited states, the HOMO is the MO in which the excited electron enters. Thus, the first excited state of LiH_2^+ will be linear (HOMO $1b_2$). The bent BH_2 molecule becomes linear in the excited state (HOMO $1b_1$, next lower MO $1b_2$). The singlet excited state and triplet ground state of CH_2 will have larger and similar bond angles, since the transitions are from $2a_1$ to $1b_1$ MOs ($^1B_1$, 140°; $^3B_1$, ~136°; Refs. 134, 135).

When electrons fill a bonding or an antibonding MO, electron density will respectively increase or decrease in the binding region (see Section 3-3-4d); therefore the bond length will, respectively, decrease or increase. For the $2a_1$ and $1b_1$ MOs, the atomic density of A attracts the two protons and decreases the bond lengths, e.g., in the series BH_2 (1.18 Å), CH_2 (1.11 Å), NH_2 (1.02 Å), H_2O (0.958 Å). In the series H_2O (0.958 Å), H_2S (1.334 Å), H_2Se (1.47 Å), H_2Te (1.7 Å), the bond length increases because the valence *s* and *p* AOs become more diffuse from O to Te and the electron density in the binding region decreases from H_2O to H_2Te.

2. AH_3 *Molecules.* The approach here is very similar to that for AH_2 molecules. From seven symmetry-adapted (*76*) valence AO functions we obtain three a_1 and two sets of e MOs for a C_{3v} molecule. The $1a_1$ MO is fully bonding, the $2a_1$ MO is a lone-pair orbital almost localized on A, and the $3a_1$ MO is antibonding between the A and H atoms. The $1e_x$ (in-phase mixing of p_{xa} and $2h_1 - h_2 - h_3$) and $1e_y$ (in-phase mixing of p_{ya} and $h_2 - h_3$) are degenerate bonding MOs, while $2e_x$ and $2e_y$ are degenerate antibonding MOs (Fig. 3-16).[39] The energy order is given by the previous rules as $1a_1 < 1e < 2a_1 < 3a_1 < 2e$.

Figure 3-16 indicates that the $1a_1$ MO concentrates more charge inside the molecular pyramid and thus causes a positive transverse force on the terminal protons (see fcd in Fig. 3-19). The $2a_1$ lone-pair MO also exerts a bending force on the protons. The $1e_x$ MO concentrates more charge outside the pyramid and thus favors a planar structure, while the $1e_y$ charge density tends to open the

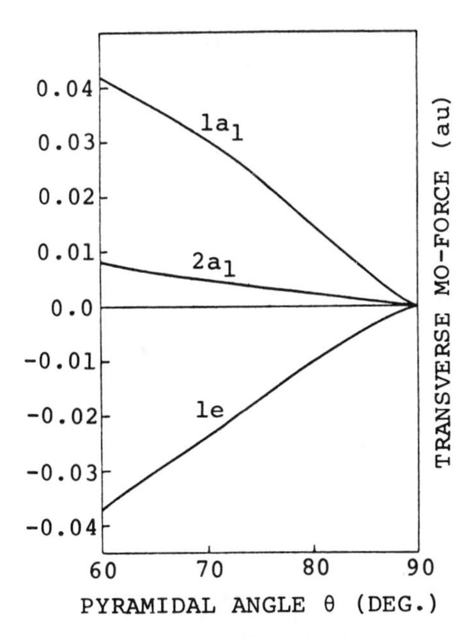

Fig. 3-19. Force correlation diagram for AH_3 molecules, constructed with the data for NH_3 molecule. (Reproduced from Ref. 76, courtesy the American Chemical Society.)

H_2AH_3 angle, favoring a T shape (planar). The overlap density in the $3a_1$ MO causes a negative transverse force (Fig. 3-18).

From the HOMO postulate, the ground states of AH_3 molecules with 1, 2, 7, or 8 valence electrons will be pyramidal, but those with 3-6, 9, or 10 valence electrons will be planar (Table 3-8). As in the Walsh rules, the model predicts 7-valence-electron CH_3, NH_3^+, BH_3^- (*285*, *286*) to be pyramidal, in disagreement with experimental and theoretical studies (Section 3-2-1). The planar Rydberg excited states of NH_3 (*134*) and PH_3 (*276*) will also constitute exceptions to the model (see Section 3-3-5).

The out-of-plane bending force constant increases in the planar series LiH_3^+, BeH_3, BH_3, because the $1e$ HOMOs are progressively filled. Because bonding and lone pair MOs are progressively filled, bond length should decrease in the series LiH_3^+, BeH_3, BH_3 (1.22 Å; Ref. 270), CH_3 (1.08 Å), NH_3 (1.02 Å). For the series NH_3 (1.02 Å, 106.6°), PH_3 (1.42 Å, 93.5°), AsH_3 (1.52 Å, 91.8°), the changes in bond lengths and angles have been attributed to the increasing diffuseness of the valence s and p AOs of the central atom.

For planar LiH_3^+ and BeH_3, the HOMO ($1e$) level is doubly degenerate, leading to an FOJT distortion. The $1e_x$ MO tends to shorten the A—H_1 bond relative to the other two (the hydrogen group orbital is $2h_1 - h_2 - h_3$), while the

$1e_y$ MO prefers a T shape and shortens the A—H_2 and A—H_3 bonds relative to the third. The resultant shapes are distinguishable, having different stabilities, but both belong to C_{2v} symmetry.

3. AB_2 *Molecules.* Using the same qualitative rules as for AH_2 and AH_3 molecules, one can pictorially construct the twelve valence MOs of C_{2v} AB_2 molecules (Fig. 3-20) and arrange them in the energy order $1a_1 < 1b_2 < 2a_1 < 1b_1 < 3a_1 < 2b_2 < 1a_2 < 3b_2 < 4a_1 < 2b_1 < 5a_1 < 4b_2$ (for details, see Ref. 76). For some AB_2 molecules, the $1a_2$, $3b_2$ and $5a_1$, $4b_2$ MOs may be reversed in energy, but this will not affect the predictions, since all these MOs prefer a linear form (see below).

The schematic MOs in Fig. 3-20 and the H-F force picture in Fig. 3-15 indicate the nature (positive or negative) of the transverse force exerted by each

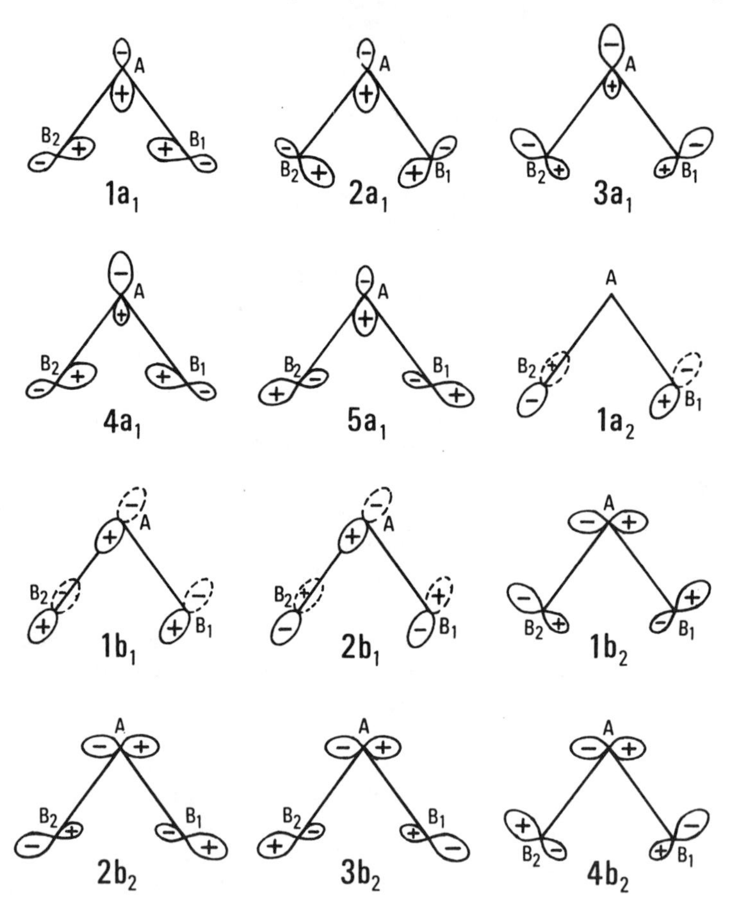

Fig. 3-20. Schematic MOs for AB_2 molecules. (Reproduced from Ref. 76, courtesy the American Chemical Society.)

MO on the terminal nuclei. The effects of the π MOs ($1a_2$, $1b_1$, $2b_1$) can be deduced from the signs of the π overlap densities between the terminal atoms. For the $5a_1$ MO, Deb explains that one atomic and one overlap term give strong negative forces that would surpass the two atomic and two overlap terms, which give smaller positive forces. Combining the resulting shape diagram with the MO energy order, the HOMO postulate leads to the following predictions for the shapes of ground-state AB_2 molecules: $1a_1$ $(1,2;b)$, $1b_2$ $(3,4;l)$, $2a_1$ $(5,6;b)$, $1b_1$ $(7,8;b)$, $3a_1$ $(9,10;l)$, $2b_2$ $(11,12;l)$, $1a_2$ $(13,14;l)$, $3b_2$ $(15,16;l)$, $4a_1$ $(17,18;b)$, $2b_1$ $(19,20;b)$, $5a_1$ $(21,22;l)$, $4b_2$ $(23,24;l)$, where the MO is the shape-determining HOMO, the numbers in parentheses refer to valence electrons, and b and l denote bent and linear forms, respectively (Table 3-8). Similar predictions can be made for excited states. These agree with known shapes. However, the 5-electron, linear He_3^+ (*245*) molecule seems to be an exception. Also, contrary to the predictions of Walsh, Deb, and Gillespie, the 16-valence-electron molecules BaF_2, $BaCl_2$, $BaBr_2$, BaI_2, SrF_2, $SrCl_2$, and CaF_2 are bent (Table 3-6). The 21-electron molecule ClF_2 might be another exception (*81*, *185*, *292*). Like Walsh's model, Deb's model does not accommodate predominantly ionic molecules such as Li_2H^+, Li_2O, and LiOH which are linear.[40]

The anchor shape of ClF_3 is explained as follows. The ClF_2 fragment in ClF_3 may be regarded as a 22-electron molecule and thus linear, giving rise to a T-shaped ClF_3. In this shape, the two ClF_2 fluorine nuclei "see" more electrons because of the remaining Cl—F bond; the resulting electron-nuclear attraction slightly bends the ClF_2 bonds (for the ESF explanation, see Section 3-3-4b).

Consider geometry changes due to ligand substitutions. The change from OF_2 (1.42 Å, 103.3°) to OCl_2 (1.70 Å, 111°) may be attributed to greater diffuseness of Cl valence AOs. Due to this the O—Cl bond will be longer than the O—F bond. Also, since the overlap between the terminal AOs is reduced for the HOMO ($2b_1$), the bond angle should be larger in OCl_2 (see the ESF account in Section 3-3-4d).

4. HAAH *Molecules.* The prediction of geometry of an HAAH molecule proceeds in two steps: (a) to decide on the linearity of the molecule by examining the in-plane transverse force on the protons in a planar *cis* (or *trans*) form, (b) to find the stable conformation for rotation about the A—A bond, by examining the transverse rotational force on the protons in a nonplanar (C_2) configuration. The present model cannot distinguish between C_2 and C_{2v} (planar *cis*) forms in geometry prediction.

Consider step a, starting from the *cis* form.[41] As before, the ten valence MOs of a C_{2v} HAAH molecule can be qualitatively constructed and arranged in energy order $1a_1 < 1b_2 < 2a_1 < 1b_1 < 3a_1 < 2b_2 < 1a_2 < 3b_2 < 4a_1 < 4b_2$ (*82*). In Fig. 3-21, the MOs which throw more electron density inside the molecular trapezeum favor a bent form, whereas those MOs which concentrate more charge outside will favor a linear form. Thus, the $1a_1$, $2a_1$, and $2b_2$ MOs

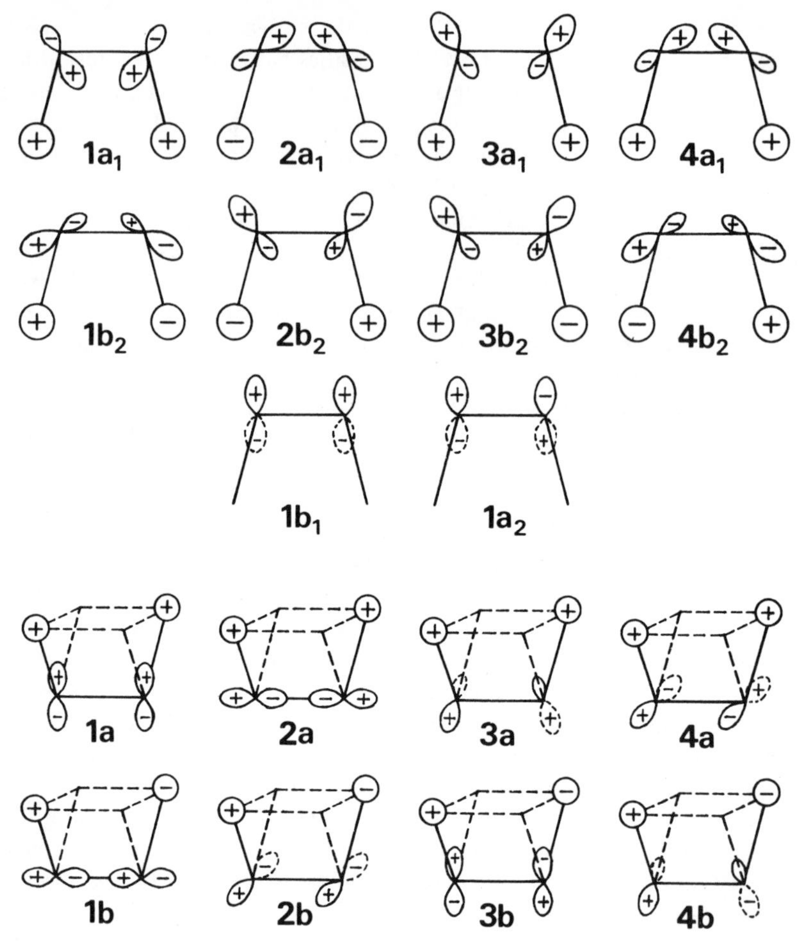

Fig. 3-21. Schematic MOs for C_{2v} HAAH (top) and C_2 HAAH (bottom) molecules. (Reproduced from Ref. 82, courtesy the American Chemical Society.)

favor a bent form, the $3a_1$, $1b_2$, and $3b_2$ MOs favor a linear form, and the $1b_1$ (π bonding) and $1a_2$ (π antibonding) MOs are almost insensitive to bending in the molecular plane. Using the shape diagram (Fig. 3-18) and the HOMO postulate, the predictions in Table 3-8 can be readily obtained. For the bent 11-14-valence-electron molecules the shape-determining MO ($2b_2$) is feebly A—A antibonding and feebly A—H bonding. This leads to an increase in the A—A bond length and a decrease in the A—H bond length in the series C_2H_2 (1.21, 1.06 Å), N_2H_2 (1.22, 1.08 Å), H_2O_2 (1.48, 0.97 Å) (*243, 283, 284*). The differences in the bond lengths and angles of H_2O_2 (0.97 Å, 105°), H_2S_2 (1.33 Å, 95°) may be attributed to greater diffuseness of sulfur valence AOs.

Consider now step b. The valence MOs of a C_2 HAAH molecule may be qualitatively constructed (*82*) and arranged in the energy order $1a < 1b < 2a < 2b < 3a < 3b < 4a < 4b$. From Fig. 3-21 we see that when an MO accumulates more electron density inside the molecular prism, it favors a C_2 or C_{2v} (*cis*) form, and when more electron density is accumulated outside the prism, a C_{2h} (*trans*) form is favored, if the MO maintains its role throughout the rotational fcd.[42] Thus, $1a$, $2a$, $1b$, and $3b$ MOs favor a C_2/C_{2v} form, while the $3a$, $4a$, $2b$, and $4b$ MOs favor a C_{2h} form. Therefore, one can readily predict rotational conformations of bent HAAH molecules (Table 3-8) using the shape diagram (Fig. 3-18) and the HOMO postulate. Since the $3b$ and $4a$ MOs cross each other at a dihedral angle of $\sim 90°$, both *cis* (or C_2) and *trans* forms are possible for 11, 12-electron molecules, whereas for 13, 14-electron molecules (e.g., H_2O_2, H_2S_2) the $3b$ MO seems to dominate over the $4a$ MO.

3-4-4c. Transferability of Geometry Predictions. The geometries of many large molecules can be predicted from the geometries of their fragments. For example, the AB_2 predictions may be employed for AB_3, AB_4, and AB_5 molecules (*76*, *78*). An example, ClF_3, has already been discussed in Section 3-4-4b. We consider here the shapes of AH_4 molecules. Since the two AH_2 planes are mutually perpendicular, each AH_2 fragment has little effect on the shape of the other. Therefore, if the $(n - 2)$-electron AH_2 fragment of a parent n-electron AH_4 is linear, the AH_4 molecule will be square planar; if the AH_2 fragment is bent, the AH_4 molecule will be tetrahedral (T_d) or D_{2d} (distorted tetrahedral). In cases of orbital degeneracy, one should also consider the FOJT effect. Using the AH_2 predictions in Table 3-8, we conclude that AH_4 molecules with 5, 6 valence electrons will be square planar and those with 3, 4, or 7–10 electrons will be T_d or D_{2d}. These predictions agree with known geometries and with predictions which are independently derived for AH_4 molecules *without* the help of AH_2 geometries. Deb (*76*, *78*, *79*) has thus noted that *shape is another property which is transferable, in quite a subtle way, from one molecule to another.*

3-4-4d. Why HOMO? The qualitative geometry predictions for many molecular classes made on the basis of the HOMO postulate are remarkably consistent with known geometries, although we occasionally come across minor exceptions as well as some difficulties in constructing the schematic MOs and shape diagrams for relatively large molecules and in the assignment of HOMO. Thus, the general validity of the HOMO postulate has been established.[43] The question therefore arises, why is only the HOMO, rather than other MOs, most important for molecular geometry? The reasoning might involve two steps. (1) In the model, the MOs are constructed at a configuration which is apart from symmetric and/or equilibrium geometry. At this geometry, the total electron density *should* have the characteristics of incomplete following or preceding (Sections 3-5-2, 3-3-8a). If the MO contributes to electron-cloud preceding, it gives a positive (accelerating) transverse force (left-hand side of Fig. 3-8b), and if it contributes to electron-cloud following, it gives a negative (resisting) transverse force (right-hand side of Fig. 3-8b). However, we note that electron-cloud incomplete following and preceding are the behaviors of *total* electron density and not of the individual MO density. (2) Among the MOs, the HOMO is most loosely bound. Hence, the density due to the lower MOs tends to follow *more completely* the nuclear motion (Ref. 78; Section 3-5-4) and thus have less influence on molecular shape. While step 1 has a firm foundation (Section 3-5), step 2 does not have such a foundation at present and needs to be further examined using more reliable MOs.[44]

3-5. TWO COMMON TYPES OF BEHAVIOR OF AN ELECTRON CLOUD DURING THE NUCLEAR REARRANGEMENT PROCESSES

3-5-1. Introduction

Imagine a process in which the nuclear coordinates of a system change from one configuration to another. If the process is favorable (or unfavorable), the system receives an internal force which accelerates (or resists) the process. An important factor giving rise to such forces should be the dynamic behavior of electron cloud during the process. In this section we will study the characteristic features of such behavior in a fairly general way within the Born–Oppenheimer approximation (see also Section 3-3-8).

Linnett and Wheatley (*179*) considered the possibility that the bond-forming orbitals are distorted (i.e., bent bond) during molecular vibrations. Such distortions were confirmed by valence-bond calculations on the out-of-plane bending of CH_3 (*51*, *150*), and for more general hydrides (*158*). The important effects of such relaxations of molecular charge distributions (see Chapter 5) were pointed out for infrared intensities (*57*, *70*, *157*, *189*), temperature dependence of ESR hyperfine splitting constants (*98*, *267*, *268*), vibrational force constants (*7–11*, *15*, *17*, *259*), chemical reactions and long-range interactions (see, e.g., Refs. 216, 219–221). Interestingly, Feynman had conjectured in 1939 (*101*) that the van der Waals force might be due to the simultaneous polarization of the individual atomic densities toward each other, their magnitudes being proportional to $1/R^7$ This was confirmed much later (see Sections 2-3-1, 7-2; *105*, *140*).

In this section we will first deduce two common features of electron-cloud reorganizations in an ad hoc manner from the H–F theorem and the integral H–F theorem (*166*, *227*).[45] These lead to a guiding principle governing the nuclear rearrangement process, which will be illustrated in the case of molecular shapes.[46] Finally, we will discuss the changes in the geometries of reactants during chemical reactions and their role in driving the reaction (*216*).

3-5-2. Common Features in the Behavior of Electron Clouds

Let Q be the nuclear displacement coordinate of a system from initial (i) to final (f) states (Fig. 3-22), with α as an intermediate configuration and β as the final configuration. In general, Q is a linear combination of displacement coordinates (*300*), with that of nucleus A as an important element. Let us write the electrostatic H-F theorem (for α) and the integral H-F theorem (Chapters 1,4) as

$$F_{\mathrm{A}}^{\alpha} = Z_{\mathrm{A}} \int r_{\mathrm{A}}/r_{\mathrm{A}}^{3}\, \rho_{\alpha}(r)\, dr - Z_{\mathrm{A}} \sum_{\mathrm{B} \neq \mathrm{A}} Z_{\mathrm{B}} R_{\mathrm{AB}}/R_{\mathrm{AB}}^{3}, \tag{3-31}$$

$$\Delta E(\alpha \to \beta) = \int \rho_{\alpha\beta}(r)\, \Delta H_{ne}(r)\, dr + \Delta V_{nn}, \tag{3-32}$$

$$\Delta H_{ne}(r) = H_{ne}^{\beta}(r) - H_{ne}^{\alpha}(r), \tag{3-33}$$

where $\Delta E(\alpha \to \beta)$ is the change in total energy for the isoelectronic process $\alpha \to \beta$; $\rho_{\alpha\beta}$ is the normalized transition density (Section 4-3) between α and β; ΔH_{ne} is the difference in nucleus-electron attraction operator between α and β, and ΔV_{nn} is the corresponding change in nuclear repulsion energy. The integrand in (3-31) is a 3-dimensional vector force density.

First, let the change $\alpha \to \beta$ satisfy the following two conditions. (a) It is a monotonous change from unstable (α) to more stable (β) configuration. (b) The displacement accompanies an increase in nuclear repulsion. Examples of these conditions are seen in (i) molecular structure and vibration, e.g., displacement from unstable linear (planar) to stable bent (pyramidal) forms of AX_n molecules; (ii) reactions forming stable molecules having no barrier between two points α and β; and (iii) approaches of two neutral atoms in the regions of dispersion forces. From condition a, the nuclei A of the system at α must receive the driving force F_{A}^{α} having components along Q (positive direction). From

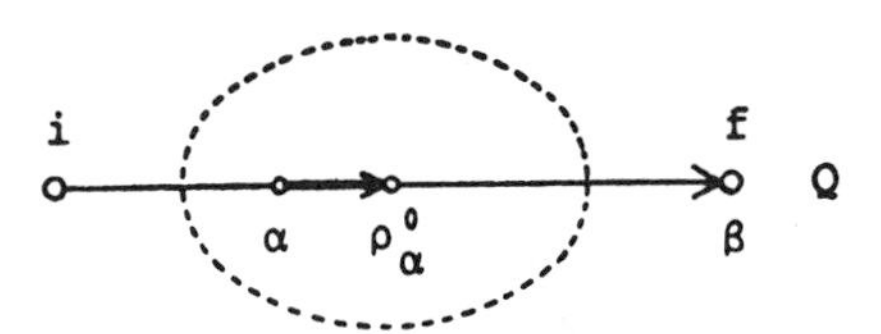

Fig. 3-22. Illustration of electron-cloud preceding in the H-F picture. The dashed ellipse denotes the electron cloud of the system in nuclear configuration α. ρ_{α}^{0} shows the projection of the centroid of weighted density on the coordinate Q and the arrow from α to p_{α}^{0} denotes the electronic part of the driving force acting at the configuration α. (Reproduced from Ref. 210, courtesy the American Chemical Society.)

condition b, the origin of the driving force must be the electronic part of (3-31), since the nuclear part is repulsive throughout. This is possible *if and only if* the center of gravity (ρ_α^0) of the weighted density $(1/r_A^3)\,\rho_\alpha(r)$ precedes the position(s) of A at α in the direction of Q. In Fig. 3-22 the arrow from α to ρ_α^0 is proportional to the electronic part of the driving force on A at α. The nuclear force is opposite to and smaller than this force (for a real system, see Ref. 219).

The preceding of the weighted density is generally expected to occur when the centroid of $\rho_\alpha(r)$ within some local region near A precedes the position of A in the Q direction (see dotted line in Fig. 3-22). The localness of such *electron-cloud preceding* comes from the $1/r_A^3$ factor in the weighted density; this is large near A but falls off rapidly with increasing distance. Note that although the preceding of weighted density is an exact consequence of a nucleus A receiving the electronic driving force in the Q direction, electron-cloud preceding is an approximate concept. However, the approximation is usually quite good, except possibly for very minute interactions, e.g., the long-range resonance force between the $1s$ and $2p$ states of two H atoms (*215*).

Similar conclusions are derived from (3-32). For conditions a and b, the destabilization energy due to ΔV_{nn} must be surpassed by the stabilization energy due to the electronic part. In (3-32) and (3-33), $H_{ne}^\beta(r)$ contributes to stabilization but $H_{ne}^\alpha(r)$ contributes to destabilization. Therefore, for the system to gain stabilization from the electronic part the transition density $\rho_{\alpha\beta}(r)$ should be denser (more positive) near the moving nuclei at β than at α (Fig. 3-23a). Such a situation will be realized if electron-cloud preceding occurs at α (Fig. 3-23b).

Now, let Q satisfy the following two conditions, which are opposite to the previous ones. (a) It is a monotonous change from stable (α) to unstable (β) configuration. (b) The displacement accompanies a decrease in nuclear repulsion. Such a displacement is realized by reversing the direction of Q in Fig. 3-22 (and Fig. 3-23). Therefore, during this process the opposite behavior, namely, incomplete following of the weighted density, $(1/r_A^3)\,\rho_\alpha(r)$, and *electron-cloud (incomplete) following* should occur.

To summarize, when electron-cloud preceding occurs during a process, the system receives a force that accelerates the process, but when electron-cloud following occurs the system receives a force that resists the process. This is the basis for the general guiding principle given below.

The primary importance of AD and EC forces in molecular shapes (Section 3-3) and chemical reactions (*16, 216, 219*) is closely related to the localness of electron-cloud preceding and following. The behavior of electron density in the atomic region of nucleus A manifests itself in the AD force on A and that in the A—B bond region manifests itself in the EC force on A. The electron density in the other regions (EGC force) should not greatly reflect the characteristics of electron-cloud preceding and following. Thus, these behaviors of electron density are reflected mainly in the AD and EC forces.[47] This is also supported

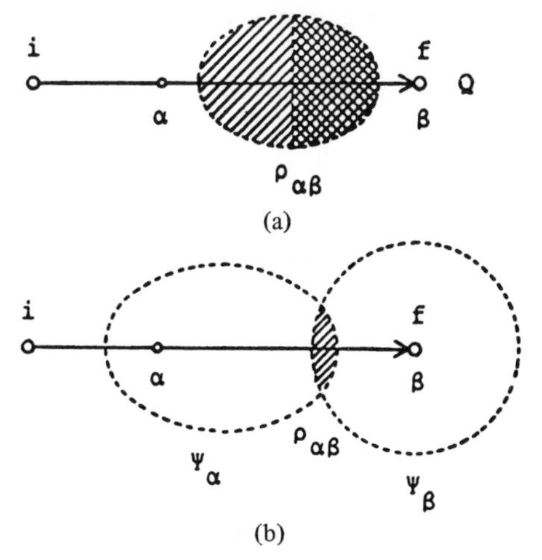

Fig. 3-23. Illustration of electron-cloud preceding in the integral H–F picture. (a) The ellipse illustrates the transition density $\rho_{\alpha\beta}(r_1)$ between the wavefunctions associated with configurations α and β. The density is higher in the cross-hatched region near β than in the singly hatched region near α. (b) The ellipse corresponds to the wavefunction associated with configuration α and is responsible for electron-cloud preceding. The circle centered on β corresonds to the wavefunction of the final state. Note that the hatched region, which shows the high-density region of $\rho_{\alpha\beta}$, is nearer to the position β than to α. (Reproduced from Ref. 210, courtesy the American Chemical Society.)

by the following relations for diatomic molecules, obtained by applying the interference partitioning of electron density *(255)* to the H–F force:

$$\left.\begin{aligned} &\text{AD force} = \text{intraatomic interference force} \\ &\text{EC force} = \text{interatomic interference force} \\ &\text{EGC force} = \text{quasi-classical force.} \end{aligned}\right\} \tag{3-34}$$

We now remark on conditions a and b. Regarding condition a, even if the process has an energy barrier like the activation energy, electron-cloud preceding in some critical regions (e.g., those between reaction sites) and due to some mechanisms (e.g., charge transfer, Ref. 108) is quite important to reduce the barrier. Condition b is obviously too strong, since electron-cloud preceding and following should occur even if this condition is relaxed as follows: (b′) the electronic part is more important than the nuclear part.

We conclude this section with the density guiding rule for nuclear rearrangement processes: *Nuclei are pulled in the direction of electron-cloud reorganization. If the energy change is monotonous throughout the process $i \to f$, electron-*

cloud preceding (following) means that configuration f is more (less) stable than configuration i.

3-5-3. Perturbational Description

The perturbational treatment of electron-cloud following and preceding indicates a connection with the SOJT theory. Taking a small nuclear displacement δQ as a perturbation, the electron density correct to first order is (*13, 14*) $\rho(Q) = \rho(Q_0) + \delta\rho$, with

$$\delta\rho = 2\delta Q \sum_{k \neq 0} \left[\int \rho_{0k}(Q_0) f(Q_0)\, dr / (E_k - E_0) \right] \rho_{0k}(Q_0), \qquad (3\text{-}35)$$

where $Q = Q_0 + \delta Q$, E_0 and E_k are ground- and excited-state energies at Q_0, ρ_{0k} is the transition density and $f(Q_0) = -(\partial H/\partial Q)_{Q_0}$ is the force operator; $\delta\rho$ represents the reorganization of electron density due to the perturbation.

It has been shown that if Q_0 is taken at the saddle point of the potential surface, $\rho(Q_0)$ and $\delta\rho$ represent, respectively, electron-cloud incomplete following and preceding. In order that preceding occur as a net effect, $\delta\rho$ should become appreciable and then the displacement along δQ will be further facilitated. If preceding does not occur, the electron cloud left behind at Q_0 (incomplete following) attracts the nuclei at Q in the $-Q$ direction and the system will revert to Q_0. Thus, the SOJT conditions (Section 3-2-3) are those under which electron-cloud preceding can occur appreciably.

When Q_0 is away from the saddle point (e.g., equilibrium geometry, geometry of highest symmetry, etc.) the unperturbed density $\rho(Q_0)$ itself would already have the characteristics of electron-cloud preceding and following. Thus, the SOJT model (employing both HOMO and LUMO) is based on $\delta\rho$ taking Q_0 at the saddle point, while Deb's model (employing only HOMO) considers basically electron-cloud following and preceding included in $\rho(Q_0)$ where Q_0 is away from the saddle point (see Section 3-4).

3-5-4. Electron-Cloud Reorganization and Molecular Shapes

3-5-4a. Examples and Roles. We shall now consider how electron-cloud following and preceding determine molecular shapes (see guiding principle in Section 3-5-2). Figure 3-8 reexpresses the ideas in Section 3-3-8 with, e.g., ethylene, from both H–F and integral H–F theorems (Figs. 3-22 and 3-23). In the H–F picture, the centroid of the C—H bond orbital precedes (π–π^* excited state) or incompletely follows (ground state) the rotational movement Q of the C—H axis. In the integral H–F picture, the overlap between the C—H bond orbitals at α and β increases (decreases) due to the preceding (following) of the C—H bond orbital at α. The resulting dominant EC force (arrow in Fig. 3-8) operates to make the π-π^* excited state of ethylene a bisected form and the ground state a planar form, in agreement with experiment. This and other examples in Section 3-3-8 validate the above guiding principle.

For the out-of-plane bending displacement of AH_3 molecules (Fig. 3-24), with Q as a linear combination of Q_A and Q_H, the guiding principle is restated as follows. If electron-cloud preceding occurs when the molecule is bent from planar shape, the constituent nuclei receive forces that accelerate this bending; but when incomplete following occurs bending is resisted and the molecule is ex-

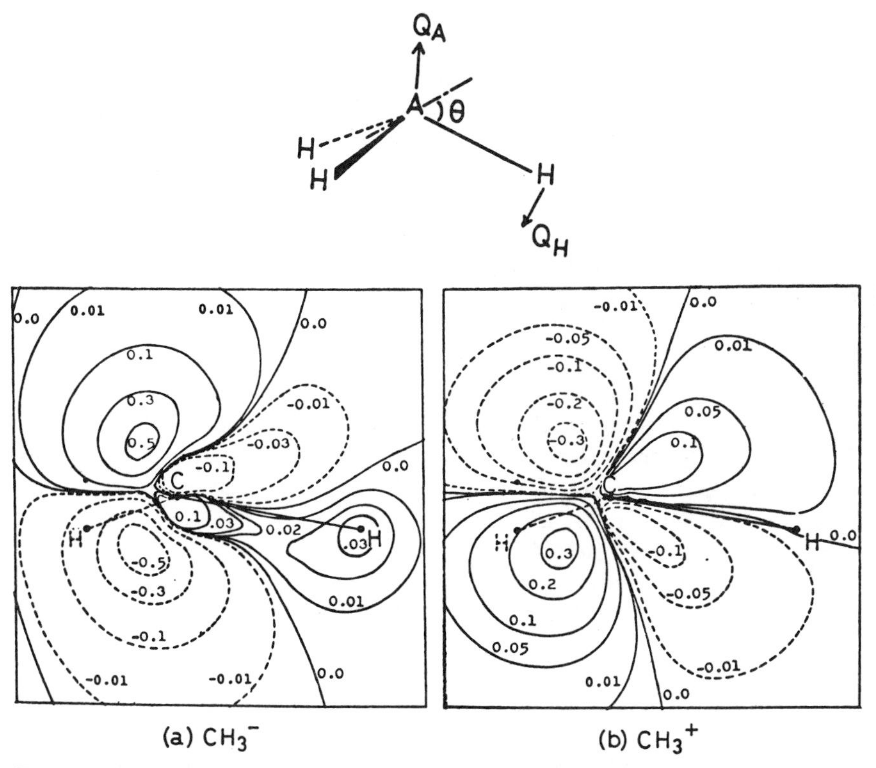

Fig. 3-24. Out-of-plane bending displacement Q (top) of AH_3 molecules from planar to pyramidal configurations. The arrows Q_A and Q_H represent, respectively, the bending movements of the central atom A and of the protons, while θ is the out-of-plane angle. The contour maps (bottom) show the changes in electron densities of CH_3^- and CH_3^+ induced by the displacement $\theta = 30°$, using nonempirical SCF–MO wavefunctions based on the minimal set of Slater orbitals. The density at $\theta = 0°$ is subtracted from that at $\theta = 30°$. The solid and dashed lines show, respectively, an increase and a decrease in density. (Reproduced from Ref. 211, courtesy the American Chemical Society.)

pected to be planar. For example, when CH_3^- is bent from planar shape, the electron density increases immediately above the C nucleus and decreases in the region immediately below it, thus producing an upward atomic dipole on the C atom. The AD force due to this electron-cloud preceding in the atomic region facilitates the displacement of the C nucleus along Q_C (Fig. 3-24a). Along the C—H axis, electron density increases in the downward region but decreases in the upward region. This electron-cloud preceding in the C—H bond region causes a positive transverse EC force on the proton that accelerates bending along Q_H. Thus, both AD and EC forces induced by electron-cloud preceding make the molecule nonplanar (Fig. 3-25) as expected from the guiding principle and in accordance with experiment (*47*). Note that Q_C does not satisfy condition b of Section 3-5-2 but satisfies b′, while Q_H satisfies both b and b′.

The changes in electron density in CH_3^+ are just the reverse of those in CH_3^- (Figs. 3-24b, 3-25). Thus, a *downward* atomic dipole is generated on the C atom (electron-cloud incomplete following). The incomplete following in the C—H bond region causes a negative transverse EC force on the proton, which resists bending along Q_H. Thus both AD and EC forces, caused by incomplete following, operate to make the molecule planar in accordance with the guiding principle above.

Consider now the electron-cloud reorganization in CH_3 (*211*), calculated through unrestricted Hartree–Fock wavefunctions (*214*, *244*). The situation here is more complex than above, making a direct application of the guiding principle difficult. The electron density near the C atom seems to show the characteristics of electron-cloud preceding, although to a much smaller extent than in CH_3^-. However, in the C—H bond region, the electron cloud follows Q_H incompletely, the extent of following increasing with increase in the out-of-plane bending angle (*51*, *150*, *211*, *267*, *268*). More accurate calculations (*221*) have shown that the EC forces on C and H make the molecule planar, exceeding the AD force on C.[48]

Interestingly, the above differences in electron-cloud reorganization among

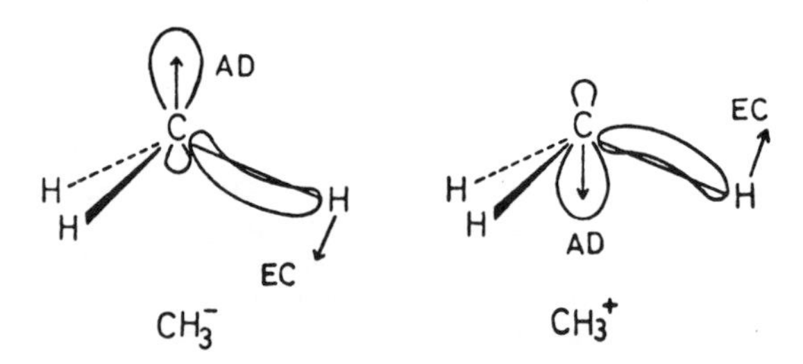

Fig. 3-25. Electron-cloud preceding in CH_3^- and incomplete following in CH_3^+ during out-of-plane bending movements. The arrows show the AD and EC forces induced by such preceding and following. (Reproduced from Ref. 211, courtesy the American Chemical Society.)

CH_3^-, CH_3^+, and CH_3 may be attributed chiefly to the singly occupied MO of CH_3. Also, electron-cloud reorganizations in isoelectronic molecules are similar (*211, 221*). These facts support the validity of the ESF model and are also important in the force reasoning behind the Walsh diagram (*66*).

3-5-4b. Bent Bond in Equilibrium Geometry. Contour maps of the electron density in NH_3 (equilibrium configuration) indicate that the bond electron cloud is slightly distorted inwardly from the N—H axis (*20, 21, 159, 223, 253, 254*). The occurrence of this "bent bond" can be explained as follows. Since the force on the protons vanishes at equilibrium geometry, the interproton repulsive force acting along $-Q_H$ (see Fig. 3-24) must exactly cancel the force of electronic origin. Considerations similar to those in Section 3-5-2 indicate that this is effectively achieved if the electron cloud near the proton and/or in the A—H bond region is distorted inwardly from the A—H axis. This is just a special ("frozen") case of electron-cloud preceding in the A—H bond region. See further Refs. 51a and 220.

3-5-4c. Localized Orbital Description. We have seen that the characteristics of electron-cloud following and preceding are expressed chiefly through the behavior of local electron clouds associated with bonds, lone pairs, etc. during nuclear rearrangement processes. One therefore expects that such characteristics would be reflected in localized MOs (*89*, *183*, *224*) obtained at non-equilibrium geometries.[49] For example, the general feature of the localized C—H bond orbital of nonplanar CH_3 may be illustrated as follows (*51*):

This clearly indicates incomplete following. In case of NH_3 (*159*), it is surprising that even in planar form the localized orbitals are *not* symmetrical with respect to the molecular plane, although the *total* electron density is symmetrical with respect to this plane. These examples show that the localized-orbital method would be a useful tool to describe intuitively electron-cloud following and preceding, and this would also hold for chemical reactions (*84, 85*). Note, however, that the force calculated from a single localized orbital has little physical meaning. Electron-cloud following and preceding are types of behavior of *total* electron density (see the definition of bond paths and bond directions in Ref. 257).

3-5-5. Change in Molecular Geometry during Chemical Reactions

The ESF concepts applied in Section 3-3 to molecular geometry can be applied to chemical reactions without much modification. For a reaction to occur, the interaction between reactants should increase electron density between the reaction sites, at the same time decreasing electron density in the region of the old breaking bond. This accumulation of electron density may occur in the overlap region common to both reaction sites, causing attractive EC forces, or in regions very close to reaction sites (e.g., the formation of *inward* atomic dipoles), causing attractive AD forces on the reaction sites. Although both of these types of electron-cloud preceding are important depending on the stages (initial, middle, final) and the classes (covalent, ionic) of reactions, the EC force seems to be more important in the significant middle stage of the reaction than the AD force, at least for homopolar reactions.

For instance, consider the dimerization of two planar CH_3 radicals to form ethane (*207*). When the two approach each other in the intermediate region (Fig. 3-26), the electrons in the $p_{\pi C}$ AO of each radical flow (precede) into the overlap region (see also Section 3-3-4d). The resultant EC force is the driving force of the reaction, since it pulls both carbons toward each other. At the same time, this EC force makes each CH_3 nonplanar as in ethane, the product, thus playing two important roles. In the final stage of the reaction, the EC(C—C) force nearly approaches the EC(C—H) force, resulting in an almost tetrahedral CH_3 fragment in ethane. Thus, in the ESF theory molecular geometry and chemical reaction are treated essentially on the same ground.

Consider also the formation of the coordinate bond in the NH_3—BH_3 system.

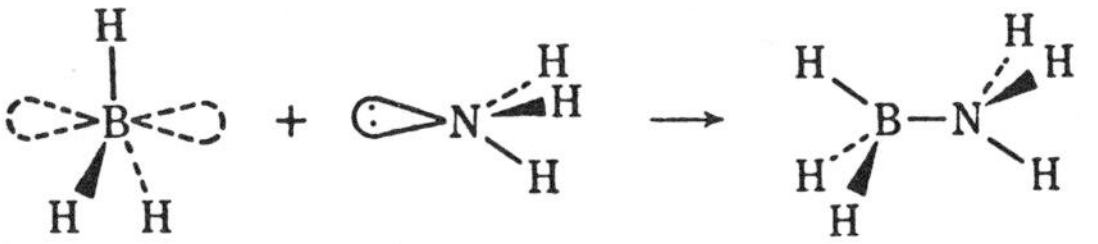

Here the p_π AO of planar BH_3 is vacant and the lone-pair orbital of NH_3 is doubly occupied. During reaction, the EC force is produced on B and the AD force on N in free NH_3 is partly transformed into the EC force. Consequently, during reaction BH_3 will gradually become pyramidal and the NH_3 valence angle will increase.[50] Deb (*76*, *78*) has also considered such changes in molecular geometry and reached similar conclusions on the basis of the model described in Section 3-4.

We now return to the dimerization of two CH_3 radicals for a more detailed

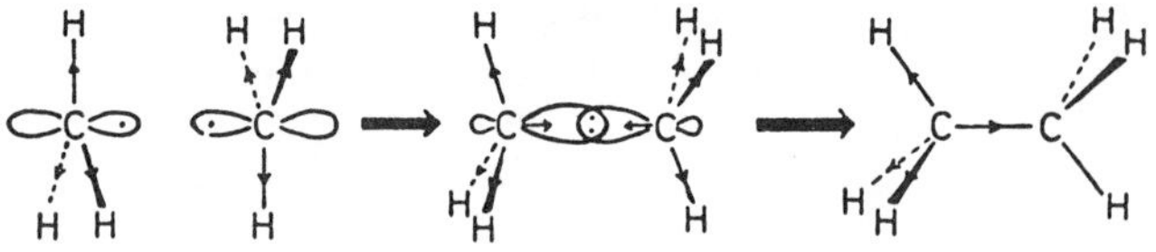

Fig. 3-26. Important forces in the reaction $CH_3 + CH_3 \rightarrow C_2H_6$. (Reproduced from Ref. 207, courtesy the American Chemical Society.)

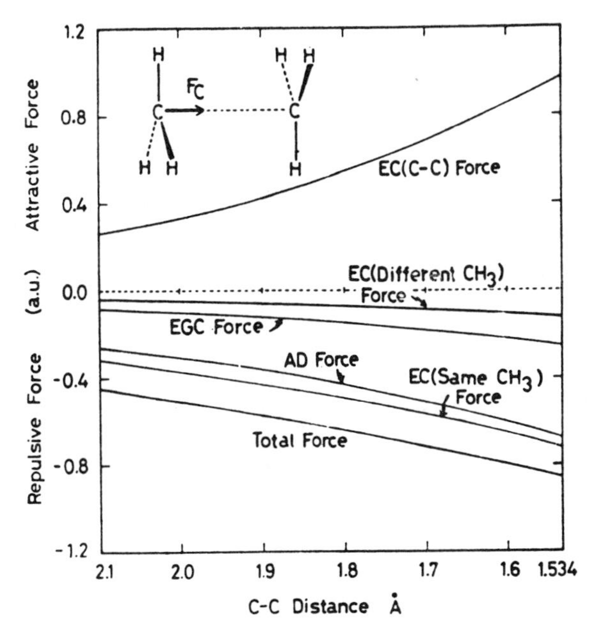

Fig. 3-27. Analysis of the force F_C acting on carbon when two *planar* methyl radicals approach each other. (Reproduced from Ref. 216, courtesy the American Chemical Society.)

study (*216*). We consider two reaction paths. (1) *planar approach*, in which the radicals are brought face to face with their configurations rigidly planar; (2) *gradually bending approach* in which the HCH angle of each radical is optimized along the path so that the bending transverse forces on protons vanish. The calculations employed extended Hückel wavefunctions (*141*) and optimized the orbital exponents in force integrals so that all forces vanish at the equilibrium geometry of ethane.

In the planar approach (Fig. 3-27), only the EC(C—C) force is attractive; all other forces, including the total force, are repulsive throughout.[51] This means that although electron density accumulates in the C—C region (see Fig. 3-28, near +0.95 Å, center of the forming C—C bond), the extent of accumulation is insufficient to drive the reaction. We also find an accumulation of electron density in the rear region along the C—H axis; this causes transverse EC(H—C) forces on protons which facilitate the *outward* bending of each radical. Clearly, this is also a kind of electron-cloud preceding, since the reaction coordinates should also include this out-of-plane bending displacement.

For the gradually bending approach (Figs. 3-29, 3-30), the AD, EC(C—C), and EGC forces are all attractive, and so the total force on carbon is attractive throughout the reaction. Comparing Figs. 3-27 and 3-30, we find that the largest effect of gradually bending each radical occurs in the AD force on car-

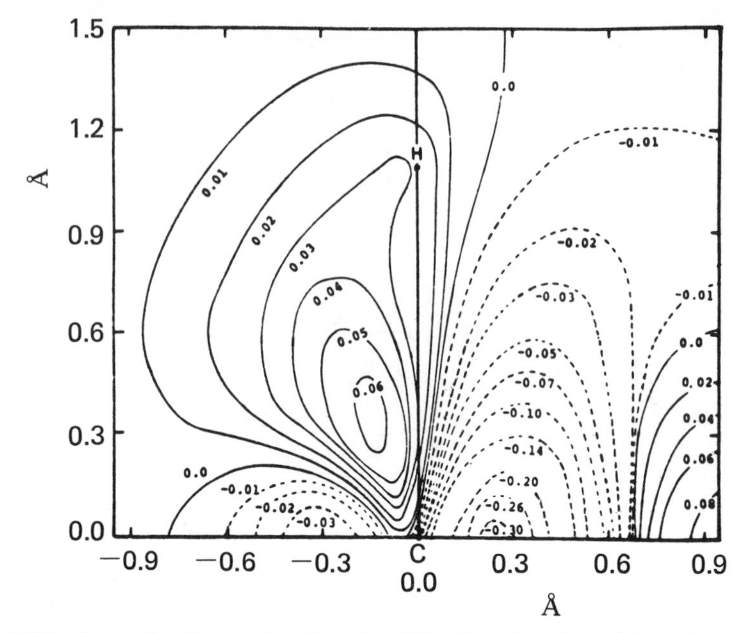

Fig. 3-28. Change in electron density induced by the interaction of two planar methyl radicals separated by 1.9 Å, obtained by subtracting the density for two free methyl radicals from that for two interacting methyl radicals at the same separation. The map is on the HCC plane and corresponds to the left-hand methyl radical. (Reproduced from Ref. 216, courtesy the American Chemical Society.)

bon. This is understandable, since out-of-plane bending induces an atomic dipole on the C atom. Thus, a change in geometry of the reactants during reaction induces electron-cloud preceding in the atomic region of the reaction site. In order to attain sufficient electron-cloud preceding in the C—C region, this secondary mechanism of preceding is quite important, and it is also induced by electron-cloud preceding in the C—H bond region (Fig. 3-28). All three kinds of preceding, manifesting themselves in the EC(C—C) force on carbon,

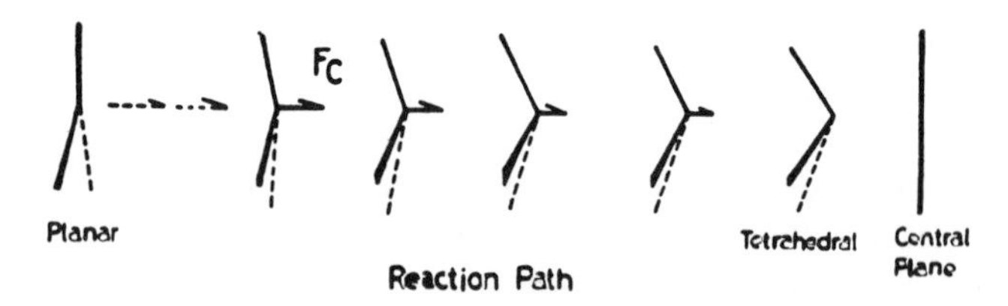

Fig. 3-29. Illustration of the gradually bending approach and the driving force F_C acting on the carbon nucleus. Another methyl radical is in the mirror-image position with respect to the central plane. (Reproduced from Ref. 216, courtesy the American Chemical Society.)

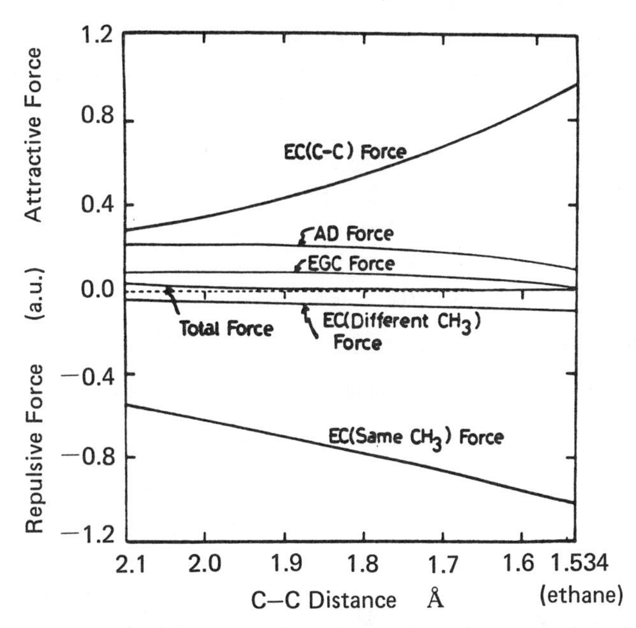

Fig. 3-30. Analysis of the driving force F_C acting on the carbon atom along the gradually bending approach of the two methyl radicals. (Reproduced from Ref. 216, courtesy the American Chemical Society.)

the EC(C—H) force on proton, and the AD force on carbon, *cooperate* to drive the reaction.

It is interesting to note (Fig. 3-30) the crossing of the curves of EC(C—C) force and AD force at C—C distance greater than 2.1 Å. This suggests that at the initial stage of the reaction the change in reactant geometry (inducing the attractive AD force) may be more important than the density accumulation in the overlap region of the two reactants. Although this is apparently similar to the dominant character of the AD force in the initial stages of H_2 formation (*16*, *140*, *215*), the mechanism is essentially different. In order to confirm this point, further study based on more accurate wavefunctions would seem to be necessary.

3-6. SUMMARY

In this chapter we have explained the force concepts applied to molecular geometry and compared them with energetic models. The two main force models, due to Nakatsuji and Deb, are quite successful for understanding and predicting the striking variety and regularity in structural chemistry. Although these two models are quite different, the force concepts used by them are very pictorial and attractive. In Section 3-5 using the force concept, we study the behavior of electron density during the nuclear rearrangement processes and obtain a general density guiding principle for such processes. These concepts are equally applicable to molecular structure and chemical reactions, as well as to the interface between them.

These successes indicate that the force concept should become a quite general and useful approach for understanding wider areas of chemical phenomena. The wide applicability of the ESF theory due to Nakatsuji and co-workers to molecular structure, chemical reactions, and long-range interactions promise such developments of the force concept.

Acknowledgment

The authors would like to acknowledge Professor B. M. Deb for kindly editing throughout this chapter.

Notes

1. Correlation diagrams were first constructed by Hund (*145*, *146*) and Mulliken (*199*, *200*). Such diagrams have been very useful in explaining, e.g., spectra of diatomic molecules, molecular shapes, spectra of transition metal complexes, reactivity of organic molecules (*302*), etc.

2. Similar rules were suggested earlier by Cassie (*50*), Penney and Sutherland (*239*), and Sidgwick and Powell (*274*).

3. A sufficient condition for a general SCF wavefunction to satisfy the H–F theorem is that the basis set includes derivative $\partial \chi_r / \partial x_r$ for any basis χ_r. A procedure which includes 'parent' and first derivative AOs seems to be useful (*221*a).

4. This may be due to terminal atoms being usually more electronegative. Walsh (*294*) had noted this for AB_2 molecules.

5. This is a modification (*210*) of the partitioning procedure of Bader et al. (*18*, *19*) for diatomic molecules.

6. This assumes that the shielding of Z_B by the electrons in χ_{sB} is complete. Although it is an overestimate, it improves as the AB distance and/or the orbital exponent of χ_{sB} increases; e.g., for a C—C distance of 1.4 Å, the error is ~8%, but for a C—H distance of 1.09 Å, the error is ~30%.

6a. See note 6.

7. According to nonempirical calculations on NH_3 (*159*) the lone-pair orbital has s–p mixing [P_{sNpN} in (3-8)] 1.063, compared with 0.866 for sp^3 hybrid and 0.943 for sp^2 hybrid.

8. The AD force is small for H atom, since here the $1s$–$2p$ mixing is small.

9. This occurs because, although the integral part of the AD/EC ratio in, e.g., the series B, C, N, O increases, the extent of s–p mixing decreases [Fig. 3-4(b)].

10. See note 6.

11. Since the force on A along A—B is always the sum of EC and EGC forces, one can reduce the EC force by incorporating the EGC force into it ('bond' force in Ref. 221).

12. Unless otherwise mentioned, experimental values are from Herzberg (*134*) or Sutton (*283*, *284*).

13. Exceptions exist from Al to Ga and As to Sb.

14. This is why inner-shell electrons are not significant for molecular shapes.

15. Since the one-center coulomb repulsion integral is large (*305*), π-donating substituents are usually σ-attracting (e.g., halogens) and vice versa.

16. An important exception occurs when both s_A and p_A are completely filled (see Section 3-3-4b).

17. The overlap effect may also be called conjugative effect in structural theory in contrast with the inductive effect given in Section 3-3-4c.

18. For the role of *d* electrons in chemical bonding, see, e.g., Ref. 63. In speaking about the importance of *d* orbitals, experimental chemists usually assume a minimal basis set (i.e., only one basis for *s*-type AO, three for *p*, five for *d*, etc.). But in a refined calculation theoreticians employ many bases for *s*-, *p*-, and *d*-type orbitals (extended basis set; see, e.g., Ref. 281). When the addition of *d* orbitals is treated as a perturbation (*217*) the first-order correction raises the Hartree–Fock orbital energy but vanishes identically for the total energy, indicating the insignificance of *d* orbitals. This result applies to the second standpoint but may not apply to the first since the minimal *s*, *p* basis set does not satisfy the requirement for the zero-order MO in perturbation theory.

19. The transfer of electrons from the lone-pair orbital of N to N—Si and SiH_3 regions is supported by the fact that $N(SiH_3)_3$ is a weak base.

20. The repulsive EC and EGC forces are responsible for steric repulsion.

21. This applies to all planar (linear)–pyramidal (bent) correlations. In the planar (linear) form the bending force vanishes by symmetry.

22. The overlap effect and the inductive-substitutent effect may or may not be cooperative (Section 3-3-4d). In case of the former, a decrease in electron density in the atomic region (decrease in AD force) leads to an increase of electron density in the bond region (increase in EC force) and vice versa.

23. Hayes (*131*) noted that for both neutral atoms and cations, the *d* levels are lower than the *p* levels for Ca, Sr, and Ba, although both levels are lowered monotonously from Be to Ba.

24. When $\theta = 0$, the electron density of C—H bond is symmetric about the bond axis.

25. Minus sign means resisting force.

26. Experimental data cited in this subsection are from Ref. 182.

27. Their sum is a good approximation to the total force.

28. From Table 3-2, for $(\pi)^2(\pi^*)^1$,

$$D(p\pi_A) = \frac{1}{1+S} + \frac{1}{2}\frac{1}{1-S} \simeq 1.5, \text{ with } S \simeq 0.25.$$

29. See Refs. 93, 104, and 299 for integration over another variable, namely, nuclear charge.

30. Both the Walsh–Allen and original Walsh diagrams were "reduced" such that orbital energies are measured relative to the linear form (compare Figs. 3-1 and 3-12).

31. Such a resemblance does not hold for unoccupied MOs indicated by the dotted curves.

32. No Walsh diagram was proposed for internal rotation.

33. For linear molecules, the Renner effect occurs (*251*).

34. For an extension to the dynamic JT effect, see Ref. 54.

35. Consideration of overlap with chlorine AOs does not alter the nature of JT distortion.

36. Such simplification may cause problems when some MO curves in the fcd cross the zero-force line (Fig. 3-13). In the reduced Walsh–Allen diagram such sign changes occur only for unoccupied orbitals.

37. (a) When the number of valence electrons is such that only one of these two MOs are occupied, two different assignments of HOMO are possible, leading to a difficulty in prediction. (b) Since it is difficult to derive the HOMO postulate on an a priori quantum mechanical basis, its validity has been asserted by testing its predictions for a wide range of molecular classes. For other details regarding this model, see Refs. 76–79, 82.

38. For BeH_2^+ (bent ground state; Ref. 246), the interlacing of occupied MO curves makes the choice of HOMO difficult (*79*).

39. The z-axis is along C_3 axis, while the x axis lies in the plane defined by z-axis and the A—H_1 bond.

40. The ESF model can explain these trends for the alkaline earths and lithium (Section 3-3-7).

41. In principle, one can also start from the *trans* form. However, in this case, the $2b_u$ and $2a_g$ MOs cross each other in the Walsh–Allen diagram (*122*), making the choice of HOMO difficult for 5–10-valence-electron molecules (see note 37). Such MO crossing is fortunately not encountered if one starts from the *cis* form (*82*).

42. Note that the MO numberings here are different from Fig. 3-13 since the latter MOs incorporate inner-shell AOs. See note 36.

43. Deb and Mahajan (*80*) have further confirmed the validity of the HOMO postulate by studying the structures of 17 quixotic hypothetical molecules according to INDO and CNDO/2 MO methods: HCLi, HBBe, $HBLi^-$, HCB, HNBe, HNB^+, HBB^-, $NaHLi^+$, LiB_2^+, $MgBe_2$, LiB_2^-, MgB_2, LiH_3^{2+} (unstable), H_3O^-, CH_5^-, HBO_2^{2+} (unstable), HBF_2^{2+} (unstable). See also notes 36, 37.

44. Relaxation calculations (see Chapter 5) by Mahajan and Deb (*184*) on BeH_2, using INDO MO densities, indicates that step 2 may not be the real reason. For a link between the SOJT model and HOMO postulate, see Ref. 25.

45. The H–F and integral H–F theorems cannot be used (*218*) to determine a priori the electron density of a system (see, however, Refs. 20, 21). Attempts (*94*, *144*, *171*, *212*, *213*, *218*, *275*, *278*, *279*) to obtain a direct deterministic equation for the electron density without using the wavefunction are not yet completely satisfactory (see, e.g., Refs. 23 and 25a).

46. Koga et al. (*168*) have recently generalized the idea of the binding–antibinding diagram of Berlin (*32*), giving an unambiguous regional definition for the role of electron density.

47. For example, in both van der Waals and chemical reaction regions of the process $H + H \rightarrow H_2$, electron-cloud preceding occurs (*16*, *140*) and causes the

driving force of the reaction, namely, AD force at long range and EC force at shorter range. Note that in, e.g., ionic reactions, electron-cloud preceding also manifests itself in the EGC force, since electron transfer from one atom to another is also "preceding."

48. The behavior of electron densities calculated from floating wavefunctions for NH_3, CH_3^+, and NH_3^+ (*221*) were very similar to those given here for CH_3^-, CH_3^+, and CH_3, respectively.

49. Note that localization of MOs is a device for interpretative purposes. It utilizes the inherent unitary ambiguity of Hartree–Fock MOs and does not affect the total wavefunction (*138*, *139*, *252*).

50. No experimental data are known for this increase.

51. For experimental evidence, see Ref. 222.

References

1. Accad, Y., Pekeris, C. L., and Schiff, B., 1971, *Phys. Rev.*, **A4**, 516.
2. Acquista, N., Abramowitz, S., and Lide, D. R., Jr., 1968, *J. Chem. Phys.*, **49**, 780.
3. Allen, L. C., 1969, *Ann. Rev. Phys. Chem.*, **20**, 315.
4. ——, 1972, *Theoret. Chim. Acta* (Berlin), **24**, 117.
5. —— and Russell, J. D., 1967, *J. Chem. Phys.*, **46**, 1029.
6. Almenningen, A., Bastiansen, O., Ewing, V., Hedberg, K., and Traetterberg, M., 1963, *Acta Chim. Scand.*, **17**, 2455.
7. Anderson, A. B., 1972, *J. Chem. Phys.*, **57**, 4143.
8. ——, 1973, *J. Chem. Phys.*, **58**, 381.
9. ——, Handy, N. C., and Parr, R. G., 1969, *J. Chem. Phys.*, **50**, 3635.
10. —— and Parr, R. G., 1970, *J. Chem. Phys.*, **53**, 3375.
11. —— and Parr, R. G., 1971, *J. Chem. Phys.*, **55**, 5490.
12. Andrews, L. and Pimentel, G. C., 1967, *J. Chem. Phys.*, **47**, 3637.
13. Bader, R. F. W., 1960, *Mol. Phys.*, **3**, 137.
14. ——, 1962, *Can. J. Chem.*, **40**, 1164.
15. —— and Bandrauk, A. D., 1968, *J. Chem. Phys.*, **49**, 1666.
16. —— and Chandra, A. K., 1968, *Can. J. Chem.*, **46**, 953; see also Chandra, A. K. and Sundar, R., 1971, *Mol. Phys.*, **22**, 369.
17. —— and Ginsburg, J. L., 1969, *Can. J. Chem.*, **47**, 3061.
18. ——, Henneker, W. H., and Cade, P. E., 1967, *J. Chem. Phys.*, **46**, 3341.
19. —— and Jones, G. A., 1961, *Can. J. Chem.*, **39**, 1253.
20. —— and Jones, G. A., 1963, *Can. J. Chem.*, **41**, 586.
21. —— and Jones, G. A., 1963, *J. Chem. Phys.*, **38**, 2791.
22. —— and Preston, H. J. T., 1966, *Can. J. Chem.*, **44**, 1131.
23. ——, Srebrenik, S., and Nguyen-Dang, T. T., 1978, *J. Chem. Phys.*, **68**, 3680.
24. Ballhausen, C. J., 1962, *Introduction to Ligand Field Theory* (McGraw-Hill, New York), pp. 193, 228.
25. Bamzai, A. S. and Deb, B. M., 1978, *Pramana*, **11**, 191.

25a. —— and Deb, B. M., 1981, *The Role of Single-Particle Density in Chemistry*, *Rev. Mod. Phys.*, forthcoming.

26. Bartell, L. S., 1963, *J. Chem. Educ.*, **40**, 295.
27. ——, 1966, *Inorg. Chem.*, **5**, 1635.
28. ——, 1968, *J. Chem. Educ.*, **45**, 754.
29. Bent, H. A., 1960, *J. Chem. Educ.*, **37**, 616.
30. ——, 1961, *Chem. Rev.*, **61**, 275.
31. ——, 1966, *J. Chem. Educ.*, **43**, 170.
32. Berlin, T., 1951, *J. Chem. Phys.*, **19**, 208.
33. Bills, J. L. and Snow, R. L., 1975, *J. Am. Chem. Soc.*, **97**, 6340.
34. Bingel, W. A., 1959, *J. Chem. Phys.*, **30**, 1250, 1254.
35. ——, 1960, *J. Chem. Phys.*, **32**, 1522.
36. ——, 1961, *Z. Naturforsch.*, **A16**, 668.
37. ——, 1964, in *Molecular Orbitals in Chemistry, Physics, and Biology*, eds. P.-O. Löwdin and B. Pullman (Academic Press, New York), p. 191.
38. Blustin, P. H., 1977, *Chem. Phys. Lett.*, **46**, 386.
39. Born, M. and Oppenheimer, R., 1927, *Ann. Physik*, **84**, 457.
40. Britton, D., 1963, *Can. J. Chem.*, **41**, 1632.
41. Büchler, A., Stauffer, J., and Klemperer, W., 1964, *J. Chem. Phys.*, **40**, 3471; *J. Am. Chem. Soc.*, **86**, 4544.
42. ——, Stauffer, J., Klemperer, W., and Wharton, L., 1963, *J. Chem. Phys.*, **39**, 2299.
43. Bünau, G. V., Diercksen, G., and Preuss, H., 1967, *Int. J. Quantum Chem.*, **1**, 645.
44. Buenker, R. J. and Peyerimhoff, S. D., 1966, *J. Chem. Phys.*, **45**, 3682.
45. —— and Peyerimhoff, S. D., 1972, *Theoret. Chim. Acta* (Berlin), **24**, 132.
46. —— and Peyerimhoff, S. D., 1974, *Chem. Rev.*, **74**, 127.
47. Bugg, C., Desiderato, R., and Sass, R. L., 1964, *J. Am. Chem. Soc.*, **86**, 3157.
48. Byers Brown, W., 1958, *Proc. Cambridge Phil. Soc.*, **54**, 251.
49. Cade, P. E., Bader, R. F. W., and Pelletier, J., 1971, *J. Chem. Phys.*, **54**, 3517.
50. Cassie, J., 1933, *Nature* (London), **131**, 438.
51. Chang, S. Y., Davidson, E. R., and Vincow, G., 1970, *J. Chem. Phys.*, **52**, 5596.
51a. Chipman, D. M., Palke, W. E., and Kirtman, B., 1980, *J. Am. Chem. Soc.*, **102**, 3377.
52. Čížek, J., 1963, *Mol. Phys.*, **6**, 19.
53. Claxton, T. A. and Benson, G. C., 1966, *Can. J. Chem.*, **44**, 157.
54. Clinton, W. L., 1960, *J. Chem. Phys.*, **32**, 626.
55. —— and Hamilton, W. C., 1960, *Rev. Mod. Phys.*, **32**, 422.
56. —— and Rice, B., 1959, *J. Chem. Phys.*, **30**, 542.
57. Cohan, N. V. and Coulson, C. A., 1956, *Trans. Faraday Soc.*, **52**, 1163.
58. Cole, T., 1961, *J. Chem. Phys.*, **35**, 1169.
59. ——, Pritchard, H. O., Davidson, N. R., and McConnell, H. M., 1958, *Mol. Phys.*, **1**, 406.
60. Cooper, J., Hudson, A., and Jackson, R. A., 1972, *Mol. Phys.*, **23**, 209.

61. Corradini, P. and Allegra, G., 1959, *J. Am. Chem. Soc.*, **81**, 5511.
62. Coulson, C. A., 1962, *Valence*, 2nd ed. (Oxford University Press, New York), p. 111.
63. ——, 1972, *d-Electrons in Chemical Bonding* (Proceedings of the Robert A. Welch Foundation Conferences on Chemical Research. XVI. Theoretical Chemistry, Texas), p. 61.
64. ——, 1974, *Israel J. Chem.*, **11**, 683.
65. —— and Deb, B. M., 1969, *Mol. Phys.*, **16**, 545.
66. —— and Deb, B. M., 1971, *Int. J. Quantum Chem.*, **5**, 411.
67. —— and Hurley, A. C., 1962, *J. Chem. Phys.*, **37**, 448.
68. —— and Luz, Z., 1968, *Trans. Faraday Soc.*, **64**, 2884.
69. —— and Nielson, A. H., 1963, *Disc. Faraday Soc.*, **35**, 71.
70. —— and Stephen, M. J., 1957, *Trans. Faraday Soc.*, **53**, 272.
71. —— and Strauss, H. L., 1962, *Proc. Roy. Soc.* (London), **A269**, 443.
72. Davidson, E. R., 1964, *J. Chem. Phys.*, **41**, 656.
73. ——, 1965, *J. Chem. Phys.*, **42**, 4199.
74. Deb, B. M., 1971, *Proc. Indian Nat. Sci. Acad.*, **37A**, 349.
75. ——, 1973, *Rev. Mod. Phys.*, **45**, 22.
76. ——, 1974, *J. Am. Chem. Soc.*, **96**, 2030.
77. ——, 1975, *J. Am. Chem. Soc.*, **97**, 1988.
78. ——, 1975, *J. Chem. Educ.*, **52**, 314.
79. ——, Bose, S. K., and Sen, P. N., 1976, *Indian J. Pure Appl. Phys.*, **14**, 444.
80. —— and Mahajan, G. D., 1981, forthcoming.
81. ——, Mahajan, G. D., and Vasan, V. S., 1977, *Pramana*, **9**, 93.
82. ——, Sen. P. N., and Bose, S. K., 1974, *J. Am. Chem. Soc.*, **96**, 2044.
83. Devaquet, A., 1972, *J. Am. Chem. Soc.*, **94**, 5626, 9012.
84. Dixon, D. A. and Lipscomb, W. N., 1973, *J. Am. Chem. Soc.*, **95**, 2853.
85. ——, Pepperberg, I. M., and Lipscomb, W. N., 1974, *J. Am. Chem. Soc.*, **96**, 1325.
86. Drago, R. S., 1973, *J. Chem. Educ.*, **50**, 244.
87. Driessler, F., Ahlrichs, R., Staemmler, V., and Kutzelnigg, W., 1973, *Theoret. Chim. Acta* (Berlin), **30**, 315.
88. Dykstra, C. E., Hereld, M., Lucchese, R. R., Schaefer, H. F., III, and Meyer, W., 1977, *J. Chem. Phys.*, **67**, 4071.
89. Edmiston, C. and Ruedenberg, K., 1963, *Rev. Mod. Phys.*, **35**, 457.
90. Ellison, F. O. and Shull, H., 1955, *J. Chem. Phys.*, **33**, 2348.
91. Ellison, G. B., Engelking, P. C., and Lineberger, W. C., 1978, *J. Am. Chem. Soc.*, **100**, 2556.
92. Englman, R., 1972, *The Jahn–Teller Effect in Molecules and Crystals* (John Wiley & Sons, New York).
93. Epstein, S. T., Hurley, A. C., Wyatt, R. E., and Parr, R. G., 1967, *J. Chem. Phys.*, **47**, 1275.
94. —— and Rosenthal, C. M., 1976, *J. Chem. Phys.*, **64**, 247.
95. Evans, J. C. and Lo, G. Y-S., 1966, *J. Phys. Chem.*, **70**, 11.
96. —— and Lo, G. Y-S., 1967, *J. Phys. Chem.*, **71**, 3697.

97. —— and Lo, G. Y-S., 1969, *J. Phys. Chem.*, **73**, 448.
98. Fessenden, R. W., 1967, *J. Phys. Chem.*, **71**, 74.
99. —— and Schuler, R. H., 1963, *J. Chem. Phys.*, **39**, 2147.
100. —— and Schuler, R. H., 1965, *J. Chem. Phys.*, **43**, 2704.
101. Feynman, R. P., 1939, *Phys. Rev.*, **56**, 340.
102. Fink, W. H. and Allen, L. C., 1967, *J. Chem. Phys.*, **46**, 2261.
103. Fletcher, R., 1970, *Mol. Phys.*, **19**, 55.
104. Frost, A. A., 1962, *J. Chem. Phys.*, **37**, 1147.
105. ——, 1966, University of Wisconsin Theoretical Chemistry Report, WIS-TCI-204, Dec.
106. ——, 1967, *J. Chem. Phys.*, **47**, 3707, 3714.
107. Fukui, K., 1966, *Bull. Chem. Soc. Jap.*, **39**, 498.
108. ——, 1970, *Theory of Orientation and Stereoselection* (Springer-Verlag, Heidelberg).
109. —— and Fujimoto, H., 1968, *Bull. Chem. Soc. Jap.*, **41**, 1989.
110. —— and Fujimoto, H., 1969, *Bull. Chem. Soc. Jap.*, **42**, 3399.
111. ——, Yonezawa, T., Nagata, C., and Shingu, H., 1954, *J. Chem. Phys.*, **22**, 1433.
112. ——, Yonezawa, T., and Shingu, H., 1952, *J. Chem. Phys.*, **20**, 722.
113. Gavin, R. M., Jr., 1969, *J. Chem. Educ.*, **46**, 413.
114. Gayles, J. N. and Self, J., 1964, *J. Chem. Phys.*, **40**, 3530.
115. Gerratt, J. and Mills, I. M., 1968, *J. Chem. Phys.*, **49**, 1719.
116. Gillespie, R. J., 1960, *J. Am. Chem. Soc.*, **82**, 5978.
117. ——, 1960, *Can. J. Chem.*, **38**, 818.
118. ——, 1963, *J. Chem. Educ.*, **40**, 295.
119. ——, 1967, *Angew. Chem. Intern. Ed.*, **6**, 819.
120. ——, 1972, *Molecular Geometry* (Van Nostrand Reinhold, New York).
121. —— and Nyholm, R. S., 1957, *Quart. Rev. Chem. Soc.*, **11**, 339.
122. Gimarc, B. M., 1970, *J. Am. Chem. Soc.*, **92**, 266.
123. ——, 1971, *J. Am. Chem. Soc.*, **93**, 593.
124. ——, 1971, *J. Am. Chem. Soc.*, **93**, 815.
125. Glidewell, C., Rankin, D. W. H., Robiette, A. G., and Sheldrick, G. M., 1970, *J. Chem. Soc.*, A318.
126. Gole, J. L., Siu, A. K. Q., and Hayes, E. F., 1973, *J. Chem. Phys.*, **58**, 857.
127. Goodisman, J., 1966, *J. Chem. Phys.*, **45**, 4689.
128. Grow, D. T. and Pitzer, R. M., 1977, *J. Chem. Phys.*, **67**, 4019.
129. Hall, G. G., 1961, *Phil. Mag.*, **6**, 249.
130. Harcourt, R. D. and Harcourt, A., 1973, *Chem. Phys.*, **1**, 238.
131. Hayes, E. F., 1966, *J. Phys. Chem.*, **70**, 3740.
132. Hedberg, K., 1955, *J. Am. Chem. Soc.*, **77**, 6491.
133. Hellmann, H., 1937, *Einführung in die Quantenchemie* (Franz Deuticke, Leipzig and Vienna), p. 285.
134. Herzberg, G., 1965, *Molecular Spectra and Molecular Structure. III. Electronic Spectra and Electronic Structure of Polyatomic Molecules* (D. Van Nostrand, Princeton, N.J.).

135. —— and Johns, J. W. C., 1971, *J. Chem. Phys.*, **54**, 2276.
136. Higuchi, J., 1956, *J. Chem. Phys.*, **24**, 535.
137. Hinze, J. and Jaffé, H. H., 1962, *J. Am. Chem. Soc.*, **84**, 540; *J. Phys. Chem.*, **67**, 1501.
138. Hirao, K., 1974, *J. Chem. Phys.*, **60**, 3215.
139. —— and Nakatsuji, H., 1973, *J. Chem. Phys.*, **59**, 1457.
140. Hirschfelder, J. O. and Eliason, M. A., 1967, *J. Chem. Phys.*, **47**, 1164.
141. Hoffmann, R., 1963, *J. Chem. Phys.*, **39**, 1397.
142. ——, 1974, *Chem. Eng. News*, **52**(30), 32.
143. ——, Gleiter, R., and Mallory, F. B., 1970, *J. Am. Chem. Soc.*, **92**, 1460.
144. Hohenberg, P. and Kohn, W., 1965, *Phys. Rev.*, **B136**, 864.
145. Hund, F., 1927, *Z. Phys.*, **40**, 742; **42**, 93.
146. ——, 1928, *Z. Phys.*, **51**, 759.
147. Hurley, A. C., 1954, *Proc. Roy. Soc.* (London), **A226**, 170, 179, 193.
148. ——, 1964, in *Molecular Orbitals in Chemistry, Physics, and Biology*, eds. P.-O. Löwdin and B. Pullman (Academic Press, New York), p. 161.
149. Ibers, J. A., 1964, *J. Chem. Phys.*, **40**, 402.
150. Itoh, T., Ohno, K., and Kotani, M., 1953, *J. Phys. Soc. Jap.*, **8**, 41.
151. Jackel, G. S., Christiansen, J. J., and Gordy, W., 1967, *J. Chem. Phys.*, **47**, 4274.
152. —— and Gordy, W., 1968, *Phys. Rev.*, **176**, 443.
153. Jahn, H. A. and Teller, E., 1937, *Proc. Roy. Soc.* (London), **A161**, 220.
154. Janoschek, R., Pruess, H., and Diercksen, G., 1968, *Int. J. Quantum Chem.*, **2**, 159.
155. Johansen, H. and Wahlgren, U., 1978, *Chem. Phys. Lett.*, **55**, 245.
156. Johannesen, R. B., Candela, G. A., and Tsang, T., 1968, *J. Chem. Phys.*, **48**, 5544.
157. Jones, W. D. and Simpson, W. T., 1960, *J. Chem. Phys.*, **32**, 1747.
158. Jordan, P. C. H. and Longuet-Higgins, H. C., 1962, *Mol. Phys.*, **5**, 121.
159. Kaldor, U., 1967, *J. Chem. Phys.*, **46**, 1981.
160. —— and Shavitt, I., 1966, *J. Chem. Phys.*, **44**, 1823.
161. —— and Shavitt, I., 1968, *J. Chem. Phys.*, **48**, 191.
162. Kari, R. E. and Csizmadia, I. G., 1969, *J. Chem. Phys.*, **50**, 1443.
163. Katriel, J., 1972, *Phys. Rev.*, **A5**, 1990.
164. ——, 1972, *Theoret. Chim. Acta* (Berlin), **23**, 309.
165. Kaufman, J. J., Sachs, L. M., and Geller, M., 1968, *J. Chem. Phys.*, **49**, 4369.
166. Kim, H. J. and Parr, R. G., 1964, *J. Chem. Phys.*, **41**, 2892.
167. Koga, T. and Nakatsuji, H., 1976, *Theoret. Chim. Acta* (Berlin), **41**, 119.
168. ——, Nakatsuji, H., and Yonezawa, T., 1978, *J. Am. Chem. Soc.*, **100**, 7522.
168a. ——, Nakatsuji, H., and Yonezawa, T., 1980, *Mol. Phys.*, **39**, 239.
169. Komornicki, A., Ishida, K., Morokuma, K., Ditchfield, R., and Conrad, M., 1977, *Chem. Phys. Lett.*, **45**, 595.
170. Kohl, D. A., 1972, *J. Chem. Phys.*, **56**, 4236.

171. Kohn, W. and Sham, L. J., 1965, *Phys. Rev.*, **A140**, 1133.
172. Kuczkowski, R. L., Lide, D. R., Jr., and Kriser, L. C., 1966, *J. Chem. Phys.*, **44**, 3131.
173. Lemberger, A. and Pauncz, R., 1969, *Acta Phys. Acad. Sci. Hung.*, **27**, 169.
174. Lide, D. R., Jr. and Kuczkowski, R. L., 1967, *J. Chem. Phys.*, **46**, 4768.
175. Liebman, J. F., 1971, *J. Chem. Educ.*, **48**, 188.
176. ——, 1974, *J. Am. Chem. Soc.*, **96**, 3053.
177. —— and Vincent, J. S., 1975, *J. Am. Chem. Soc.*, **97**, 1373.
178. ——, Politzer, P., and Sanders, W. A., 1976, *J. Am. Chem. Soc.*, **98**, 5115.
179. Linnett, J. W. and Wheatley, P. J., 1949, *Trans. Faraday Soc.*, **45**, 33, 39.
180. Löwdin, P.-O., 1955, *Phys. Rev.*, **97**, 1474.
181. ——, 1955, *Phys. Rev.*, **97**, 1490.
182. Lowe, J. P., 1968, *Progr. Phys. Org. Chem.*, **6**, 1.
183. Magnasco, V. and Perico, A., 1967, *J. Chem. Phys.*, **47**, 971.
184. Mahajan, G. D. and Deb, B. M., 1981, forthcoming.
185. Mamantov, G., Vasini, E. J., Moulton, M. C., Vickroy, D. G., and Maekawa, T., 1971, *J. Chem. Phys.*, **54**, 3419.
186. Matsumura, C. and Lide, D. R., Jr., 1969, *J. Chem. Phys.*, **50**, 71, 3080.
187. McDiarmid, R. and Charney, E., 1967, *J. Chem. Phys.*, **47**, 1517.
188. McIver, J. W., Jr. and Komornicki, A., 1971, *Chem. Phys. Lett.*, **10**, 303.
189. McKean, D. C. and Schatz, P. N., 1956, *J. Chem. Phys.*, **24**, 316.
190. Merer, A. J. and Mulliken, R. S., 1969, *Chem. Rev.*, **69**, 639.
191. —— and Schoonveld, L., 1968, *J. Chem. Phys.*, **48**, 522.
192. —— and Schoonveld, L., 1969, *Can. J. Phys.*, **47**, 1731.
193. Messmer, R. P. and Birss, F. W., 1969, *J. Phys. Chem.*, **73**, 2085.
194. Millie, P. and Berthier, G., 1968, *Int. J. Quantum Chem.*, **S2**, 67.
195. Milligan, D. E. and Jacox, M. E., 1967, *J. Chem. Phys.*, **47**, 5146.
196. Moccia, R., 1967, *Theoret. Chim. Acta* (Berlin), **8**, 8.
197. Morehouse, R. L., Christiansen, J. J., and Gordy, W., 1966, *J. Chem. Phys.*, **45**, 1751.
198. Morokuma, K., Pedersen, L., and Karplus, M., 1968, *J. Chem. Phys.*, **48**, 4801.
199. Mulliken, R. S., 1928, *Phys. Rev.*, **32**, 186.
200. ——, 1932, *Rev. Mod. Phys.*, **4**, 1.
201. ——, 1932, *Phys. Rev.*, **41**, 751.
202. ——, 1933, *Phys. Rev.*, **43**, 279.
203. ——, 1934, *J. Chem. Phys.*, **2**, 782.
204. ——, 1942, *Rev. Mod. Phys.*, **14**, 204.
205. ——, 1955, *J. Chem. Phys.*, **23**, 1833.
206. ——, 1959, *Tetrahedron*, **5**, 253.
207. Nakatsuji, H., 1973, *J. Am. Chem. Soc.*, **95**, 345.
208. ——, 1973, *J. Am. Chem. Soc.*, **95**, 354.
209. ——, 1973, *J. Am. Chem. Soc.*, **95**, 2084.
210. ——, 1974, *J. Am. Chem. Soc.*, **96**, 24.

211. ——, 1974, *J. Am. Chem. Soc.*, **96**, 30.
212. ——, 1976, *Phys. Rev.*, **A14**, 41.
213. ——, 1977, *J. Chem. Phys.*, **67**, 1312.
214. ——, Kato, H., and Yonezawa, T., 1969, *J. Chem. Phys.*, **51**, 3175.
215. —— and Koga, T., 1974, *J. Am. Chem. Soc.*, **96**, 6000.
216. ——, Kuwata, T., and Yoshida, A., 1973, *J. Am. Chem. Soc.*, **95**, 6894.
217. —— and Musher, J. I., 1974, *Chem. Phys. Lett.*, **24**, 77.
218. —— and Parr, R. G., 1975, *J. Chem. Phys.*, **63**, 1112.
219. ——, Koga, T., Kondo, K., and Yonezawa, T., 1978, *J. Am. Chem. Soc.*, **100**, 1029.
220. ——, Matsuda, K., and Yonezawa, T., 1978, *Chem. Phys. Lett.*, **54**, 347; *Bull. Chem. Soc. Jap.*, **51**, 1315.
221. ——, Kanayama, S., Harada, S., and Yonezawa, T., 1978, *J. Am. Chem. Soc.*, **100**, 7528.
221a. ——, Kanda, K., and Yonezawa, T., 1980, *Chem. Phys. Lett.*, **75**, 340.
221b. ——, Hayakawa, T., and Yonezawa, T., 1981, forthcoming.
222. Neunhoeffer, O. and Hasse, H., 1958, *Chem. Ber.*, **91**, 1801.
223. Newton, M. D., Switkes, E., and Lipscomb, W. N., 1970, *J. Chem. Phys.*, **53**, 2645.
224. Niessen, W. von, 1972, *J. Chem. Phys.*, **56**, 4290.
225. Orgel, L. E., 1960, *An Introduction to Transition-Metal Chemistry* (John Wiley & Sons, New York).
226. Palke, W. E. and Pitzer, R. M., 1967, *J. Chem. Phys.*, **46**, 3948.
227. Parr, R. G., 1964, *J. Chem. Phys.*, **40**, 3726.
228. ——, 1974, *CIT-JHU-UNC Quanta*, Third Issue (Dept. of Chemistry, University of North Carolina, Chapel Hill).
229. Parsons, A. E. and Searcy, A. W., 1959, *J. Chem. Phys.*, **30**, 1635.
230. Patton, J. W. and Hedberg, K., 1968, *Bull. Am. Phys. Soc.*, **13**, 831.
231. Pauling, L., 1960, *The Nature of the Chemical Bond*, 3rd ed. (Cornell University Press, Ithaca, New York).
232. Pearson, R. G., 1969, *J. Am. Chem. Soc.*, **91**, 1252, 4947.
233. ——, 1970, *J. Chem. Phys.*, **52**, 2167; **53**, 2986.
234. ——, 1970, *Theoret. Chim. Acta* (Berlin), **16**, 107.
235. ——, 1971, *Chem. Phys. Lett.*, **10**, 31.
236. ——, 1971, *Acc. Chem. Res.*, **4**, 152.
237. ——, 1976, *Symmetry Rules for Chemical Reactions* (John Wiley & Sons, New York).
238. Pedersen, L. and Morokuma, K., 1967, J. *Chem. Phys.*, **46**, 3941.
239. Penney, W. G. and Sutherland, G. B., 1936, *Proc. Roy. Soc.* (London), **A156**, 654.
240. Peyerimhoff, S. D., Buenker, R. J., and Allen, L. C., 1966, *J. Chem. Phys.*, **45**, 734.
241. Politzer, P., 1967, *J. Phys. Chem.*, **71**, 3068.
242. ——, 1976, *J. Chem. Phys.*, **64**, 4239.
243. Pople, J. A. and Beveridge, D. L., 1970, *Approximate Molecular Orbital Theory* (McGraw-Hill, New York).

244. —— and Nesbet, R. K., 1954, *J. Chem. Phys.*, **22**, 571.
245. Poshusta, R. D., Haugen, J. A., and Zetik, D. F., 1969, *J. Chem. Phys.*, **51**, 3343.
246. ——, Klint, D. W., and Liberles, A., 1971, *J. Chem. Phys.*, **55**, 252.
247. Preuss, H. and Diercksen, G., 1967, *Int. J. Quantum Chem.*, **1**, 641.
248. Pulay, P., 1969, *Mol. Phys.*, **17**, 197.
249. Ray, N. K., 1970, *J. Chem. Phys.*, **52**, 463.
250. Reichman, S. and Schreiner, F., 1969, *J. Chem. Phys.*, **51**, 2355.
251. Renner, R., 1934, *Z. Phys.*, **92**, 172.
252. Roothaan, C. C. J., 1951, *Rev. Mod. Phys.*, **23**, 69.
253. Rothenberg, S., 1969, *J. Chem. Phys.*, **51**, 3389.
254. ——, 1971, *J. Am. Chem. Soc.*, **93**, 68.
255. Ruedenberg, K., 1962, *Rev. Mod. Phys.*, **34**, 326.
256. ——, 1977, *J. Chem. Phys.*, **66**, 375.
257. Runtz, G. R., Bader, R. F. W., and Messer, R. R., 1977, *Can. J. Chem.*, **55**, 3040.
258. Salem, L., 1961, *Proc. Roy. Soc.* (London), **A264**, 379.
259. ——, 1963, *J. Chem. Phys.*, **38**, 1227.
260. ——, 1968, *J. Am. Chem. Soc.*, **90**, 543.
261. ——, 1969, *Chem. Phys. Lett.*, **3**, 99.
262. —— and Wright, J. S., 1969, *J. Am. Chem. Soc.*, **91**, 5947.
263. Schmidtke, H. H., 1962, *Z. Naturforsch*, **17a**, 121.
264. ——, 1963, *Z. Naturforsch*, **18a**, 496.
265. —— and Preuss, H., 1961, *Z. Naturforsch*, **16a**, 790.
265a. Schmiedekamp, A., Cruickshank, D. W. J., Skaarup, S., Pulay, P., Hargittai, I., and Boggs, J. E., 1979, *J. Am. Chem. Soc.*, **101**, 2002.
266. Schnuelle, G. W. and Parr, R. G., 1972, *J. Am. Chem. Soc.*, **94**, 8974.
267. Schrader, D. M., 1965, *J. Chem. Phys.*, **46**, 3895.
268. —— and Karplus, M., 1964, *J. Chem. Phys.*, **40**, 1593.
269. Schwartz, M. S., 1969, *J. Chem. Phys.*, **51**, 4182.
270. Schwartz, M. E. and Allen, L. C., 1970, *J. Am. Chem. Soc.*, **92**, 1466.
271. Searcy, A. W., 1958, *J. Chem. Phys.*, **28**, 1237.
272. ——, 1959, *J. Chem. Phys.*, **31**, 1.
273. Shull, H. and Ebbing, D., 1958, *J. Chem. Phys.*, **28**, 866.
274. Sidgwick, N. V. and Powell, H. M., 1940, *Proc. Roy. Soc.* (London), **A176**, 153.
275. Simons, J., 1973, *J. Chem. Phys.*, **59**, 2436.
276. Smith, W. L. and Warsop, P. A., 1968, *Trans. Faraday Soc.*, **64**, 1165.
277. Sovers, O. J., Kern, C. W., Pitzer, R. M., and Karplus, M., 1968, *J. Chem. Phys.*, **49**, 2592.
278. Srebrenik, S. and Bader, R. F. W., 1975, *J. Chem. Phys.*, **63**, 3945.
279. ——, Bader, R. F. W., and Nguyen-Dang, T. T., 1978, *J. Chem. Phys.*, **68**, 3667.
280. Stenkamp, L. Z. and Davidson, E. R., 1973, *Theoret. Chim. Acta* (Berlin), **30**, 283.
281. Stewart, R. F., 1970, *J. Chem. Phys.*, **52**, 431.

282. Surratt, G. T. and Goddard, W. A., 1977, *Chem. Phys.*, **23**, 39.
283. Sutton, L. E. (ed.), 1958, Chemical Society Special Publication No. 11.
284. —— (ed.), 1965, Chemical Society Special Publication No. 18.
285. Symons, M. C. R., 1969, *Nature* (London), **222**, 1123.
286. —— and Wardale, H. W., 1967, *Chem. Commun.*, 758.
287. Takahata, Y., Schnuelle, G. W., and Parr, R. G., 1971, *J. Am. Chem. Soc.*, **93**, 784.
288. Thompson, H. B., 1968, *Inorg. Chem.*, **7**, 604.
289. ——, 1971, *J. Am. Chem. Soc.*, **93**, 4609.
290. —— and Bartell, L. S., 1968, *Inorg. Chem.*, **7**, 488.
291. Tsao, P., Cobb, C. C., and Classen, H. H., 1971, *J. Chem. Phys.*, **54**, 5247.
292. Ungemach, S. R. and Schaefer, H. F., III, 1976, *J. Am. Chem. Soc.*, **98**, 1658.
293. Urey, H. C. and Bradley, C. A., 1931, *Phys. Rev.*, **38**, 1969.
294. Walsh, A. D., 1953, *J. Chem. Soc.*, 2260, 2266, 2288, 2296, 2301, 2306, 2321.
295. Watson, R. E., 1960, *Phys. Rev.*, **118**, 1036.
296. ——, 1960, *Phys. Rev.*, **119**, 1934.
297. White, D., Seshadri, K. S., Deves, D. F., Mann, D. E., and Leneusky, M. J., 1963, *J. Chem. Phys.*, **39**, 2463.
298. Wilkinson, R. G. and Mulliken, R. S., 1955, *J. Chem. Phys.*, **23**, 1895.
299. Wilson, E. B., Jr., 1962, *J. Chem. Phys.*, **36**, 2232.
300. ——, Decius, J. C., and Cross, P. C., 1955, *Molecular Vibrations* (McGraw-Hill, New York).
301. Winnewisser, G., Winnewisser, M., and Gordy, W., 1968, *J. Chem. Phys.*, **49**, 3465.
302. Woodward, R. B. and Hoffmann, R., 1970, *The Conservation of Orbital Symmetry* (Verlag Chemie GmbH, Weinheim, Germany, and Academic Press, New York).
303. Wulfman, C. E., 1959, *J. Chem. Phys.*, **31**, 381.
304. ——, 1960, *J. Chem. Phys.*, **33**, 1567.
305. Yonezawa, T., Nakatsuji, H., and Kato, H., 1968, *J. Am. Chem. Soc.*, **90**, 1239.
306. ——, Nakatsuji, H., Kawamura, T., and Kato, H., 1969, *Bull. Chem. Soc. Jap.*, **42**, 2437.

4
Energies, Energy Differences, and Mechanisms of Internal Motions

Michael T. Marron
Department of Chemistry
University of Wisconsin–Parkside
Kenosha, Wisconsin

Contents

4-1. INTRODUCTION

The energy of a system is viewed by most scientists as an extremely useful concept. However, it is not energy itself, but rather the difference in energy between two states of a system that forms the basis for understanding physical phenomena. Two states of a system may differ in the position of two or more nuclei, in electron distribution, or in a change in the external constraints. Knowledge of the difference in energy between two states of a system is necessary and sometimes sufficient for a full understanding of changes observed in nature. In this chapter we discuss methods for computing energy differences and we focus upon one method in particular, the integral Hellmann-Feynman (iHF) theorem, which is a generalization of the Hellmann-Feynman (H–F) theorem (*7,20*).

The conventional procedure for computation of energy differences is to find wavefunctions and energies for each state using the Schrödinger equation and then to take the difference in energies. When approximate wavefunctions are employed (the usual case) this procedure calls for taking the difference of expectation values

$$\Delta\tilde{E}_{ed} = \langle\tilde{\psi}_B|H_B|\tilde{\psi}_B\rangle\langle\tilde{\psi}_B|\tilde{\psi}_B\rangle^{-1} - \langle\tilde{\psi}_A|H_A|\tilde{\psi}_A\rangle\langle\tilde{\psi}_A|\tilde{\psi}_A\rangle^{-1}. \qquad (4\text{-}1)$$

Here we have adopted the notation that a tilde placed above a quantity indicates that it is approximate. There are at least two other methods for computing energy differences and both are based on the H–F theorem: the *integrated* H–F theorem and the iHF theorem (see Chapter 1). For the discussion here we presume that the wavefunction for the system is known as a function of some change parameter or parameters, λ, where $\lambda = A$ corresponds to ψ_A and $\lambda = B$ corresponds to ψ_B. We further limit our consideration to isoelectronic processes. This is not a serious limitation, for by proper definition of the system most of biology, chemistry, and physics is included in this class of processes. Later, in Section 4-7, we will discuss how this restriction may be relaxed for treatment of ionization processes. The integrated H–F theorem is written

$$\Delta E_d = \int_A^B \langle\psi(\lambda)|\partial H/\partial\lambda|\psi(\lambda)\rangle\langle\psi(\lambda)|\psi(\lambda)\rangle^{-1}\, d\lambda, \qquad (4\text{-}2)$$

and the iHF theorem is

$$\Delta E_I = \langle \psi_B | H_B - H_A | \psi_A \rangle \langle \psi_B | \psi_A \rangle^{-1}. \tag{4-3}$$

When the exact wavefunctions are known, i.e., when $\psi(\lambda)$ is known for the full range from A to B, these three methods all produce the same result. When approximate wavefunctions are substituted into the equations they will generally produce different results. Epstein et al. (*5*) refer to (4-2) and (4-3) as formulas instead of theorems when used with approximate wavefunctions not among the small class of approximate wavefunctions that do provide agreement among (4-1), (4-2), and (4-3). Their rationale is that (4-2) and (4-3) may be considered to define energy differences which may not agree with $\Delta\tilde{E}_{ed}$ but which also rest upon the Schrödinger equation for their validity. The pros and cons of which formula is best to use in a given situation have been discussed at considerable length in the literature. Various viewpoints are based on several widely differing considerations, including the amount of work involved for computation of ΔE, the accuracy of the ΔE, the independence or lack of independence of ΔE on the path taken in going from A to B, physical insight provided into a given process, and the potential for classical modeling of a process.

The three formulas all require that wavefunctions be known for the initial and final states; (4-2) also requires knowledge of the wavefunction for all intermediate states. If the wavefunctions are determined in the conventional fashion, namely, by means of the variation principle, the ΔE_{ed} is easily obtained as a byproduct of the computation. Once the wavefunctions have been obtained the computation of ΔE by (4-2) or (4-3) is often a simple matter involving evaluation of only one-electron integrals much fewer in number than those required for determination of the wavefunction itself.

Since there is no general, a priori way for selecting a method of computation that will provide the most accurate result for a given situation, selection of a method must rest upon considerations other than the amount of work involved. In Sections 4-2 and 4-3 we outline the method of computation using (4-2) and (4-3) and provide details of the methods. In Sections 4-4 and 4-5 a discussion of accuracy and path dependence for formula (4-3) is given. The remaining three sections are devoted to applications of (4-3) to different types of processes. Comments on interpretation and modeling of processes may be found there.

No introduction to applications of H–F-type formulas to molecular systems is complete without reference to the pioneering work of Berlin (*2*). Both Hellmann and Feynman noted that their theorem provided a theoretical basis for interpreting quantum phenomena in terms of classical electrostatics. Berlin showed how to do this for chemical binding in molecules (see Chapters 2 and 3).

Much of the work with iHF theorem was motivated and inspired by the search for a simple model for understanding such inherently quantum phenomena as barriers to rotation, dissociation of molecules, and ionization energies. The work of Berlin and many other careful workers who followed his lead have pointed the way.

4-2. THE INTEGRATED HELLMANN-FEYNMAN THEOREM AND MISCELLANEOUS RELATED FORMULAS

4-2-1. The Integrated Hellmann-Feynman Theorem

The H-F theorem (*7, 20*) gives the forces in a molecular system as

$$F_\lambda = \frac{\partial E}{\partial \lambda} = \left\langle \psi(\lambda) \middle| \frac{\partial H}{\partial \lambda} \middle| \psi(\lambda) \right\rangle \langle \psi(\lambda) | \psi(\lambda) \rangle^{-1}, \tag{4-4}$$

where λ is a parameter or parameters. If λ is the internuclear distance in a diatomic molecule, F_λ would be the slope of the potential energy if one makes the usual Born-Oppenheimer approximation. Integration of the force over λ yields (4-2), which expresses the energy expended in the process carrying the system from the initial value to the final value of λ. If one chooses λ to be the internuclear distance with an initial value of ∞, then the energy computed is the binding energy of the diatomic molecule. Note that for (4-2) to be workable both $\psi(\lambda)$ and $\partial\psi/\partial\lambda$ must be bounded.

The conditions under which $\Delta\tilde{E}_d$ will equal $\Delta\tilde{E}_{ed}$ have been worked out very carefully (*5, 21, 22*). Hurley has termed this class of functions *floating* functions. When the condition of equality exists rigorously (as opposed to accidental agreement for selected values of λ) the integrated H-F theorem is satisfied; otherwise we refer to (4-2) as the integrated H-F formula. Two examples of floating functions are (1) wavefunctions constructed from basic orbitals whose locations have been optimized and which need not reside at atomic centers (i.e., "float"), and (2) the true Hartree-Fock wavefunctions for a system. Note that it is possible to obtain a function of this class without optimizing orbital locations.

The principal and in most instances overwhelming impediment to application of (4-2) for computation of energy differences is the requirement that ψ be known for all values of λ throughout the range of integration. It is for this reason, for example, that (4-2) has usually been applied only in theoretical developments (see, e.g., Ref. 23) or as a basis for forming classical models for energy differences (see, e.g., Ref. 11). These are important applications, sometimes more important than the actual computation of numbers; we mention the impediment however, by way of explaining the small number of numerical studies based on the integrated H-F theorem. Examples of such computations are given by Epstein et al. (*5*).

A novel application of (4-2) is one proposed by Foldy (*9*) for atoms, and Wilson (*60*) and Frost (*11*) for molecules. These authors imagine an artificial process of constructing an atom or molecule by giving the electron density as the square of a wavefunction that depends explicitly on some parameter λ, where λ varies from 0 to 1 as the nuclear charges vary between 0 and their final values. The energy of the system is obtained by integrating according to (4-2) between the limits 0 and 1. The energy difference computed by this technique is the difference between free electrons and the final atomic or molecular state. This energy is commonly referred to as the molecular electronic energy. For an N-electron system undergoing this process we have (using atomic units)

$$H(\lambda) = T + \lambda^2 \sum_{\alpha<\beta} Z_\alpha Z_\beta R_{\alpha\beta}^{-1} - \lambda \sum_{\alpha,i} Z_\alpha r_{\alpha i}^{-1} + \sum_{i<j} r_{ij}^{-1} \tag{4-5}$$

$$\frac{\partial H}{\partial \lambda} = 2\lambda \sum_{\alpha<\beta} Z_\alpha Z_\beta R_{\alpha\beta}^{-1} - \sum_{\alpha,i} Z_\alpha r_{\alpha i}^{-1} \tag{4-6}$$

and

$$\Delta E_d = V_{nn} - \int_0^1 \left\langle \psi(\lambda) \middle| \sum_{\alpha,i} Z_\alpha r_{\alpha i}^{-1} \middle| \psi(\lambda) \right\rangle d\lambda, \tag{4-7}$$

where α and β refer to nuclei, i and j refer to electrons, T is the total electronic kinetic energy operator, and the term V_{nn} in (4-7) is the result of integrating over the first term in (4-6). Frost simplified (4-7) by inverting some of the integrations indicated by the bracket notation in (4-7) to arrive at

$$\Delta E_d = V_{nn} - \sum_\alpha Z_\alpha \int P(r_1)\, r_{\alpha 1}^{-1}\, d\tau_1, \tag{4-8}$$

where

$$P(r_1) = N \int_0^1 d\lambda \int \cdots \int |\psi(\lambda)|^2\, d\tau_2 \cdots d\tau_N \tag{4-9}$$

is the one-electron density for the molecule. Foldy, Frost, and Wilson have pointed out that an advantage to computing atomic and molecular energies in this fashion is that one may interpret the energy using only classical concepts. Equation (4-8) tells us that the electronic energy ΔE_d can be obtained by evaluating the potential energy of interaction between the one-electron charge density (4-9) and the array of nuclei having charges Z_α.

Frost illustrated this process by computing ΔE_d for the hydrogen atom:

$$\psi(\lambda) = \left(\frac{\lambda^3}{\pi}\right)^{1/2} \exp(-\lambda r)$$

$$P(r) = \left(\frac{3}{8\pi}\right) \left[1 - \exp(-2r) \left(1 + 2r + 2r^2 + \frac{4}{3} r^3\right)\right],$$

from which one may compute

$$\Delta E_d = -4\pi \int_0^\infty P(r)\, r^{-1} r^2\, dr = -\tfrac{1}{2}.$$

The requirement that ψ must be known throughout the range of integration is difficult to meet in most instances, so approximation schemes would seem to be called for. Ruedenberg (*51*) analyzed the barrier to rotation in ethane using formula (4-2) by decomposing ψ into localized orbitals and then presuming that the orbitals were fixed on the unrotated portion of the molecule. His computation provides a classical basis for interpreting the barrier. This model is discussed further in Section 4-8.

4-2-2. Miscellaneous Related Formulas

Several clever, fairly specific studies using the H–F theorem have been published wherein direct application of the theorem yields an energy difference. For example, there is a method for computing barriers to internal rotation in ethane, hydrogen peroxide, and similar molecules that relies on the symmetry of the molecule (*13*, *14*, *65*). The central idea is to compute the torque for a rotational conformation midway between the stable and most unstable forms. Using the molecular symmetry it is possible to relate the torque at one point to the height of the barrier. This method is discussed more fully in Section 4-8, which is devoted to rotational barriers.

The H–F theorem may also be used for estimating error in approximate computations of molecular energies. Schwartz (*52*) has proposed a technique for estimating the correlation energy E_{corr}, defined as the difference between the exact energy and the energy given by the true Hartree–Fock wavefunction. If ψ is taken to be the exact wavefunction and $\tilde{\psi}$ the Hartree–Fock wavefunctions, the variation of E_{corr} with a parameter Z is given by the H–F theorem as

$$\frac{\partial E_{\text{corr}}}{\partial Z} = \left\langle \psi \left| \frac{\partial H}{\partial Z} \right| \psi \right\rangle - \left\langle \tilde{\psi} \left| \frac{\partial H}{\partial Z} \right| \tilde{\psi} \right\rangle.$$

If the one-electron part of the Hamiltonian H depends on Z through its nuclear-attraction parts, $V_1 = V_1(Z)$, one may conclude that $\partial E_{\text{corr}}/\partial Z \approx 0$ because $\partial H/\partial Z$ will contain only one-electron operators and the Hartree–Fock wavefunction is known to give good values for such quantities. By appropriate constructions of $V_1(Z)$ one may compare a system of interest to another system for which E_{corr} is known and conclude that E_{corr} is the same. Consider, for example, the two-electron diatomic molecule AB and let $V_1(Z) = -(2 - Z) r_{\text{A1}}^{-1} - r_{\text{B1}}^{-1}$. For $Z = 0$ we have HeH^+, while $Z = 1$ corresponds to H_2. One thus expects nearly equal E_{corr}, as is indeed the case: at $R = 1.6$ au, $E_{\text{corr}}(H_2) = -0.04219$ au and $E_{\text{corr}}(HeH^+) = -0.04580$. Note that by another choice of V_1 one can compare HeH^+ and the united atom Li^+ ($E_{\text{corr}} = -0.0435$ au).

Gopinathan and Whitehead (*15*) have recently proposed a correction scheme for improving energy estimates made by the SCF-$X\alpha$ method. They make use of the H–F theorem to derive conditions under which their technique may be expected to apply. The method itself is derived from the virial theorem.

Marron and Weare (*35*) have derived a variational principle for energy differences that is stationary around the iHF result (4-3). This method of computation is more complicated to apply than (4-3), in that two new auxiliary functions are introduced and integrals over both one- and two-electron operators are required. Only one realistic application of their technique, by Trindle and George (*58*), has appeared and it employed very simple wavefunctions; its authors concluded that this method offers no advantages over direct application of the iHF theorem.

A time-dependent iHF theorem has also been derived (*17*). No applications are given by its authors, but they note that the theorem is the point of departure for Golden's (*12*) quantum mechanical theory of gas-phase kinetics. Epstein (*6*) has applied the theorem to the examination of the average dipole moment of a molecule in an external oscillating electric field. Heinrichs (*19*) has also examined this problem using time-dependent perturbation theory and derived a perturbation expansion for the time-dependent iHF theorem in the course of this treatment.

4-3. THE INTEGRAL HELLMANN–FEYNMAN THEOREM

The formula we refer to as the iHF theorem, (4-3), has appeared in the literature without that name in a variety of quantum mechanical analyses. The earliest published use of this formula seems to be by James (*24*) for valence-bond computations on the Li_2 molecule. Similar formulas arise in perturbation-theoretic computations for atomic (*64*) and molecular wavefunctions (*16*), and in scattering theory (*3, 41*). Its usefulness in computing possible large energy differences for quantum chemical processes was first recognized by Parr (*44*), who named the theorem, and by Richardson and Pack (*48*). Within three years of Parr's first publication, he and his associates had published a number of papers exploring implications of the theorem in several different applications (*5, 25, 26, 45, 62, 63*). The formulation of the theorem by Richardson and Pack differs from that of Parr in that the wavefunction for the final state, ψ_B, is not antisymmetric with respect to exchange of electrons between all segments of the whole system. Because of this the two-electron operators in ΔH do not cancel and one of the principal simplifying features of Parr's treatment is lost, namely, the reduction of a problem so that it involves only one-electron operators and a first-order transition density. Physical interpretation and modeling of two-electron, nonlocal interactions is considerably more difficult than is the case for a one-electron situation.

Details of Parr's formulation of the theorem are given most easily if we consider only isoelectronic processes in which A and B correspond to different (static) nuclear configurations for a system. It was mentioned earlier that the restriction to isoelectronic processes may be relaxed; this situation will be discussed in Section 4-7. The difference between the Hamiltonians for the system in the two states may be written

$$\Delta H = \Delta V_{nn} + \Delta V_{ne},$$

where ΔV_{nn} is the difference in nuclear-nuclear repulsion energy for the two configurations

$$\Delta V_{nn} = \sum_{\alpha<\beta} Z_\alpha Z_\beta (R^B_{\alpha\beta})^{-1} - \sum_{\alpha<\beta} Z_\alpha Z_\beta (R^A_{\alpha\beta})^{-1}$$

and ΔV_{ne} is the difference in nuclear attraction operators

$$\Delta V_{ne} = \sum_\mu H'(\mu) \tag{4-10}$$

$$H'(\mu) = \sum_\alpha Z_\alpha (r^B_{\alpha\mu})^{-1} - \sum_\alpha Z_\alpha (r^A_{\alpha\mu})^{-1}.$$

Equation (4-3) may be rewritten using this notation as

$$\Delta E_l = \Delta V_{nn} + \langle\psi_B|\Delta V_{ne}|\psi_A\rangle\langle\psi_B|\psi_A\rangle^{-1}. \tag{4-11}$$

Given an N-electron system, we may define the first-order (spinless) electron transition density between states as *(63)*

$$\rho_{\mathrm{AB}} = \frac{N}{S} \int \psi_{\mathrm{B}}^{*}(1,2,\ldots,N)\, \psi_{\mathrm{A}}(1,2,\ldots,N)\, ds(1)\, d\tau(2) \cdots d\tau(N), \tag{4-12}$$

in which S is the overlap integral, $ds(1)$ implies summation over spin functions for electron number 1, and $d\tau(i)$ implies integration over space and spin coordinates for the ith electron. Using (4-12), (4-11) can be written as

$$\Delta E_l = \Delta V_{nn} + \int \rho_{\mathrm{AB}} H'(1)\, dv(1), \tag{4-13}$$

where $dv(1)$ is the configuration space volume element. Note that the definition of ρ_{AB} is such that the integral over all space equals the number of electrons. This property is one exhibited by a charge density which may be obtained from (4-12) by setting A = B. The reader should see Ref. 25 for detailed derivations of these formulas.

If one considers a closed-shell system, then approximate wavefunctions may be employed that are in the form of single determinants

$$\tilde{\psi}_{\mathrm{A}} = |\chi_1^{\mathrm{A}}(1)\, \chi_2^{\mathrm{A}}(2) \cdots \chi_N^{\mathrm{A}}(N)|,$$

$$\tilde{\psi}_{\mathrm{B}} = |\chi_1^{\mathrm{B}}(1)\, \chi_2^{\mathrm{B}}(2) \cdots \chi_N^{\mathrm{B}}(N)|,$$

and an orbital expansion for the transition density ρ_{AB} is easily obtained. If each spin orbital χ_i is expressed as a space orbital ϕ_i times a spin factor, the following orthonormality properties may be imposed upon the spatial components:

$$\langle \phi_i^{\mathrm{A}} | \phi_j^{\mathrm{A}} \rangle = \langle \phi_i^{\mathrm{B}} | \phi_j^{\mathrm{B}} \rangle = \delta_{ij}, \quad \langle \phi_i^{\mathrm{A}} | \phi_j^{\mathrm{B}} \rangle = d_{ij}.$$

The $(N/2)^2$ overlap elements d_{ij} define a square matrix $\boldsymbol{d}$. If $\tilde{\psi}_{\mathrm{A}}$ and $\tilde{\psi}_{\mathrm{B}}$ are substituted into (4-12) one obtains

$$\tilde{\rho}_{\mathrm{AB}} = 2 \sum_{i=1}^{N/2} \sum_{j=1}^{N/2} \phi_i^{\mathrm{A}}\, d_{ij}^{-1}\, \phi_j^{\mathrm{B}*}, \tag{4-14}$$

where d_{ij}^{-1} is an element of the inverse of $\boldsymbol{d}$.

Wyatt and Parr *(63)* give similar formulas employing both molecular and localized orbitals. The interpretation of (4-13) is considerably simplified if (4-14) can be brought into diagonal form. King et al. *(27)* describe a technique for determining a pair of transformations among each set $\{\phi_i^A\}$ and $\{\phi_i^B\}$ such that the matrix $\boldsymbol{d}$ is diagonal. They refer to the transformed orbitals as *corresponding orbitals.* Corresponding orbitals permit the simplest interpretation of (4-13). If the corresponding orbital transformation is made, (4-13) reduces to the following equation in diagonal form:

$$\Delta E_l = \Delta V_{nn} + 2 \sum_{k=1}^{N/2} \hat{d}_{kk}^{-1} \langle \hat{\phi}_k^B | H' | \hat{\phi}_k^A \rangle, \tag{4-15}$$

where corresponding orbitals are indicated by a caret.

A simple illustration may help to clarify the discussion so far. Consider the process wherein a hydrogen atom is converted to a He^+ atom. The Hamiltonian for the system is $H = T - Zr^{-1}$ where $Z = 1$ for hydrogen and $Z = 2$ for the helium ion. The eigenfunctions and eigenvalues for this sytem are well known.

$$\psi = \left[\frac{Z^3}{\pi}\right]^{1/2} \exp(-Zr) \text{ and } E = -\frac{Z^2}{2}.$$

It is easily verified that computation of the energy difference by either formula (4-1) or by (4-3), and for that matter by (4-2), gives the value $\Delta E = -1.5$ au precisely. It is instructive to perform these computations using approximate wavefunctions, even though the exact wavefunctions are known. We choose the simplest approximations imaginable, namely, the linear functions

$$\tilde{\psi}_A = C_A(1 - r/k_A) \text{ and } \tilde{\psi}_B = C_B(1 - r/k_B),$$

for $0 \leqslant r \leqslant k$, and $\tilde{\psi}_A = \tilde{\psi}_B = 0$ elsewhere. C_A and C_B are normalization constants; k_A and k_B are variational parameters selected to minimize the energy. The variation principle gives $k_A = 4$ and $k_B = 2$ with $C^2 = 30/k^3$. Application of formula (4-1) gives

$$\Delta \tilde{E}_{ed} = -20/16 + 5/16 = -0.9375.$$

This is a fairly good value, considering the degree of approximation in linear wavefunctions. Computation using (4-3) proceeds as follows:

$$\Delta \tilde{E}_l = \langle \tilde{\psi}_B | -r^{-1} | \tilde{\psi}_A \rangle \langle \tilde{\psi}_B | \tilde{\psi}_A \rangle^{-1}$$

$$= -\int_0^2 (1 - r/2)(1 - r/4)\, r\, dr \Bigg/ \int_0^2 (1 - r/2)(1 - r/4)\, r^2\, dr$$

$$= -15/14 = -1.0714.$$

In this example, the result found by using the iHF formula is superior to that found using the expectation value difference. This will not be the case every time. If a different pair of approximate functions were selected the situation might well reverse itself. Epstein et al. (*5*) treat this same example using different approximate wavefunctions.

Because formula (4-1) has been the standard method of computing energy differences for many years, one would like to know when $\Delta\tilde{E}_l$ and $\Delta\tilde{E}_{ed}$ are likely to agree. If they do not agree, then one would like to know which estimate is more likely to be accurate. Epstein et al. (*5*) give the best discussion of conditions necessary and sufficient for $\Delta\tilde{E}_l = \Delta\tilde{E}_{ed}$; this discussion has been expanded by others (*23, 35, 53*). The class of functions that ensure $\Delta\tilde{E}_l = \Delta\tilde{E}_{ed}$ have been termed *superfloating* functions by Epstein et al., in a generalization of the floating concept used by Hurley (*21*) to describe functions that satisfy the H-F theorem. The only way known to construct such functions requires that both $\tilde{\psi}_A$ and $\tilde{\psi}_B$ be variationally selected from the same (linear) basis set. The simplest way to satisfy this condition is to mix $\tilde{\psi}_A$ and $\tilde{\psi}_B$ together to form new functions

$$\tilde{\tilde{\psi}}_A = a\tilde{\psi}_A + b\tilde{\psi}_B \quad \text{and} \quad \tilde{\tilde{\psi}}_B = c\tilde{\psi}_A + d\tilde{\psi}_B.$$

The parameters a, b, c, and d are determined by the conventional variational method which results in the familiar secular equation containing some not-so-familiar integrals, e.g., $\langle\tilde{\psi}_B|H^A|\tilde{\psi}_A\rangle$. This "Epstein mixing" may be illustrated for the process $H \rightarrow He^+$ using our simple linear functions. We find

$$a = 0.74232, \; b = 0.35305, \; c = 0.36413, \text{ and } d = 0.66103,$$

with $\Delta\tilde{\tilde{E}}_l = \Delta\tilde{\tilde{E}}_{ed} = -0.9823$. Note that in this example the expectation value difference has been improved and the new iHF value is worse than before mixing. This is not a general result. It is possible for both values to improve or both values may even become worse.

Before we proceed to a discussion of the accuracy of these formulas, it should be mentioned that situations may arise where formula (4-3) yields an indeterminate result. Two examples may be given that possess this behavior: dissociation and electronic excitation. Dissociation entails transporting one or more nuclei an infinite distance from the remaining nuclei to construct state B. For systems with more than one electron, this results in a 0/0 indeterminacy in (4-3), even for exact wavefunctions. Electronic excitation also causes (4-3) to become indeterminate if approximate wavefunctions are used; indeterminacy is caused in this case by orthogonality between $\tilde{\psi}_A$ and $\tilde{\psi}_B$. There are two ways of circumventing this difficulty. One is to translate one or more nuclei at the same time electronic excitation occurs, thus removing orthogonality.

This technique introduces a new concern referred to as the path problem and dealt with below in Section 4-5. The second method entails a limiting procedure. This procedure and a similar one for dissociation are discussed in Section 4-7.

4-4. THE ACCURACY OF THE iHF FORMULA

Shortly after the first papers using the iHF theorem began to appear, several authors noted a shortcoming in the theorem (*31, 35, 39, 42, 49, 58*). When approximate functions are employed the error in the energy difference is of first order in errors in the wavefunctions $\tilde{\psi}_A$ and $\tilde{\psi}_B$, whereas the expectation value difference formula (4-1) produces a difference good to second order in the errors. Before we embark on a discussion of error (see also Sections 1-10 and 1-11), it may be useful to point out that theory serves two proper and sometimes separate roles. The first is to provide a computational basis for purposes of prediction and the second is to provide a framework for interpretation of experimental findings. Discussions of the utility of the iHF theorem generally fall into one or the other of these two classes. Discussions of accuracy usually come under the first heading, although there have been some novel attempts to bridge the two (*31, 53*). A fair statement of affairs at this writing is that (4-1) is the preferred technique of most investigators for the computation of energy differences. There are, however, certain types of processes (and wavefunctions) for which the iHF formula can be expected to produce superior results in absolute terms. The utility of the iHF theorem for understanding processes involving energy differences is much easier to establish. It rests upon the fact that there is a classical interpretation for every term in (4-3). The tautology between these two schools brings to mind a quote by Toffler (*56*):

> Theories do not have to be 'right' to be enormously useful. Even error has its uses. The maps of the world drawn by the medieval cartographers were so hopelessly inaccurate, so filled with factual error, that they elicit condescending smiles today when almost the entire surface of the earth has been charted. Yet the great explorers could never have discovered the New World without them. Nor could the better, more accurate maps of today been drawn until men, working with limited evidence available to them, set down on paper their bold conceptions of worlds they had never seen.

Epstein et al. (*5*) state that in their opinion the utility of the iHF formula for computing energy differences "is very much a problem for numbers, not formal analysis. . . . '[O]rder' is to some extent a theoretical construct, and its use implies numerical coefficients decreasing with 'order,' which they may not do." There are numerous examples of computations in the literature which serve to make this point; one was given above. The iHF theorem can produce results superior to those found using approximate wavefunctions in (4-1). One reason that an order-of-error analysis cannot be wholly correct is that the error also depends on the size of ΔH (*53*). Calculations described below indicate that for small changes, such as stretching vibrations in diatomic molecules where ΔH is small, the iHF formula gives very accurate results. Results are obtained that are closer to the true energy difference than those obtained by taking the difference of expectation values.

Marron and Weare (*35*) have attempted to overcome the accuracy problem by deriving a variation principle for energy differences that is stable about formula (4-3). Their method provides a means of computing energy differences directly in terms of ΔH which has an error that is bilinear in the errors involved in the wavefunctions for the two states. That is, the error term is second order in the sense that it involves a product of errors associated with each wavefunction. These authors present several simple examples in which error is introduced into the wavefunctions and comparisons are made between the error in energy differences computed using their method, e.g., (4-1), and using (4-3). As expected, the iHF formula produces a large, nearly linear error term in ΔE as a function of the errors in $\tilde{\psi}_A$ and $\tilde{\psi}_B$, one that is much larger than the error in ΔE for the variational method or (4-1). The fact that accuracy analyses are "very much a problem for numbers" was convincingly demonstrated recently by Trindle and George (*58*). These authors tested the Marron–Weare formula for a number of 5- and 6-atom systems undergoing reasonable motions (but using very simple wavefunctions) and concluded that, although there were instances where the variational method was superior to any other, there were also instances in which it provided a most ridiculous estimate of ΔE, far from the reasonable estimates provided by either (4-1) or (4-3). They did not perform a complete analysis of their results, but they do present an argument to account for the deterioration in a second-order estimate of ΔE relative to a first-order estimate. They note that if the error in the wavefunction is large, the parabola associated with a second-order error may well exceed the linear, first-order error in ΔE. In a sense they extended the plot of Marron and Weare into a region where second-order estimates are worse than first-order estimates.

There have been several computational studies involving the iHF aimed at determining the accuracy of the formula. These compare the iHF results to the expectation value difference. Hayes and Parr (*18*) used one-center wavefunctions for computing stretching energies for HeH^+. Other workers (*5*, *36*, *50*) have reported similar computations for the stretching of the hydrogen molecular ion using a variety of wavefunctions and positioning the initial and final wavefunctions in different ways. Marron and Parr provide an extensive error analysis. Cooney (*4*) computed the energy of the He atom using a homogeneous electron gas for the initial state. The final state was described by a variety of wavefunctions, including several which had explicit correlation terms. Recently a series of computations was performed (*59*) for energy differences between pairs of isoelectronic molecules using extremely simple, floating Gaussian wavefunctions.

The studies listed in the previous paragraph all conclude that iHF estimates of energy differences are not well behaved under certain conditions and that it is difficult to predict when the iHF formula will produce reasonable values. Several of these authors have gone further and suggested that the iHF theorem is useless for computations and should be applied only for purposes of interpretation of energy differences.

Several studies have also been published which show the iHF formula can produce results for certain processes that are far superior to estimates by other methods. In one of the earliest studies, dissociation energies were computed using the iHF formula (*48*). The iHF estimates were almost always far better than the $\Delta\tilde{E}_{ed}$. This study fostered a number of other barrier computations that are discussed in Section 4-8. Studies of bond stretching energies (*18, 39*) indicated that the iHF method may be the method of choice, especially for small stretches and for cases where one-center wavefunctions are available. Simons (*55*) has recently developed a formulation for ionization and electron affinity computations using the iHF theorem. His initial results indicate that the iHF estimate will be extremely accurate.

Clearly there are instances when the iHF formula provides exceptionally good estimates of ΔE and others where it is poor. When may one expect the iHF formula to produce a good result? Lowe and Mazziotti (*31*) address this question using the concept of local energy. Their ideas were tested (*36, 39, 59*). The central idea underlying their error analysis is that the transition density (4-12) should be constructed in such a way as to emphasize those regions of configuration space for which $\tilde{\psi}_A$ and $\tilde{\psi}_B$ provide good descriptions and to deemphasize the regions where the functions are poor. This prescription is not always easy to follow. The transition density may be constructed in different ways by shifting positions of the nuclei in the A and B states or by varying one's choice of reference state. Location of good and poor regions depends upon, among other things, the type of wavefunction basis set (STO, Gaussian, single-center), the refinement of the wavefunctions, and the type of system. Even if one does not desire to examine the effects of various factors important in constructing a transition density, the concept is a useful one for analyzing the accuracy of a particular iHF ΔE estimate.

Analysis of the computation reported by Richardson and Pack (*48*) will help illustrate the Lowe-Mazziotti technique. Richardson and Pack calculated dissociation energies for diatomic molecules using as a starting point a $\tilde{\psi}_A$ which is unsymmetrized product of the atomic wavefunctions centered at the atomic nuclei. The final state $\tilde{\psi}_B$ is an MO wavefunction for the diatomic molecule.

Using H_2 as an example one can see that this selection of a reference state emphasizes configurations in H_2 where one electron is near one nucleus while the second electron is near the other nucleus. It also deemphasizes ionic configurations corresponding to small interelectronic separations. This is desirable because MO wavefunctions are uncorrelated and provide poor descriptions of configurations where the electrons are in the same region of space. One can be certain that had Richardson and Pack chosen to place the two reference state H atoms in superposition on the same nucleus, their ΔE estimate would have been terrible. This reference state emphasizes configurations with small r_{12}. Placement of the atom in regions where the molecular wavefunction is very small would also be a poor choice because variationally determined wavefunctions tend to be poorest in these regions.

One can understand the poor results reported by Cooney (*4*) who used a homogeneous two-electron gas as his reference state for computing the He atom energy. This reference state gives equal weight to regions of the wavefunctions far from and close to the nuclei. A reference state emphasizing regions near the nuclei would certainly produce better results. Mazziotti and Lowe (*39*) have provided a detailed analysis of stretching calculations using one-center wavefunctions to represent the states and determine quite precise conditions under which one might expect very accurate $\Delta\tilde{E}_l$ estimates.

Lowe and Mazziotti have also derived an upper bound to the difference $|\Delta\tilde{E}_{ed} - \Delta\tilde{E}_l|$ and find that it is inversely proportional to the overlap integral $\langle\tilde{\psi}_B|\tilde{\psi}_A\rangle$. This premise has been tested (*36, 59*). Both studies conclude that as a rule of thumb, when one has a choice in constructing the transition density, it should be done in such a way as to maximize the overlap between the two states. Trindle and George find the relationship between the error in $\Delta\tilde{E}_l$ and the overlap integral is quite nonlinear. They also discovered that the iHF estimate may be improved substantially by the simple expedient of mixing basis functions into $\tilde{\psi}_A$ that are appropriate to the B state and vice versa. Since the entire basis sets for states A and B do not overlap, this does not correspond to Epstein mixing of the two functions $\tilde{\psi}_A$ and $\tilde{\psi}_B$, and so $\Delta\tilde{E}_l$ need not equal $\Delta\tilde{E}_{ed}$. The authors cite a proverb of theoretical chemistry folklore to explain the improved $\Delta\tilde{E}_l$ estimate. They state: "to describe a process or a change in a system in a given basis, it is appropriate to include in that basis functions appropriate to the system at several points along the transformation."

4-5. THE PATH PROBLEM

In the previous section we alluded to a problem that might be termed the path problem. If we consider a set of processes $A \to B \to C$, then the energies computed by formula (4-1) obey the following associativity relation:

$$\Delta E(A \to B) + \Delta E(B \to C) = \Delta E(A \to C). \qquad (4\text{-}16)$$

This is true even when approximate wavefunctions are used. This relation does not hold in general for iHF energy differences. For cyclic processes this is analogous to a hysteresis effect.

The path problem was noted by Wyatt and Parr (*62, 63*) in their study of the barrier to rotation in ethane using the iHF theorem. Twisting one methyl group by 60° to convert between staggered and eclipsed conformations produces a different transition density and $\Delta \tilde{E}_l$ than that obtained by twisting each methyl group 30° in opposite directions. Hayes and Parr (*18*) reported a series of computations for stepwise stretching of HeH^+ that exhibit the same difficulty. Computations for stretching of the H_2^+ molecule were reported (*5*, *36*, *50*) and analyzed in detail (*31*, *36*). The concepts described in the previous section provide a basis for understanding these results. If a process is carried out so as to maximize the overlap between the initial-state and final-state wavefunctions, one usually obtains the best results. Lowe and Mazziotti have also noted that nuclei may be positioned in such a way as to emphasize the best portion of wavefunction in the transition density. They give rules that apply to several different classes of wavefunctions.

These concepts have been employed to advantage (*37, 38, 55*) in computing molecular binding and electronic ionization energies. By careful selection of reference states, these authors are able to obtain estimates for ΔE by means of the iHF formula superior to those found by the difference of expectation values. Simons (*55*) gives a recipe for choosing a satisfactory reference state.

Note that in the case where superfloating functions are employed, ΔE estimates using (4-1), (4-2), and (4-3), all agree and the associativity relation (4-16) is satisfied for all cases.

4-6. NUCLEAR CHARGING PROCESSES

The charging process converting $H \rightarrow He^+$ described in Section 4-2 is one example of a class of isoelectronic processes characterized by a change in the charge of one or more nuclei. Isoelectronic processes such as $H_2O \rightarrow O^{2-}$ or $N_2 \rightarrow CO$ are two more examples. These processes are of course, artificial; they are constructed for the purpose of comparing the energy of one system with that of another. One of the systems is usually selected because it is well understood or because wavefunctions for it are easily obtainable. In these cases this state may be thought of as a reference state. Nontrivial examples of charging processes have been analyzed using the iHF theorem (*38, 40, 59*).

Since the early years of quantum mechanics, chemists have attempted to understand molecular binding in diatomic hydrides by viewing the molecular species as a perturbation of the united atom. Pauling (*46*) suggested that, for example, the molecule LiH may be viewed as a Be atom in which a proton has been removed from the nucleus. He proposed that the equilibrium internuclear distance in the LiH molecule should be the radius of a sphere centered at the

Be nucleus that contains three of the four electrons, i.e., at which the proton "sees" an effective charge on the Be nucleus of +1. If the proton were placed outside this radius, it would be attracted by the excess negative charge. Inside the radius the proton is repelled. Platt (*47*) extended these ideas to include the computation of force constants (see Section 5-6) and applied the model to a large number of molecules, with unexpected success. Platt's model has been extended (*40*) by application of the iHF formula to the computation of equilibrium bond lengths and force constants in metal carbonyl hydrides, a typical process being $Fe(CO)_5 \rightarrow Mn(CO)_5H$. In developing their model these authors make two simplifying assumptions. They begin by writing down the energy difference expression (4-13), and then presume that all interactions of the CO groups with one another and with the central metal are unchanged in the process. This reduces ΔV_{ne} to $-\Sigma_\alpha r_{H\alpha}^{-1}$; experimental data are given to justify this assumption. The second assumption follows closely the spirit of Pauling and Platt. They replace ρ_{AB} by the one-electron charge density on the central metal atom. These authors conclude that the results obtained from this simple model are quite good. The approach can provide insight into how relatively rare transition metal hydride structures relate to known organometallic structures. It also provides a nice rationale for the observation that M—H stretching frequencies are nearly the same as those for the corresponding diatomic metal hydrides.

Marron (*38*) has analyzed diatomic binding for H_2 and LiH using both neutral and negative atomic reference states according to the two processes $He \rightarrow H_2 \rightarrow H^-$ and $Be \rightarrow LiH \rightarrow Li^-$. This analysis brings out one of the advantages of the iHF method of computing energy differences, namely, a classical electrostatic interpretation of the process can be supplied. Let us consider in more detail the computation for $Be \rightarrow LiH$. Let the two states be described by single-determinant wavefunctions represented as $(1s)^2(2s)^2$ for Be and as $(1\sigma)^2(2\sigma)^2$ for LiH. The transition density (4-12) may then be written:

$$\rho_{AB} = 2\,\frac{1s\cdot 1\sigma\langle 2s\cdot 2\sigma\rangle + 2s\cdot 2\sigma\langle 1s\cdot 1\sigma\rangle - 1s\cdot 2\sigma\langle 2s\cdot 1\sigma\rangle - 2s\cdot 1\sigma\langle 1s\cdot 2\sigma\rangle}{\langle 1s\cdot 1\sigma\rangle\langle 2s\cdot 2\sigma\rangle - \langle 1s\cdot 2\sigma\rangle\langle 2s\cdot 1\sigma\rangle}. \tag{4-17}$$

Note that the integral of the transition density over all space is 4, the number of electrons. The cross terms that appear in this expression complicate the analysis and may be removed by means of the corresponding orbital transformation of King et al. (*27*) to give

$$\rho_{AB} = 2\{1\hat{s}\cdot 1\hat{\sigma}\langle 1\hat{s}\cdot 1\hat{\sigma}\rangle^{-1} + 2\hat{s}\cdot 2\hat{\sigma}\langle 2\hat{s}\cdot 2\hat{\sigma}\rangle^{-1}\}. \tag{4-18}$$

The caret over an orbital indicates a corresponding orbital. Corresponding orbitals have the additional virtue of maximizing the trace of the overlap matrix.

This property has been interpreted by various authors to mean that the orbitals are those that are most transferable from one molecular conformation to another (*54, 57*). For the process Be → LiH this means that the $1\hat{s} \cdot 1\hat{\sigma}$ term in the transition density remains nearly unchanged for positions of the proton slightly greater or less than the equlibrium internuclear distance. Any variation in the transition density is expressed in the $2\hat{s} \cdot 2\hat{\sigma}$ distribution. The force constant may thus be computed by taking the derivative of the energy equation with respect to internuclear distance and ignoring contributions due to the $1\hat{s} \cdot 1\hat{\sigma}$ distribution.

The explicit form for the iHF energy difference is (atomic units)

$$\Delta\tilde{E}_l(\text{Be} \to \text{LiH}) = 3R_{12}^{-1} + 2\langle 1\hat{s} \cdot 1\hat{\sigma}|r_1^{-1} - r_2^{-1}\rangle\langle 1\hat{s} \cdot 2\hat{\sigma}\rangle^{-1} + 2\langle 2\hat{s} \cdot 2\hat{\sigma}|r_1^{-1} - r_2^{-1}\rangle\langle 2\hat{s} \cdot 2\hat{\sigma}\rangle^{-1}, \quad (4\text{-}19)$$

where position 1 is the position associated with the Be or Li nucleus and position 2 is the position of the H atom in the final state. If we note that the transition density is very small outside a sphere centered at position 1 having a radius equal to the LiH bond distance, it is a simple matter to make a semiempirical estimate for each term in (4-19). This exercise highlights the underlying intuitive features of this approach.

The $1\hat{s} \cdot 1\hat{\sigma}$ distribution is *s*-type in character, so we guess that

$$\langle 1\hat{s} \cdot 1\hat{\sigma}|r_2^{-1}\rangle\langle 1\hat{s} \cdot 1\hat{\sigma}\rangle \approx R_{12}^{-1}. \quad (4\text{-}20)$$

The actual value of this integral product is 0.32 au, whereas $R_{12}^{-1} = 0.33$ au. Upon rearrangement of (4-19) and substitution of (4-20) we have

$$\Delta\tilde{E}_l \approx R_{12}^{-1} + 2\langle \rho_{\text{AB}}|r_1^{-1}\rangle - 2\langle 2\hat{s} \cdot 2\hat{\sigma}|r_2^{-1}\rangle\langle 2\hat{s} \cdot 2\hat{\sigma}\rangle^{-1}. \quad (4\text{-}21)$$

The second term in this equation represents the energy required to remove a proton centered at position 1 within the transition density cloud to infinity. This term may thus be approximated by the empirical energy difference $\Delta E(\text{Be} \to \text{Li}^-) = 7.17$ au. The final term in (4-21) may be estimated in the same fashion as (4-20), giving

$$\langle 2\hat{s} \cdot 2\hat{\sigma}|r_2^{-1}\rangle\langle 2\hat{s} \cdot 2\hat{\sigma}\rangle^{-1} \approx R_{12}^{-1}. \quad (4\text{-}22)$$

The actual value of this integral product is 0.31 au. This estimate is expected to be worse because the $2\hat{s} \cdot 2\hat{\sigma}$ distribution is relatively diffuse. A two-term multipole expansion would probably improve matters considerably. The final result of all these approximations is

$$\Delta\tilde{E}_l(\text{Be} \to \text{LiH}) \approx -R_{12}^{-1} + \Delta E(\text{Be} \to \text{Li}^-) = 6.84 \text{ au}. \quad (4\text{-}23)$$

This should be compared with the experimental value of 6.60 au. The physical interpretation of the binding energy of LiH using the Be atom as a reference state is crystal clear: it is the energy needed to move a proton at position 1 to position 2 within a fixed Be-LiH transition density plus the nuclear-nuclear repulsion energy for LiH. Our analysis considered movement of the proton from position 1 to position 2 in two parts: first, removal of the proton to infinity, followed by returning it to position 2 from infinity. One feature of this analysis that sets it apart from previous simple treatments such as Platt's is that the presumption of a fixed transition density during movement of the proton is not an approximation.

At this point it is probably worthwhile repeating a caveat of Epstein et al. (*5*): "It is important not to carry a physical discussion appropriate for one formula over to another one." When one employs the iHF formula, the transition density must be analyzed and understood. It is possible that by using different formulas for energy differences one might, for example, understand energy differences entirely in terms of variations of kinetic energy of the electrons.

Trindle and George (*59*) have employed the iHF formula to examine energy differences in the isoelectronic series CH_3N, CH_3NH^+, CH_3CH, $CH_3CH_2^+$, CH_3BH_2, CH_2NH, and $CH_2NH_2^+$. One purpose of their study was to determine if reasonable results could be obtained when very simple wavefunctions are used to describe the initial and final states of the system. Their findings are encouraging. They also discovered a means of improving the accuracy of the iHF results by (sometimes) several orders of magnitude. This work is discussed in more detail in the previous two sections.

4-7. STRETCHING, DISSOCIATION, AND IONIZATION

Applications of the iHF formula to the computation of energy differences resulting from stretching of the diatomic molecules H_2^+ (*5*, *36*, *50*) and HeH^+ (*18*, *39*) have been reported in fairly extensive detail (see Section 4-5). The iHF formula generally provides good estimates for small variations in the bond distance and also for larger variations in the bond distances when one-center wavefunctions are used (*31*, *36*, *39*). Kim (*26*) derives general formulas for the higher-order derivatives of the energy with respect to internuclear separation of a diatomic molecule by using the H-F and iHF theorems.

Marron and Parr (*36*) have examined the limit of the stretching process, namely, dissociation of a molecule. H_2^+ was studied and considered to dissociate either unsymmetrically by the removal of one of the nuclei to infinity or symmetrically by simultaneous removal of both the nuclei to infinity. The final separated-atom state was represented by the exact wavefunction, while the initial molecular state was represented by a variety of wavefunctions, including variational wavefunctions in both AO bases and elliptic MO bases, wavefunctions constrained to have the correct cusp behavior, wavefunctions having correct long-range behavior, and superfloating wavefunctions formed by Epstein-mixing initial-state and final-state wavefunctions. The iHF energy differences $\Delta\tilde{E}_l$ are compared to the expectation value differences $\Delta\tilde{E}_{ed}$; the iHF results are generally different for the two pathways (symmetrical versus unsymmetrical) while the $\Delta\tilde{E}_{ed}$ values are not. For the case of symmetric dissociation of H_2^+ one obtains an indeterminate transition density; Marron and Parr describe a limiting procedure that resolves this indeterminacy. Plots of the transition densities for stretching and dissociation are given.

Marron (*37*) later applied the iHF formula to dissociation of H_2, LiH, and Li_2 molecules. For these molecules a limiting procedure must be applied in all instances regardless of path for dissociation. The results of these calculations are very sensitive to the accuracy of the wavefunctions used. It is argued that approximate functions can give a qualitatively correct transition density and so the iHF method can be used as an interpretational tool. The dissociation of LiH is analyzed using classical concepts and a model is suggested for the dissociation process.

Simons (*55*) has extended and applied the iHF formula to nonisoelectronic processes of electronic ionization and electron attachment, and to electronic excitation processes. Excitation processes can produce indeterminate transition densities similar to those found in molecular dissociation. He derives formulas for calculating single or multiple ionization potentials and in the course of this treatment develops the notion of a local ionization potential. Computations for the He atom using his technique and various approximate wavefunctions, are significantly more accurate than those estimated by (4-1). The sign of the electron affinity of the hydrogen atom is correctly predicted by the iHF formula, whereas the expectation value difference estimate gives the wrong sign. Unfortunately, Simons's method has not been applied to any other but these two test systems.

4-8. BARRIERS TO INTERNAL ROTATION

We end our review of applications of H–F-type formulas to the computation of energy differences on a topic where much of the activity began. Computation of barriers to internal rotation has absorbed the interest of chemists for a long time. The problem is a difficult one because the barriers are usually a very small fraction of the total molecular energy. Lowe (*32, 34*) has reviewed the numerous theoretical approaches to computing and understanding barriers to rotation. In this section we focus on three methods based on the H–F theorem: one (*13, 14, 65*) which employs the (differential) H–F theorem directly; an application of the integrated H–F theorem (*51*); and methods deriving from the iHF theorem (see also Section 3-3-8).

The first method for computing barriers to internal rotation (*13*, *14*, *65*) relates the torque for a rotational conformation midway between the most stable and unstable conformations to the barrier height. The idea is really quite simple. Imagine that the potential energy for rotation is given by a sine curve $V = A \sin \theta$ where $\theta = -90^\circ$ is the minimum of the curve and $\theta = +90^\circ$ is the maximum. The quantity $2A$ is the barrier height. If one knows the torque at the midpoint, $\theta = 0$, the constant A is determined because $dV/d\theta\big|_{\theta=0} = A \cos \theta\big|_{\theta=0} = A$. Using similar reasoning the torque at any point can be related to the barrier height; the torque at the midpoint is simply a convenient choice for highly symmetric barrier potentials.

By using Hartree-Fock wavefunctions for a system with hindered rotation and a barrier potential that can be fit by a sine curve, one can determine the barrier height from the torque with second-order accuracy in the error in the wavefunctions (*10*). Allen and Arents (*1*) examined several experimentally determined barrier potentials to see if they may reasonably be presumed to be sinelike for purposes of calculating a barrier height by this method. By invoking Freed's theorem (*10*) they conclude that if one uses Hartree-Fock wavefunctions and one computes the barrier from the torque at the midpoint using the H-F theorem, the results should be accurate to within ±5%. The catch here is that true Hartree-Fock functions are available only for the simplest molecules.

Goodisman performed computations for ethane using several approximate wavefunctions and obtained barriers ranging between 0.7 and 2.1 kcal/mole, compared to the experimental value of 2.9 kcal/mole. The total energy for the stable, staggered conformer is 50,000 kcal/mole. Note that the barrier is a very small fraction of the total energy. Extensive analysis by Goodisman of this technique led him to conclude that barriers computed using reasonable wavefunctions will generally be too small. He points to incomplete consideration of polarization effects as the main source of error, and also observes that the method is insensitive to small variations in the wavefunction. One advantage of this method is that it should be useful for investigating the nature and origin of barriers, because one has some physical intuition for charge densities and forces. Goodisman suggests a few appropriate classical models based on this method but does not develop them in any detail.

Let us examine more carefully the barrier to rotation in ethane by means of this method. Imagine that one end of the molecule is fixed and the other is rotated. The H-F theorem gives the torque experienced by the rotating end as a sum of two terms: a repulsive term caused by repulsive forces between the nuclei and an attractive term representing attraction between the protons and the molecular electrons. Following Goodisman we write the following equation for torque:

$$-\frac{\partial E(\theta)}{\partial \theta} = 3R \left\{ F_1 + \int \psi^* \mathfrak{F}_1 \psi \, d\tau \right\},$$

where R is the perpendicular distance of a proton from the C—C axis, F_1 is the force on proton 1 on the rotating end, in the direction of rotation (and normal to R) due to the other nuclei, and $\mathfrak{F}_1$ is the electronic force operator in this same direction. The force F_1 is a simple matter to compute and visualize. The two carbon atoms do not contribute, contributions from the two other protons on the rotating end cancel, and only the protons on the fixed end contribute a rotational force and thus to the torque. The effect of the electronic term is to moderate or attenuate the force due to nuclear-nuclear repulsion. The observed barrier is about 60% of that given by considering nuclear-nuclear repulsion alone. Goodisman has analyzed the ethane molecule in great detail and he concludes that the major attractive contribution to the barrier arises from attraction between the proton and the electrons in the C—H bonds in the fixed methyl group. A fair approximation to the barrier is obtained using a point-charge model to derive the torque. Protons on the fixed end are modeled as point charges having a value somewhat less than one to represent the shielding of the electrons in the bond. Goodisman estimates these charges from atomic electron populations.

A similar interpretation of the ethane rotation barrier has been given by Ruedenberg (*51*) who bases his discussion on the integrated H-F formula. By integrating the H-F expression for the torque on the rotating methyl group between limits defined by the staggered and eclipsed conformers he obtains a formula for the energy difference or the barrier height. Two terms result. The first term is due to nuclear-nuclear repulsion and is simply the difference between the nuclear-nuclear repulsion energies in the two configurations. It is identical to ΔV_{nn} given in Section 4-3. This term is positive (proceeding from the stable staggered conformer to the eclipsed conformer) and about 40% too large. The second term is negative and arises from the nuclear-electronic attraction terms in the Hamiltonian. Ruedenberg decomposes the charge density into a sum of localized molecular orbital squares to facilitate interpretation. To simplify the computation he makes the following approximation: all localized molecular orbitals remain unchanged as the rotation occurs except the three C—H bond orbitals on the rotating end.[1] Because of the symmetry of the ethane molecule the only orbitals that furnish nonvanishing contributions to the nuclear-electronic term are the C—H bond orbitals on the fixed end of the molecule. This situation parallels the previous one where a similar result occurs for nuclear-electronic contributions to the torque in ethane. The reasons are the same in both cases.

The interpretation of the rotational barrier in ethane based on Ruedenberg's method is that it arises because of nuclear-nuclear repulsion (ΔV_{nn}) attenuated by nuclear-electronic attraction for the rotating protons by the C—H bond electrons on the fixed end of the molecule. Ruedenberg modeled the bond orbitals as line charges–a very rough approximation–and finds a barrier about 50% too small. The qualitative features of his exposition are, however, on firm ground, and a number of different methods of modeling the nuclear-electronic attraction term suggest themselves. The key here is that the method is quantum mechanical, but it results in terms–electrostatic interactions–that admit a classical interpretation.

The nuclear-electronic (NE) contribution to the barrier can be examined more closely by focusing on a single proton. Ethane's symmetry requires that the total NE contribution be three times the NE contribution of a single proton. The NE contribution equals the NE attraction of the electrons in the C—H bonds for a proton in the eclipsed conformation minus the staggered conformation attraction. The NE energy is greater (more negative) in the eclipsed conformation because the proton is closest to a C—H bond in that conformation.

By way of introduction to the third and last H-F method for analyzing rotational barriers, we note that Ruedenberg's formula may be derived by making similar, though not identical, approximations (*25*) to the iHF formula (4-3). The iHF formula shares the feature with the previous two methods of providing a means for computing energy differences directly rather than taking the difference of two large energies. It is no wonder that one of the first applications of the iHF method was to analyze the barrier to rotation in ethane (*61-63*). The iHF analysis of the barrier proceeds in many respects in the same fashion as it does by Ruedenberg's method. The barrier to rotation is given as a sum of two terms [see (4-13)]. The nuclear-nuclear repulsion term favors the staggered conformer by 5.1 kcal/mole and the electronic term of 2.2 kcal/mole favors the eclipsed form. The principal and important difference is that the electronic term involves a transition density instead of a charge density as it did in the previous two descriptions. Understanding the ΔV_{nn} term is trivial, so an analysis of a barrier in this formulation becomes an analysis of the electronic term $\int \rho_{AB} H'(1)\, dv(1)$. Wyatt and Parr find that the regions near the protons on the fixed end of the molecule contribute most to the electronic portion of the barrier. They emphasize that this component includes changes in electronic kinetic energy, electron-electron repulsion energy, and nuclear-electron attraction energy but that these various components need not be computed separately in this method of analyzing the barrier. Their success in applying the iHF formula to computation of the barrier in ethane suggested the possibility that the iHF method might be the method of choice for computation of barriers. This possibility was immediately tested by three independent groups (*8*, *43*, *49*), all of whom selected H_2O_2 for analysis. Each group employed different wavefunctions, and all found poor agreement with experiment. It should be noted that H_2O_2 poses a much more rigorous test for a theory of barriers than does a calculation for a system with a methyl group because the symmetry of the methyl group may result in significant error cancellation.

Wyatt and Parr (*62*) suggested modeling the rotation process using point-charge models for the transition density. The iHF formula offers the advantage for modeling that approximations that are weaker than those for other methods can be made to arrive at workable models. For example, in Ruedenberg's method one must assume both frozen and perfect following for localized orbitals to reduce the integrals to tractable form. The iHF does not require this assumption. One pays a price for this advantage in that transition densities instead of charge densities must be modeled. Most people have a better intuitive grasp of charge densities than of transition densities.

Following this suggestion a semiempirical electrostatic model was developed (*28-30*, *32*, *33*) based on the iHF formula. According to the iHF formula the problem of constructing a model for an internal rotation barrier reduces to one of making a suitable model for the transition density. Equation (4-13) assures us that a sufficiently detailed description, e.g., several terms in a multipole expansion of the transition density, will certainly produce accurate results. However, to keep the model mathematically and conceptually tractable a simple model involving only a few parameters is desirable. Lowe and Parr (*29*, *30*) considered rotation about a single bond where one end is fixed. As the other end rotates there is a change in nuclear-nuclear repulsion and also a change in the attraction between nuclei and the transition density. They considered H_2O_2 at first and assumed only that no gross changes in local electron distribution occur as one end of the HOOH molecule is rotated. This assumption means that the transition density on the rotated end will be symmetrically disposed between the initial and final proton positions and will thus make no contribution to the energy difference. The transition density on the rest of the molecule will look very much like the ordinary charge density with a cylindrically symmetric portion located along the O—O rotation axis and a portion along the (fixed) O—H bond. Only the non-cylindrically symmetric part of the transition density contributes to the barrier. Lowe and Parr generated the H_2O_2 potential curve as a function of rotation angle by representing the O—H bond portion of the transition density as a quantity of negative charge at a point on the O—H bond axis. The size and position of the charge were selected for optimum agreement with the experimental curve (Fig. 4-1). The shape of the curve, and hence the torque predicted by this model, agrees quite well with experiment. For such a model to be useful the information derived empirically for one system must be transferable to another. In this case one hopes that for similar types of molecules the O—H segment of the model would remain unchanged. Multiple bonds or highly electronegative atoms could alter this situation drastically, however. Lowe and Parr looked at rotation in methyl alcohol, CH_3—OH, using the same parameters for the fixed OH end, and obtained extremely good agreement with experiment (Fig. 4-2). Subsequent papers by these authors examined situations involving very electronegative atoms and multiple bonds. Using this model, the problem of computing and understanding barriers to rotation resolves itself into questions such as changes in electron distribution in going from an OH

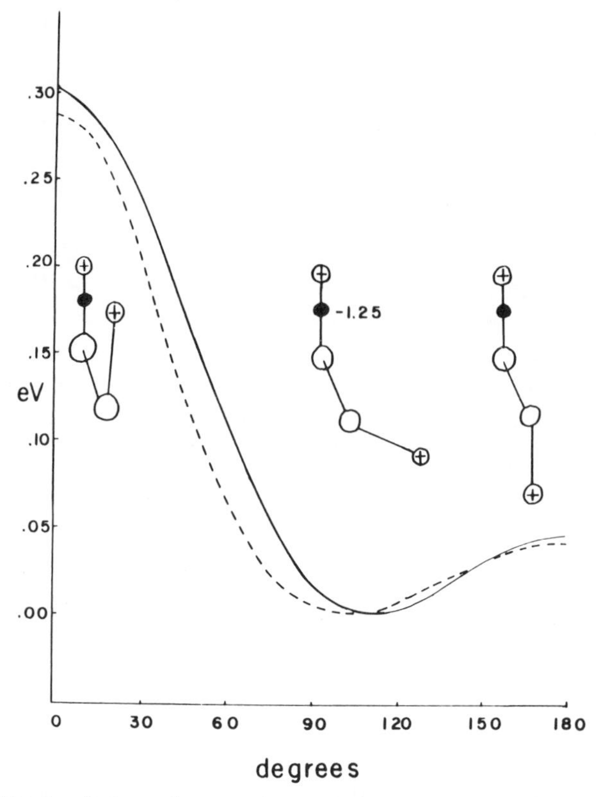

Fig. 4-1. Predicted and observed energy changes for internal rotation in hydrogen peroxide. The model has a negative charge of 1.25 au at a point 61.2% of the distance along the O—H bond. $V(O)$: exper. = 0.303 eV, model = 0.289 eV; $V(\pi)$: exper. = 0.048 eV, model = 0.043 eV; angle of minimum energy: exper. = 111.5°, model = 102°. —— Observed, --- model. (Reproduced from Ref. 28, courtesy the American Institute of Physics.)

to an SH bond or when fluorine is substituted for H in ethane. These questions may be answered in terms of concepts familiar to the chemist, namely, electronegativity, induction, and resonance.

This model has been applied by Lowe and Parr to a wide variety of molecules with remarkable success both in predicting barriers and in understanding previously unexplained relationships between barriers for different molecules. In spite of the attractiveness of the model it has not enjoyed wide acceptance, leading one of its authors (*34*) to discontinue his use of it and turn to other techniques, ones that "are couched in terms and concepts currently in the mainstream of chemical thinking." Perhaps as techniques are developed for direct determination of electron densities, and, it is to be hoped, transition densities, the iHF theorem will become widespread in use and enter the mainstream of chemical thinking.

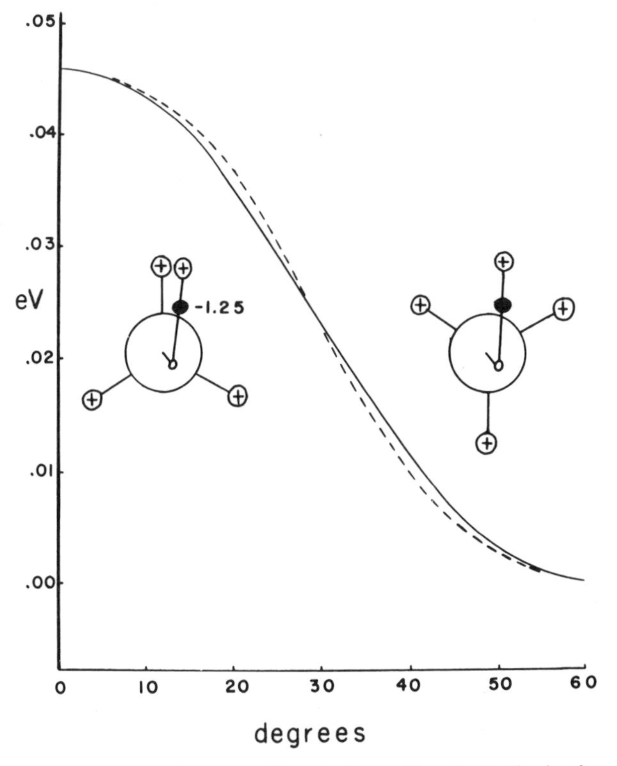

Fig. 4-2. Predicted and observed energy changes for methanol. Both give barrier of 0.046 eV. Model deviates slightly from cos (3θ) behavior. —— Observed, --- model. (Reproduced from Ref. 28, courtesy the American Institute of Physics.)

4-9. SUMMARY

We have discussed methods for computing energy differences that are based on the H–F theorem. We described the integrated H–F theorem and miscellaneous related methods for computing energy differences by integrating the force necessary to take a system from some initial state to some different final state. Several examples were given in Sections 4-2, 4-6, and 4-8 that employed this method. The principal advantages seen of this and other methods based on the H–F theorem is that they admit a rigorous physical interpretation based on classical electrostatics.

We have also described, at some length, methods based on the iHF theorem. Theoretical questions relating to agreement between this method and others, the accuracy of the method, and the so-called path problem were taken up in Sections 4-3 through 4-5. The last three sections of this chapter were devoted to applications of the iHF. Special emphasis was placed on the physical insight gained by this method into the various processes considered.

Acknowledgments

I wish to thank Prof. Robert Parr for a bibliography of H–F papers and Prof. Saul Epstein for a copy of the first chapter of this book. Profs. Lowe and Parr, and the American Institute of Physics were kind enough to grant permission to reproduce Figs. 4-1 and 4-2.

Note

1. Silverstone (*53*) calls such orbitals frozen, perfect following orbitals and discusses this approximation more fully.

References

1. Allen, L. C. and Arents, J., 1972, *J. Chem. Phys.*, **57**, 1818.
2. Berlin, T., 1951, *J. Chem. Phys.*, **19**, 208.
3. Breit, G., 1951, *Rev. Mod. Phys.*, **23**, 238.
4. Cooney, W. A., 1970, *Int. J. Quantum Chem.*, **IIIS**, 381.
5. Epstein, S. T., Hurley, A. C., Wyatt, R. E., and Parr, R. G., 1967, *J. Chem. Phys.*, **47**, 1275.
6. ——, 1974, *The Variational Method in Quantum Chemistry*, (Academic Press, New York), Appendix C.
7. Feynman, R. P., 1939, *Phys. Rev.*, **56**, 340.
8. Fink, W. H. and Allen, L. C., 1967, *J. Chem. Phys.*, **46**, 3270.
9. Foldy, L. L., 1951, *Phys. Rev.*, **83**, 397.
10. Freed, K. F., 1968, *Chem. Phys. Lett.*, **2**, 255.
11. Frost, A. A., 1962, *J. Chem. Phys.*, **37**, 1147.
12. Golden, S., 1949, *J. Chem. Phys.*, **17**, 620.
13. Goodisman, J., 1966, *J. Chem. Phys.*, **44**, 2085; **45**, 4689.
14. ——, 1967, *J. Chem. Phys.*, **47**, 334.
15. Gopinathan, M. S. and Whitehead, M. A., 1976, *J. Chem. Phys.*, **65**, 196.
16. Gray, B. F., 1958, *J. Chem. Phys.*, **29**, 1246.
17. Hayes, E. F. and Parr, R. G., 1965, *J. Chem. Phys.*, **43**, 1831.
18. —— and Parr, R. G., 1966, *J. Chem. Phys.*, **44**, 4650.
19. Heinrichs, J., 1968, *Phys. Rev.*, **172**, 1315; errata, 1968, *Phys. Rev.*, **176**, 2168.
20. Hellmann, H., 1937, *Einführung in die Quantenchemie*, (Franz Deuticke and Co., Leipzig and Vienna), p. 285.
21. Hurley, A. C., 1954, *Proc. Roy. Soc.* (London), **A226**, 170, 179.
22. ——, 1964, in *Molecular Orbitals in Chemistry, Physics, and Biology*, eds. P.-O. Löwdin and B. Pullman (Academic Press, New York), p. 161.
23. ——, 1967, *Int. J. Quantum Chem.*, **IS**, 677.
24. James, H. M., 1934, *J. Chem. Phys.*, **2**, 794.
25. Kim, H. and Parr, R. G., 1964, *J. Chem. Phys.*, **41**, 2892.
26. ——, 1968, *J. Chem. Phys.*, **48**, 301.

27. King, H. F., Stanton, R. E., Kim, H., Wyatt, R. E., and Parr, R. G., 1967, *J. Chem. Phys.*, **47**, 1936.
28. Lowe, J. P. and Parr, R. G., 1965, *J. Chem. Phys.*, **43**, 2565.
29. —— and Parr, R. G., 1966, *J. Chem. Phys.*, **44**, 3001.
30. ——, 1966, *J. Chem. Phys.*, **45**, 3059.
31. —— and Mazziotti, A., 1968, *J. Chem. Phys.*, **48**, 877.
32. ——, 1968, in *Progress in Physical Organic Chemistry*, vol. 6, eds. A. Streitwieser and R. W. Taft (Wiley-Interscience, New York), p. 1.
33. ——, 1969, *J. Chem. Phys.*, **51**, 832.
34. ——, 1973, *Science*, **179**, 527.
35. Marron, M. T. and Weare, J. H., 1968, *Int. J. Quantum Chem.*, **2**, 729.
36. —— and Parr, R. G., 1970, *J. Chem. Phys.*, **52**, 2109.
37. ——, 1970, *J. Chem. Phys.*, **52**, 3600.
38. ——, 1970, *J. Chem. Phys.*, **52**, 3606.
39. Mazziotti, A. and Lowe, J. P., 1969, *J. Chem. Phys.*, **50**, 1153.
40. McDugle, W. G., Jr., Schreiner, A. F., and Brown, T. L., 1967, *J. Am. Chem. Soc.*, **89**, 3114.
41. McIntosh, J. S., Park, S. C., and Rawitscher, G. H., 1964, *Phys. Rev.*, **134B**, 1010.
42. Musher, J. I., 1965, *J. Chem. Phys.*, **43**, 2145.
43. Melrose, M. P. and Parr, R. G., 1967, *Theoret. Chim. Acta* (Berlin), 8, 150.
44. Parr, R. G., 1964, *J. Chem. Phys.*, **40**, 3726.
45. ——, Borkman, R. F., and Marron, M. T., 1968, *J. Chem. Phys.*, **48**, 1425.
46. Pauling, L., 1927, *Proc. Roy. Soc.* (London), **A114**, 181.
47. Platt, J. R., 1950, *J. Chem. Phys.*, **18**, 932.
48. Richardson, J. W. and Pack, A. K., 1964, *J. Chem. Phys.*, **41**, 897.
49. Rothstein, S. M. and Blinder, S. M., 1967, *Theoret. Chim. Acta* (Berlin), **8**, 427.
50. —— and Blinder, S. M., 1968, *J. Chem. Phys.*, **49**, 1284.
51. Ruedenberg, K., 1964, *J. Chem. Phys.*, **41**, 588.
52. Schwartz, M. E., 1966, *J. Chem. Phys.*, **45**, 4754.
53. Silverstone, H. J., 1965, *J. Chem. Phys.*, **43**, 4537.
54. Shchembelov, G. A. and Ustynyuk, Y. A., 1974, *J. Am. Chem. Soc.*, **96**, 4189.
55. Simons, G., 1975, *J. Chem. Phys.*, **63**, 2206.
56. Toffler, A., 1970, *Future Shock* (Random House, New York), p. 6.
57. Trindle, C., 1970, *J. Am. Chem. Soc.*, **92**, 3251.
58. —— and George, J. K., 1975, *Theoret. Chim. Acta* (Berlin), **40**, 119.
59. —— and George, J. K., 1976, *Int. J. Quantum Chem.*, **10**, 21.
60. Wilson, E. B., Jr., 1962, *J. Chem. Phys.*, **36**, 2232.
61. Wyatt, R. E. and Parr, R. G., 1964, *J. Chem. Phys.*, **41**, 3262.
62. —— and Parr, R. G., 1965, *J. Chem. Phys.*, **43**, S217.
63. —— and Parr, R. G., 1966, *J. Chem. Phys.*, **44**, 1529.
64. Wertheim, M. S. and Igo, G., 1955, *Phys. Rev.*, **98**, 1.
65. Zulicke, L. and Spangenberg, H. J., 1966, *Theoret. Chim. Acta* (Berlin), **5**, 139.

5
Calculation and Interpretation of Force Constants

Jerry Goodisman
Department of Chemistry
Syracuse University
Syracuse, New York

Contents

5-1. INTRODUCTION AND DEFINITIONS

The force constants of a molecule are defined in the context of the Born-Oppenheimer approximation which makes the "electronic" energy U (including internuclear repulsion) depend on the parameters specifying the nuclear configuration. Letting λ' and λ'' represent such parameters, we define a corresponding quadratic force constant (FC) as

$$k_{\lambda'\lambda''} = (\partial^2 U/\partial\lambda'\partial\lambda'')_e. \tag{5-1}$$

λ', λ'', and the remaining $\{\lambda\}$ may be nuclear positions in a space-fixed coordinate system, internuclear distances, bond angles, or other coordinates which define the configuration of the nuclei. Cubic and higher FCs are correspondingly defined in terms of third and higher derivatives.

Of course, not all the $3N$ coordinates of an N-atom molecule are independent with respect to changes of U, since overall translations or rotations of the molecule leave U unchanged (Sections 5-2 to 5-4). For a diatomic molecule, the only nonzero FC corresponds to changing the internuclear distance R, i.e.,

$$k_{RR} = (\partial^2 U/\partial R^2)_e. \tag{5-2}$$

The subscript e in (5-1) and (5-2) refers to evaluation at the equilibrium nuclear configuration, the values of the nuclear coordinates being those which minimize U.

Most commonly, one calculates an FC by calculating U for a series of nuclear configurations, generating a potential-energy surface $U(\{\lambda\})$ and evaluating the required derivatives of U numerically, e.g., by fitting the surface to some analytical function which can then be differentiated. The present chapter is concerned with alternative procedures which involve calculation of the force

$$F_{\lambda'} = -\partial U/\partial\lambda' \tag{5-3}$$

followed by numerical or analytical differentiation of $F_{\lambda'}$. If the force is calculated using the Hellmann-Feynman (H-F) theorem, simplified formulas for FCs result (although they are not quite as simple as those the H-F theorem gives for the forces themselves). We will discuss the FC formulas in Sections 5-2 (using the virial form of the H-F theorem) and 5-3 (using the electrostatic form). In Section 5-4, derivations using perturbation theory are considered; they give additional information about the terms in the formulas.

Of course, most approximate wavefunctions do not satisfy the H-F theorem; it would seem to be compounding the error to differentiate a formula which itself is invalid. With respect to actual calculation of FCs, this is certainly true. However, one obtains expressions for FCs which are exact for the exact wavefunction (as well as for certain approximate ones), and these serve to aid our understanding of why FCs have the values they do, and provide starting points for simplifications and approximations. Attempts at analysis of FCs are discussed in Section 5-5, and simple models for the calculation of forces in Section 5-6.

The remaining sections deal with the calculation of FCs from forces rather than energies. Whether or not the H–F theorem holds, there may be advantages in calculating $k_{\lambda'\lambda''}$ by first evaluating $F_{\lambda'}$. As discussed in Section 5-7, for certain kinds of wavefunctions the force may be calculated analytically from the wavefunction at a single nuclear configuration; one does not require a series of calculations to obtain wavefunctions and energies at several configurations. This economy becomes important for polyatomic molecules because the number of coordinates λ and force constants $k_{\lambda'\lambda''}$ grows with the number of nuclei. The calculation of FCs from forces will be discussed in Section 5-8.

A number of extensive discussions of the calculation of FCs have appeared recently, dealing mostly with the usual approach of generating the energy surface and finding its derivatives (*18, 50, 52, 114*). At the same time, there continues to be discussion in the literature about the derivation of FCs from spectroscopic and other measurements (*21, 65, 77, 123, 143*).

By way of introduction, we consider briefly one expression for FCs afforded by the H–F theorem. For calculation of the force itself, the theorem tells us that it is not necessary to know how the wavefunction varies with nuclear displacement, i.e.,

$$\partial U/\partial \lambda' = \int \psi^*(\partial H/\partial \lambda')\, \psi \, d\tau. \tag{5-4}$$

The force $F_{\lambda'}$, is just the expectation value of the force operator, $-(\partial H/\partial \lambda')$, over the wavefunction which is an eigenfunction of the Hamiltonian H. In differentiating H with respect to a nuclear coordinate, the values of electronic coordinates are to be held fixed in whatever coordinate system is chosen. Differentiating again we have

$$\partial^2 U/\partial \lambda' \partial \lambda'' = \int \psi^*(\partial^2 H/\partial \lambda' \partial \lambda'')\, \psi \, d\tau + \int \psi^*(\partial H/\partial \lambda')(\partial \psi/\partial \lambda'')\, d\tau + \int (\partial \psi^*/\partial \lambda'')(\partial H/\partial \lambda')\, \psi \, d\tau. \tag{5-5}$$

The terms involving the derivative of the wavefunction with respect to λ'' cannot be shown to vanish, since ψ is not an eigenfunction of $\partial H/\partial \lambda'$. Indeed, these terms are extremely important, as the discussion of Sections 5-2, 5-3, and 5-5 will show, often all but canceling off the first term (expectation value of $\partial^2 H/\partial\lambda'\partial\lambda''$). They are often referred to as *relaxation terms*, since they represent the effect of changes in the wavefunction due to the change in the parameter λ''. Since the wavefunction is to remain normalized as the nuclear configuration of a molecule changes, the internuclear repulsion V_{nn} does not contribute to these terms:

$$\int \psi^*(\partial V_{nn}/\partial\lambda')(\partial\psi/\partial\lambda'')\,d\tau + \int (\partial\psi^*/\partial\lambda'')(\partial V_{nn}/\partial\lambda')\,\psi\,d\tau$$

$$= (\partial V_{nn}/\partial\lambda')\left[\partial\left(\int \psi^*\psi\,d\tau\right)\Big/\partial\lambda''\right] = 0.$$

A simple example will show how the relaxation terms can be of the greatest importance. Let the nucleus of an isolated atom located on the x-axis at $x = x_A$ have its position moved to $x_A + \delta x$ (δx small). Using (5-5), the change in energy, invoking the H–F theorem to drop out first-order terms, is

$$\Delta U = \left[\int \psi^*(\partial^2 H/\partial x_A^2)\,\psi\,d\tau + \int (\partial H/\partial x_A)(\partial|\psi|^2/\partial x_A)\,d\tau\right](\delta x)^2.$$

With a coordinate system for the electrons fixed in space,

$$\partial^2 H/\partial x_A^2 = -Z_A e^2 \sum_i \partial^2(1/r_{Ai})/\partial x_A^2,$$

where $Z_A e$ is the nuclear charge, $-e$ the electronic charge, and the sum is over electrons. In differentiating $1/r_{Ai}$, the position of electron i is to be held fixed while x_A changes. Now ΔU must be zero, so the relaxation term must exactly cancel off the expectation value of $\partial^2 H/\partial x_A^2$. A similar result obtains for the translation or rotation of a molecule as a whole. The origin of the relaxation is easily understood, since the electron density of the atom completely follows the displacement of its nucleus. As seen in a coordinate system fixed in space, the change $\partial|\psi|^2/\partial x_A$ is large.

Thus, the relaxation terms are a priori important, even if they do not necessarily represent what we should call relaxation. However, as will be seen in more detail in Sections 5-2 and 5-3, the actual meaning of these terms depends very much on what nuclear displacements are chosen for λ' and λ'', and on what coordinate systems are used. In the present example, suppose the electrons are fixed in a coordinate system whose origin is always on the nucleus. Then $\partial\psi/\partial x_A$ vanishes, and so does $\partial^2 H/\partial x_A^2$. Our example shows that the sizes of the individual terms in (5-5) depend very much on the coordinate system used, although their sum does not.

An early discussion (*34*) of the relaxation terms referred to them as nonclassical terms, since the expectation value of $\partial^2 H/\partial\lambda'\partial\lambda''$ would give the FC in a classical electrostatic picture in which the nuclei move through a fixed charge density due to the electrons and the other nuclei, whereas the perturbation-theory expressions for the relaxation terms involve transition densities. We note, however, that keeping electronic coordinates fixed does not necessarily mean the electronic wavefunction is unchanged, since the coordinate system used may depend parametrically on the nuclear configuration.

We will learn more about these terms by considering (5-5) further for a particular choice of λ' and λ''. Supposing both λ' and λ'' represent the x-coordinate of nucleus A in a molecule, the force constant $k_{\lambda'\lambda''}$ is given by

$$\frac{\partial^2 V_{nn}}{\partial x_A^2} - Z_A \int \psi^*\psi \frac{\partial^2}{\partial x_A^2}\left[\sum_i (1/r_{Ai})\right] d\tau - Z_A \int \frac{\partial(\psi^*\psi)}{\partial x_A}\frac{\partial}{\partial x_A}\left[\sum_i (1/r_{Ai})\right] d\tau$$

$$= \sum_{B \neq A} Z_A Z_B \frac{\partial^2(1/r_{AB})}{\partial x_A^2} - Z_A \sum_i \left[\int |\psi|^2 \frac{\partial}{\partial x_A}\left(\frac{\cos\theta_{Ai}}{r_{Ai}^2}\right) d\tau\right.$$

$$\left. + \int \frac{\partial |\psi|^2}{\partial x_A}\frac{\cos\theta_{Ai}}{r_{Ai}^2}\, d\tau\right], \tag{5-6}$$

where r_{Ai} is the distance from electron i to nucleus A and θ_{Ai} is the angle between the x-axis and the vector from A to i, so that $\cos\theta_{Ai} = (x_i - x_A)/r_{Ai}$ (see Fig. 5-1); atomic units are used. $\partial|\psi|^2/\partial x_A$ arises primarily (*17*) from the part of the electron density which follows the movement of A (as for the atomic example above). The result is that the second term in the square brackets on the right-hand side of (5-6) is negative and tends to cancel the other electronic

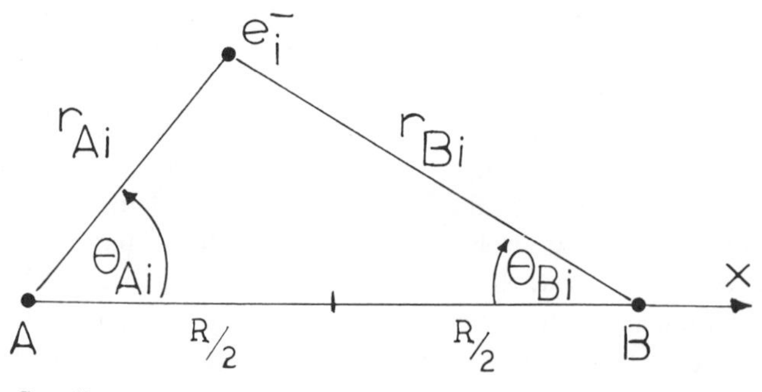

Fig. 5-1. Coordinate system for a diatomic molecule AB, with internuclear distance R.

term. It is reasonable for the relaxation term to give a negative contribution to the FC in most coordinate systems, since electronic relaxation should facilitate the movement of a nucleus, and decrease $\partial^2 U/\partial x_A^2$. Clearly the electronic charge that follows the nucleus rather than remaining fixed in space makes no contribution to the restoring force on the nucleus, and so lessens $\partial^2 U/\partial x_A^2$. Furthermore, a calculation of $\partial^2 U/\partial x_A^2$ by perturbation theory (Section 5-4) shows the relaxation term is always negative for ground electronic states.

If the entire electronic charge density follows A rigidly, the relaxation term exactly cancels the other electronic term, and only the internuclear repulsions contribute to the FC. For a diatomic molecule, this would make the stretching force constant equal to $2Z_A Z_B e^2/R_e^3$. Since actual FCs are much smaller than this, the cancellation within the right-hand square brackets of (5-6) is far from complete. One should not assume that all of the electronic density follows nucleus A and leaves nucleus B behind, since this would be inconsistent with the Born–Oppenheimer approximation (*104*) and the virial theorem (*39*). Although a simple model (Section 5-6) based on just such an assumption was much used for predicting FCs (*104*), its success was in large part accidental.

The simplification afforded by the H–F theorem is still evident in (5-6). Since in a space-fixed coordinate system only the electron–nuclear attraction part of the electronic Hamiltonian depends on nuclear coordinates, all the terms in (5-6) are calculable from the one-electron density

$$\rho = N \int \psi^* \psi \, d\tau', \tag{5-7}$$

where $d\tau'$ means integration over all electronic coordinates but one and $\int \psi^* \psi \, d\tau = 1$. For H_2, the electronic force was calculated, as a function of R (internuclear distance), using a good approximation to the Hartree–Fock wavefunction (*54*). Values of R_e, quadratic and cubic FCs obtained by polynomial fits to the calculated force curve were as good as or even better than those from the energy curve, and the force, containing as it does only one-electron integrals, is considerably easier to calculate than the expectation value of the Hamiltonian.

The problem, of course, is that the latter has to be evaluated to obtain the wavefunctions for force calculation, since as yet we do not have a satisfactory method for independent, direct calculation of molecular one-electron density (*139*). Also, since the H–F forces are very sensitive to small errors in the wavefunction near nuclei (*55*, *63*; Chapter 1), when simple MO wavefunctions are used to evaluate the electric field at a nucleus the equilibrium condition of zero field is generally violated. Thus, although reasonable quadratic and cubic FCs can be obtained using approximate Hartree–Fock wavefunctions for second-row diatomics (*17*), different values result from forces calculated on different nuclei, indicating that the functions do not satisfy the H–F theorem as Hartree–Fock functions should (Section 5-7).

For a function which does not obey the H–F theorem, energy derivatives calculated from the $U(R)$ surface are usually better than those from (5-5). To get $k_{\lambda'\lambda''}$ for such functions, one adds to the right-hand side of (5-5) terms due to deviations from the H–F theorem, namely,

$$\frac{\partial}{\partial \lambda''}\left(\int \frac{\partial \psi^*}{\partial \lambda'} H\psi \, d\tau + \int \psi^* H \frac{\partial \psi}{\partial \lambda'} \, d\tau\right)$$

$$= \int \frac{\partial^2 \psi^*}{\partial \lambda' \partial \lambda''} H\psi \, d\tau + \int \frac{\partial \psi^*}{\partial \lambda'} \frac{\partial H}{\partial \lambda''} \psi \, d\tau + \int \frac{\partial \psi^*}{\partial \lambda'} H \frac{\partial \psi}{\partial \lambda''} \, d\tau$$

$$+ \int \frac{\partial \psi^*}{\partial \lambda''} H \frac{\partial \psi}{\partial \lambda'} \, d\tau + \int \psi^* \frac{\partial H}{\partial \lambda''} \frac{\partial \psi}{\partial \lambda'} \, d\tau + \int \psi^* H \frac{\partial^2 \psi}{\partial \lambda' \partial \lambda''} \, d\tau. \quad (5\text{-}8)$$

The derivatives with respect to nuclear coordinates in (5-4), (5-5), and (5-8) hold the electronic coordinates fixed in some as yet unspecified coordinate system. The difference between H–F theorems in different coordinate systems can be written as a hypervirial theorem (*44*, *49*, *62*). A recent study on long-range forces (*75*) has emphasized that if the coordinate system is properly chosen the H–F theorem is not such a poor approach to numerical calculations as had been believed.

For a diatomic molecule, one can move either nucleus or both nuclei. Although the change in R is the same, the force expressions differ. Also, the electronic positions (fixed during differentiation) may be given in a space-fixed coordinate system, a coordinate system fixed on either nucleus, or confocal ellipsoidal coordinates. In the last case the electron–nuclear distances depend on R and the H–F theorem is the virial theorem. Formulas for FCs from both these theorems are discussed in Sections 5-2 and 5-3.

5-2. FORCE CONSTANTS FROM THE VIRIAL THEOREM

For the diatomic molecule AB the confocal ellipsoidal coordinates of electron 1 are

$$\xi_1 = (r_{A1} + r_{B1})/R, \quad \eta_1 = (r_{A1} - r_{B1})/R, \quad \phi$$

and depend on R. It follows (*46*) that

$$(\partial H/\partial R)_{\xi\eta} = -2TR^{-1} - VR^{-1},$$

where $H = T + V$, T and V being the operators for kinetic and potential energy.

Equation (5-4) now becomes

$$R(dU/dR) = -2\langle T\rangle - \langle V\rangle, \tag{5-9}$$

where the brackets represent expectation values over the normalized wavefunction. This is the virial theorem. The quantity $-R^{-1}(2\langle T\rangle + \langle V\rangle)$ is sometimes referred to (*135*) as the *virial force.* It is not a force on a nucleus or a particular coordinate. Using $E = \langle T\rangle + \langle V\rangle$, the virial theorem can be rewritten in either of the two forms:

$$d(R^2E)/dR = R\langle V\rangle, \tag{5-10a}$$

$$d(RE)/dR = -\langle T\rangle. \tag{5-10b}$$

According to (5-10), potential curves may be discussed by considering expectation values of either the potential or kinetic energy alone.

An advantage of using the kinetic energy is that it is a one-electron operator and simple to calculate. Also, it seems to be less sensitive to details of the wavefunction than, say, the electrostatic force at a nucleus, which means it may be possible to calculate it accurately with an approximate wavefunction. At an arbitrary value of R, we have from (5-10b)

$$R(d^2U/dR^2) + 2(dU/dR) = -d\left(\int \psi^* T\psi\, d\tau\right)\Big/ dR. \tag{5-11}$$

Therefore the quadratic FC is given by

$$k = (d^2U/dR^2)_e = (-R^{-1}\, d\langle T\rangle/dR)_e, \tag{5-12a}$$

since dU/dR vanishes at $R = R_e$. This shows that the existence of a positive FC requires the electronic kinetic energy to be decreasing with R for R near R_e. Several authors (*25*, *26*, *48*, *114*, *118*, *135*) have used the variation of kinetic energy for theoretical calculation of FCs. The calculations are simple when used with MO wavefunctions, and produce accurate quadratic and higher FCs for diatomic hydrides (*117*, *118*). Of course, a wavefunction which is not properly scaled does not obey the virial theorem, but achieving optimum scaling (*55*, *135*) is not difficult. Furthermore, the forces calculated via the virial theorem seem relatively insensitive to errors in the wavefunction (*25*, *26*).

For diatomics, the virial force is derived by expansion and rearrangement of (5-10b):

$$dE/dR = -R^{-1}(E + \langle T\rangle).$$

This was calculated (*116–118, 138*) from SCF wavefunctions for diatomics and polyatomics, and differentiated to get harmonic and higher FCs. One requires knowledge of how the matrix of molecular orbital coefficients changes with nuclear displacements. The assumption of a complete basis set simplified the equations, but also led to inaccuracies. Instead of being used to calculate forces, the virial theorem could be used to obtain relations between FCs. For H_2O, using the three internuclear coordinates, the virial theorem reads

$$\sum_{\tau=1}^{3} R_\tau (\partial E/\partial R)_\tau = 0.$$

Differentiation of this provided two independent relations which could be combined with two experimental data to predict four FCs. Only on using a wavefunction constrained to satisfy the virial theorem did the method lead to acceptable results.

An alternative expression for the FC is

$$\frac{d^2U}{dR^2} = \frac{d^2E}{dR^2} = -\frac{E}{R}\left(\frac{d(\langle T\rangle/E)}{dR}\right)_e, \qquad (5\text{-}12b)$$

where E is the total electronic energy $[U = E(R) - E(\infty)]$. Bishop (*24*) advocated use of (5-12b) in conjunction with the determination of R_e by the virial theorem: it is more convenient to find R_e as the point for which $\langle T\rangle = E$ (vanishing of the so-called virial force) than as the minimum in a plot of E or U against R. Thus from calculated values of $\langle T\rangle/E$ for a series of R values, one can find the point for which $\langle T\rangle/E = 1$ and obtain k from the slope at this point.

If one takes the operator T in a space-fixed coordinate system, the right-hand side of (5-11) requires knowledge of how ψ changes with R. Calculating $d\psi/dR$ by perturbation theory (Section 5-4) gives (*40, 41*)

$$R^2(d^2U/dR^2) + 4R(dU/dR) + 2U = 2S,$$

where S, a sum over excited states, comes from the change in electronic charge distribution with R. From a study of S one can derive information about U.

Formulas for higher FCs of a diatomic molecule in terms of $\langle T\rangle$ result from differentiating (5-11):

$$R(d^3U/dR^3) + 3(d^2U/dR^2) = -d^2\langle T\rangle/dR^2.$$

At $R = R_e$, this reduces to

$$R_e l + 3k = -d^2\langle T\rangle/dR^2.$$

Higher derivatives of U at R_e can be evaluated similarly (*31, 94, 125*).

Now, electrons in a spherical box of radius a would have $\langle T \rangle$ proportional to a^{-2} (*39*). Supposing a proportional to R implies $d\langle T \rangle/dR$ proportional to R^{-3}, so that (5-12a) makes k proportional to R_e^{-4}. The approximate constancy of kR_e^4 is known to spectroscopists. The result, kR_e^4 = constant, holds also if $\langle T \rangle$ is of the more general form $T_0 + T_2 R^{-2}$. Assuming that $Q = -R^{-1}\, d(R^2 \langle T \rangle)/dR$ was constant led to accurate predictions about relations between the FCs of a diatomic molecule (*31, 94–96*). The constancy of Q means

$$\langle T \rangle = T_0 + T_2 R^{-2},$$

since then Q becomes $-2T_0$. Inserting $\langle T \rangle$ from (5-10b) and $d\langle T \rangle/dR$ from (5-11) into the expression for Q, we have

$$R^2 (d^2 U/dR^2) + 4R(dU/dR) + 2E = -2T_0.$$

This may be integrated (*95*) to give

$$E = -T_0 + U_1 R^{-1} + U_2 R^{-2},$$

with U_1 and U_2 integration constants. The three-term expansion in R^{-1} works well for U, and leads to predictions of higher FCs from lower ones. It also suggests simple models (Section 5-6) for diatomic molecules.

This *Fues potential* can be derived by a perturbation theory (*99*) in which the parameter is $\gamma = 1 - R_e/R$ (Section 5-4), and the series is truncated after the term in γ^2. But, the virial theorem is not satisfied to this order (*89*); indeed, if the kinetic energy is correctly calculated to second order, it is in better agreement with experiment than that derived from the second-order energy via the virial theorem. It was suggested that $\langle T \rangle$ be computed to second order in γ, and that the virial theorem be used to obtain the energy from $\langle T \rangle$. If

$$\langle T \rangle = T_0 + a\gamma + b\gamma^2,$$

then (5-10b) implies that U includes, in addition to terms in R^{-1} and R^{-2}, a term in $R^{-1} \ln R$. The parameters can be chosen to fit quadratic and cubic FCs, and higher FCs can then be predicted. The results are usually an improvement on the Fues potential.

Calculation of FCs from the *potential* energy alone has the attractive feature that the latter may be thought of as a classical electrostatic interaction, once the charge distribution is chosen. Differentiating (5-10a) with respect to R and noting that $(dU/dR)_e$ vanishes, we have

$$(d\langle V \rangle/dR)_e = R_e k \tag{5-13}$$

for the quadratic FC and

$$(d^2\langle V\rangle/dR^2)_e - 4R_e^{-1}(d\langle V\rangle/dR)_e = R_e l \tag{5-14}$$

for the cubic FC, where

$$l = (d^3 U/dR^3)_e.$$

The left-hand side of (5-13) may be simplified (*119*) using the electrostatic H–F theorem:

$$(d\langle V\rangle/dR)_e = \left(\int \psi^*(\partial V/\partial R)\psi\, d\tau + \int V(\partial|\psi|^2/\partial R)\, d\tau\right)_e$$

$$= \left(\int V(\partial|\psi|^2/\partial R)\, d\tau\right)_e,$$

because $\langle \partial V/\partial R\rangle$ is the electrostatic force, which vanishes at R_e.

Thus the FC involves *only* the "relaxation" of the electron density, $\partial|\psi|^2/dR$. The importance of the relaxation term was previously noted (*39*) in a slightly different way. Differentiating (5-10a) and inserting (5-9), (5-10a), and the electrostatic H–F theorem in the result, one has

$$R^2 d^2U/dR^2 = R\int V(\partial|\psi|^2/dR)\, d\tau - 2R(dU/dR). \tag{5-15a}$$

Neglecting the relaxation term gives $U = a + b/R$, which represents no binding. Invoking perturbation theory (*119*) in space-fixed coordinates for $\partial|\psi|^2/\partial R$,

$$k = -\frac{1}{R_e}\sum_n{}' \frac{(\partial V/\partial R)_{0n}(V)_{n0} + (V)_{0n}(\partial V/\partial R)_{n0}}{E_n - E_0}, \tag{5-15b}$$

where the prime indicates that the sum is over excited electronic states. It is an unpleasant fact that V (in contrast with $\partial V/\partial R$) is the full potential-energy operator, including two-electron terms.

The above formulas use a space-fixed coordinate system in differentiating $\langle V\rangle$ with respect to R, since the expectation value of $(\partial V/\partial R)$ was taken equal to dU/dR (electrostatic H–F theorem). One could equally well perform this differentiation in the ellipsoidal coordinate system (*103*), from which we derived the virial theorem (5-9) in the first place. Then the differentiation of (5-9) yields

$$R(d^2U/dR^2)+(dU/dR)=-2\left[\langle -2T/R\rangle+\int(\partial\psi/\partial R)^*T\psi\,d\tau+\int\psi^*T(\partial\psi/\partial R)\,d\tau\right]$$

$$-\left[\langle -V/R\rangle+\int(\partial\psi^*/\partial R)\,V\psi\,d\tau+\int\psi^*V(\partial\psi/\partial R)\,d\tau\right]$$

$$(d^2U/dR^2)=R^{-2}\left[6\langle T\rangle+2\langle V\rangle\right]-2R^{-1}\,\mathrm{Re}\left[\int(\partial\psi^*/\partial R)(2T+V)\psi\,d\tau\right], \tag{5-16}$$

where we have used (5-9) again and Re means "real part of."

Now $(\partial\psi/\partial R)$ refers to fixed values of electronic coordinates in the ellipsoidal system. Since the coordinates scale with R in this system, one can say roughly that $(\partial\psi/\partial R)$ corresponds to the change in the wavefunction with shape but not size; if the entire wavefunction expanded uniformly with R, $\partial\psi/\partial R$ would vanish. Though we saw above that neglecting completely the change of electron density with R led to no FC at all, neglecting $\partial\psi/\partial R$ is not as serious. At $R=R_e$ we obtain $k=(d^2U/dR^2)_e=R_e^{-2}[2\langle T\rangle-2R_e(dU/dR)_e]=R_e^{-2}[-2\langle H\rangle]=-2E/R_e^2$. Since the last term (relaxation) in (5-16) is negative by perturbation theory (*46*), $d^2U/dR^2<\langle\partial^2H/\partial R^2\rangle=R^{-2}(6\langle T\rangle+2\langle V\rangle)$. A similar result is valid for any approximate wavefunction for which all variational parameters are optimized for each R. But, the inequality $R_e^2k<-2E(R_e)$ is too weak to be useful (*46*).

For polyatomic molecules, various forms of the virial theorem correspond to different choices of internal coordinates to specify the nuclear configuration. Thus, the relation (*97*)

$$2\langle T\rangle+\langle V\rangle+\sum_{K<L}R_{KL}(\partial U/\partial R_{KL})=0,$$

where the sum runs over *all* internuclear distances, can be simplified (*90*) since all these distances are not independent:

$$2\langle T\rangle+\langle V\rangle+\sum_i R_i(\partial U/\partial R_e)_{R_j}=0. \tag{5-17}$$

Here the sum runs over any collection of internuclear distances which together with angular coordinates $\{\theta_j\}$ suffice to specify the nuclear configuration. Thus, as in diatomics, U can be treated through either kinetic energy or potential energy alone. Note that $\langle V\rangle=2E=-2\langle T\rangle$ for all configurations that maintain $\{R_i\}$, which could be bond lengths, at their equilibrium values. These facts can be used to study molecular shapes (e.g., bond angles) or to develop a localized description of bonding (*128*; Section 3-2-4).

Differentiating (5-17) twice with respect to an angular coordinate gives a formula (*129*) for bending constants,

$$\sum_i R_i(\partial^3 U/\partial\theta^2 \partial R_i) = \partial^2(2E - \langle V\rangle)/\partial\theta^2 .$$

Neglecting the left-hand side, one has $\partial^2 U/\partial\theta^2$ equal to $\frac{1}{2}\partial^2\langle V\rangle/\partial\theta^2$; the FC is $R_j^{-2}(\partial^2 U/\partial\theta^2)$. A localized-charge model was used to treat bending FCs of symmetric AB_2 molecules (*129*). Differentiating (5-17) with respect to any of the A—B and B—B distances gives an equation that the interatomic FCs must satisfy, e.g.,

$$\partial(\langle T\rangle + E)/\partial R_1 = \partial[R_1(\partial U/\partial R_1) + R_2(\partial U/\partial R_2) + R_3(\partial U/\partial R_3)]/\partial R_1,$$

and $(\partial E/\partial R_1)_e = 0$. H_2O was studied in this way (*117*), although using an unsatisfactory wavefunction.

The polyatomic virial theorem, because it involves several distances, generally does not lead to individual FCs of interest. An exception, however, occurs with methane, where by taking $R_1, \ldots, R_4$ in (5-17) as the C—H distances one obtains an expression for the breathing FC (*24*) which looks very much like (5-12b) for a diatomic and provides an alternative to taking the second derivative of the energy.

5-3. FORCE CONSTANTS FROM THE ELECTROSTATIC H-F THEOREM

In (5-4), $\partial H/\partial\lambda$ is $\partial V/\partial\lambda$ if electronic coordinates are held fixed in a space-fixed coordinate system. If λ is a nuclear displacement, $\langle -\partial V/\partial\lambda\rangle$ is proportional to the electrostatic field at the nucleus. For FCs, the second derivative of U may also be taken in another coordinate system.

One can reasonably expect (*60, 105*) that a wavefunction satisfying the H-F theorem will give accurate forces and energies. Wavefunctions giving poor forces can give good FCs if the calculated force surface is parallel to the true one (*23*). Wavefunctions for which all parameters changing with nuclear displacement have been variationally determined satisfy the H-F theorem, e.g., floating wavefunctions, Hartree-Fock functions, and certain single-center expansions (*55*). Note that a wavefunction not satisfying the H-F theorem can give different values for an FC, depending on how internal coordinates are defined with respect to overall translations and rotations of the molecule (*17, 105, 106, 148*).

For a diatomic molecule,

$$\frac{d^2U}{dR^2} = \left\langle \frac{d^2V}{dR^2} \right\rangle + 2\,\mathrm{Re}\left\{ \int \psi^* \left(\frac{dV}{dR} - \frac{dE}{dR} \right) \frac{d\psi}{dR}\, d\tau \right\},$$

where electronic positions are unchanged in differentiation. If ψ satisfies the H–F theorem, dE/dR can be dropped in this equation. More generally, let λ be any parameter in the Hamiltonian H (coordinate of a nucleus, charge of a nucleus, etc.) and E the eigenvalue of H, with $H\psi = E\psi$. Then

$$d^2E/d\lambda^2 = \langle\psi|d^2H/d\lambda^2|\psi\rangle + \langle d\psi/d\lambda|dH/d\lambda|\psi\rangle + \langle\psi|dH/d\lambda|d\psi/d\lambda\rangle,$$

and the sum of the last two terms is negative (*47*, *128*). This holds for optimized variational functions (all variational parameters chosen to minimize the expectation value of $H(\lambda)$ for each $\{\lambda\}$) as well as for the exact wavefunction. Thus we will have for such a function ϕ (which depends on λ through these parameters)

$$d^2\langle\phi|H|\phi\rangle/d\lambda^2 \leqslant \langle\phi|d^2H/d\lambda^2|\phi\rangle.$$

In the case of an FC, this means the relaxation terms are negative. The condition on ϕ is essentially the condition for ϕ to satisfy the H–F theorem.

Even for the diatomic, and even keeping the electrons in a space-fixed coordinate system, arbitrariness remains as to the meaning of d/dR, since R can be changed by δR by an infinite number of combinations of displacements of nuclei A and B. Let us first consider the result of moving nucleus A only, so that dV/dR is $-\partial V/\partial x_A$ which (see equation (5-6), Fig. 5-1) is $\Sigma_i \cos\theta_{Ai}/r_{Ai}^2$. In differentiating this again with respect to x_A we use

$$\frac{\partial}{\partial x_A}\left(\frac{\cos\theta_{Ai}}{r_{Ai}^2}\right) = \frac{3\cos^2\theta_{Ai} - 1}{r_{Ai}^3} - \frac{4\pi}{3}\delta(r_i - r_A), \tag{5-18}$$

which follows because the Laplacian of r_{Ai}^{-1} is $-4\pi\delta(r_i - r_A)$. Using this in (5-6) yields

$$\frac{d^2U}{dx_A^2} = \frac{d^2U}{dR^2} = 2\frac{Z_AZ_B}{R^3} - Z_A\int|\psi|^2\sum_i\left(\frac{3\cos^2\theta_{Ai} - 1}{r_{Ai}^3}\right)d\tau + \frac{4\pi}{3}Z_A\rho(A)$$
$$- Z_A\int\left(\frac{\partial|\psi|^2}{\partial x_A}\right)\sum_i\left(\frac{\cos\theta_{Ai}}{r_{Ai}^2}\right)d\tau \tag{5-19}$$

for the FC.

Kim (*67*) has derived a general formula for the *n*th derivative of U with respect to x_A and given explicit expressions for the first three derivatives, that for the second derivative being (5-19). Schwendeman (*125*, *126*) had earlier considered such expressions, using the relation

$$\int\rho(\cos\theta_A/r_A^2)\,d\tau = \int r_A^{-1}(\partial\rho/\partial x_A)\,d\tau$$

to rearrange the formulas before performing subsequent differentiations. He conjectured that, because ρ is not singular, the importance of the nth-order derivatives of ρ should decrease rapidly with n. Since these enter the relaxation terms, the ratio of $d^n U/dR^n$ to $\langle \partial^n V_{nn}/\partial R^n \rangle$ should decrease rapidly (V_{nn} = internuclear repulsion), but the experimental results indicated only a slow decrease.

Use of the equation above is an important simplification, because otherwise (*92*) the FC expressions which require successive differentiation with respect to nuclear coordinates will involve integrals over high inverse powers of $r_{1\mathrm{A}}$ (distance of the electron from nucleus A). The manipulations needed to obtain this equation can be performed on the expression for the force itself, i.e., write $\nabla_{\mathrm{A}}(1/r_{1\mathrm{A}}) = -\nabla_1(1/r_{1\mathrm{A}})$ and integrate by parts in electronic coordinates. Then the following expression for the Cartesian FC results (*92*):

$$k_{\mathrm{A}p,\mathrm{B}q} = \frac{N}{2}\int d\tau\, \phi^* \left[\delta_{\mathrm{AB}}\left(\frac{Z_{\mathrm{A}}}{r_{1\mathrm{A}}} + \frac{Z_{\mathrm{B}}}{r_{1\mathrm{B}}}\right)\frac{\partial^2 \phi}{\partial x_{1p}\,\partial x_{1q}} + \frac{Z_{\mathrm{A}}}{r_{1\mathrm{A}}}\frac{\partial^2 \phi}{\partial x_{\mathrm{B}q}\partial x_{1p}} + \frac{Z_{\mathrm{B}}}{r_{1\mathrm{B}}}\frac{\partial^2 \phi}{\partial x_{\mathrm{A}p}\partial x_{1q}}\right].$$

Here, p and q specify Cartesian coordinates, A and B specify nuclei, and N is the number of electrons. Formulas for higher FCs can be derived similarly. Applying these to Hartree–Fock wavefunctions constructed from Slater or Gaussian bases, one encounters integrals not much different from those needed for the calculation of the wavefunctions. Other schemes for simplification of the integrals appearing in force calculations have been investigated (*1*, *2*).

Salem (*119*) wrote (5-19) as

$$k = \frac{d^2 U}{dR^2} = Z_{\mathrm{A}}\left[q_{\mathrm{A}} + \frac{4\pi}{3}\rho(\mathrm{A}) - \int\left(\frac{\partial \rho}{\partial x_{\mathrm{A}}}\right)\left(\frac{\cos\theta_{\mathrm{A}}}{r_{\mathrm{A}}^2}\right) d\tau\right]_{R=R_e}, \quad (5\text{-}20)$$

where

$$q_{\mathrm{A}} = 2\frac{Z_{\mathrm{B}}}{R^3} - \int \rho \frac{3\cos^2\theta_{\mathrm{A}} - 1}{r_{\mathrm{A}}^3} d\tau \quad (5\text{-}21)$$

is the electrostatic field gradient at nucleus A due to nucleus B and the electrons. Salem interpreted this as the contribution charges outside nucleus A make to the second derivative of the electrostatic potential at A with displacement of nucleus A in the x-direction. The contact term, $(4\pi/3)\,\rho(\mathrm{A})$, can be related to the magnetic hyperfine interaction, and represents the contribution to this second derivative due to charges inside nucleus A. It may be thought of (*17*) as a restoring force due to charge density at nucleus A which, after the nuclear displacement, is centered off A. The last term in (5-20) is evidently the relaxation or readjustment term, which can be shown (*34*) to be negative. It flattens out the energy curve and reduces the force constant. Individual terms in (5-20) or (5-19) are large compared to k, and there is much cancellation: electron density rigidly following A (e.g., core electrons) contributes to several terms but makes no contribution to the FC.

If both nuclei are moved simultaneously to change R, the expression for d^2U/dR^2 is a combination of expressions like (5-20), involving the field gradients and electron densities at both nuclei. When each moves by the same distance in opposite directions, for instance, one obtains (*119*)

$$k = \frac{2Z_A Z_B}{R^3} - \int \rho \left[Z_A \frac{3\cos^2\theta_A - 1}{4r_A^3} + Z_B \frac{3\cos^2\theta_B - 1}{4r_B^3} \right]_e d\tau$$

$$+ \frac{4\pi}{3} \frac{Z_A \rho(A) + Z_B \rho(B)}{4}$$

$$+ \frac{1}{4} \int \left(\frac{\partial\rho}{\partial x_B} - \frac{\partial\rho}{\partial x_A} \right) \left(\frac{Z_A \cos\theta_A}{r_A^2} + \frac{Z_B \cos\theta_B}{r_B^2} \right) d\tau. \tag{5-22}$$

Here, $\partial\rho/\partial R$ is $\partial\rho/\partial x_B - \partial\rho/\partial x_A$. This is not just the average of (5-19) and the expression obtained from (5-19) by putting B for A (moving nucleus B while keeping A fixed). In fact, if this average is subtracted from twice the expression (5-22), the result is an expression for the FC from which the electronic field gradient terms, including $\rho(A)$ and $\rho(B)$, are eliminated:

$$k = \frac{2Z_A Z_B}{R_e^3} + \frac{1}{2} \int \left[\left(\frac{\partial\rho}{\partial x_B} \right) \frac{Z_A \cos\theta_A}{r_A^2} - \left(\frac{\partial\rho}{\partial x_A} \right) \frac{Z_B \cos\theta_B}{r_B^2} \right]_e d\tau.$$

Note that for a wavefunction satisfying the H–F theorem,

$$Z_A \int (\partial\rho/\partial x_B)(\cos\theta_A/r_A^2)\, d\tau = -Z_B \int (\partial\rho/\partial x_B)(\cos\theta_B/r_B^2)\, d\tau.$$

A host of other relations are possible, e.g., it is possible (*7*) to eliminate the terms in $\rho(A)$ and $\rho(B)$.

One can obtain dU/dR in space-fixed coordinates and then differentiate it in confocal ellipsoidal coordinates to obtain quadratic and higher FCs (*23*, *103*). Using a particular expression for k, one expects significant cancellation between terms. Provided the wavefunction employed satisfies the H–F theorem, one can choose any expression (from several equivalent formulas) for k that emphasizes the terms one is interested in. For a polyatomic molecule, the FCs for the normal modes can be computed (*92*), by linearly combining FCs for Cartesian displacements of individual nuclei, using (5-19).

The Cartesian FCs for displacement of nucleus A may be expressed (*7*) in terms of the corresponding components of the field gradient at nucleus A,

$$k_{xx}^{(A)} = Z_A \left(q_{xx}^{(A)} + \frac{4\pi}{3} \rho(A) - \int \frac{x - x_A}{r_A^3} \left(\frac{\partial \rho}{\partial x_A} \right) d\tau \right),$$

$$k_{xy}^{(A)} = Z_A \left(q_{xy}^{(A)} - \int \frac{x - x_A}{r_A^3} \left(\frac{\partial \rho}{\partial y_A} \right) d\tau \right).$$

For a diatomic molecule these reduce to (5-20). The terms in $\rho(A)$ are large compared to the individual FCs. To eliminate them one can compute combinations such as $k_{xx} - \frac{1}{2}(k_{yy} + k_{zz})$ or $k_{xx} - k_{yy}$. For a diatomic molecule with internuclear axis along the x-axis, the resulting expression for the FC is (*7*)

$$\begin{aligned} k^{(A)} &= k_{xx}^{(A)} - \tfrac{1}{2}(k_{yy}^{(A)} + k_{zz}^{(A)}) \\ &= Z_A \left(\frac{3}{2} q_A - \frac{1}{2} \int \left[\frac{2(x - x_A)}{r_A^3} \frac{\partial \rho}{\partial x_A} - \frac{(y - y_A)}{r_A^3} \frac{\partial \rho}{\partial y_A} \right.\right. \\ &\quad \left.\left. - \frac{(z - z_A)}{r_A^3} \frac{\partial \rho}{\partial z_A} \right] d\tau \right). \end{aligned} \tag{5-23}$$

On the other hand, the average of $k_{xx}^{(A)}$, $k_{yy}^{(A)}$, and $k_{zz}^{(A)}$ is

$$\begin{aligned} \nabla_A^2 U = \left(\frac{d^2U}{dX_A^2} + \frac{d^2U}{dY_A^2} + \frac{d^2U}{dZ_A^2} \right) &= \sum_{B \neq A} \nabla_A^2 \frac{Z_A Z_B}{R_{AB}} \\ &\quad + 4\pi Z_A \rho(A) - Z_A \int \nabla_A \rho \cdot \nabla_A \frac{1}{r_A} d\tau. \end{aligned}$$

This has been used to discuss "atomic" FCs in molecules (*22*, *51*, *71*, *72*).

The sum of the squares of the vibrational frequencies ν_k in a polyatomic molecule may be shown (*51*) to obey

$$\sum \nu_k^2 = C \sum_B \mu_B \nabla_B^2 U, \tag{5-24}$$

where C is a constant and μ_B is the reciprocal mass of atom B. It is possible to find a set of positive values for the quantities $\nabla_B^2 U$ for different atoms such that, for many molecules, the sum on the right-hand side of (5-24) reproduces the observed value of $\Sigma \nu_k^2$ to a few percent. The value for a given atom is independent of the hybridization or bonding around it. Improvement can be obtained if different values are used for $\nabla_H^2 U$, depending on the atom to which H is bonded. An analysis of the density into atomic orbitals allowed $\nabla_B^2 U$ to be analyzed into atomic and overlap contributions.

5-4. FORMULAS FROM PERTURBATION THEORY

The formulas for FCs in Section 5-3 were obtained by differentiating the energy or the electronic Schrödinger equation. An alternative route to differentiation is perturbation theory (*55*). Although the resulting formulas may be identified with those of Section 5-3, perturbation theory adds insight into their meaning and, in some cases, gives additional information.

The first-order part of the perturbation (see e.g. Ref. 132) corresponding to a change of internuclear distance for a diatomic may be taken as

$$H_1 = [C_A(\partial H/\partial x_A)_0 + C_B(\partial H/\partial x_B)_0]\,\delta R \tag{5-25}$$

and the second-order part as

$$H_2 = [C_A^2(\partial^2 H/\partial x_A^2)_0 + 2C_A C_B(\partial^2 H/\partial x_A \partial x_B)_0 + C_B^2(\partial^2 H/\partial x_B^2)_0]\,(\delta R)^2\,. \tag{5-26}$$

The subscript 0 refers to some reference nuclear configuration, and space-fixed coordinates are used for the electrons. The perturbation corresponds to increasing the x-coordinate of nucleus A by $C_A \delta R$ and the x-coordinate of nucleus B by $C_B \delta R$. Various expressions for the first- and second-order energies are possible, but they must all give the same result as long as $C_B - C_A = 1$. The standard formulas of perturbation theory give

$$E^{(2)} = \frac{1}{2}(\partial^2 U/\partial R^2)\,(\delta R)^2 = \langle 0|H_2|0\rangle - \sum_{n\neq 0} \frac{\langle 0|H_1|n\rangle\langle n|H_1|0\rangle}{E_n^{(0)} - E_0^{(0)}}. \tag{5-27}$$

We recognize the terms discussed previously: the expectation value of H_2 over the unperturbed or zero-order wavefunction corresponds to moving the nuclei without allowing the electrons to readjust, while the second is the relaxation term, written as a sum of contributions over excited states, n. The excited electronic states generate different potential surfaces $E_n(R)$; $E_0^{(0)}$ and $E_n^{(0)}$ are the energies on ground and excited surfaces for $R = R_0$. The negative relaxation term tends to cancel off part of the effect of the first term, which is large and positive. The internuclear repulsion does not contribute to the relaxation term because of the orthogonality of the excited electronic states to the ground state. The operators $(\partial H/\partial x_A)$ and $(\partial H/\partial x_B)$ are one-electron operators so that, if $\psi_0^{(0)}$ is taken as a single determinant, only states $\psi_k^{(0)}$ differing in one orbital from $\psi_0^{(0)}$ contribute to the sum.

If the reference nuclear configuration is chosen as the equilibrium configuration, the expression for the second-order energy becomes an expression for the force constant. In the general case of a polyatomic molecule (*55*) in the *m*th electronic state,

$$k_{\lambda\lambda'} = \frac{\partial^2 U_m}{\partial\lambda\partial\lambda'} = \left\langle m \left| \frac{\partial^2 H}{\partial\lambda\partial\lambda'} \right| m \right\rangle + 2 \sum_{p\neq m} \frac{\langle m|(\partial H/\partial\lambda)|p\rangle\langle p|(\partial H/\partial\lambda')|m\rangle}{E_p - E_m},$$

where λ and λ' are nuclear coordinates. The relations between expressions for FCs derived from the H–F theorem and those from perturbation theory have been discussed by Salem (*119*), who also used experimental results to evaluate the sizes of the terms in the case of H_2.

For the diatomic molecule, the choice $C_A = -1$, $C_B = 0$ in (5-25) and (5-26) gives for the stretching FC the expression

$$k = \langle 0|\partial^2 H/\partial x_A^2|0\rangle - 2\sum_n{}' |\langle 0|\partial H/\partial x_A|0\rangle|^2/(E_n^{(0)} - E_0^{(0)})$$

(the prime means $n = 0$ is excluded from summation) and a similar expression obtains for $C_A = 0$, $C_B = 1$. If we take $C_B = \frac{1}{2}$, $C_A = -\frac{1}{2}$, we obtain

$$k = \frac{1}{4}\langle 0|(\partial^2 H/\partial x_A^2) - 2(\partial^2 H/\partial x_A \partial x_B) + (\partial^2 H/\partial x_B^2)|0\rangle$$

$$- \frac{1}{4}\sum_n{}' \frac{|\langle 0|\partial H/\partial x_A|n\rangle|^2 - 2\langle 0|\partial H/\partial x_A|n\rangle\langle n|\partial H/\partial x_B|0\rangle + |\langle 0|\partial H/\partial x_B|n\rangle|^2}{E_n^{(0)} - E_0^{(0)}}.$$

In contrast, the choice $C_A = C_B = \frac{1}{2}$ should produce no change in the energy, representing as it does a uniform translation:

$$0 = \frac{1}{4}\langle 0|(\partial^2 H/\partial x_A^2) + 2(\partial^2 H/\partial x_A \partial x_B) + (\partial^2 H/\partial x_B^2)|0\rangle$$

$$- \frac{1}{4}\sum_n{}' \frac{|\langle 0|\partial H/\partial x_A|n\rangle|^2 + 2\langle 0|\partial H/\partial x_A|n\rangle\langle n|\partial H/\partial x_B|0\rangle + |\langle 0|\partial H/\partial x_B|n\rangle|^2}{E_n^{(0)} - E_0^{(0)}}.$$

Subtracting from the previous expression for k, a new expression is obtained:

$$k = -\langle 0|(\partial^2 H/\partial x_A \partial x_B)|0\rangle - \sum_n{}' \langle 0|(\partial H/\partial x_A)|n\rangle \langle n|(\partial H/\partial x_B)|0\rangle/(E_n^{(0)} - E_0^{(0)}). \quad (5\text{-}28)$$

The terms in this sum are not necessarily all of the same sign, whereas the terms in previous sums were all negative.

The advantages of this expression (*86*) over others are: (i) the individual terms are relatively smaller, (ii) there can be cancellation between terms in the sum over excited states, and (iii) the matrix elements in the sum involve only electron-nuclear interaction operators while the "expectation value" is just the inter-nuclear repulsion [cf. (5-22)]. Explicitly,

$$k = \mathrm{Re}\left[\sum_{n>0} \frac{\left\langle 0\left| Z_A \sum_i \cos\theta_{Ai}/r_{Ai}^2 \right| n\right\rangle \left\langle n\left| Z_B \sum_i \cos\theta_{Bi}/r_{Bi}^2 \right| 0\right\rangle}{E_0^{(0)} - E_n^{(0)}}\right] + \frac{2Z_A Z_B}{R_e^3}. \quad (5\text{-}29)$$

Murrell (*86*) plotted $k/2Z_A Z_B$ against R_e^{-3} for all diatomic molecules for which the information was available. Families of points were apparent, according to the groups of the periodic table of the atoms from which the molecules were formed. For the homonuclear alkali diatomics, a line of slope close to unity was obtained; for all other molecules, the points lay to the right of this line, meaning that the first term of (5-29) is negative. Attempts to calculate the actual values of the integrals appearing led to an analysis in terms of atomic orbitals.

An expression like (5-29) was also derived (*86*) for the stretching-interaction FC of linear triatomics ACB. In this case, the energy due to a displacement of A by δR_A and B by δR_B changes the energy by $\frac{1}{2}k_{AC}(\delta R_A)^2 + k_{AB}(\delta R_A)(\delta R_B) + \frac{1}{2}k_{BC}(\delta R_B)^2$, assuming the system remains linear. If the A—B bond is compressed while B—C is stretched by displacing A by $-\delta R_A$ and B by δR_B, the energy change is $\frac{1}{2}k_{AC}(\delta R_A)^2 + k_{AB}(-\delta R_A)(\delta R_B) + \frac{1}{2}k_{BC}(\delta R_B)^2$. By subtracting the second-order perturbation theory expression for this change from that for the previous one, one obtains an expression for k_{AB} which is the same as (5-29).

Perturbation theory for change of R can actually become a tool for computing $U(R)$. A variational principle can be used to obtain approximations to the first-order wavefunction, from which the second- and third-order energies can be obtained, leading to quadratic and cubic FCs. However, calculations on H_2 by such a method (*23*) did not produce very good results unless a correction, employing experimental information, was introduced. Mills (*84*) earlier performed approximate calculations for larger molecules with approximate MO functions as zero-order functions, approximating the first-order wavefunction by the zero-order determinant times a sum of one-electron functions to be optimized, since the perturbation for change of R is one-electron.

Instead of carrying out perturbation theory starting from the equilibrium nuclear configuration, one can start from the united atom (UA) and calculate (*33*) first- and second-order energies with the united atom as zero-order nuclear configuration, the perturbations being nuclear displacements from the united-atom origin. The changes in the electronic wavefunction and electronic energy are written as expansions in this perturbation, and the energy expansion differentiated with respect to nuclear coordinates to yield expressions for the forces and force constants. Besides yielding Platt's united-atom model (Section 5-6), the method calculates interaction FCs of molecules like CH_4 with good success. The nuclear and electronic contributions to the FC, which have opposite signs, are each much larger than the FC itself, the electronic contribution all being due to relaxation or rearrangement of the electron distribution during perturbation. Related to the UA theory is a perturbation theory starting from the spherically symmetric molecular puff (*70*), which corresponds to replacing the light nuclei in a molecule like CH_4 by a shell of positive charge. The energy of the puff (unperturbed energy) is closer to that of the molecule than is the united-atom energy. Calculations in this formalism for 10-electron MH_k molecules determined the breathing FCs, as well as FCs for bend–stretch interactions, with generally good success.

Byers Brown (*34*) considered perturbations involving changes in scale parameters, such as those which lead to the virial theorem in first order, instead of nuclear displacements. The second-order perturbation energy led to the equation

$$\frac{1}{2}\sum_{\mathrm{A,B}} R_{\mathrm{A}} \cdot \nabla_{\mathrm{A}}(\nabla_{\mathrm{B}} U) \cdot R_{\mathrm{B}} = 3\langle T\rangle + \langle V\rangle + \sum_{m}{}' \frac{|\langle 0|T|m\rangle|^2}{E_0^{(0)} - E_m^{(0)}}.$$

The sums on the left-hand side are over nuclei, and T is the kinetic energy operator for the electrons. Since $H = T + V$ and the unperturbed wavefunctions are orthogonal eigenfunctions of H, one can put V for T in the sum. A related perturbation treatment (*99*) for diatomic systems wrote $H = R^{-2}t + R^{-1}v$, where t and v were the kinetic and potential energy operators in confocal ellipsoidal coordinates, to derive the energy in a series in R^{-1}. The perturbation parameter was $\mu = (R - R_e)/R$, the energy being derived as R_e/R times the power series $\Sigma_{i=0}^{\infty}\, \omega_i \mu^i$. The quadratic FC is $2R_e^{-2}(\omega_2 - \omega_1)$, and expressions for higher FCs are derived. If $\omega_n - \omega_{n-1} = 0, n \geqslant 3$, one gets $U = U_0 + U_1 R^{-1} + U_2 R^{-2}$, the Fues potential.

This treatment was extended to polyatomic systems (*37*) by multiple perturbation theory. Using confocal ellipsoidal coordinates for each pair of nuclei i, U was expanded in the parameters $\lambda_i = 1 - R_{ie}/R_i$, the coefficients being obtained through experimental FCs. Solution of the first-order perturbation equation can give quadratic and cubic FCs.

The perturbation treatment of the long-range interaction between H atoms yields the energy as a series in R^{-1}. Each term can be obtained through either an energy or an H-F force calculation, using the perturbed wavefunction. Unfortunately, for a given term in the energy, it was necessary (*63*) to go to higher terms in the wavefunction if the H-F route is taken. More precise wavefunctions are needed to get $U(R)$ from forces (cf. Section 5-7). Some apparent contradictions between the perturbation treatments of force and energy vanish if one chooses a proper coordinate system (*75*; Section 7-2).

A quite different perturbation theory for FCs and other properties treats interelectronic repulsion as the perturbation (*131*). Although the total energy is correct to within 10%, this gives poor FCs unless the wavefunctions are constrained to satisfy the virial theorem, $\langle T \rangle + E = 0$, $[d\langle T \rangle/dR]_e + R_e k_e = 0$ [see (5-9) and (5-12)]. E_e, R_e, and k_e are the experimental values.

We now return to the FC expressions to see how their evaluation with accurate wavefunctions helps our understanding of the various terms. Sections 5-6 and 5-7 discuss calculations with more approximate wavefunctions.

5-5. ANALYSIS OF CALCULATED FORCE CONSTANTS

By calculating individual terms in the FC expressions Bader and co-workers (*17, 18*) gave a detailed analysis of FCs for small diatomics: Using atomic population analysis to make the nonunique division of ρ into $\rho_A + \rho_B + \rho_{AB}$, and moving nucleus A, the FC becomes

$$k = (2Z_A/R_e^3)(Z_B - K_A^{e1}), \tag{5-30}$$

where K_A^{e1} has contributions from the three parts of ρ. With approximate Hartree-Fock wavefunctions, these contributions are all about the same size, but different signs appear, making for cancellation.

The contribution of ρ_B (K_A^{BB}), a relaxation term when viewed in a coordinate system fixed on A, is equivalent to a static field-gradient contribution, since

$$-\int \left(\frac{\partial \rho_B}{\partial x_A}\right)\left(\frac{\cos\theta_A}{r_A^2}\right) d\tau = \frac{4\pi}{3}\rho_B(A) - \int \rho_B \frac{3\cos^2\theta - 1}{r_A^3} d\tau.$$

The FC becomes

$$k = (2Z_A/R_e^3)[(Z_B - K_A^{BB}) - (K_A^{AA} - K_A^{AB})],$$

where $Z_B - K_A^{BB}$ is the "total unshielded charge on B" and the rest "true relaxation terms." K_A^{AA} is positive (decreasing k) if, when A is moved, ρ_A is polarized in the direction of motion.[1] The polarization direction could be predicted from polarization already present at R_e. For second-row covalent diatomics the relaxation of the atomic density ρ_A tends to oppose nuclear motion and increase k ($K_A^{AA} < 0$) while the relaxation of the overlap density facilitates nuclear motion. The latter dominates in covalent molecules, the former in ionic ones. However, for ionic molecules with A as anion, relaxation of anionic density makes k less than the static contribution, $(2Z_A/R_e^3)(Z_B - K_A^{BB})$ while k for covalent molecules is made large by deshielding (*17*).

For second-row ionic hydrides (*18*), the relaxation terms K_A^{AB} and K_A^{AA} tend to oppose displacement of the heavy nucleus; the reverse is true for covalent hydrides. Still, FCs are larger for covalent molecules because of larger field-gradient (nuclear deshielding) contributions. A similar analysis (*35*, *36*) for H_2 and Li_2 found that both core and overlap densities relaxed to facilitate nuclear motion and lower k, i.e., density moved from the overlap and atomic regions to a region outside the moving nucleus. However, the conclusion (*36*) that core density on the moving nucleus does not follow rigidly has been challenged (*6*).

FCs for polyatomic molecules have been analyzed by Nakatsuji et al. (*88*), using extended Hückel wavefunctions in the context of ESF theory (Chapter 3). These wavefunctions do not satisfy the H-F theorem, but parameters were introduced to remedy this and other defects. The force on a nucleus was separated into three parts, AD, EC, and EGC (or GC) forces (Section 3-3-2). It was found that for the C—C stretch in ethane the largest contributions to the FC were the C—C exchange and the $C—H_3$ exchange forces, but these cancel leaving no dominant contribution. For the bending FC the AD force on C, which is due to hybridization, is most important. An alternative analysis for polyatomic molecules, based on maximum overlap of central atomic orbitals with ligand bonding orbitals, is available (*83*).

By decomposing ρ into localized orbitals for ethane, ethylene, and acetylene, the signs of interaction FCs (*109*) can be determined from the displacements of localized electron pairs during geometry distortion. Other analysis (*85*) of interaction FCs is in terms of the change in hybridization due to orbital following of bending displacements. For example, increasing *s*-orbital content gives a shorter bond and a positive stretch–bend interaction constant; this is related to the overlap method. Other interpretations of signs and magnitudes of interaction FCs have also been presented (*16*, *45*, *78*). For a displacement along the normal mode q_i, the energy changes by (Sections 5-4, 3-2-3)

$$\Delta U_0 = \frac{1}{2}\langle 0|\partial^2 V/\partial q_i^2|0\rangle\, q_i^2 + \sum_k{}' \frac{|\langle 0|\partial V/\partial q_i|k\rangle|^2}{U_0 - U_k}\, q_i^2,$$

where the sum in the relaxation term (negative) is over excited states. Assuming that the sum is dominated by the lowest-lying excited state, Bader (*16*) searched out the q_i which couples this state to the ground state, i.e., $\langle 0|\partial V/\partial q_i|k\rangle \neq 0$. Thus, the interaction FC for two valence displacements having the same signs in q_i will be negative (see also Refs. 145, 146).

5-6. SIMPLE MODELS

In an expression like (5-19) the most difficult term is that due to the relaxation of electron density as the nuclei move. Platt (*104*) proposed a model for diatomic hydrides which neglects this term. Although this violates the virial theorem and the Born–Oppenheimer approximation, the results were surprisingly good because of a cancellation of errors. The electron density is taken, independent of internuclear distance, as that of the corresponding UA, centered on the heavy atom. The outward force on the proton is

$$F = -ZR^{-2} + \int (x - R)|r - R|^{-3} \rho^{\mathrm{UA}}\, d\tau, \tag{5-31}$$

where Z is the heavy nuclear charge and the proton is on the positive x-axis at a distance R from the origin. Since ρ^{UA} is spherically symmetric, the electronic force is

$$4\pi \frac{d}{dR}\left[\int_0^R \rho^{\mathrm{UA}} \left(\frac{r^2}{R}\right) dr + \int_R^\infty \rho^{\mathrm{UA}}\, r\, dr\right] = -\frac{4\pi}{R^2}\int_0^R \rho^{\mathrm{UA}}\, r^2\, dr. \tag{5-32}$$

The total force on the proton is $C(R)\,R^{-2}$ where $C(r)$ is the total charge (nuclear plus electronic) in a sphere of radius r. At $R = R_e$, the number of electrons inside a sphere[2] of radius R equals Z. The force constant becomes

$$-dF/dR = 2C(R)\,R^{-3} + 4\pi R^{-2} \rho^{\mathrm{UA}}(R)\,R^2$$

evaluated at $R = R_e$. Since $C = 0$ here,

$$k = 4\pi\rho^{\mathrm{UA}}(R_e). \tag{5-33}$$

In (5-33) one can use either experimental or calculated (putting $C = 0$) R_e, the former giving slightly better values of k. The calculated values of both R_e and k were in good agreement with experiment (*104*).

The use of the spherically symmetric united-atom density is more reasonable for polyatomic hydrides XH_2, XH_3, XH_4 than for diatomics. Taking R_A as the distance of nucleus A (proton) from the center and R_{AB} an internuclear distance, the potential surface is (*79*)

$$U(R) = \sum_{A} \phi(R_A) + \sum_{A<B} (R_{AB})^{-1},$$

where the potential $\phi(R)$ obeys, as in (5-32), the constraint

$$-d\phi/dR = R^{-2}\left[Z - 4\pi \int_0^R r^2 \rho^{UA}\, dr\right] = R^{-2}C(R),$$

Z being the united-atom nuclear charge. The calculation of equilibrium nuclear configurations and FCs now becomes simply a problem of geometry. For example, in case of tetrahedral hydrides, $R_{AB} = R_A(2\sqrt{6}/3)$ and

$$\partial U/\partial R_A = 4C(R_A)/R_A^2 - (3\sqrt{6}/2R_A^2).$$

At $R_A = (R_A)_e$, $C(R) = 3\sqrt{6}/8$. Taking $R_A = (R_A)_e$ and n as the number of protons, the breathing FC for a spherical electron density is (*24*)

$$k_V = 4\pi n\rho(R_A) + nR_A^{-2} \int_0^{R_0} 4\pi r^2 \,[d\rho/dr]_{r=R_A}\, dr.$$

U may be expanded to second order in ΔR_A and ΔR_{AB} to obtain FCs in terms of $\rho(R_A)$ which can be calculated from experimental k_V. Overall, these calculated FCs are not very satisfactory (*79*). Furthermore, where dipole moments could be calculated from the model, requiring additional assumptions for the heavy-atom density, results are poor, casting doubt on the validity of the model.

The reasons for the success of Platt's model with diatomic hydrides have been investigated in some detail. It is clear that some cancellation of errors is involved since [see (5-15)] neglecting electron relaxation leads to $U = a + bR^{-1}$, i.e., an unbound molecule (*39*). Writing $\tilde{F}_A$ for the deviation of the force on nucleus A (proton) from the force, (5-32), due to the UA density, the correct expression for the FC is

$$k = \{4\pi\rho^{UA} + 2Z_B\tilde{F}_A/R + Z_B(\partial\tilde{F}_A/\partial R)\}_e.$$

Only the first (Platt's) term remains if $\tilde{F}_A$ is proportional to R^{-2} (*39*).

Bratož and co-workers (*3*, *32*, *33*) calculated $U(R)$ for various molecules with configuration–interaction (CI) wavefunctions built from determinants of UA orbitals, one of which was the ground-state UA wavefunction. The calculated FC was analyzed into electrostatic ($\langle \partial^2 H/\partial R^2 \rangle$) and electron relaxation parts, the former having contributions from the UA and other configurations. Considerable cancellation occurred between the relaxation contribution to k and that of the "excited" configurations to the electrostatic part (*3*), so that k was coincidentally almost equal to that obtained by ignoring relaxation and assuming that near R_e the molecular density is well represented by the UA density when, in fact, it is not. More recent work (*19*) also confirms the latter, showing that the error is canceled by neglecting relaxation. On the other hand, for stretch–stretch interaction FCs, the relaxation term cancels the UA term, so Platt's model works poorly. Analysis of k into orbital contributions (*33*) shows that UA core electrons act like point charges, each contributing $-2/R^3$ to k, while valence electrons contribute much less. Neglecting valence electrons, $k = 2Z'/R^3$ where Z' is the net charge of the core. This formula works to ~10% for alkali hydrides and homonuclear diatomics.

In order to remove the inconsistency of Platt's model with the virial theorem, the electronic wavefunction can be made R-dependent by taking the scaled UA density at each R, the scale factor η being chosen variationally (*61*). Let $T(1)$, $V(1, R/\eta)$ represent the kinetic and potential energies computed with the unscaled wavefunction at nuclear configuration R/η. The scaled energy for configuration R is $\eta^2 T(1) + \eta V(1, R/\eta)$. Then the optimum scale factor is determined by

$$\frac{\partial E(\eta, R)}{\partial \eta} = 2\eta T(1) + V(1, R/\eta) + \frac{R}{\eta} \frac{dV(1, R/\eta)}{d(R/\eta)} = 0$$

for each R. Explicit formulas for R_e and k result from wavefunctions and energies for different R. This model, which is almost as simple as Platt's and gives better results for hydrides, includes part of the variation of ρ with R in that ρ varies in size but not in shape (*61, 80*). An attempt to use separated-ion electron densities instead of UA densities led to considerably worse results. The scaled UA model has also been applied (*80*) to metal carbonyl hydrides, using the integral H–F theorem (Chapters 1 and 4) and assuming that the energy change on removing a proton from a metal nucleus of charge $Z + 1$ and placing it at a distance r_{MH} is equal to

$$Zr_{\text{MH}} - \int \rho_{\text{XY}}^{\text{M}}(1)\, r^{-1}\, d\tau,$$

with the transition density $\rho_{\text{XY}}^{\text{M}}(1)$ equal to the electron density of atom M of charge Z, the CO groups surrounding M being unchanged. Equilibrium M—H distances and stretching frequencies have been discussed.

The electron density of a diatomic AB can be divided into three parts depending on how it behaves as R changes, namely, a part which rigidly follows nucleus A, a part which remains rigidly attached to nucleus B, and a part which follows neither, i.e.,

$$\rho(r) = \rho_A(r) + \rho_B(r) + \rho_{NPF}(r), \tag{5-34}$$

where NPF means "not perfectly following" (8). Here ρ_A depends on r relative to nucleus A, independent of the position of nucleus B, while ρ_B depends on r in a coordinate system centered on B. For a polyatomic molecule (8),

$$\rho(r; R_A, \{R_N\}) = \rho_A(r - R_A) + \sum_{N \neq A} \rho_N(r - R_N) + \rho_{NPF}(r; R_A, \{R_N\}). \tag{5-35}$$

Now, the FC for displacement of nucleus A is (Sections 5-1 and 5-2)

$$\nabla_A^2 U = 4\pi Z_A \rho(A) - Z_A \int \nabla_A \rho \cdot \nabla_A |r - R_A|^{-1}\, d\tau, \tag{5-36}$$

and there is significant cancellation between the terms on the right-hand side because the part of density which moves rigidly with A gives no contribution to $\nabla_A^2 U$. Using (5-35), ρ_A does not enter and since in a diatomic molecule $\nabla_A \rho_B = 0$ (B is fixed while A moves),

$$\nabla_A^2 U^e = d^2U/dR^2 + (2/R)(dU/dR) = 4\pi Z_A \rho_B(A) + \{4\pi Z_A \rho_{NPF}(A) - Z_A \langle \nabla_A \rho_{NPF} \cdot \nabla_A |r - R_A|^{-1} \rangle\}, \tag{5-37}$$

which is an exact equation. If one assumes that $\rho_{NPF}(A)$ is negligible compared to $\rho_B(A)$ and drops the last terms, (5-37) leads to a classical Poisson equation for nuclear motion. The FCs are

$$k = (\nabla_A^2 U)_{R=R_e} \simeq [4\pi Z_A \rho_B(A)]_{R_e},$$

$$l = \left[\left(\frac{d}{dR} - \frac{2}{R}\right)\nabla_A^2 U\right]_{R=R_e} \simeq 4\pi Z_A \left[\left(\frac{d}{dR} - \frac{2}{R}\right)\rho_B(A)\right]_{R_e},$$

and so on. Taking $\rho_B(A)$ as the total electron density at a distance R from B, opposite to A, leads to good agreement for k and higher FCs (8).

This implies that the terms in curly braces in (5-37) add up to a small contribution. Requiring this to vanish defines the quantity $\rho_B^e(A)$ such that

$$d^2U/dR^2 + (2/R)(dU/dR) = 4\pi Z_A \rho_B^e(A). \tag{5-38}$$

Similarly, $\rho_A^e(B)$ may be defined so that the left member of this equation is equal to $4\pi Z_B \rho_A^e(B)$. If the effective following densities ρ_B and ρ_A are assumed to be spherically symmetric, they can be completely mapped out from the potential curve by using (5-38), e.g., for H_2^+, H_2, and LiH *(6, 98)*. This is related *(72)* to Platt's and other models. For hypothetical cases where $\nabla_A^2 U$ exactly equals $4\pi Z_A \rho_B(A)$, atomic densities may be used *(11)* to obtain $\rho_B(A)$. With B as the heavy atom, predictions of k for several diatomics were accurate, in contrast to results with B as the light atom. Presumably the light atom density is less rigidly held to its nucleus.

Integration of (5-37) gives *(5, 9, 10, 12)*

$$U(R) - U(R_e) = \int_{R_e}^{R} s(1 - s/R)\, F(s)\, ds,$$

where $F(s)$ is an effective-density function which, according to the success of the approximate formulas for k and l, is approximately $4\pi Z_B \rho_A(B)$ or $4\pi Z_A \rho_B(A)$. Simple parametrized forms for $F(s)$, e.g., exponential decay, were used to generate $U(R)$, by choosing parameters to reproduce some experimental quantities (R_e, k, l). Higher FCs and dissociation energies were predicted with mixed success. In a similar approach *(4)* for triatomics, the rigidly following densities for related diatomics were used, along with interaction terms. Note that using LCAO–MOs, $\{\phi_i\}$,

$$\phi_i = \sum_k C_k^{(i)} X_k, \tag{5-39}$$

where X_k is an AO, the electron density is

$$\rho = \sum_i \sum_{k,l} C_k^{(i)*} C_l^{(i)} X_k X_l. \tag{5-40}$$

This can be divided into a part rigidly following each nucleus and a not-perfectly-following part, if desired.

Kim *(67)* derived FCs by using the partitioning of ρ in conjunction with the integral H–F formula for $U - U_0$. Extensive cancellation occurred between terms in ρ and its derivatives. The neglect of ρ_{NPF} or its parts was also investigated. One can also partition ρ as $\rho_A^s + \rho_B^s + \rho_D$, where ρ_A^s and ρ_B^s are spherically symmetric densities about nuclei A and B. Neglecting the remainder, ρ_D, and its derivatives yields approximate harmonic and anharmonic FCs, but this neglect is not justified (*125*).

Assumptions about electron density have been used to simplify other exact expressions for FCs. King (*72*) considered his atomic FCs [see (5-24)] in conjunction with the expression

$$\rho = \sum_{\mathrm{A,B}} \sum_{i,j} C_{ij}^{(\mathrm{AB})} \chi_i^{(\mathrm{A})} \chi_j^{(\mathrm{B})},$$

where $C_{ij}^{(\mathrm{AA})}$ contributes to atomic population density and $C_{ij}^{(\mathrm{AB})}$ ($\mathrm{A} \neq \mathrm{B}$) to the bond-order matrix. $\nabla_{\mathrm{A}}^2 U$ was divided into two parts which were studied by CNDO wavefunctions: the interaction of a bare nucleus with fixed densities on other atoms, and a term due to charge redistribution. If core electrons are treated as nucleus-fixed negative charges, it appeared that the relaxation terms may be ignored and the approximate atomic FCs become

$$\nabla_{\mathrm{A}}^2 U \simeq 4\pi Z_{\mathrm{A}} \sum_{\mathrm{B} \neq \mathrm{A}} \sum_{i} C_{ii}^{(\mathrm{BB})} [\phi_i^{(\mathrm{B})}(r_{\mathrm{A}})]^2 .$$

However, all the observed trends in FCs could not be understood. Actually, King's atomic FCs follow from exact expressions if one invokes the partitioning of ρ and neglects the NPF contributions. The generalization of (5-37) to polyatomics is

$$\nabla_{\mathrm{B}} \cdot \nabla_{\mathrm{A}} U = -4\pi Z_{\mathrm{A}} \rho_{\mathrm{B}}(\mathrm{A}) - \int \nabla_{\mathrm{B}} \rho_{\mathrm{NPF}} \cdot \nabla_{\mathrm{A}} r_{\mathrm{A}}^{-1} \, d\tau .$$

Conditions for invariance to translations and rotations (Sections 5-1 and 5-4) give some FCs in terms of others. For triatomics, $\partial^2 U/\partial x_{\mathrm{A}} \partial x_{\mathrm{B}}$ is expressible in terms of $\partial^2 U/\partial x_{\mathrm{A}}^2$. With neglect of ρ_{NPF} and models for $\rho_{\mathrm{B}}(\mathrm{A})$, these are in turn expressible in terms of diatomic FCs and are relatively insensitive to the bonding environment and local point-charge distributions (see Ref. 22).

Murrell (*86*) considered only valence electrons $\{i\}$ explicitly and included core electrons in the effective nuclear charges $\overline{Z}_{\mathrm{A}}$ and $\overline{Z}_{\mathrm{B}}$, to obtain

$$\frac{k}{2\overline{Z}_{\mathrm{A}}\overline{Z}_{\mathrm{B}}} = \frac{1}{R_e^3} + Re \left[\sum_{k>0} \frac{\left\langle 0 \left| \sum_i \cos\theta_{\mathrm{A}i}/r_{\mathrm{A}i}^2 \right| k \right\rangle \left\langle k \left| \sum_i \cos\theta_{\mathrm{B}i}/r_{\mathrm{B}i}^2 \right| 0 \right\rangle}{E_0^{(0)} - E_k^{(0)}} \right].$$

A plot of this against R_e^{-3}, using observed k's and R_e's, revealed interesting regularities.

The occurrence of the independently measurable field gradient q_{A} in

$$k = Z_A \left[q_A + \frac{4\pi}{3}\rho(A) - \int \left(\frac{\partial \rho}{\partial x_A} \right) \frac{\cos \theta_A}{r_A^2} d\tau \right]_{R=R_e} \tag{5-41}$$

suggests the possibility of predicting FCs from other measurements. Indeed, the cancellation of the second and third terms is almost complete in certain cases,[3] leading to a proportionality of k to q. $\partial\rho/\partial x_A = 0$ if most of the density follows rigidly nucleus B (*119*), e.g., when B is the heavy atom in M^-H^+ ionic hydrides. Putting $Z_A = 1$ in (5-41), $k = q_A + 4\pi\rho(A)/3$ which, coupled with $k = 4\pi\rho(A)$ from (5-33), gives

$$q_A = 2k/3. \tag{5-42}$$

For M^+H^- molecules, a simple model is a fixed unit point charge for M^+, the remainder of electron density following A rigidly and hence not contributing to k. Thus $k = q_A = 2/R_e^3$, calculating q_H from the point-charge model. If q_H is calculated from unit point charges and k is calculated as $\frac{3}{2}q_H$, then k becomes $3R_e^{-3}$, which works better (*11, 12*) than $2R_e^{-3}$. Alternatively (*7*), the invariance of energy to overall translations and rotations yields other equations for k than (5-41), e.g., (5-23). Neglecting the integral, $k = \frac{3}{2}Z_A q_A$, which works well for both M^-H^+ and M^+H^- compounds (*7, 119*). However, Schwendeman (*125*) has cast doubt on the existence of a real physical connection between q and k. By expanding the electron density in spherical harmonics about nucleus A he showed that while the $l = 2$ (d-like) component determined q, k involved the $l = 1$ (p-like) component. Nevertheless, cancellation between different components of the density and their derivatives is actually extensive (*67*).

We shall now discuss certain simple electrostatic models for electron density, taking point-charge or equally simple models first. Such distributions may give realistic electronic potential energy but *not* the kinetic energy. However, according to the virial theorem, the energy of a diatomic and its FCs can be generated from potential energy alone. By allowing the point charges to move as the nuclei move, the model can also include electron relaxation.

Now, the potential function

$$E = U_0 + U_1 R^{-1} + U_2 R^{-2} \tag{5-43}$$

describes many diatomics very well, as evidenced by relations between k and higher FCs. Parr and Brown (*97*) generalized this to polyatomics, via the virial theorem, e.g., for CO_2 they derived relations between twelve valence FCs and the five parameters in their proposed potential functions. Choosing these parameters to reproduce five experimental valence FCs, they were able to predict most of the others correctly (see also Refs. 124 and 144).

Applying the virial theorem to (5-43) gives $\langle V \rangle = 2U_0 + U_1 R^{-1}$ and $\langle T \rangle =$

$-U_0 + U_2R^{-2}$. In the latter, $-U_0$ is the contribution of core electrons and U_2R^{-2} is that of the valence electrons (*31*).[4] Writing $qh^2/8m\nu^2$ for U_e and choosing the effective number of electrons q and the effective box length νR to fit experimental R_e and k, one can predict anharmonicities and various relations between potential constants of diatomic molecules (*56*). Using (5-43) one obtains the following relations at $R = R_e$:

$$0 = -U_1R_e^{-2} - 2U_2R_e^{-1},$$

$$R_e^4U_e'' = 2U_2,\; R_e^5U_e''' = -12U_2,\; R_e^6U^{\mathrm{IV}} = 72U_2,$$

so that $U_e''U_e^{\mathrm{IV}}/(U_e''')^2 = 1$, compared to $\frac{7}{9}$ from the Morse potential. Both 1 and $\frac{7}{9}$ represent reality equally well.

The form of $\langle V\rangle$ suggests interaction of point charges and led to the bond charge model for homo- and heteronuclear diatomics (*31*, *94-96*), linear triatomics (*97*, *114*, *129*), nonlinear triatomics (adding a point dipole for the central-atom lone pair), and excited electronic states (*112*). For homonuclear diatomics, a charge $-qe$ was placed at the bond center and a charge $+\frac{1}{2}qe$ on each atom; the values of bond charges are often physically reasonable and transferable from one molecule to another. However, for heteropolar diatomics, the charges and their positions have less physical significance (*96*). Using the diatomic (AB) bond charges for triatomics ABA allowed prediction of $k_{RR'}$ or at least its sign, from k_{RR} and the diatomic FC. Similarly, one can relate FCs of ground and excited states (*112*).

By writing the total energy as a truncated expansion in inverse powers of internuclear separations, one can derive relations between various FCs of a polyatomic molecule (*129*). Thus, for linear symmetric triatomics (*113*), $k_{\theta\theta} \simeq 0.026(k_{RR} + k_{RR'})$ where $k_{RR'}$ is the stretch-stretch interaction constant; for nonlinear triatomics (*113*), $k_{\theta\theta} \simeq 0.10(k_{RR} + k_{RR'}) + 0.04k_{R\theta}$. These rules, previously known to spectroscopists, work well with some exceptions (*130*).

An apparently more sophisticated electrostatic model (*15*) employs a quantum statistical pseudopotential to represent the potential of valence electrons and invokes the H-F theorem to choose certain parameters. In a molecule AB, each atom is an ionic core plus valence electrons, and part of the latter is localized at a distance bR_e from atom A, b being determined from covalent radii as $R_A/(R_A + R_B)$. The pseudopotential includes the effective density of the valence electrons as a free parameter whose value is chosen to give the best fit with experimental results. There are several additional parameters, some of which are chosen to achieve electrostatic equilibrium (zero field) at each nucleus. While FCs were reasonable (10-20% error), dipole moments were poor because of an extreme sensitivity to the value of b.

A more sophisticated electrostatic model regards certain diatomics as constructed from polarizable ions (*56*, *115*) and has also been extended to alkaline earth dihalides (*58*). Although successful for heats of formation and bond angles, it works poorly for FCs. The problem lies with the electrostatic terms in the energy.

5-7. FORCE CONSTANTS FROM PARTICULAR TYPES OF WAVEFUNCTION

We shall now discuss a priori calculations of FCs that start by calculating forces. In Section 5-7 we consider formulas for forces and FCs which apply to particular types of wavefunction, and in Section 5-8 we discuss the results and their accuracy.

Formulas for forces and FCs are simpler for MO wavefunctions because the force operator is one-electron. However, such functions often involve expansions in basis functions centered on origins which change as the nuclear configuration changes (see, e.g., Ref. 145). If the origins are not chosen variationally for each nuclear configuration and if the basis set is not very large, such functions will not satisfy the H–F theorem, i.e., the force $-\partial U/\partial\lambda$ is not equal to $\langle -\partial H/\partial\lambda\rangle$. One-center expansions are free from this problem, and we shall consider such cases first.

For di- and polyatomic hydrides, one-center calculations (*20*, *59*) can give fair FCs by the conventional technique of fitting calculated points on the surface $U(\lambda)$ to a power series. The nonlinear parameters in basis functions are determined variationally for each nuclear configuration, rather than simply using UA orbitals as basis functions. Bishop (*25*, *26*, *28*) has long contended that the conventional technique for getting FCs is uneconomical and numerically inaccurate, and that it gives no insight into their interpretation. His formalism (see below) for obtaining FCs directly, applied to a one-center, single-determinant function for H_2O (*28*), shows the advantages of his method. His work extended earlier formulas (*3*, *32*) which did not take into account changes in nonlinear parameters. A CI wavefunction is taken, employing configurations built from one-center basis orbitals, which are independent of nuclear configuration. If the CI coefficients for the ith state are denoted by $\{C_{is}\}$, its energy at a particular nuclear configuration is

$$U_i = \sum_{s,t=1}^{N} C_{is}C_{it}H_{st}, \tag{5-44}$$

where $H_{st} = \langle s|H|t\rangle$. H includes internuclear repulsion, and s and t refer to configurations. By hypothesis, only the C_{is} change with nuclear position, so that

$$dU_i/d\lambda_j = \sum_{s,t=1}^{N} [(dC_{is}/d\lambda_j)C_{it}H_{st} + C_{is}(dC_{it}/d\lambda_j)H_{st} + C_{is}C_{it}(\partial H_{st}/\partial\lambda_j)].$$

Since the coefficients are determined from the eigenvalue equation

$$\sum_{t=1}^{N} H_{st}C_{it} = U_i S_{st} C_{it}, \tag{5-45}$$

and normalization is conserved independently of nuclear displacements,

$$\sum_{s,t=1}^{N} C_{is}C_{it}S_{st} = \sum_{s,t=1}^{N} C_{is}C_{it}\langle s|t\rangle = 1, \tag{5-46}$$

the terms involving $dC_{is}/d\lambda_j$ vanish. Thus we obtain

$$dU_i/d\lambda_j = \sum_{s,t=1}^{N} C_{is}C_{it}(\partial H_{st}/\partial\lambda_j), \tag{5-47}$$

which is the H–F theorem for a linear variational wavefunction (*55*). Here,

$$\partial H_{st}/\partial\lambda_j = \langle s|\partial H/\partial\lambda_j|t\rangle.$$

Note that if we were discussing Hartree–Fock or a similar equation for determining MOs things would be less simple, because H would be an effective Hamiltonian which depends on the coefficients C_{is}.

Differentiating the force expression (5-47), we find

$$d^2U_i/d\lambda_j d\lambda_k = \sum_{s,t=1}^{N} C_{is}C_{it}(\partial^2 H_{st}/\partial\lambda_j\partial\lambda_k)$$

$$+ \sum_{s,t=1}^{N} [(\partial C_{is}/\partial\lambda_k)C_{it}(\partial H_{st}/\partial\lambda_j) + C_{is}(\partial C_{it}/\partial\lambda_k)(\partial H_{st}/\partial\lambda_j)].$$

One recognizes immediately the division into classical electrostatic and relaxation terms. In the linear variation framework with fixed basis functions (orthonormal), the latter can be calculated by expansion in the eigenvectors with energies U_h for the initial nuclear positions. After some manipulation, one finds (*56*)

$$d^2U_i/d\lambda_j d\lambda_k = \sum_{s,t=1}^{N} C_{is}C_{it}(\partial^2 H_{st}/\partial\lambda_j\partial\lambda_k) + 2\sum_{h=1}^{N}{}'$$

$$\cdot\left[\sum_{s,t=1}^{N} C_{is}C_{ht}(\partial H_{st}/\partial\lambda_j)\right]\left[\sum_{s,t=1}^{N} C_{is}C_{ht}(\partial H_{st}/\partial\lambda_k)\right][U_i - U_h]^{-1}, \tag{5-48}$$

where the prime means $h = i$ is excluded. Thus the CI calculation need be done only once. On the other hand, one requires the integrals $\langle s|\partial H/\partial\lambda_j|t\rangle$ and $\langle s|\partial^2 H/\partial\lambda_j\partial\lambda_k|t\rangle$. If a space-fixed coordinate system is used for the electrons, the only parts of H that enter are the nuclear-electronic attraction and internuclear repulsion terms.

For a molecule like NH_3, λ_j's may be individual nuclear coordinates or symmetry coordinates (*32*). The calculated FCs can be split into a zero-order contribution (from a single configuration), a deformation contribution (due to mixing of configurations), and a relaxation contribution (arising from $dC_{is}/d\lambda_k$). Except in the case of $k_{RR'}$, there was appreciable cancellation between the last two terms for FCs of NH_3 and H_2O (*32*).

If the configurations contain nonlinear parameters whose values are determined variationally for each nuclear configuration, the H-F theorem still holds in the form of (5-47). In the second derivatives, extra terms arise (*28*) because of changes in these parameters, so that calculation of $d^2U_i/d\lambda_j d\lambda_k$ requires $\partial^2 U_i/\partial p_m \partial p_n$ at the equilibrium nuclear configuration (p_m and p_n are nonlinear parameters), as well as quantities like $dp_m/d\lambda_k$. The former can be obtained variationally, while the latter can be obtained through simultaneous equations set up from quantities $\partial^2 U_i/\partial\lambda_j\partial p_n$ and $\partial^2 U_i/\partial p_m \partial p_n$.

Multicenter basis functions, centered on various nuclei of a molecule, change with nuclear displacement if they are viewed in a space-fixed coordinate system. The H-F theorem is not satisfied unless the origins are chosen variationally ("floating functions") or a very large basis is used (*55*). FC formulas which take into account the changes with nuclear positions have been developed for use with approximate functions in common use (*25, 26, 53*). We will be concerned first with MO wavefunctions whose basis functions are AOs fixed on various nuclei. The coefficients are given by the Roothaan equations

$$FC_i = \epsilon_i SC_i, \tag{5-49}$$

where C_i is the vector of expansion coefficients for the ith molecular orbital, S is the overlap matrix over the basis functions, and the matrix operator F consists of the matrix of the one-electron part of the Hamiltonian, h, plus G. Here

$$G_{jk} = \sum_{l=1}^{N} \sum_{m=1}^{p} \sum_{n=1}^{p} C_{ml} C_{nl} g_{jm,kn}, \tag{5-50a}$$

with

$$g_{jm,kn} = \int \phi_j(1)\,\phi_m(2)(r_{12})^{-1}\,[\phi_k(1)\,\phi_n(2) - \phi_n(1)\,\phi_k(2)]\,d\tau_1\,d\tau_2. \tag{5-50b}$$

There are N occupied MOs and p basis functions ϕ_i, and all functions have been assumed real. The energy is

$$E = \sum_{i=1}^{N} \frac{1}{2} \left(\epsilon_i + \sum_{m=1}^{p} \sum_{n=1}^{p} C_{mi} C_{ni} h_{mn} \right).$$

We are interested in $\partial^2 E/\partial\lambda_j \partial\lambda_k$. The problem is that at least some of the basis functions change with λ_j and λ_k, so that the elements of $\boldsymbol{S}$, $\boldsymbol{h}$, and $\boldsymbol{G}$ also change. Gerratt and Mills calculated the changes in Roothaan equations and their solutions by perturbation theory (*53, 55*). From the first-order correction to the wavefunction (including changes in coefficients and basis functions) the second-order energy and hence FCs can be obtained.

The force of course corresponds to the first-order change in energy. Letting superscripts refer to orders of perturbation theory, one finds

$$E^{(1)} = \sum_{i=1}^{N} \left[C_i^{(0)\dagger} \boldsymbol{h}^{(1)} C_i^{(0)} - \epsilon_i^{(0)} C_i^{(0)\dagger} \boldsymbol{S}^{(1)} C_i^{(0)} + \frac{1}{2} \sum_{j=1}^{N} \sum_{k,l,m,n=1}^{p} C_{ki}^{(0)} C_{lj}^{(0)} C_{mi}^{(0)} C_{nj}^{(0)} g_{kl,mn}^{(1)} \right]. \quad (5\text{-}51)$$

Here, we write

$$h_{ij}^{(1)} = \delta\lambda_j \frac{\partial}{\partial\lambda_j} \int \phi_i h \phi_j \, d\tau,$$

where h is the one-electron part of the Hamiltonian, $E^{(1)} = \delta\lambda_j(\partial E/\partial\lambda_j)$, and so on; $-\partial E/\partial\lambda_j$ is of course the force. In calculating $h_{ij}^{(1)}$ one has to take into account the changes in h with λ_j as well as the changes in ϕ_i and ϕ_j; for $S_{ij}^{(1)}$ and $g_{kl,mn}^{(1)}$ one need consider only the change in basis functions (*100*). The force can be calculated from the unperturbed coefficients $C_i^{(0)}$ alone (zero-order wavefunction), as shown in (5-51). However, (5-51) shows that the function is not assumed to satisfy the H–F theorem because the change in the wavefunction, as reflected in $\boldsymbol{h}^{(1)}$, $\boldsymbol{S}^{(1)}$, and $\boldsymbol{g}^{(1)}$, must be taken into account. In the Gerratt–Mills perturbation method, $E^{(2)}$ is calculated from the first-order wavefunction rather than from the force. In any case, one requires evaluation of integrals involving $\partial\phi_i/\partial\lambda$. If ϕ_i is an s-type Slater orbital centered on nucleus A,

$$\frac{\partial \phi_i^{(1)}}{\partial x_A} = \frac{\partial}{\partial x_A} r_{A1}^n \exp(-\zeta r_{A1}) = [r_{A1}^n \exp(-\zeta r_{A1})] \left(\frac{n}{r_{A1}} - \zeta \right) \left(\frac{x_A - x_1}{r_{A1}^2} \right).$$

In general, differentiation of an s atomic orbital yields a sum of p's, differentiation of a p a sum of d's, etc. By invoking the H–F theorem, one could avoid calculation of certain derivatives of atomic orbitals, but this would be a poor approximation for most functions. It may be noted that the origins of basis functions are nonlinear parameters in the sense of Bishop, but whose values are fixed for each nuclear configuration instead of determined variationally. In general, the wavefunction depends on the origins (represented by q) and other, variational, parameters (represented by p). In calculating the force, one writes symbolically,

$$dE/d\lambda = \partial E/\partial\lambda + (\partial E/\partial q)(\partial q/\partial\lambda) + (\partial E/\partial p)(\partial p/\partial\lambda),$$

and notes that the variational condition makes $\partial E/\partial p$ vanish (*110*).

An analog of the Gerratt–Mills perturbation formulation directly differentiates the Roothaan equations to calculate FCs from MO wavefunctions; this was tested for H_2 (*25–27*). Using the force expression of Gerratt and Mills in conjunction with semiempirical MO methods to calculate equilibrium molecular geometries was more efficient than that generating the energy surface by evaluating energy for many geometries (*81*).

The calculation of forces is further simplified (*110*) if one uses CNDO wavefunctions. Since the matrix elements of the Hamiltonian depend analytically on nuclear separations, evaluating their derivatives with respect to λ becomes simple. For the overlap matrix elements, calculated directly, one can differentiate explicitly:

$$\partial S_{ij}/\partial\lambda = \int (\partial X_i/\partial\lambda) X_j \, d\tau + \int X_i (\partial X_j/\partial\lambda) \, d\tau.$$

The CNDO energy is written as a sum of monatomic and diatomic contributions

$$E = \sum_{\mathrm{A}} E_{\mathrm{A}} + \sum_{\mathrm{A}<\mathrm{B}} E_{\mathrm{AB}}.$$

E_{A} are independent of nuclear positions while E_{AB} is written (*74, 110*) in terms of overlap populations, coulomb integrals, and resonance integrals. The coulomb and resonance integrals between AOs X_i and X_j are given as analytic functions of the separation of the centers of X_i and X_j. Thus calculating forces by differentiating E requires little additional computation. Similar conclusions hold (*81, 148*) for other approximation schemes, e.g., extended Hückel, INDO, and MINDO.

FCs calculated in this way are identical to those obtained by applying numerical procedures to the energy surface. For diatomic and small polyatomic molecules, CNDO and INDO generally give too high stretching FCs, although bending FCs (fixed bond lengths) are often well reproduced. Extended Hückel calculations can also give reasonable predictions of bond angles and bending FCs. In comparing this with Hartree-Fock results, one requires an understanding of how the total energies can be written as a sum of molecular orbital energies, which can be done for the extended Hückel but not for true Hartree-Fock (see Section 3-2-1). A set of assumptions which permit such a separation was presented (*93*) in connection with the potential surface of C_2H_4 (see Ref. 141 for B_2H_6). Stretching FCs were poor, bending FCs good, and interaction FCs fair. It was argued that chemical binding in the extended Hückel model occurs only because the proper distance behavior of the overlaps is included, not because the energy is correctly described. Thus, only for bending FCs can one expect good results. However, a separation of total energy into MO contributions could be made (*43*) from the H-F theorem. The energy was obtained as an integral of the force, and FCs by differentiating the force.

The works discussed in this section show the practicality of calculating forces analytically rather than by numerical differentiation of the energy $U(\lambda)$. We will now discuss the following two questions: how accurate are FCs obtained from forces calculated in this way, and are there other advantages to deviating from the conventional procedure of double numerical differentiation of $U(\lambda)$? Several authors answer the second question affirmatively.

5-8. FORCE CONSTANTS CALCULATED FROM FORCES

In some early work (*25, 26*) calculation of FCs by double analytical differentiation of energy was shown to be superior, in the case of H_2, to single analytical differentiation to get the force at several internuclear distances followed by numerical differentiation. Calculating quadratic FCs for H_2 and several diatomic hydrides using one-center functions (*67–69*) gave good agreement with experiment, but not when the origin of coordinates was chosen as the center of expansion of the wavefunction. Still earlier calculations (*84*) on CH_4 using one-center wavefunctions in a perturbation formalism (Section 5-5) did not give consistently good agreement with experimental FCs because of approximations made in expressing the potential of the nuclei. Also, the formalism could not be used for the breathing vibration, which maintains the molecular symmetry.

Further discussions of the application of double analytical differentiation are available (*28, 105, 133, 134*). Since all methods give the same results when exact formulas are used, the choice is a matter of convenience and numerical accuracy. Comparisons with Gaussian wavefunctions for H_2 led to the conclusion that the dependence of orbital centers (floating orbitals) on nuclear configuration must be taken into account and that good FCs can be obtained from some wavefunctions not close to the Hartree-Fock limit (*54*). Pulay (*105*) concluded that the most efficient method was analytical differentiation to get the forces, from which equilibrium geometry can be obtained, followed by numerical differentiation to obtain FCs. His force formulas (Chapter 9) assumed an expansion in basis orbitals rigidly following the nuclei.

The above discussion uses the word *force* to mean either the H-F force ($\langle \partial H/\partial\lambda \rangle$) or the energy derivative with nuclear displacement, calculated directly instead of numerically. Pulay (*105*) divided his calculated forces into the H-F force and the "wavefunction force," the latter vanishing when the H-F theorem is satisfied. The same results are obtained if $\partial U/\partial\lambda$ is calculated directly or by numerical differentiation; the choice depends on convenience, numerical accuracy, or interpretability of results. An early analysis (*121*) led to the conclusion that FCs calculated from $\langle \partial H/\partial\lambda \rangle$ are generally less reliable than those obtained conventionally, by computing energy at a series of internuclear distances; examples are known where energy calculations gave reasonable results, force calculations nonsense. The error analysis may be made as follows.

Let the deviation of the normalized approximate wavefunction ϕ from the exact normalized one, ψ, be represented in terms of a parameter ϵ,

$$\psi = (1 - \epsilon^2)^{1/2}\phi + \epsilon\Gamma, \tag{5-52}$$

where the function Γ is normalized and orthogonal to ϕ. The error in the energy is

$$\langle\psi|H|\psi\rangle - \langle\phi|H|\phi\rangle = -\epsilon^2\langle\phi|H|\phi\rangle + 2\epsilon(1 - \epsilon^2)^{1/2}\langle\phi|H|\Gamma\rangle + \epsilon^2\langle\Gamma|H|\Gamma\rangle.$$

Since the difference between ϕ and ψ is of size ϵ or smaller and $H\psi = E\psi$, $\langle\phi|H|\Gamma\rangle$ differs from $E\langle\psi|\Gamma\rangle$ by terms of size ϵ or smaller, and the second term on the right-hand side is of order ϵ^2. Thus each point on the potential energy surface calculated as $\langle\phi|H|\phi\rangle$ differs from the corresponding point on the exact surface $\langle\psi|H|\psi\rangle$ by second-order terms. It seems that energy derivatives with respect to parameters in H, calculated for several values of these parameters, should also be correct to second order. On the other hand, the expectation value of the force operator, kinetic energy operator, or other operators not commuting with H would be

$$\langle\psi|Q|\psi\rangle = (1 - \epsilon^2)\langle\phi|Q|\phi\rangle + 2\epsilon(1 - \epsilon^2)\langle\phi|Q|\Gamma\rangle + \cdots$$

The second term on the right-hand side is first order in ϵ because ψ is not an eigenfunction of Q; thus $\langle\psi|Q|\psi\rangle$ differs from $\langle\phi|Q|\phi\rangle$ in first order. It is thus expected that, in general, errors in calculated force, kinetic energy, etc. will be larger than errors in the energy itself, so that calculating FCs from these quantities would lead to more inaccurate results than calculating them from the energy.

Several exceptions must be taken to this statement. First, Hartree-Fock wavefunctions give rise to second-order errors in the expectation values of one-electron operators (*57*) such as force or kinetic energy, permitting accurate results to be obtained from the H-F or the virial force. Of course, Hartree-Fock functions satisfy the H-F and virial theorems, so that they must yield exactly the same results for FCs when either of the theorems is invoked as when the energy is used directly. For a "stable" wavefunction (*60*) the error in the generalized force $\langle \phi | \partial H/\partial \alpha | \phi \rangle$ is second order rather than first order; α may be any parameter in H, including a nuclear displacement. A stable wavefunction is one for which $\mathrm{Re}\,\langle \phi | H | \partial\phi/\partial\alpha \rangle$ vanishes. This seems to imply that wavefunctions satisfying the H-F theorem should lead to better FCs as well as better equilibrium geometries (*105*). Other authors do not agree (*134*) and state similarly that scaling to make a function satisfy the virial theorem does not much improve values of k calculated via the virial theorem. Wavefunctions built from spherical floating Gaussians have been given (*134*) for some small diatomics and used for calculation of k by various methods (perturbation theory, direct differentiation of Roothaan equations, virial theorem, H-F theorem, etc.), sometimes forcing the basis functions to follow the nuclei and sometimes determining their origins variationally. It was concluded that the most reliable method was to use the full energy derivative and not worry about satisfying the H-F theorem by allowing the basis orbitals to float.

A second exception to the statement on the accuracy of calculated FCs was raised by Schwendeman (*126*), who displayed explicitly the difference between the equilibrium internuclear distances for a diatomic calculated from exact (R_e) and approximate (R_e') wavefunctions,

$$R_e' - R_e = \epsilon^2 \langle \Gamma | \partial H/\partial R | \Gamma \rangle / (d^2 U/dR^2), \tag{5-53}$$

with $U = E + V_{nn}$. The error in E is second order, formally like that in U, but d^2E/dR^2 and d^2V_{nn}/dR^2 are of opposite signs and cancel each other to a large extent. The smallness in the denominator of (5-53) means that $R_e' - R_e$ is much larger than the factor ϵ^2 would lead us to expect and should perhaps be considered (*126*) as "pseudo-first-order." The error in the FC is

$$k' - k = (d^2U'/dR^2)_{R_e'} - (d^2U/dR^2)_{R_e}$$

$$= \left(\frac{d^2[E + \epsilon^2 X]}{dR^2}\right)_{R_e'} - \left(\frac{d^2E}{dR^2}\right)_{R_e} + \left(\frac{d^2V_{nn}}{dR^2}\right)_{R_e'} - \left(\frac{d^2V_{nn}}{dR^2}\right)_{R_e},$$

where X is $\langle\phi|H|\phi\rangle - \langle\psi|H|\psi\rangle$, the error in electronic energy. Thus

$$k' - k = (d^2U/dR^2)_{R_e'} - (d^2U/dR^2)_{R_e} + [d^2(\epsilon^2 X)/dR^2]_{R_e'}$$

and the difference between the second derivatives of the exact potential curve at R_e and R_e' is proportional to $R_e' - R_e$. It appears that the error in k will be larger than second order. The remedy (*126*) for this is to evaluate derivatives of U' at the experimentally determined internuclear distance rather than at the minimum of the calculated $U'(R)$. Other authors (*76*, *122*, *133*) have concurred, but the effect of changing the origin is not always important (*21*, *111*). There is still cancellation between electronic and nuclear contributions to k, so errors in the former are magnified (*134*).

The discussion of the error in calculated forces and FCs can be carried out by double perturbation theory, one perturbation being the change in R and the other representing the difference between the approximate and exact wavefunctions. Kirtman et al. (*73*) carried out a double perturbation treatment for a diatomic molecule to show the relation between "exact" forces, obtained by numerical differentiation of the approximate potential energy curve (variational), and the forces calculated by perturbation theory. They concluded that while the errors in the former were second order in the error in the wavefunction [ϵ in (5-52)], the forces from perturbation theory contain a first-order error, like the H–F forces. The variational method calculates the FC as

$$k = \left[\left\langle\psi_0\left|\frac{\partial^2 H}{\partial R^2}\right|\psi_0\right\rangle + 4\left\langle\frac{\partial\psi_0}{\partial R}\left|\frac{\partial H}{\partial R}\right|\psi_0\right\rangle + 2\left\langle\frac{\partial\psi_0}{\partial R}|H - E|\frac{\partial\psi_0}{\partial R}\right\rangle + 2\left\langle\frac{\partial^2\psi_0}{\partial R^2}|H - E|\psi_0\right\rangle\right]_{R_e}, \quad (5\text{-}54)$$

where ψ_0 is the approximate wavefunction. In perturbation theory, $H(R)$ is written as $H^{(0)}(R) + \mu V^{(1)}(R)$, where $H^{(0)}(R)\psi_0(R) = E^{(0)}(R)\psi_0(R)$. The parameter μ defines order of perturbation theory for the perturbation which represents the deviation of exact from approximate function. The operators $H^{(0)}$ and $V^{(1)}$, and the wavefunction $\psi_0(R)$, are expanded about $R = R_e$; the change in R is the second perturbation. The expression derived (*73*) by double perturbation theory for the FC agrees with (5-54) to terms which are second order in μ, but that obtained from single perturbation theory (the perturbation being the change in R from R_e) differs by terms proportional to μ (first-order errors), as expected. Similar results obtain for higher FCs, or for FCs corresponding to normal coordinate displacements in a polyatomic molecule.

Calculations with single-determinant MO wavefunctions are susceptible to expansion error (due to an insufficient basis set) and correlation error. The expansion error has been discussed (*118*) in terms of the off-diagonal hypervirial theorem. FCs calculated by the force method (*82*) were not very sensitive to the basis set. However, the work ignores the change of orbital exponents with nuclear displacement (*38*).

Since Hartree-Fock functions for closed-shell systems tend to dissociate incorrectly, this raises calculated energies for large internuclear distances and tends to make FCs too high (*56*). Using extended Hartree-Fock functions which are correct through first order in configuration interation for H_2 and Li_2 to calculate k from energy and force curves, better accuracy was found for the former (*36*). Thus, it may be concluded that for most approximate wavefunctions more accurate FCs can be obtained from the energy than from the H-F force. The latter is of interest when it avoids calculation of the complete energy expression (see below).

Since calculation of FCs from the energy surface is time-consuming, the possibility of curtailing computation time by approximating integrals in SCF calculations has been investigated (*2*, *100–102*). The use of semiempirical methods may be advantageous, although some methods (e.g., CNDO) which are acceptable for energies and other properties give unreliable FCs (*14*, *137*). The difficulties connected with numerical differentiation are also receiving attention (*64*, *91*, *114*, *136*). One needs a scheme requiring the smallest number of energies,[5] and maximizing the accuracy and reliability of the results (see Ref. 140).

For polyatomic molecules, it is advantageous to calculate FCs via forces from semiempirical or approximate Hartree-Fock functions (*105*, *106*, *122*). The force here is $\partial\langle H\rangle/\partial\lambda$, so that it includes both H-F force ($\langle\partial H/\partial\lambda\rangle$) and relaxation terms (wavefunction force). One does not get good results by considering the variation of $\langle\partial H/\partial\lambda\rangle$ alone (*38*, *105*). From a wavefunction for a single nuclear configuration a number of forces (corresponding to different coordinates λ) may be calculated by direct differentiation of $\langle H\rangle$ in less time than needed to carry out energy calculations for the various series of configurations needed to get all the $\partial\langle H\rangle/\partial\lambda$. The second differentiation of $\langle H\rangle$ to get FCs is done numerically by getting wavefunctions for a series of configurations. From calculations of forces at two configurations differing by the values of λ_i one can get (*110*) all FCs $k_{ij} = \partial^2 U/\partial\lambda_i\partial\lambda_j$. For example, for the five FCs of CH_4 the wavefunction and its derivatives had to be calculated (*110*) at five nuclear configurations: the equilibrium configuration, two for which two C—H distances were changed by ±0.03 Å, and two for which one HCH angle was changed by ±4°. For formaldehyde (*82*) nine configurations sufficed to get two values (which could be compared to check accuracy) for each of ten quadratic FCs. Note that a coupling or interaction FC $k_{ij} = \partial^2 U/\partial\lambda_i\partial\lambda_j$ can be calculated (*110*) from the force $\phi_i = -\partial U/\partial\lambda_i$ according to $k_{ij} = -\Delta\phi_i/\Delta\lambda_j$, or from the force $\phi_j = -\partial U/\partial\lambda_j$ as $k_{ij} = -\Delta\phi_j/\Delta\lambda_i$. This procedure is especially efficient when one wants to calculate interaction as well as diagonal FCs for polyatomics, but has advantages for evaluation of k in a diatomic molecule as well (see Chapter 9 for details). An algorithm has been proposed (*127*) for calculating k using calculated values for forces as well as energy.

Pulay found that FCs, especially interaction FCs, came out surprisingly well with approximate Hartree-Fock functions. The derivatives of the forces were evaluated at the experimental geometry rather than the theoretical one. This procedure should give more accurate results (formally correct to second order) according to the argument of Schwendeman (*126*) discussed above. Molecules for which FCs have been calculated by Pulay and his co-workers include (see also Table 9-6): ONF and NF (*106*); HCN, FCN, $(CN)_2^+$, and N_2F^+ (*111*); CH_4, H_2O (*106, 110*); CH_2O (*82*); C_2H_6, C_2H_4, and C_2H_2 (*108, 109*); HF, NH_3, and BH_4^- (*106*). The wavefunctions used were limited-basis approximations to Hartree-Fock, employing Gaussian bases, except that CNDO was used in some earlier work. Stretching FCs tend to be 10-20% high (they are worse from CNDO calculations), but bending and interaction FCs, as well as trends in FCs, are well represented (*110*). The interaction FCs are often the most difficult to extract unambiguously from spectroscopic data. In fact, calculated interaction FCs are sometimes (*109*) more reliable than those derived from experiment because of assumptions and neglected terms in the interpretation of experimental data, so the theoretical values could be used to distinguish between values proposed by experimenters (*108*). Stretch-stretch interaction FCs predicted on the basis of theoretical potential functions were less reliable. In some cases cubic FCs were evaluated. For reviews of Pulay's method, see Chapter 9, and Refs. 107 and 147.

If the simplifications associated with wavefunction calculation by semiempirical methods are combined with the simplifications of the force method itself, one can generate the complete quadratic force field for larger polyatomics. Such attempts have met with mixed success. FCs from a maximum overlap method (*13*) were not satisfactory. CNDO/2 bending and interaction FCs for H_2CO, F_2CO, CF_4, etc. were in reasonable agreement with experiment, as were some trends in stretching FCs, although the values of the latter are twice the experimental ones (*66*). Although CNDO/2 and MINDO/2 FCs for a series of polyatomics including C_2H_2 and CH_3CHO are not quite satisfactory, it is possible to formulate a set of empirical multipliers for different types of force constants, giving the scheme some predictive value (*76*).

This approach is not unique to the force method. Calculations of quadratic FCs (from energies) for small hydrocarbons gave C—H and C—C stretching constants differing by fairly constant ratios from experimental values (*29*). Thus, one might derive a set of scale factors, transferable from molecule to molecule, to use in conjunction with FC calculations. Choosing such scale factors from results for propane, ethane, and cyclopropane, satisfactory FCs for related molecules can be obtained (*30*). It appears that calculations can yield information about FCs which, used "judiciously" (*77*) with experimental data, can lead to construction of molecular force fields even where neither experiment nor theory alone suffices (see Ref. 142).

One may also calculate quadratic and anharmonic FCs from nonvariational approximate wavefunctions which may be constructed from the H-F theorem itself, by satisfying the constraint of electrostatic equilibrium. Such results for H_2, Li_2, and Na_2 (*120*) were in fair agreement with experiment, although the forces are quite sensitive to changes in the wavefunction. In a related treatment, Nakatsuji and co-workers (*87, 88*) chose parameters in approximate wavefunctions so that H-F theorem was satisfied and other defects of the wavefunction remedied. FCs were given by force derivatives (at zero force). Using parameters chosen from ethane, twelve stretching FCs and one bending FC were calculated for a number of molecules. Results were better than those of extended Hückel energies, but were not good in all cases. Extensive cancellation between nuclear and electronic contributions was emphasized (for details of this ESF theory, see Chapter 3). The bending FC of H_2O was studied (*43*) by differentiating the perpendicular force on a proton with respect to the HOH angle α, using minimum-basis MO functions. This gave a $k_{\alpha\alpha}$ closer to the experimental value than any previous calculation. The FC could also have been calculated from forces on the oxygen nucleus, but the wavefunction was inadequate to represent polarization and hence the force satisfactorily.

5-9. SUMMARY AND CONCLUSION

While force constants can be calculated from the energy as a function of nuclear configuration, there are good reasons to first calculate the force by analytical differentiation of the energy expression. For the exact wavefunction, various expressions are obtained for the force constant, according to the way the differentiation is carried out (Sections 5-2 and 5-3). These expressions permit physical interpretation, and suggest approximations and simple models (Section 5-6). On the other hand, if individual terms in the force and force-constant expressions are evaluated using accurate wavefunctions, we add to our understanding of the process of molecular binding (Section 5-5).

There are also reasons connected with numerical accuracy and convenience for performing the first differentiation of energy analytically (Section 5-8). These become particularly cogent for polyatomic molecules. For certain kinds of approximate wavefunctions, the analytical differentiation of energy is not difficult (Section 5-7). As a result of much experimentation with different methods, calculations of a variety of force constants to a useful accuracy are now possible (Sections 5-4 and 5-8). The force concept has played an important role in the research leading to this success.

Notes

1. The part of ρ_A which does not change with R makes no contribution to k.
2. The total electronic charge of the united atom exceeds Z.
3. In diatomic hydrides the field gradient at H is ~85% of k (*119*) and that at ^{14}N in benzonitrile derivatives is about 90% of the C—N stretching FC (*42*).
4. Note that kinetic energy of particles in a one-dimensional box varies as the inverse square of box length.
5. Calculated derivatives depend on the choice of points for energy values.

References

1. Adamov, M. N. and Bulychev, V. P., 1966, *Teor. Eksp. Khim.*, **2**, 685.
2. —— and Ivanov, A. I., 1977, *Zh. Strukt. Khim.*, **18**, 245 (English transl.: *J. Struct. Chem.*, **18**, 199).
3. Allavena, M. and Bratož, S., 1963, *J. Chim. Phys.*, **60**, 1199.
4. Anderson, A. B., 1972, *J. Chem. Phys.*, **57**, 4143.
5. ——, 1972, *J. Mol. Spectrosc.*, **44**, 411.
6. ——, 1973, *J. Chem. Phys.*, **58**, 381.
7. ——, Handy, N. C., and Parr, R. G., 1969, *J. Chem. Phys.*, **50**, 3634.
8. —— and Parr, R. G., 1970, *J. Chem. Phys.*, **53**, 3375.
9. —— and Parr, R. G., 1971, *J. Chem. Phys.*, **55**, 5490.
10. —— and Parr, R. G., 1971, *Chem. Phys. Lett.*, **10**, 293.
11. —— and Parr, R. G., 1972, *Theoret. Chim. Acta* (Berlin), **26**, 301.
12. —— and Parr, R. G., 1972, *J. Chem. Phys.*, **56**, 5204.
13. Arce, F., Casado, J., Ríos, M. A., and Tato, V., 1976, *An. Quim.*, **72**, 505.
14. ——, Casado, J., Ríos, M. A., and Tato, V., 1977, *Rev. Roum. Chim.*, **22**, 335, 345.
15. Aslaksen, E. W., 1972, *Phys. Rev.*, **A6**, 1367.
16. Bader, R. F. W., 1960, *Mol. Phys.*, **3**, 137.
17. —— and Bandrauk, A. D., 1968, *J. Chem. Phys.*, **49**, 1666.
18. —— and Gangi, R. A., 1975, *Theor. Chem.* (Chem. Soc. Spec. Per. Rep.), **2**, 1.
19. —— and Ginsburg, J. L., 1969, *Can. J. Chem.*, **47**, 3061.
20. Banyard, K. E. and Hake, R. B., 1964–66, *J. Chem. Phys.*, **41**, 3221; **43**, 2684; **44**, 3523.
21. Bartell, L. S., Fitzwalter, S., and Hehre, W. J., 1975, *J. Chem. Phys.*, **63**, 4750.
22. Bartlett, R. J. and Parr, R. G., 1977, *J. Chem. Phys.*, **67**, 5828.
23. Benston, M. L. and Kirtman, B., 1966, *J. Chem. Phys.*, **44**, 119, 126.
24. Bishop, D. M., 1963, *Mol. Phys.*, **6**, 305.
25. —— and Macias, A., 1969, *J. Chem. Phys.*, **51**, 4997.
26. —— and Macias, A., 1970, *J. Chem. Phys.*, **53**, 3515.
27. —— and Macias, A., 1972, *J. Chem. Phys.*, **56**, 999.
28. —— and Randić, M., 1966, *J. Chem. Phys.*, **44**, 2480.
29. Blom, C. E., Slingerland, P. J., and Altona, C., 1976, *Mol. Phys.*, **31**, 1359.

30. —— and Altona, C., 1976, *Mol. Phys.*, **31**, 1377.
31. Borkman, R. F. and Parr, R. G., 1968, *J. Chem. Phys.*, **48**, 1116.
32. Bratož, S. and Allavena, M., 1962, *J. Chem. Phys.*, **37**, 2138.
33. ——, Daudel, R., Roux, M., and Allavena, M., 1960, *Rev. Mod. Phys.*, **32**, 412.
34. Brown, W. B., 1958, *Proc. Cambridge Phil. Soc.*, **54**, 250.
35. Chandra, A. K. and Sundar, R., 1971, *Mol. Phys.*, **22**, 369.
36. —— and Sundar, R., 1972, *Chem. Phys. Lett.*, **14**, 577.
37. Chang, S. Y., 1972, *J. Chem. Phys.*, **56**, 2161.
38. Chong, D. P., Gagnon, P. A., and Thorhallson, J., 1971, *Can. J. Chem.*, **49**, 1047.
39. Clinton, W. L., 1960, *J. Chem. Phys.*, **33**, 1603.
40. ——, 1963, *J. Chem. Phys.*, **38**, 2339.
41. —— and Frattali, S. D., 1963, *J. Chem. Phys.*, **39**, 3316.
42. Colligiani, A., Ambrosetti, R., Angelone, R., and Oja, T., 1975, *Proc. 2nd Int. Symp. Nucl. Quadrup. Res.*, ed. A. Colligiani (Augusto Vallerini, Pisa, Italy), p. 239.
43. Coulson, C. A. and Deb, B. M., 1971, *Int. J. Quantum Chem.*, **5**, 411.
44. —— and Hurley, A. C., 1962, *J. Chem. Phys.*, **37**, 448.
45. —— and Longuet-Higgins, H. C., 1948, *Proc. Roy. Soc.* (London), **A193**, 456.
46. Davidson, E. R., 1962, *J. Chem. Phys.*, **36**, 2527.
47. Deb, B. M., 1972, *Chem. Phys. Lett.*, **17**, 78.
48. Empedocles, P., 1967, *J. Chem. Phys.*, **46**, 4474.
49. Epstein, S. T., 1963, *J. Chem. Phys.*, **42**, 3813.
50. Fadini, A., 1976, *Wiss. Forschungsber.*, Ser. 1, Vol. 75, Part A.
51. Gaughan, R. R. and King, W. T., 1972, *J. Chem. Phys.*, **57**, 4530.
52. Gans, P., 1977, *Adv. Infrared Raman Spectrosc.*, **3**, 87.
53. Gerratt, J. and Mills, I. M., 1968, *J. Chem. Phys.*, **49**, 1719, 1730.
54. Goodisman, J., 1963, *J. Chem. Phys.*, **39**, 2397.
55. ——, 1973, *Diatomic Interaction Potential Theory. Vol. I: Fundamentals* (Academic Press, New York).
56. ——, 1973, *Diatomic Interaction Potential Theory. Vol. II: Applications* (Academic Press, New York).
57. —— and Klemperer, W. A., 1963, *J. Chem. Phys.*, **38**, 721.
58. Guido, M. and Gigli, G., 1976, *J. Chem. Phys.*, **65**, 1397.
59. Hake, R. B. and Banyard, K. E., 1966, *J. Chem. Phys.*, **45**, 3199.
60. Hall, G. G., 1961, *Phil. Mag.* (London), **6**, 249.
61. —— and Rees, D., 1962, *Mol. Phys.*, **5**, 279.
62. Hirschfelder, J. O. and Coulson, C. A., 1962, *J. Chem. Phys.*, **36**, 941.
63. —— and Eliason, M. A., 1967, *J. Chem. Phys.*, **47**, 1164.
64. Ilin, V. V., Pinchuk, V. M., and Sorochinskaya, V. E., 1977, *Opt. Spektrosk.*, **43**, 993 (English transl.: *Opt. Spectrosc.*, **43**, 587).
65. IUPAC Phys. Chem. Div., 1977, *Appl. Spectrosc.*, **31**, 569.
66. Kanakavel, M., Chandrasekhar, J., Subramanian, S., and Singh, S., 1976, *Theoret. Chim. Acta* (Berlin), **43**, 185.

67. Kim, H., 1968, *J. Chem. Phys.*, **48**, 301.
68. ——, Cho, U. I. and Choi, S., 1972, *Daehan Hwahak Hwojee*, **14**, 133.
69. —— and Kim, H., 1972, *Daehan Hwahak Hwojee*, **16**, 214, 261.
70. —— and Parr, R. G., 1968, *J. Chem. Phys.*, **49**, 3071.
71. King, W. T., 1968, *J. Chem. Phys.*, **49**, 2866.
72. ——, 1972, *J. Chem. Phys.*, **57**, 4535.
73. Kirtman, B., Chang, S. Y., and Scott, W. R., 1974, *J. Chem. Phys.*, **61**, 3700.
74. Klopman, G. and O'Leary, B., 1970, *Fortsch. Chem. Forsch.*, **15**, 445.
75. Koga, T. and Nakatsuji, H., 1976, *Theoret. Chim. Acta* (Berlin), **41**, 119.
76. Kozmutza, K., 1976, *Acta Phys. Acad. Sci. Hung.*, **40**, 245.
77. Levin, I. W. and Pearce, R. A. R., 1975, *Vib. Spectra Struct.*, **4**, 101.
78. Linnett, J. W. and Hoare, M. F., 1949, *Trans. Faraday Soc.*, **45**, 844.
79. Longuet-Higgins, H. C. and Brown, D. A., 1955, *J. Inorg. Nucl. Chem.*, **1**, 60.
80. McDugle, W. G., Jr., Schreiner, A. F., and Brown, T. L., 1967, *J. Am. Chem. Soc.*, **89**, 3111, 3114.
81. McIver, J. W. and Komornicki, A., 1971, *Chem. Phys. Lett.*, **10**, 303.
82. Meyer, W. and Pulay, P., 1974, *Theoret. Chim. Acta* (Berlin), **32**, 253.
83. Mezei, M. and Pulay, P., 1968, *Acta Chim. Acad. Sci. Hung.*, **56**, 167, 331.
84. Mills, I. M., 1958, *Mol. Phys.*, **1**, 107.
85. ——, 1963, *Spectrochim. Acta*, **19**, 1585.
86. Murrell, J. N., 1960, *J. Mol. Spec.*, **4**, 446.
87. Nakatsuji, H., 1973, *J. Am. Chem. Soc.*, **95**, 345, 354, 2084.
88. ——, Kuwata, T., and Yoshida, A., 1973, *J. Am. Chem. Soc.*, **95**, 6894.
89. Nalewajski, R. F., 1977, *Chem. Phys.*, **22**, 257.
90. Nelander, B., 1970, *J. Chem. Phys.*, **51**, 469.
91. Noor Mohammed, S., 1977, *Int. J. Quantum Chem.*, **12**, 813.
92. Novosadov, B. K., 1975, *Zh. Strukt. Khim.*, **16**, 967.
93. Paldus, J. and Hrabe, P., 1968, *Theoret. Chim. Acta* (Berlin), **11**, 401.
94. Parr, R. G. and Borkman, R. F., 1967, *J. Chem. Phys.*, **46**, 3683.
95. —— and Borkman, R. F., 1968, *J. Chem. Phys.*, **49**, 1055.
96. —— and Borkman, R. F., 1969, *J. Chem. Phys.*, **50**, 58.
97. —— and Brown, J. E., 1968, *J. Chem. Phys.*, **49**, 4849.
98. ——, Finlan, J. M., and Schnuelle, G. W., 1972, *Chem. Phys. Lett.*, **14**, 72.
99. —— and White, R. J., 1968, *J. Chem. Phys.*, **49**, 1059.
100. Perevozchikov, V. I. and Gribov, L. A., 1975, *Vopr. Sborn. Kvant. Khim.*, p. 192.
101. —— and Gribov, L. A., 1976, *Opt. Spektrosk.*, **41**, 332 (English transl.: *Opt. Spectrosc.*, **41**, 190).
102. —— and Gribov, L. A., 1977, *Opt. Spektrosk.*, **42**, 203 (English transl.: *Opt. Spectrosc.*, **42**, 113).
103. Phillipson, P., 1963, *J. Chem. Phys.*, **39**, 3010.
104. Platt, J. R., 1950, *J. Chem. Phys.*, **18**, 932.

105. Pulay, P., 1969, *Mol. Phys.*, **17**, 197.
106. ——, 1971, *Mol. Phys.*, **21**, 329.
107. ——, 1977, in *Modern Theoretical Chemistry*, vol. 4, ed. H. F. Schaefer (Plenum Press, New York), p. 153.
108. —— and Meyer, W., 1971, *J. Mol. Spectrosc.*, **40**, 59.
109. —— and Meyer, W., 1974, *Mol. Phys.*, **27**, 473.
110. —— and Torok, F., 1973, *Mol. Phys.*, **25**, 1153.
111. ——, Ruoff, A., and Sawodny, W., 1975, *Mol. Phys.*, **30**, 1123.
112. Ramaswamy, K. and Balasubramanian, V., 1972, *Indian J. Pure Appl. Phys.*, **10**, 853.
113. Ray, N. K. and Parr, R. G., 1973, *J. Chem. Phys.*, **59**, 3934.
114. Richards, W. G., Raftery, J. and Hinkley, R. K., 1974, *Theor. Chem.* (Chem. Soc. Spec. Per. Rep.), **1**, 1.
115. Rittner, E. S., 1951, *J. Chem. Phys.*, **19**, 1030.
116. Rossikhin, V. V., Marozov, V. P., and Bezzub, L. I., 1968, *Teor. Eksper. Khim.*, **4**, 37, (English transl.: *Theor. Exper. Chem. USSR*, **4**, 22).
117. ——, Marozov, V. P., and Tsaune, A. Ya., 1968, *Teor. Eksper. Khim.*, **4**, 42 (English transl.: *Theor. Exper. Chem. USSR*, **4**, 25).
118. ——, Bolotin, A. B., and Zaslavskaya, L. I., 1972, *Liet. Fiz. Rinkinys*, **12**, 753.
119. Salem, L., 1963, *J. Chem. Phys.*, **38**, 1227.
120. —— and Alexander, M., 1963, *J. Chem. Phys.*, **39**, 2994.
121. —— and Wilson, E. B., Jr., 1962, *J. Chem. Phys.*, **36**, 3421.
122. Sawodny, W. and Pulay, P., 1974, *J. Mol. Spectrosc.*, **51**, 135.
123. Sawodny, W. 1976, *Kem. Kozl.*, **46**, 263.
124. Schnuelle, G. W. and Parr, R. G., 1972, *J. Am. Chem. Soc.*, **94**, 8974.
125. Schwendeman, R. H., 1966, *J. Chem. Phys.*, **44**, 556.
126. ——, 1966, *J. Chem. Phys.*, **44**, 2115.
127. Shipman, L. L. and Christoffersen, R. E., 1971, *Chem. Phys. Lett.*, **11**, 101.
128. Silverman, J. N. and Van Leuven, J. C., 1970, *Chem. Phys. Lett.*, **7**, 37.
129. Simons, G., 1972, *J. Chem. Phys.*, **56**, 4310.
130. —— and Choc, C. E., 1974, *Chem. Phys. Lett.*, **25**, 413.
131. Sorochinskaya, V. E. and Morozov, V. P., 1973, *Teor. Eksper. Khim.*, **9**, 795 (English transl.: 1975, *Theor. Exper. Chem. USSR*, **9**, 625).
132. —— and Morozov, V. P., 1975, *Teor. Eksper. Khim.*, **11**, 613 (English transl.: 1976, *Theor. Exper. Chem. USSR*, **11**, 555).
133. Swanstrom, B., Thomsen, K., and Yde, P. B., 1971, *Mol. Phys.*, **20**, 1135.
134. ——, Thomsen, K., and Yde, P. B., 1972, *Mol. Phys.*, **23**, 691.
135. Thorhallson, J. and Chong, D. P., 1969, *Chem. Phys. Lett.*, **4**, 405.
136. Van Duijneveldt-van de Rijdt, J. G. C. M. and Van Duijneveldt, F. B., 1976, *J. Mol. Struc.*, **35**, 263.
137. Zakharyan, R. Z., 1975, *Vopr. Sborn. Kvant. Khim.*, G-64, p. 3.
138. Zaslavskaya, L. I., Rossikhin, V. V., and Bolotin, A. B., 1976, *Fiz. Mol.*, **2**, 16.

Supplementary References

139. Bamzai, A. S. and Deb, B. M., 1981, *The Role of Single-Particle Density in Chemistry, Rev. Mod. Phys.*, forthcoming.
140. Beran, Z. S. and Zidarov, D. C., 1978, *Int. J. Quantum Chem.*, **13**, 227.
141. Blom, C. E. and Müller, A., 1978, *J. Chem. Phys.*, **69**, 3397.
142. Creswell, R. A. and Robiette, A. G., 1978, *Mol. Phys.*, **36**, 869.
143. Fieck, G., 1978, *Theoret. Chim. Acta* (Berlin), **49**, 211.
144. Gazquez, J. L., Ray, N. K., and Parr, R. G., 1978, *Theoret. Chim. Acta* (Berlin), **49**, 1.
145. Lakdor, T. B., Suard, M., Taillandier, E., and Berthier, G., 1978, *Mol. Phys.*, **36**, 509.
146. Liu, B., Sando, K. M., North, C. S., Friederich, H. B., and Chipman, D. M., 1978, *J. Chem. Phys.*, **69**, 1425.
147. Pulay, P., Meyer, W., and Boggs, J. E., 1978, *J. Chem. Phys.*, **68**, 5077.
148. Smit, W. A. and Roos, F. A., 1978, *Mol. Phys.*, **36**, 1017.

6
Models for Chemical Reactivity

Peter Politzer and Kenneth C. Daiker
Department of Chemistry
University of New Orleans
New Orleans, Louisiana

Contents

6-1. INTRODUCTION

One of the continuing themes of theoretical chemistry has been the effort to develop a better basis for understanding and predicting the reactive properties of molecules. This has led to the introduction of a number of proposed indices of reactivity, designed to provide some quantitative measure of the chemical activities of the various sites and regions of a molecule. Among these indices are such quantities as atomic charges, bond orders, free valencies, frontier electron densities, localization energies, and others. The values of such indices are generally determined by quantum chemical calculations.

It is not the purpose of this chapter to survey these older approaches to the problem of chemical reactivity; detailed analyses are available elsewhere (*52*,

101–103, *113*, *158*, *163*, *292*). However, one of these techniques, the use of calculated atomic charges, does fit into the present discussion, since it can be regarded as a forerunner of the concepts and methods with which this chapter deals. We will therefore begin with a brief account of some aspects of the problem of estimating atomic charges. A more complete treatment has recently been given (*145*).

6-2. ATOMIC CHARGES

If it were possible to associate a physically meaningful positive or negative charge with each atom in a molecule, then these charges might provide some indication of the relative susceptibilities of these atoms to electrophilic or nucleophilic attack. But while atomic charge is a familiar and intuitive concept, it is not a physical observable—i.e., it is not a physically measurable property. If atomic charge were a physical observable, there would be associated with it a quantum mechanical operator, and there would be a rigorous and well-defined procedure for calculating it. Instead, the actual situation is that the quantitative basis for the concept of atomic charge is completely arbitrary. This is of course because the concept itself is an ambiguous one; it refers to a property of an atom in a molecule, which is not a well-defined entity. Many different formulas for computing atomic charges have been proposed (*145*); unfortunately, the results sometimes have little physical meaning.

The most widely used procedure for computing atomic charges is probably the population analysis suggested by Mulliken (*195*; see also Ref. 194). This can be illustrated in terms of a simple example, a molecular orbital which is a linear combination of atomic orbitals on atoms A and B:

$$\psi = C_A \psi_A + C_B \psi_B. \tag{6-1}$$

If ψ is normalized, then

$$C_A^2 + 2C_A C_B S_{AB} + C_B^2 = 1, \tag{6-2}$$

where S_{AB} is the overlap integral between ψ_A and ψ_B, $S_{AB} = \int \psi_A \psi_B \, d\tau$. Letting N be the number of electrons in molecular orbital ψ, Mulliken proposed that the quantity

$$C_A^2 + C_A C_B S_{AB}$$

be used to represent the fraction of these electrons that is associated with atom A. The electronic charge on A is therefore

$$Q_A = N(C_A^2 + C_A C_B S_{AB}). \tag{6-3}$$

Q_B is defined in analogous fashion. Equation (6-3) can easily be generalized as follows:

$$Q_r = \sum_i N_i \left[\sum_m \left(C_{im}^2 + \sum_s \sum_n C_{im} C_{in} S_{mn} \right) \right]_{s \neq r} . \tag{6-4}$$

For the case of a polyatomic molecule described by a set of molecular orbitals ψ_i,

$$\psi_i = \sum_r \sum_m C_{im} \psi_m . \tag{6-5}$$

The subscripts m and n refer to atomic orbitals on atoms r and s, respectively; N_i is the number of electrons in ψ_i. C_{im} is the coefficient of atomic orbital ψ_m in molecular orbital ψ_i; it is determined by minimizing the total energy of the molecule. The net charge on atom r is then $(Z_r - Q_r)$, Z_r being its nuclear charge.

Equation (6-4) has an intuitive appeal and it can easily be programmed. It is therefore applied very extensively as a formula for atomic charge, and with a certain degree of success. Indeed, charges calculated by using (6-4) are a standard feature of many wavefunction computations, both ab initio and semiempirical.

This formula does have some serious weaknesses, however (*89, 145, 230, 232, 271*). An obvious one is that the overlap term, $2C_A C_B S_{AB}$ in (6-2), is apportioned equally between A and B, even when these are two different atoms. Several modifications of (6-4) have been proposed which either attempt to treat this term more realistically (*178, 244*), or simply to eliminate it by orthogonalizing the atomic orbitals of the basis set and thereby making the overlap integrals, and overlap terms, equal to zero (*80, 291*). While some of these methods do indeed give more satisfactory results than (6-4), they cannot overcome a second problem, which is that they assign the entire contribution of the C_A^2 term in (6-2) to atom A, and the entire C_B^2 contribution to atom B. In reality, however, the function ψ_A, to which C_A corresponds, may have its greatest concentration at some significant distance from nucleus A, perhaps even closer to nucleus B than to A (*230, 271*). The same is of course possible for ψ_B. It has been demonstrated that reasonably good wavefunctions for such molecules as methane and ammonia can be written solely in terms of atomic basis orbitals belonging to the central atom (see, e.g., Ref. 29). It is clear that a portion of the electronic charge represented by such a wavefunction must be associated with the peripheral atoms. Yet (6-4) assigns all of the electrons to the central atom, since all of the atomic basis orbitals belong to it. Thus the carbon in methane would be depicted as having a charge of -4.

Another very serious weakness in (6-4) is its sensitivity to the atomic-orbital basis set used in writing the wavefunction. It can happen that two different wavefunctions for a given molecule will yield nearly identical values for the various physical observables, indicating that they are virtually equivalent descriptions of the molecule, but nevertheless will predict, according to (6-4), quite different atomic charges, because of some minor difference in their basis sets (*216*, *232*). This is clearly an artifact, which reflects the fact that the physical observables of a molecule are calculated from the total wavefunction, which has a fundamental significance as a quantum mechanical description of the molecule, whereas atomic charges computed with (6-4) are obtained from just certain of the terms in the wavefunction, which in themselves have no fundamental significance. Thus, for example, two different near-Hartree-Fock wavefunctions for hydrogen fluoride, which are essentially identical descriptions of the molecule as far as all of its physically measurable properties are concerned, nevertheless yield quite different atomic charges (*232*) by (6-4): H(+0.23), F(−0.23) versus H(+0.48), F(−0.48).

In order to avoid this artificial dependence of calculated charges upon the mathematical form of the wavefunction, atomic charges have also been defined in terms of the electronic density function,

$$\rho(r) = N \int \psi^*(r_1, r_2, \cdots, r_N)\, \psi(r_1, r_2, \cdots, r_N)\, dr_2 \cdots dr_N. \tag{6-6}$$

$\psi(r_1, r_2, \cdots, r_N)$ is the wavefunction describing an N-electron molecule, and $\rho(r)$ is the average number of electrons in a unit volume element at the point r. One approach to obtaining atomic charges from $\rho(r)$ is by integrating it over regions of space "belonging" to the various atoms in the molecule. The net charge on atom r is then $(Z_r - Q_r)$, where

$$Q_r = \int_r \rho(r)\, dr, \tag{6-7}$$

the integration being carried out only over the region associated with atom r. Two possibilities for defining the region associated with an atom have been proposed (*18*, *230*) and a comparison of some of the charges resulting from these as well as other approaches has been given (*240*).

It certainly seems reasonable to use the electronic density function as a basis for obtaining atomic charges, and there is the very desirable feature that $\rho(r)$ is a physical observable. Thus, two molecular wavefunctions which differ in form (for instance they may have different basis sets) but which give the same physical description of the molecule, including its charge distribution, will predict the same atomic charges. For example, the two hydrogen fluoride wavefunctions mentioned earlier, which yield such different charges when these are computed using (6-4), give the values H(+0.26), F(−0.26) and H(+0.27), F(−0.27) when

(6-7) is used in conjunction with the Politzer–Harris procedure for defining the atomic regions (*232*).

Even atomic charges determined using the electronic density function contain a significant element of arbitrariness, however, because of the inherent ambiguity of the concept of an atom in a molecule. The weakness of (6-7) is, of course, the fact that there are no rigorous and unique guidelines that establish a region of space associated with each atom in a molecule. Very recently, Hirschfeld (*128*) has proposed a most promising related approach that avoids this problem. For the molecules compared, Hirshfeld's charges are quite similar to those obtained by the Politzer–Harris method (see also Section 2-7).

6-3. THE ELECTROSTATIC POTENTIAL

When calculated atomic charges are used as an index of chemical reactivity, the rationale is that an approaching electrophile will favor those sites on a molecule that have the more negative charges, while an approaching nucleophile will favor the more positively charged sites. From this point of view, which admittedly oversimplifies the situation, the atomic charges are really being used as a point-charge representation of the electrostatic potential that is generated at all points in the neighborhood of the molecule by its electrons and nuclei, since in reality it is this potential that acts upon any nearby entity. Clearly, a much more accurate and more detailed picture could be obtained by directly calculating this electrostatic potential. It could then be used instead of the atomic charges as an indicator of the reactive regions and sites in the molecule. The basic principle would be the same—an electrophile would seek regions of negative potential, a nucleophile would seek positive ones—but the level of application would be considerably higher and more realistic. This very important development, the use of the calculated electrostatic potential as a guide to the reactive regions of molecules, was pioneered by Bonaccorsi and co-workers (*30–32*).

The electrostatic potential (EP) at a point r in the space around a molecule is (in atomic units)

$$V(r) = \sum_{A} \frac{Z_A}{|R_A - r|} - \int \frac{\rho(r')\,dr'}{|r' - r|}. \qquad (6\text{-}8)$$

Z_A is the charge on nucleus A, located at R_A, and $\rho(r')$ is the electronic density function for the molecule. The first term on the right-hand side of (6-8) represents the effect of the nuclei, the second represents that of the electrons. As seen, the two have opposite signs, and therefore opposite effects. $V(r)$ is their resultant at each point r; it is an indication of the net electrostatic effect produced at the point r by the total charge distribution (electrons plus nuclei) of the molecule.

Unlike the traditional indices of reactivity mentioned in Section 6-1, the EP is a real physical property; in quantum mechanical terms, it is a physical observable. In addition to its definition, (6-8), it can be given any of several other interpretations, all of which are rigorously exact:

1. Given a point charge $\pm Q$ located at r', then $\pm QV(r)$ is exactly equal to the electrostatic interaction energy between the undistorted molecule and the point charge.
2. In a perturbation-theory treatment of the total (not just electrostatic) interaction between the molecule and the point charge, $\pm QV(r)$ is exactly equal to the first-order term in the total interaction energy.
3. The negative gradient of the quantity $\pm QV(r)$ is exactly equal to the electrostatic force that is exerted by the molecule's charge distribution upon the point charge $\pm Q$.

$$F(r) = -\nabla[\pm QV(r)].$$

Some applications of this property of the EP will be discussed in Section 6-8.

Note that the EP $V(r)$ is *not* an energy quantity; however $\pm QV(r)$ *is* an energy. When $V(r)$ is in atomic units, as in (6-8), then the energies $\pm QV(r)$ corresponding to interpretations 1 and 2 above are in atomic units of energy. These can be converted to kcal/mole by multiplying by 627 (see note 4). It is customary, in using the EP to study molecular reactivity, to express $V(r)$ in kcal/mole, which means that the quantity being given is actually $QV(r)$ with $Q = +1$.

The EP is a property of a molecule in a definite state; it reflects the particular geometry and electronic density distribution of the molecule in that state. The EP does not, in itself, take any account of the specific nature of any entity which may be approaching the molecule, all three interpretations given above are in terms of a point charge, $\pm Q$. The potential also does not reflect changes that occur in the molecule itself as it begins to interact with some other entity; it takes no account, for example, of polarization, charge transfer, or exchange effects, which can, of course, be very significant factors in the overall interaction.

For these reasons, it is not in general proper to use the EP as a measure of the total energy of interaction of the molecule with some entity, not even with a point charge. It does sometimes happen that $V(r)$ can be directly related to the total interaction energy; such situations will be treated in Section 6-6-1. $V(r)$ can also be used to compute exactly the electrostatic interaction energy between the molecule and the other entity, provided that an appropriate charge distribution function is available for the latter. This will be discussed in Sections 6-4-4 and 6-6-2.

However, the real role of the molecular EP in the study of the reactive properties of molecules is essentially a qualitative one. The sign and magnitude of the EP in various regions around a molecule is taken as an indication of the tendency for an approaching electrophile to attack the molecule in these regions. The electrophile will presumably tend to go where $V(r)$ is negative (meaning that the electrostatic effect of the molecule's electrons is dominant over that of its nuclei) and to avoid those regions where $V(r)$ is positive. These tendencies should be stronger, the greater the magnitudes of the potentials (see, e.g., Fig. 6-1).

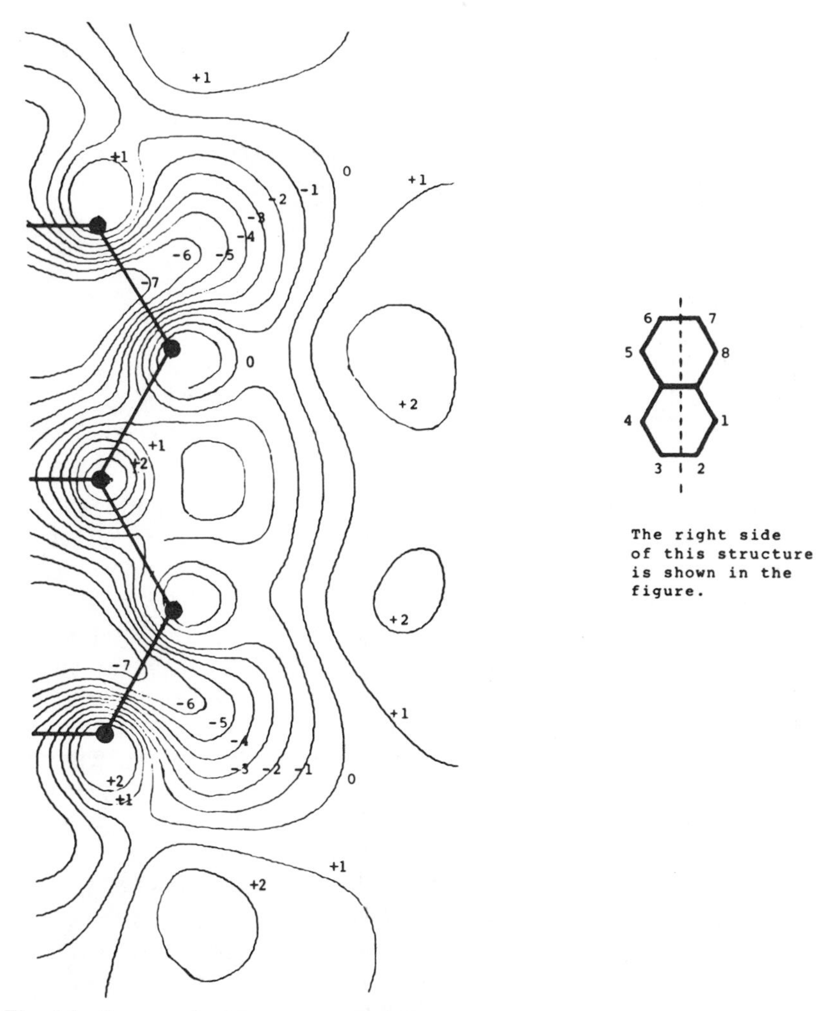

Fig. 6-1. Contour diagram showing electrostatic potential of naphthalene, kcal/mole, in plane 1.32 Å above molecular plane. Only half of plane is shown, as indicated in structure above. Note channels of negative potentials, showing favored paths for electrophilic attack. Naphthalene is not perfectly planar; this accounts for the slight lack of symmetry in the contours. (Reproduced from Ref. 241, courtesy John Wiley & Sons, Inc.)

Thus, it is reasonable to anticipate that a knowledge of the EP around a molecule will help considerably in developing an understanding of which sites and regions on it are the most attractive for an approaching electrophile. Similarly, a knowledge of the potentials around several molecules will allow predictions as to their relative order of reactivity toward an electrophile. Such interpretations and predictions will generally be more valid as the electrostatic contribution is a greater portion of the total interaction. This means that the usefulness of the EP as a guide to a molecule's reactive sites and regions is greater in the early stages of an interaction, before an approaching electrophile comes very close to it, since such effects as polarization, charge transfer, and exchange become more important at smaller separations.

Some of the points that have been made in these last two sections are brought out very well in a study (*10*) of the reactivity of aceheptylene (I).

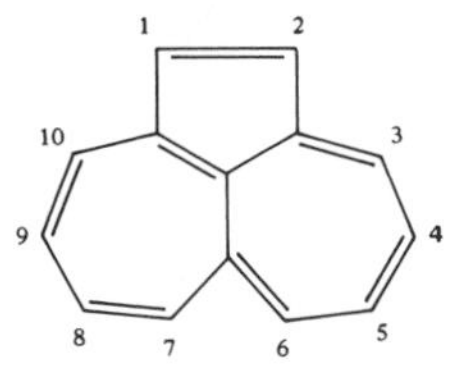

I

The authors computed wavefunctions for this molecule by an ab initio SCF procedure and several different semiempirical methods, and then calculated the charges on the various atoms, using (6-4) (with $S_{mn} = 0$ for the semiempirical wavefunctions). These sets of charges differed very significantly, indeed even qualitatively from each other, and predicted, on the whole, that C4 is the most favorable protonation site. This is contrary to the experimental observation (for the 3,5,8,10-tetramethyl derivative) that protonation initially occurs exclusively at C1, followed by the gradual migration of the proton to first C4 and then C6 (see Ref. 10). At thermodynamic equilibrium only C6 is protonated. When the EP around aceheptylene was computed using the ab initio wavefunction, it showed the most negative region to be near C1 and C2, followed by C4 and C9 and then the C6, C7 pair. Thus the EP correctly predicts that the most favored approach for a proton is toward the C1, C2 region. However this example also illustrates the point that the potential indicates where an electrophile is most likely to approach a molecule, and not necessarily where it will undergo the thermodynamically most favored interaction.

The preceding discussion of the use of EP in studying the reactive properties of molecules has referred only to their interactions with electrophiles, although it would seemingly be possible to treat nucleophilic reactions in analogous fashion; the attack would now be expected to occur preferentially at regions of positive potential. The fact is, however, that positive potentials do not necessarily reflect an affinity for nucleophiles. Sometimes they are due to hydrogen atoms, which are very often positive without being sites for nucleophilic attack; other times it may be that the positive charge of a nucleus, being so highly concentrated, produces an electrostatic effect greater than the actual tendency to interact with a nucleophile. It is relevant to point out, in this connection, that the EP around free neutral atoms has invariably been found to be positive everywhere (*236, 242*), reflecting the concentrated charge of the nucleus. Thus, while calculated molecular EPs have been utilized extensively to interpret the reactive properties of the molecules toward electrophiles, they have been little used in conjunction with nucleophilic attack. Among the few examples of such an application are investigations of the interaction of Cl^- with the nucleic acid bases (*112, 263*).

The limitations upon the information that can be conveyed by the EP, which have been discussed in this section, should be kept in mind. They can be quite significant. For example, both the bromination and the nitration of toluene involve electrophilic attack upon the toluene, yet the percentages of the ortho and para products are 33 and 67, respectively, in the former case, and 58 and 38 in the latter (*193*). The EP of toluene certainly could not have predicted *both* sets of results! In a great many instances, however, the EP does provide very useful information and insight into the reactive properties of molecules. Some examples will be discussed in Section 6-5.

6-4. CALCULATION OF THE ELECTROSTATIC POTENTIAL

6-4-1. General Procedure

In principle, the evaluation of $V(r)$ as defined by (6-8) is straightforward. As pointed out earlier, (6-8) is perfectly general and there is no restriction upon the mathematical form in which $\rho(r)$ is expressed. However, the type of wavefunction that is most likely to be available for a molecule is one that is in terms of molecular orbitals, these being written as linear combinations of atomic orbitals, as in (6-5). Then,

$$\rho(r) = \sum_i N_i \sum_r \sum_s \sum_m \sum_n C_{im} C_{in} \psi_m(r)\, \psi_n(r). \tag{6-9}$$

As in (6-4) and (6-5), the subscripts m and n refer to atomic orbitals on atoms r and s, respectively, while N_i is the number of electrons in molecular orbital ψ_i. Inserting this expression for $\rho(r)$ into (6-8),

$$V(r) = \sum_A \frac{Z_A}{|R_A - r|} - \sum_i N_i \sum_r \sum_s \sum_m \sum_n C_{im} C_{in} \int \frac{\psi_m(r')\, \psi_n(r')\, dr'}{|r - r'|}. \tag{6-10}$$

The integrals that appear in (6-10) are of a familiar type; they have the form of nuclear-electronic attraction integrals such as are evaluated in the course of computing a wavefunction. The point $\mathbf{r}$ now takes the place of the position of the nucleus. Thus the calculation of $V(\mathbf{r})$ involves a large number of two- and three-center nuclear-electronic attraction integrals.

The specific techniques used to compute these integrals depend upon the nature of the atomic orbitals ψ_m and ψ_n. The computational procedure is particularly efficient when Gaussian-type orbitals are involved. These have the property that any product of two Gaussian functions can be expressed as a single function, even if the original two were on different centers (in which case the new function is centered at an intermediate point). This permits a considerable simplification of the integrals in (6-10), which can then be evaluated in terms of the error function (*44, 45, 283*).[1]

There are also analytical formulas available for treating the integrals in (6-10) when ψ_m and ψ_n are Slater-type orbitals (see, e.g., Refs. 19, 20, 269, and 282). In many instances, however, these are rather time-consuming to use—an important consideration in view of the large number of integrals that must be calculated. In fact, it turns out to be computationally more efficient to expand each Slater-type function in terms of Gaussian orbitals and then to follow the procedure described above, rather than to calculate the integrals directly in terms of Slater-type orbitals. This may, of course, introduce some degree of uncertainty into the results, depending upon how many Gaussians are used to represent each Slater-type function. It has been found, however, that the latter can be described quite accurately by six-term expansion (*123*). Indeed, replacing each Slater-type orbital by such a six-Gaussian (6G) expansion changes the calculated electrostatic potential "in the most delicate regions" by less than 0.5 kcal/mole; even a 3G expansion, for a group of three-membered ring molecules, yields potentials which are within 1-3 kcal/mole of those obtained directly from the Slater-type orbitals (*224*). To show how the calculated value converges as the number of Gaussians in the expansion increases, we quote the following figures (*220*), representing the calculated EP at a particular point in the neighborhood of oxaziridine (CH_2ONH): 2G, $V = -78$ kcal/mole; 3G, $V = -63$; 4G, $V = -64$, 5G, $V = -65$; 6G, $V = -65$ kcal/mole. It should be emphasized that even the use of a six-Gaussian expansion to replace each Slater-type function, while very much increasing the total number of integrals to be calculated, still results in a great saving of computational time, because the integrals over Gaussian orbitals can be evaluated so much more readily.

In general, the absolute accuracy of the calculated EP depends upon the quality of the wavefunction, since the latter determines $\rho(\mathbf{r})$, and upon the degree of rigor with which the integrals in (6-10) are computed. Although these are not difficult in principle, the large number that is required for any reasonably sized molecule can lead to computational problems. Hence, they are sometimes approximated, as shall be seen. So far, this discussion has emphasized the exact treatment of these integrals. It is true that some degree of error can enter when Slater-type orbitals are represented by Gaussian expansions, but this error can be made essentially insignificant by using 6G expansions. This means that the calculation of $V(r)$ as discussed up to this point is, for all practical purposes, rigorously correct in the sense that, although the accuracy of the result necessarily depends upon the quality of the wavefunction, no further significant approximation is involved in going from the wavefunction to the EP.

There are two matters that must now be considered with respect to the calculation of $V(r)$. First, what sorts of results are obtained with various types of molecular wavefunctions? Second, what approximations have been used in evaluating the integrals in (6-10), and how good are they?

6-4-2. Relationship of Potential to Quality of Wavefunction

If the exact wavefunction were available for a molecule, then the true EP could be obtained, provided that all integrals were computed rigorously. In reality, however, an ab initio self-consistent-field (SCF) wavefunction[2] is the best that can presently be achieved for most molecules of chemical interest (and even that is not practical in many cases in which some semiempirical procedure has to be used; see, e.g., Ref. 144). Fortunately, the operator $1/|r - r'|$ used in computing the EP is a one-electron operator, and properties associated with such operators are usually given relatively accurately by SCF wavefunctions. For a Hartree–Fock wavefunction, properties computed with one-electron operators are correct through first-order; any corrections must be at least second-order effects (*116*, *189*). This indicates [but does not guarantee, since the second-order and higher corrections may conceivably be quite large (*247*)] that Hartree–Fock wavefunctions should yield fairly accurate values for such properties. However it is only for Hartree–Fock wavefunctions that there are no first-order corrections to one-electron properties, and the SCF functions that can be computed for most molecules of chemical interest fall considerably short of the Hartree–Fock limit. While there are indications that even such SCF wavefunctions do give at least some one-electron properties relatively accurately (see, e.g., Refs. 228 and 317), it is important in terms of the present discussion that this point be examined in the particular case of the EP.

A number of studies have compared potentials obtained with SCF functions having basis sets of different sizes and approaching the Hartree–Fock limit to varying extents. The overall conclusion that emerges is that increasing the basis set, and thereby generally improving the quality of the wavefunction, does not greatly alter the overall form of the EP for a given molecule, although there are certainly changes in detail. The locations of the potential minima (the points at which the EP reaches its most negative values) are likely to change somewhat, and the numerical values of the potentials at these points may be affected, perhaps significantly.[3]

For example, Ghio and Tomasi (*105*) computed $V(r)$ for aziridine, $(CH_2)_2NH$, using a small Gaussian basis set (molecular energy = -132.40 hartrees)[4] and compared it to previous results obtained with a minimum basis of Slater-type orbitals (STO) with best-atom exponents (molecular energy = -132.66 hartrees). The EPs were examined in the symmetry plane of the molecule, which is perpendicular to the ring and contains the N—H bond. In overall appearance, the results are nearly identical. Both EPs show two large negative regions, one associated with the nitrogen lone pair and one outside the C—C bond. Each of these regions has its own minimum, and they are in approximately the same locations. However, those obtained with the STO wavefunction are closer to the molecule (by about 0.24 Å and 0.36 Å for the nitrogen and the C—C minima, respectively) and they are more negative (-92.6 versus -76.9 kcal/mole, and -17.2 versus -10 kcal/mole, respectively.

Just recently, using a basis set of roughly double-zeta quality (molecular energy = -132.96 hartrees), a very similar potential map was again produced (*58*), with an even more negative nitrogen minimum (-108 kcal/mole) but a less negative one by the C—C bond (-7.5 kcal/mole). With the same type of wavefunction, Catalan and Yanez (*58*) also found a more negative minimum for the oxygen in oxirane than did Ghio and Tomasi (*105*) with their small Gaussian basis set: -80 versus -44.8 kcal/mole. These various differences are undoubtedly due to the use of different geometries as well as to the disparity in basis sets in the two works.

Almlöf and Stogard (*9*) compared the EPs computed for ethylene from five different SCF wavefunctions having various basis sets and energies ranging from -77.76 to -78.05 au (the Hartree-Fock energy was estimated to be -78.08 au). The values of the potential minimum were all between -18.0 and -22.9 kcal/mole, showing no correlation with the calculated total energy. The sensitivity of the computed potential to the quality of the wavefunction was found, in general, to increase with distance from the nuclei.

A particularly significant study (*216*) examined the effect of including phosphorus $3d$ basis functions upon the atomic charges [(6-4)] and the EP of the dimethylphosphate anion, $(CH_3)_2PO_4^-$. It was found that the charges were changed very considerably by the inclusion of $3d$ basis functions (the worst case was the phosphorus atom, which went from +1.35 to +0.52); the potential, on the other hand, was relatively little affected. The minima became less negative by 7-9% and moved about 0.05 Å farther away from the nearest atoms; they also shifted by angles of about 25° with respect to these atoms. These changes are comparatively minor ones, however, and do not alter the qualitative picture; they are certainly much less than the changes in atomic charges. These results, as well as comparisons of other properties, show that the essential features of the charge distribution in $(CH_3)_2PO_4^-$ are represented fairly accurately even without the inclusion of $3d$ basis orbitals, and that the effect observed upon the calculated charges is an artifact, reflecting one of the weaknesses (see Section 6-2) of (6-4). The EP is less sensitive to the form of a molecular wavefunction and more indicative of its substance.

A similar conclusion regarding the effect of d-type basis functions emerges from a recent study (*188*). For a group of five molecules, using roughly double-zeta-level basis sets, they found that the general result of including d-type "polarization" functions was to diminish the magnitudes of the negative minima by some 20–25%, and to move them slightly further away from the atoms with which they were associated.

The important point brought out by these and related investigations (*36*, *75*, *85*, *226*, *280*) is that a generally reliable qualitative picture of the EP can be obtained with an SCF wavefunction, even if it is only of minimum-basis-set quality. The major positive and negative regions will be apparent, and the principal minima will be in approximately the correct locations. Furthermore, the available evidence indicates that when there are two minima within the same molecule, their order will generally not be inverted unless their magnitudes are quite similar (*277*).[5] On the other hand, the numerical values of the calculated potentials and the precise positions of the minima should not be taken too seriously unless the wavefunction is very close to the Hartree–Fock limit; even then there is the possibility of appreciable second-order corrections to the potential. In view of this uncertainty in the numerical values of the computed potential, it is clear that potentials near zero should be regarded with particular caution, since they may be even qualitatively incorrect. For instance, the EP calculated for pyridine using a double-zeta wavefunction (with polarization basis functions on the hydrogens) showed a weak minimum of −3.3 kcal/mole near the β-carbon (*6*). When a minimum-basis STO wavefunction was used, each Slater-type orbital being replaced by three Gaussian functions in computing the potential (STO-3G basis set), a minimum was found in approximately the same location but it had a positive value: +0.1 kcal/mole (*223*). The difference in magnitude is therefore only 3.4 kcal/mole, but it is associated with a qualitative difference, a change in the sign of the EP.

The preceding discussion has dealt entirely with SCF calculations, because these have been the basis for nearly all EPs obtained from ab initio wavefunctions. An important recent study (*85*), however, examined the effects of including configuration interaction (CI) upon the EP of formaldehyde in its ground state and in two low-lying excited states (see Section 6-5-9). Rather reassuringly, the inclusion of CI produced very little change in the general pattern of the ground-state potential. Its primary effect was to gradually diminish somewhat the magnitudes of the negative minima. Thus, as the number of determinants included in the wavefunction increased from 1 (SCF) to 261, 634, 1188, and 2474 (using an enriched double-zeta basis set), the oxygen lone-pair minima correspondingly went from -60.4 to -51.2, -50.3, and -50.2 kcal/mole, the last value being for both the 1188- and 2474-determinant computations. It is encouraging to note how distinctly and rapidly the magnitudes of the potential minima level off and reach stable values. This property is definitely not shown by the total molecular energy; for example, the decrease in energy between the 1188- and 2474-determinant functions is greater (-0.1356 hartrees) than between the 1- and 261-determinant functions (-0.0804 hartrees).

The general conclusion concerning the meaningfulness of EPs computed from ab initio SCF wavefunctions is a favorable one. It appears that they are reasonably reliable for qualitiative purposes, and for comparisons between molecules, provided that the wavefunctions for all of the molecules are in terms of the same basis sets and are otherwise equivalent. However, many of the molecules which would be most interesting to investigate are too large for ab initio calculations; a semiempirical approach is necessary.[6]

A detailed investigation of the accuracy of EPs computed with CNDO/2 wavefunctions[7] (*246*) has been carried out (*54*, *108*, *109*, *111*). Because of the approximations inherent in this semiempirical technique, there is some ambiguity as to the best approach to use in evaluating (6-10). For example, since a key element of the CNDO/2 method is that integrals involving a product of two different atomic orbitals, $\psi_m(r)\,\psi_n(r)$ where $m \neq n$, are set equal to zero, it would be fully consistent to similarly set such terms equal to zero in calculating $V(r)$. Or, since another element of the CNDO/2 procedure is to treat integrals over p orbitals as though they involved the corresponding s orbitals, it might be argued that the same should be done in computing $V(r)$. Giessner-Prettre and Pullman tested various combinations of these and other possible approaches, comparing the resulting EPs with those obtained with SCF functions for the same molecules; these latter were taken as the standards for comparison. It was found that the best results were achieved when the CNDO/2 wavefunctions were first "deorthogonalized," after which the potentials were calculated with all terms included and all integrals evaluated accurately. The deorthogonalization was accomplished by treating the CNDO/2 wavefunctions as though they were written in terms of rigorously orthogonal atomic orbitals,[8] the supposed orthogonalization having been carried out by a Löwdin transformation of an original STO basis set (*177*). If this view is adopted, then the CNDO/2 wavefunction can similarly be transformed "back" to a basis of nonorthogonal STO (*107*, *279*):

$$\tilde{C}_{\mathrm{STO}} = \tilde{S}^{-1/2}\,\tilde{C}_{\mathrm{CNDO/2}}, \tag{6-11}$$

where $\tilde{C}_{\mathrm{STO}}$ is the matrix of the coefficients of the STO in the deorthogonalized molecular wavefunction, $\tilde{S}$ is the STO overlap matrix, and $\tilde{C}_{\mathrm{CNDO/2}}$ is the matrix of the coefficients in the CNDO/2 wavefunction. Once the latter wavefunction has been deorthogonalized, the rigorous evaluation of the integrals involved in determining $V(r)$ can be carried out by methods described in Section 6-4-1. Since these integrals are over STOs, a likely first step is to represent them as expansions in terms of Gaussian-type functions.

Giessner-Prettre and Pullman found, for the molecules H_2O and H_2CO, that this procedure produced EPs that were in good agreement with those obtained from SCF functions, a conclusion that was also reached subsequently for a group of seven three-membered ring molecules (*220*). In the latter study, a good linear relationship was found between the values of the potential minima calculated with the deorthogonalized CNDO/2 wavefunctions and those corresponding to minimum-basis-set SCF wavefunctions. These results are very encouraging, since they demonstrate the possibility of obtaining meaningful EPs by a semiempirical technique, one which can be applied to molecules that are too large to be handled by ab initio methods.

This optimism concerning the use of CNDO/2 wavefunctions to determine the EPs of large molecules must be tempered, however, by a realization of some important shortcomings that have been observed despite deorthogonalization and rigorous evaluation of integrals. One of the most serious is that CNDO/2 potentials for aromatic molecules such as benzene, fluorobenzene, etc. fail to show the negative regions above and below the ring(s) that are revealed by potentials calculated with SCF wavefunctions (*224*, *226*, *242*). CNDO/2 potentials are found to be positive above and below aromatic rings. Another problem is that CNDO/2 wavefunctions sometimes significantly distort the extent of charge transfer to heteroatoms in organic molecules, with obvious consequences for the EPs. For example, there is a tendency to indicate an exaggerated transfer of electronic charge to oxygen atoms, at the expense of nitrogens in the same molecule (*220*). Thus, the CNDO/2 potentials for both guanine (II) and cytosine

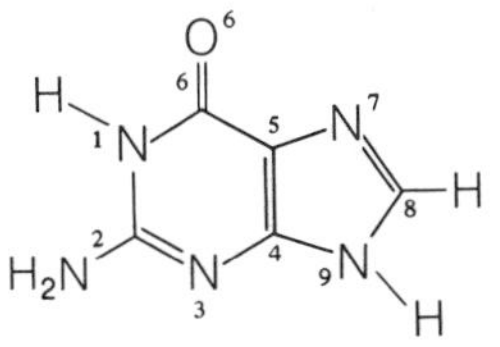

II

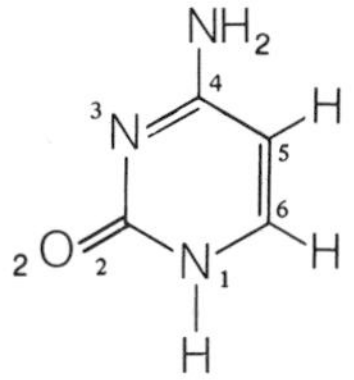

III

(III) predict the minima to be near the oxygens, contrary to both SCF potentials and also experimental observations, which indicate them to be by the N7 and N3 nitrogens, respectively (*110, 111*).

CNDO/2 potentials have also sometimes been found to be deficient in showing secondary minima that are revealed by SCF results. In some instances the secondary minimum is there, but with a positive rather than a negative value

(*220, 277*); sometimes, however, a secondary minimum is not even shown (*238*). In summary, while the CNDO/2 method does afford a practical and useful approach to the EPs of large molecules, it is clear that the results must be interpreted cautiously.

The potentials obtained with other semiempirical techniques, such as the INDO (*246*) and the extended-Hückel (*183*) methods, appear not to have been subjected to such detailed analysis and comparison to SCF results as have CNDO/2 potentials. The situation with respect to the INDO procedure seems to be basically the same as for CNDO/2 (*110, 224*); again deorthogonalization appears to be desirable,[9] the negative regions above and below aromatic rings are not properly indicated, and the N3 minimum in cytosine is less negative than that of the oxygen. There seem to have been rather few calculations of EPs using extended-Hückel wavefunctions. The regions above and below aromatic rings are again found to be a problem; in addition, the results in general are reported to be greatly exaggerated (*224*). For example, some minima are overestimated by a factor of about four. However, extended-Hückel wavefunctions are quite sensitive to the parameters used in computing them and to certain variable details of the procedure (*229, 233, 291*). Thus it may be that with a suitable choice of parameters and procedure, more accurate potentials could be obtained.

6-4-3. Integral Approximations

In the preceding discussion, the emphasis has been on comparing and assessing the EPs calculated with wavefunctions of various levels of approximation, both SCF and semiempirical. Whatever inaccuracies there might have been in the computed potentials could be assumed to be due primarily to the approximate nature of the wavefunctions, rather than to any lack of rigor in calculating the potential from the wavefunction, i.e., in evaluating the integrals that appear in (6-10). The main inaccuracy arising from the latter source would have been in some instances in which Slater-type basis orbitals were represented by relatively small Gaussian expansions in computing the integrals. However this would be expected to produce an error of only 1-3 kcal/mole (see Section 6-4-1).

This rigorous evaluation of all integrals may be extremely time consuming, since there can be a very large number of them. For instance, to compute $V(\mathbf{r})$ for benzene at just one point in space, using a CNDO/2 wavefunction and expanding each STO as a sum of six Gaussians, requires computing 16,740 integrals.[10] If three-Gaussian expansions are used, the number of integrals per point is 4,185. This can be very costly in terms of computer time. There has accordingly been some investigation of ways in which the time required to treat the integrals in (6-10) could be shortened. One possibility in those cases in which the integral evaluation involves expanding STOs as sums of Gaussian functions is to decrease the number of terms in the expansion. This has a significant effect upon the number of integrals, since the latter is proportional to the square of the number of terms in the expansion; the effect upon the accuracy of the results has been discussed in Section 6-4-1.

In the course of their analysis of potentials computed from CNDO/2 wavefunctions, Giessner-Prettre and Pullman examined two possible integral approximations, both suggested by the CNDO/2 method itself (*108*, *109*, *111*, *217*). One of these is to replace the nuclear-electronic attraction integrals that appear in (6-10) by the negative of the Coulomb electronic repulsion integral, $\langle \psi_m^*(1)\, \psi_m(1) | 1/r_{12} | 1s_H^*(2)\, 1s_H(2) \rangle$, in which ψ_m is taken to be spherically symmetric and $1s_H$ represents a hydrogen $1s$ orbital located at the point in space at which the potential is being calculated. The second integral approximation tested is one in which the nuclear-electronic attraction integrals in (6-10) are evaluated rigorously, but both ψ_m and ψ_n are taken to be spherically symmetric. These two approximations were tested in conjunction with the further simplification that only those integrals were considered in which $\psi_m = \psi_n$; all others were set equal to zero. With both approximations, the resulting potentials were found to be roughly similar to those obtained by more accurate procedures. There were some considerable differences in detail, however; both positions and also numerical values of potential minima were predicted incorrectly. Indeed, both approaches failed to even show a minimum near the N7 position in guanine (II), which is in reality the preferred position for protonation (*39*, *99*, *165*). For uracil (IV), in contrast, the first integral approximation did

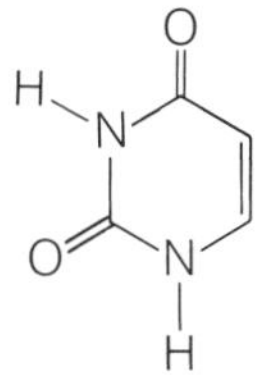

IV

produce negative regions and minima near both oxygens, consistent with a more accurate treatment, and the minima were properly ranked (*217*).

Another proposed procedure (*54*, *57*) for simplifying the integral evaluation associated with (6-10) is to apply the Mulliken approximation, which in this case takes the form:

$$\int \frac{\psi_m^*(r')\, \psi_n(r')\, dr'}{|r - r'|} = \frac{S}{2} \left[\int \frac{\psi_m^*(r')\, \psi_m(r')\, dr'}{|r - r'|} + \int \frac{\psi_n^*(r')\, \psi_n(r')\, dr'}{|r - r'|} \right]. \tag{6-12}$$

S is the overlap integral $\int \psi_m^*(r')\,\psi_n(r')\,dr'$. The two integrals on the right-hand side of (6-12) would then be computed rigorously, as discussed in Section 6-4-1. This approximation is in the same spirit as the equal partitioning of the overlap charge in Mulliken's population analysis procedure (Section 6-2). The advantage of applying (6-12) is that it replaces a large number of integrals in terms of other integrals which would have to be calculated anyway. For a wavefunction written in terms of a basis set of n atomic orbitals, the use of (6-12) would reduce (*57*) the number of integrals to be evaluated from $\frac{1}{2}n(n+1)$ to n, certainly a dramatic change.

Carbo and Martin (*57*) have used (6-12) in computing the EPs of several molecules. Unfortunately they did not present the results that would be obtained with the same wavefunctions if the integrals were calculated rigorously: this would have permitted a more meaningful assessment of the usefulness of (6-12). It appears at present that this approximation produces at least a rough qualitative picture of the EP. However, both the positions and also the values of the minima may be significantly in error, and it is necessary to be cautious in drawing conclusions from the results. For example, with (6-12) NH_3 was found to have a negative region in the outer portion of the cone formed by the three N—H bonds. No trace of this is detected in a rigorously computed $V(\mathbf{r})$ for NH_3 (*277*).

On the whole, it seems to be advisable to calculate all integrals rigorously. This can include expanding STOs in terms of Gaussian functions. If computer time is a problem, the safest time-saving step in the integral evaluation process is to limit Gaussian expansions of STOs to three terms.

6-4-4. Multipole Expansion Methods

This discussion of various approaches to the calculation of $V(\mathbf{r})$ has been limited to the direct evaluation of (6-10). However, it is possible to rewrite (6-10) in another form, in which the terms to be computed are quite different in nature from the nuclear-electronic attraction type of integrals which are the basis of (6-10). This alternative form may be derived by expanding the quantities $1/|\mathbf{R}_A - \mathbf{r}|$ and $1/|\mathbf{r} - \mathbf{r}'|$ in (6-10) in Neumann expansions, whereby (6-10) becomes (*129, 180*)

$$V(\mathbf{r}) = V(r,\theta,\phi) = \sum_{n=0}^{\infty} \sum_{m=-n}^{n} \frac{(n-|m|)!\,Q_n^m}{(n+|m|)!\,r^{n+1}} P_n^{|m|}(\cos\theta)\exp(-im\phi), \tag{6-13}$$

where $P_n^{|m|}(\cos\theta)$ represents the associated Legendre polynomials, and

$$Q_n^m = \sum_A Z_A R_A^n P_n^{|m|}(\cos\theta_A)\exp(im\phi_A) - \iiint \rho(r',\theta',\phi')\, r'^n P_n^{|m|}(\cos\theta') \times \exp(im\phi')\, d\tau, \quad (6\text{-}14)$$

where (R_A, θ_A, ϕ_A) are the coordinates of nucleus A, $\rho(r') = \rho(r', \theta', \phi')$ is the electronic density function for the molecule [cf. (6-6)], and $d\tau = r'^2 \sin\theta'\, dr'\, d\theta' d\phi'$. The expression for $V(r)$ given in (6-13) is valid only if the point r is outside of the molecular charge distribution; in practice, this requirement can be satisfied reasonably well by computing the potential only at points outside of the van der Waals radius of the molecule.

Equation (6-13) can easily be given a physical interpretation by working out the initial terms of the series. For $n = 0$, $m = 0$, the integration in (6-14) yields simply the total number of electrons N, and therefore $Q_0^0 = \Sigma_A Z_A - N$. The first term in (6-13) is therefore

$$\frac{\sum_A Z_A - N}{r},$$

which is the EP at the point r that would be produced by the total net charge[11] of the molecule if it were a point charge located at the origin. For a neutral molecule, this is of course equal to zero.

The next three terms in (6-13) are

$$\frac{Q_1^0 \cos\theta}{r^2} + \frac{Q_1^{-1}\sin\theta\exp(im\phi)}{2r^2} + \frac{Q_1^1 \sin\theta\exp(-im\phi)}{2r^2}$$

$$= \frac{Q_1^0\cos\theta}{r^2} + \frac{\sin\theta}{2r^2}[\cos\phi(Q_1^1 + Q_1^{-1}) - i\sin\phi(Q_1^1 - Q_1^{-1})]. \quad (6\text{-}15)$$

From (6-14),

$$Q_1^0 = \sum_A Z_A R_A \cos\theta_A - \iiint \rho(r',\theta',\phi')\, r'\cos\theta'\, r'^2 \sin\theta'\, dr'\, d\theta'\, d\phi',$$

which is simply the component of the molecular dipole moment μ along the z-axis (Fig. 6-1). Thus, $\mu_z = Q_1^0$. Similarly, $\mu_x = \frac{1}{2}(Q_1^1 + Q_1^{-1})$ and $\mu_y = (1/2i)(Q_1^1 - Q_1^{-1})$. The three terms on the right-hand side of (6-15) can therefore be written

$$\frac{\mu_z\cos\theta}{r^2} + \frac{\mu_x \sin\theta\cos\phi}{r^2} + \frac{\mu_y\sin\theta\sin\phi}{r^2}.$$

Since the numerators of these three terms are the projections of μ_z, μ_x, and μ_y upon the vector $\boldsymbol{r}$, they may be interpreted as a contribution to the potential at the point $\boldsymbol{r}$ that arises from the component of the molecular dipole moment, or the *dipole strength*, in the direction of the point $\boldsymbol{r}$.

In similar fashion, the next five terms in (6-13), corresponding to $n = 2$, $m = -2, \ldots, +2$, can be regarded as being due to the molecular quadrupole strength in the direction of $\boldsymbol{r}$; the seven $n = 3$ terms represent the octopole strength, and so forth. Thus, (6-13) gives the EP as a sum of contributions from the various multipoles of the molecular charge distribution; the equation can therefore be described as a multipole expansion of the potential and written in the form

$$V(\boldsymbol{r}) = \sum_{n=0}^{\infty} \frac{Q_n}{r^{n+1}}, \tag{6-16}$$

in which Q_n is the *strength* of the 2^n-pole of the molecular charge distribution in the direction of $\boldsymbol{r}$. The more multipoles that are included, the longer will be the expansion and the more rigorously correct will be the equation for $V(\boldsymbol{r})$. As before, however, the accuracy of $V(\boldsymbol{r})$ is limited by that of the electronic density function $\rho(\boldsymbol{r})$.

Equations (6-13)–(6-16) are written in terms of one center, or origin, to which all of the coordinates refer. The choice of that center is a question of convenience; for example, it might be taken to be the center of mass of the molecule. It is also possible, however, to write $V(\boldsymbol{r})$ as a sum of several multipole expansions, each defined with respect to a different center. To do this, it is necessary to divide up the molecular charge distribution (both electronic and nuclear) into segments associated with the various centers. These segments must satisfy the requirements,

$$\rho(\boldsymbol{r}) = \sum_i \rho_i(\boldsymbol{r}) \text{ and } \sum_{\mathrm{A}} Z_{\mathrm{A}} = \sum_i Z_i, \tag{6-17}$$

where ρ_i is the electronic density function of the ith segment, and Z_i is the total nuclear charge in that segment. These segments may be based upon any one of several possible criteria. For instance, each segment might correspond to the nuclear and electronic charge associated with one of the atoms in the molecule (the latter being determined by one of the procedures for computing atomic charges that were mentioned in Section 6-2) (*90*, *265*, *294*). Another possibility is to transform the molecular wavefunction into a set of localized orbitals, using one of the techniques that have been proposed for this purpose,[12] and then to let each term $\rho_i(\boldsymbol{r})$ in (6-17) correspond to one of the localized orbitals (*37*, *90*, *277*). The nuclear charges must be apportioned between these segments in some manner. The origins for these multisegment multipole expansions might be chosen to be the center of charge in each segment. A particularly interesting approach is to divide the molecule into groups such as CH_3, NH_2, OH, etc. and to center a multipole expansion on each of these. This might lead to "group potentials" that are transferable from one molecule to another (*40*, *277*). The expansion origins could again be taken to be the centers of total charge, or alternatively the centers of mass, or of only nuclear charge, of the segments.

One of the advantages of writing $V(r)$ as a multipole expansion, whether in terms of one or many centers, is that it leads to an analytical expression for $V(r)$. This is in contrast to the two-dimensional map that is the usual form in which $V(r)$ is expressed. It was shown above that Q_1^0, Q_1^{-1}, and Q_1^1 are related in a simple fashion to the components of the dipole moment vector, μ_x, μ_y, and μ_z. In an analogous manner, Q_2^{-2}, Q_2^{-1}, Q_2^0, Q_2^1, and Q_2^2 can be shown to be related to the components of the quadrupole moment tensor (*129*, *180*), $Q_3^{-3}, \ldots, Q_3^3$ to the components of the octopole moment, and so forth. Thus from a knowledge of the components of the multipole moments, all of the Q_n^m can be determined. Once these are known, then (6-13) readily yields an analytical form for $V(r)$. The components of the multipole moments can be calculated from molecular wavefunctions (*90*, *265*). In the case of a many-center multipole expansion, separate multipole moments and sets of components must be computed for each segment of the molecular charge distribution.

One of the instances in which an analytical expression for $V(r)$ is particularly useful is in calculating the energy of electrostatic interaction between two molecules, treated as two static charge distributions. This is given by

$$E_{es} = \sum_i \sum_j \frac{Z_i Z_j}{|R_i - R_j|} + \iint \frac{\rho_A(r)\, \rho_B(r')\, dr\, dr'}{|r - r'|} - \sum_i Z_i \int \frac{\rho_B(r')\, dr'}{|R_i - r'|} - \sum_j Z_j \int \frac{\rho_A(r)\, dr}{|R_j - r|}, \quad (6\text{-}18)$$

in which $\rho_A(r)$ and $\rho_B(r')$ are the electronic density functions of the two molecules, Z_i is the charge on the ith nucleus of molecule A, situated at R_i, and Z_j and R_j have analogous meanings with respect to molecule B. The second term on the right-hand side of (6-18) represents the interelectronic repulsion; it requires integration over both r and r', and therefore involves relatively complicated two-electron-type integrals. However, if an analytical expression for the EP due to either molecule A, $V_A(r)$, or B, $V_B(r)$, were available, then (6-18) would become, in the former case,

$$E_{es} = \int V_{\mathrm{A}}(r')\, \rho_{\mathrm{B}}(r')\, dr' + \sum_j Z_j V_{\mathrm{A}}(R_j). \tag{6-19}$$

Equation (6-19) contains only one-electron-type integrals, and the computations are therefore easier and faster than for (6-18).

The integrals that must be calculated in order to determine the components of the multipole moments, and hence to obtain $V(r)$, are of the type $\int \psi_m(r)\, r^k \cdot \psi_n(r)\, dr$, where k is a positive integer (*90, 265, 294*). These can be treated more easily and more rapidly than the integrals which appear in the rigorous expressions for $V(r)$, (6-8) and (6-10), which contain the operator r^{-1}.[13] It should be recalled, however, that (6-13) is, in practice, an approximation, since the series expansion is generally truncated after a limited number of terms (usually after $n = 3$ or 4), and also because the point $\mathbf{r}$ is not really outside of the entire molecular charge distribution.

Thus, if the EP is to be calculated by means of a multipole expansion, one of the key questions to consider is how many terms should be included in the expansion. Another important point concerns the relative merits of single-center and many-center expansions. These questions have been examined in several studies (*37, 90, 210, 288*). It is invariably found that many-center expansions are considerably more accurate than single-center ones, and the many-center series converges more rapidly as well. It should be pointed out here that the single-center expansion is at a disadvantage in that its $n = 0$ term equals zero, and is therefore useless for any neutral molecule. In a many-center expansion, in contrast, the $n = 0$ terms make definite contributions, unless the segments into which the molecular charge distribution has been divided are such that they each have a net charge of zero. The latter situation is not usually the case. Thus, for example, it could be argued that a many-center multipole expansion that includes the $n = 3$ terms should really be compared with a one-center expansion that includes the $n = 4$ terms. Even then, however, the many-center result is more accurate. For example (*210*), at a point 2.63 Å from the nitrogen atom in pyridine, where the rigorously calculated EP from (6-10) was -0.734 eV, a single-center expansion which included octopole terms gave a value of -0.606 eV, while a many-center expansion based upon "atomic" segments predicted a potential of -0.708 eV with only quadrupole contributions being taken into account. With the addition of octopole terms, the latter result improved to -0.748 eV. As expected, the accuracy of both types of expansions increased with greater distance from the nitrogen (and the main concentration of molecular charge). Thus, at a distance of 3.63 Å the results were: rigorous, -0.368 eV; single-center (octopole), -0.331 eV; many-center (dipole), -0.342 eV; many-center (quadrupole), -0.363 eV. In general, it appears that a single-center expansion must be carried at least through octopole terms in order to come reasonably close to a rigorously calculated potential, while a many-center expansion need go only through dipole terms to achieve a similar result. The question as to what sort of segments (atomic, localized orbital, etc.) are most effective in the many-center approach has not been thoroughly investigated as yet; some preliminary indications favor atomic segments (*90*).

The most extensive study and application of multipole expansions in computing electrostatic effects has been by Rein and co-workers (*265, 288, 294*). The main computation involved is the calculation of the components of the multipole moments, and they have compared the accuracies of the CNDO/2 and the iterative extended-Hückel methods in this regard. Both approaches give reasonably good dipole moments, but the iterative extended-Hückel method appears to be superior for quadrupole and octopole moments. It is possible, however, that the CNDO/2 moments could be improved somewhat by "de-orthogonalizing" the wavefunctions, as was recommended earlier for computing their electrostatic potentials using (6-10) (Section 6-4-2). Such deorthogonalization does, in general, slightly improve (*279*) the dipole moments obtained with CNDO/2 wavefunctions. Dovesi et al. (*90*) used deorthogonalized CNDO/2 functions in their study of the multipole expansion technique for computing potentials. This might conceivably be a factor in their having obtained results which seem to be better than would have been expected from the experience of Rein with the CNDO/2 method.

6-4-5. Point-Charge Approximations

The simplest possible multipole expansion of the EP is one which terminates after the $n = 1$, or monopole, term in (6-13). For a neutral molecule, this would have to be a many-center expansion; if it were in terms of just one center, then $V(r)$ would be zero everywhere, as pointed out in Section 6-4-4. A many-center monopole expansion, on the other hand, corresponds to a point-charge model. The resulting EP arises from replacing the actual molecular charge distribution with a set of point charges. The accuracy of this potential depends, of course, upon the values and locations assigned to the point charges.

One obvious possibility is to put them at the positions of the nuclei, and to give them magnitudes and signs corresponding to the respective atomic charges, as calculated by some suitable procedure (Section 6-2). Rein (*265*) tested this approach for pyridine, using three different sets of atomic charges. One set consisted of charges obtained by the CNDO/2 method, in which the CNDO/2 coefficients are inserted into (6-4) but S_{mn} is set equal to zero. The other two sets of charges were calculated using the iterative extended-Hückel procedure, first in conjunction with (6-4) and then with a modified form of this equation in which the overlap term is apportioned in a manner determined by the position of the centroid of the charge distribution $\psi_m \psi_n$ (*178*). Only the second of these sets of point charges, that corresponding to (6-4) unmodified, yielded a reasonably good potential in this particular case; the other two gave results that were often in error by factors of two to four. Somewhat greater success was achieved (*190*) for formamide, methanol, and formic acid using SCF wavefunctions and (6-4). Of course the use of (6-4) introduces the problem of the sensitivity of the computed charges to the atomic orbital basis set used for the molecular wavefunction (Section 6-2). Thus the calculated atomic charges in *N*-methylacetamide change considerably (*121*, *122*) in going from one Gaussian basis set to another; for instance, the oxygen charge went from -0.30 to -0.64. Almlöf et al. (*10*) observed even qualitative differences in atomic charges computed for aceheptylene using (6-4) from one ab initio and a variety of semiempirical ($S_{mn} = 0$) wavefunctions. All of these sets of charges led to rather poor reproductions of the EP, although they did correctly indicate the most negative region. Another study, of two different point-charge approximations to the EPs of some chlorophyll-related molecules, had a roughly similar outcome (*204*). On the other hand, the atomic point-charge model based on (6-4) was found (*106*) to be reasonably effective in describing the potential in the plane of the NO_2^- ion, except in the lone-pair region of the nitrogen, where it significantly underestimated the magnitude of the minimum.

Another possible approach to establishing a point-charge model of a molecular charge distribution is to require that the point charges have such values and locations as to permit certain conditions to be satisfied. For example, Hall (*117*) suggested that the values of the point charges be fixed by the requirement that they give the correct molecular dipole moment and total charge. Their position would be determined by reference to an SCF wavefunction for the molecule, written in terms of a minimum basis set of floating spherical Gaussian functions. One point charge would be placed at each nucleus, one at the center of each spherical basis function, and one at the center of the function corresponding to each of their products. Hall showed for LiH that the EP calculated from these point charges and the nuclear charges was remarkably close to that obtained rigorously from the aforementioned SCF molecular wavefunction. For example, at 0.63 Å from the lithium nucleus (perpendicular to the Li—H axis) the point-charge potential was +302 kcal/mole, while the presumably more accurate value was +308 kcal/mole. Hall also demonstrated mathematically that with increasing distance from the atoms the SCF potential would gradually reduce to an asymptotic form which was exactly the point-charge potential. Tait and Hall (*297*) subsequently extended this approach to H_2O and CH_4, with similar good results. They also found the point-charge model to be effective for determining other one-electron properties, such as the electric field and its gradient, and multipole moment components. Formulas were derived which make it possible to evaluate the errors in the point-charge values of these properties relative to the results that would be obtained with the SCF wavefunction. Tait and Hall pointed out that these properties can be calculated more easily with the point-charge model plus corrections than directly from the SCF function. It is interesting to note that their point charges can have quite unphysical magnitudes. While most of them are in the reasonable range of 0-6, the point charges corresponding to the oxygen lone pairs in H_2O had values of -197, while the oxygen nucleus was represented by +398.

Thinking along similar lines, Ghio et al. (*106*) attempted to improve their atomic point-charge description of NO_2^-, mentioned above, by adding six more point charges designed to help reproduce the SCF dipole and quadrupole moments. This did produce a potential that more closely resembled the SCF, although there was some overcompensation. In another example of this general approach, the electrostatic effect of the enzyme carboxypeptidase *A* upon a substrate (*N*-methylacetamide, serving as a model for glycyltyrosine) was simulated (*121*, *122*) simply by means of a set of point charges distributed over the enzyme.

Of course the procedure could be reversed, and instead of approximating the EP from a set of point charges, a rigorously calculated potential could be used to determine the point charges; these could then be utilized for some other purpose. Momany (*190*) obtained atomic charges for formamide, methanol, and formic acid in this manner. A somewhat more elaborate approach has been applied to the H_2O molecule (*33*). The nuclei were given their correct positions and charges, while the valence electrons were represented by point charges that were required to duplicate as closely as possible the EP map computed from an SCF wavefunction, and also to give the SCF dipole moment (the oxygen core electrons were incorporated into the nucleus, giving it an effective charge of +6). A point-charge distribution satisfying these conditions was obtained when each localized lone pair was replaced by three $-\frac{2}{3}$ charges positioned symmetrically around its centroid, and each localized bond orbital by two -1 charges on the bond axis, symmetrical with respect to its centroid. This simple model was able to reproduce the SCF potential remarkably accurately outside of the atoms' van der Waals radii. These point charges, together with the SCF potential, were then used to calculate the electrostatic interaction energy between two H_2O molecules and thereby to investigate possible conformations of the dimer. The results were in good agreement with direct SCF computations of the $(H_2O)_2$ total energy. This will be discussed in more detail in Section 6-6-2.

Finally, the possibility of using a point charge to represent a monatomic cation that is associated with a polyatomic anion has been tested (*258*). The ions involved were Na^+ and $(CH_3)_2PO_4^-$. Good agreement was found between the EP computed rigorously for the system $Na(CH_3)_2PO_4$ (using an SCF molecular wavefunction) and that calculated equally accurately for just $(CH_3)_2PO_4^-$

plus a +1 point charge in place of the Na^+. In view of this success, a +2 point charge was subsequently used to represent the Mg^{2+} ion in $[Mg(CH_3)_2PO_4]^+$.

6-4-6. The Use of Group Potentials

There has been some limited investigation of the use of group potentials to obtain approximate molecular EPs (*40, 42, 277*). The basic concept is that certain clearly identifiable two-electron groups such as lone pairs, core orbitals, σ or π bonds between a given pair of atoms, etc., may have associated with them potentials that change relatively little in going from one molecule to another. If this is the case, then one could estimate the potential due to a chemical grouping, such as CH_3 or NH_2, by combining the appropriate lone pair, core, and bond potentials, and eventually the total molecular potential could be produced by combining the potentials of the various chemical groupings. (Following Bonaccorsi et al. we will use the term *group* in this section to represent each of the two-electron entities plus two nuclear unit charges associated with it to provide electrical neutrality. Elsewhere in this chapter, however, the word *group* will have its customary meaning of a chemical grouping, such as CH_3.)

The approach described above is, of course, an application of the old concept that some of the properties of portions of molecules may often be transferable from one molecule to another; this approach has been found to be successful, at least in some degree, on both empirical and rigorous levels (see Ref. 173 and references cited therein).

Bonaccorsi et al. have tested the validity of this concept in the present context, and have found it very promising. Thus, in four three-membered ring molecules having the formulas CH_2NHX, where X = CH_2, O, and NH (both *cis* and *trans*), the potential at a particular point in space due to the constant CH_2NH portion varied only over the interval −154.1 to −156.0 kcal/mole. In a series of eight three-membered-ring molecules, the potential due to a common CH_2 grouping at a given point in space had a mean value of 21.66 kcal/mole, with a maximum deviation from this mean of only 1.15 kcal/mole. On the other hand, the potentials at the same point due to the C—C σ bond in cyclopropane (C_3H_6), oxirane $[(CH_2)_2O]$, and aziridine $[(CH_2)_2NH]$ ranged from 14.7 to 20.4 kcal/mole. This reflects the differing degrees of bending of this bond in these molecules, which has the effect of making its potential not transferable.

Bonaccorsi et al. (*42*) tested two techniques for calculating the group contributions, one involving transferable localized orbitals (TLO) and the other being a point-charge model, the point charges not necessarily being restricted to the nuclei. EPs were calculated by these various procedures for formamide, acetamide, *N*-methyl-formamide and 2-formyl-aminoacetamide, and compared to potentials obtained from SCF wavefunctions. On the whole, the results were very satisfactory, especially when the TLO were used. The authors suggest that in computing the potential in the vicinity of any given chemical grouping within a molecule, the group potentials corresponding to that grouping be computed in terms of TLO, but that a point-charge model be used to determine the effects of more distant chemical groupings. These approximate EPs were also found to be quite effective for calculating hydration energies via (6-19).

6-5. APPLICATIONS

6-5-1. Introduction

Calculated EPs have now been published, in more or less detail, for better than 100 molecular and ionic species. In some cases the purpose was primarily analytical, to investigate some computational technique, while in others the intention was to gain understanding of the reactive properties of the molecule or ion. The emphasis in this section will be upon the latter studies and results; the former have been surveyed in Section 6-4.

For the benefit of the reader, Table 6-1 presents an alphabetical list of some systems for which EPs have been computed and published. This list is not claimed to be complete, but it should certainly include the great majority of published potentials (see also Ref. 277a).

6-5-2. Formamide, $H_2N—CHO$

This molecule provides a good initial test of the usefulness of the EP, because it contains two distinct possibilities for protonation sites, the amine nitrogen and the carbonyl oxygen, the relative importance of which has been a matter of much controversy in the past. The relevant experimental data, which have been summarized (*36*, *133*), have been subjected to conflicting interpretations. The general opinion at present seems to be that oxygen is the primary site for protonation. The same conclusion has been reached on the basis of recent SCF calculations of the total energies of formamide and its protonated forms (*12*, *133*). While these take no account of solvation effects, arguments have been presented (*133*) claiming that interaction with water molecules would stabilize oxygen-protonated formamide relative to the nitrogen-protonated form (see, e.g., Ref. 3).

The EP of formamide, computed using a Gaussian-basis SCF wavefunction (*35*, *251*), is shown in Fig. 6-2. The very large, strong negative region near oxygen, contrasted to the small and weak one above nitrogen, clearly indicates oxygen as the preferred site for protonation, in agreement with total energy calculations and recent evaluations of experimental evidence. It is interesting to note (Fig. 6-2) that the combined effect of the two oxygen lone pairs is to produce what might be described as an extended minimum of −67 kcal/mole, in the N—C—O plane and partially encircling the oxygen atom. This is pleasingly consistent with the finding, by means of total energy calculations, that the added proton

Table 6-1. Alphabetical List of Systems for Which Electrostatic Potentials Have Been Published.

aceheptylene (*10*)
acetamide (*207*)
acetonitrile (*307*)
acetylcholine · NH_2^- (*286, 307, 313*)
acetylene (*68*)
adenine (*21, 35, 39, 108, 110, 250, 251, 277, 308*)
adenine · thymine (*219*)
allene (*57*)
alphaprodine (*49*)
ammonia (*41, 48, 57, 188, 261, 272, 277*)
ammonia · $xH_2O, x = 3, 4, 5$ (*261*)
ammonium ion (*48*)
anisole (*223*)
2-azirene (*58*)
aziridine, i.e., ethyleneimine (*31, 32, 58, 105, 277*)
azulene (*27*)
benzaldimine, ground and excited states (*67*)
benz[*a*]anthracene (*240*)
benzene (*5, 6, 223*)
benzofuran (*27*)
carbodiimide (*119*)
catechol (*223*)
chlorpromazine (*226*)
codeine (*174*)
cyanamide (*119*)
cyclopropane (*30, 277*)
cyclopropene (*30*)
cytosine (*35, 110, 251, 260*)
diaziridine (*34*)
diazirine (*31, 277*)
diazomethane (*55, 57, 119, 168*)
diazomethane, excited states (*55*)
diazomethyl anion (*119*)
dicyanobenzene anion (*75, 76*)
3,4-dihydroxy-benzyl alcohol (*225*)
diimide (*40*)
dimethylamine (*41*)
dimethylether (*1*)
dimethylether · HF (*1*)
dimethylphosphate anion (*24, 216, 258*)
ethane (*238*)
ethylene (*9, 168*)
ethylene oxide (*31, 32, 58, 277*)
ethynyl radical (*77*)
fluoroacetylene (*68*)
fluorobenzene (*7, 223*)
formaldehyde (*48, 54, 56, 57, 85, 109, 167, 188, 274*)
formaldehyde, excited states (*56, 85*)
formaldehyde · HF (*274*)
formamide (*36, 110, 190, 249-251, 274, 277*)
formamide dimer (*42*)
formamide · HF (*274*)
formic acid (*190*)
2-formylaminoacetamide (*42*)
formyl fluoride (*61, 312, 313*)
furan (*64, 235, 237*)
α-glycine (*6-8*)
guanine (*39, 108, 111, 250, 251*)
guanine · cytosine (*219*)
heroin (*174*)
histamine (*310*)
histamine · H^+ (*310*)
hydrogen chloride (*160, 188*)
hydrogen fluoride (*160, 171, 188, 280, 287, 304*)
hydrogen fluoride dimer (*171*)
hydrogen cyanide (*41*)
hydrogen sulfide (*170*)
hydromorphone base (*175*)
5-hydroxyindole (*311*)
5-hydroxytryptamine (*309, 311*)
6-hydroxytryptamine (*209, 309*)
imidazole (*226, 277*)
imidazolium cation (*311, 314*)
iminoxy radical (*96*)
indole (*27*)
isocyanamide (*119*)
isoxazole (*23, 277*)
ketene (*55, 57*)
ketene, excited states (*55*)
ketyl radical anion (*96*)
lithium fluoride (*287, 304*)
lithium hydride (*117, 287, 304*)
magnesium dimethylphosphate cation (*258*)
meperidine (*49*)
methane (*57, 238*)
methanol (*1, 190*)
methanol · HF (*1*)
N-methylacetamide (*252*)

Table 6-1. *(Continued)*

7-methyladenine (*21*)	pyrazine (*5, 277*)
2-(*m*-methylphenoxy) ethanol (*223*)	pyrazine anion (*75, 76*)
N-methylpyrrole (*237*)	pyrazole (*23, 226, 277*)
6-monoacetylmorphine (*174*)	pyridazine (*5*)
2-monoaminomorphine (*174*)	pyridine (*5, 6, 210, 223, 261, 265, 277*)
morphine (*49, 174*)	pyridine · H_2O (*261*)
nitrilimine (*119*)	pyrimidine (*5, 90*)
nitrite anion (*106*)	pyrrole (*64, 226, 237, 277*)
p-nitrobenzyl alcohol (*223, 225*)	ribose (*26, 259*)
4-nitropyridine anion (*75, 76*)	scopine acetate · OH^- (*313*)
nitrogen (*277*)	sodium dimethylphosphate (*258*)
nitrogen trifluoride (*48*)	*s*-tetrazine (*5*)
nitrosyl fluoride (*277*)	thiirane (*31*)
p-nitrotoluene (*307*)	thiophene (*104*)
nitrous acid (*82, 95*)	thymine (*35, 38, 251*)
nitroxide radical (*96*)	thymine, excited states (*38*)
oxaziridine (*34, 220*)	toluene (*60, 307*)
oxazole (*23, 277*)	2-(*m*-tolyloxy)ethanol (*225*)
oxirane (*31, 32, 58, 277*)	*s*-triazine (*5*)
oxymorphone base (*175*)	tropine acetate · OH^- (313)
ozone (*277*)	tryptamine (*64*)
phenanthrene (*240*)	uracil (*217, 259*)
phenethylamine (*182, 223*)	urea (*208*)
phenol (*223*)	uridine (*259*)
N-phenylformaldimine, ground and excited states (*67*)	water (*1, 33, 57, 109, 145, 167, 169, 188, 261, 277*)
promazine (*226*)	water dimer (*261*)
propane (*238*)	water · HF (*1*)
propyne (*307*)	

on oxygen is located in the N—C—O plane and can move relatively easily from one side of the oxygen to the other, as long as it stays in this plane; an out-of-plane O—H rotation, however, would require significantly more energy (*12*).

It should be pointed out that the atomic charges computed for formamide from several different wavefunctions, using the population analysis procedure of (6-4), predict nitrogen to be more negative than oxygen (*66, 133, 134*). On the basis of this criterion, therefore, nitrogen should be the primary protonation site, which would be contrary to both theoretical and experimental evidence, as discussed above (see also Ref. 268).

The interaction of the Li^+ cation with formamide appears to be qualitatively similar to that of H^+. According to total energy calculations (*12*), the preferred position for Li^+ is by the oxygen, essentially in the N—C—O plane and along the C—O axis. The next most stable site, however, is not nitrogen but rather

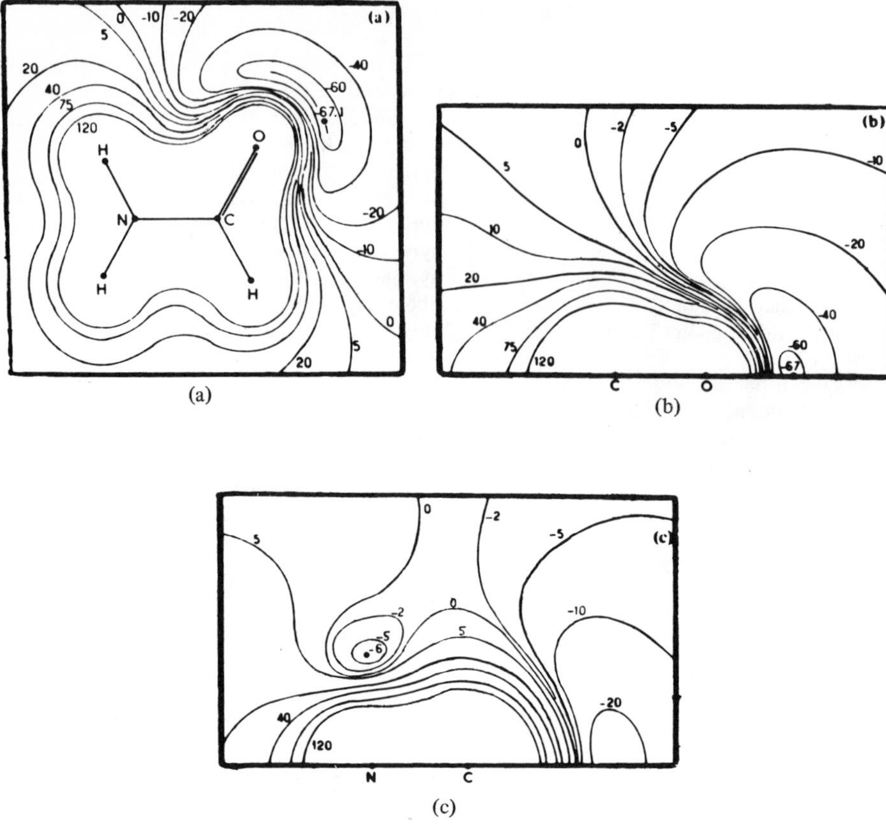

(a)

(b)

(c)

Fig. 6-2. Electrostatic potential of formamide, kcal/mole. (a) Molecular plane. (b) Perpendicular plane containing C—O bond. (c) Perpendicular plane containing N—C bond. (Reproduced from Ref. 36, courtesy The North-Holland Publishing Co.)

a bridged position in which Li^+ is only slightly above the N—C—O plane and is nearly equidistant from the nitrogen and the oxygen atoms. The formation of this latter complex was found to be accompanied by a significant rotation of the NH_2 group, away from the Li^+ ion, and a lengthening of the C—N bond. It was suggested that this complex may be involved in the *cis–trans* isomerization which is induced in some polypeptides by salts.

6-5-3. Nucleic Acid Bases

An understanding of the reactive properties of these molecules–adenine (V), guanine (II), cytosine (III), and thymine (VI)–is of obvious importance, in view of their roles as the building blocks of deoxyribonucleic acid (DNA). Each of them has several likely sites for attack by electrophiles; EP calculations can help to provide insight into the orders of preference of these sites.

V　　　　VI

6-5-3a. Adenine (V). The EP of adenine, computed from an SCF wavefunction (*39*), is shown in Fig. 6-3. There are strong negative regions by each of the pyri-

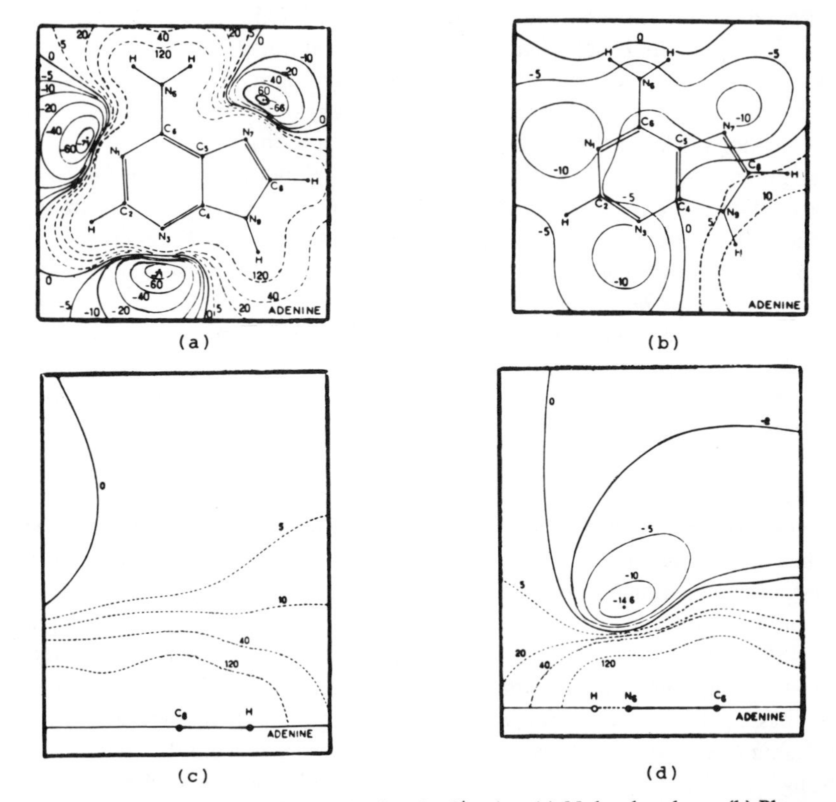

Fig. 6-3. Electrostatic potential of adenine, kcal/mole. (a) Molecular plane. (b) Plane parallel to molecule, at distance of 2 Å. (c) Plane perpendicular to molecule, passing through C8—H bond. (d) Plane perpendicular to molecule, passing through positions C6 and N6. (Reproduced from Ref. 39, courtesy Springer-Verlag.)

dinelike nitrogens (N1, N3, and N7), which extend above and below the plane of the ring but have their minima in the plane: -71, -71, and -66 kcal/mole, respectively. A much weaker minimum, -14.6 kcal/mole, is located above and below the amine nitrogen, N6 [Fig. 6-3(b)].[14]

These results suggest that the N1 and N3 positions should be the primary sites for electrophilic attack, but do not permit a distinction between them. Experimentally, it has been found that N1 is the main protonation site on adenine (*59*, *65*, *154*, *155*), although N3 is more reactive toward alkylation (*211*), which is also believed to occur via electrophilic attack. Once N1 has been protonated, the favored site for a second proton is not N3, but N7 (*154*).

There is some evidence that these protonation properties of adenine are maintained when there is an alkyl substituent on position N9 (*53*, *303*). However, the presence of a methyl group on N7 leads to an interesting change in this pattern. The primary protonation position on 7-methyladenine is N3, while a second proton (to form the 2+ ion) will favor N9 (*156*, *157*). The preference for N3 over N1 in 7-methyladenine has been interpreted by taking note of the fact that in this molecule N9 does not carry a hydrogen. The resulting lone pair at N9 can be expected to greatly enhance the already strongly negative EP near N3, making the latter the most attractive site for a proton (*157*).

A noteworthy aspect of Fig. 6-3 is the striking contrast between the EPs associated with the three types of nitrogens in adenine (*35*, *39*). There are three pyridinelike nitrogens, N1, N3, and N7; these all have large regions of strong negative potential near them. There is one aromatic amine nitrogen, N6, which also has its own negative region, although a much weaker one. Finally, there is pyrrolelike N9, with which there is associated no negative potential. This trend in the nitrogen potentials is in the same direction as the changes in basic strengths (in aqueous solution) of pyridine, aniline (an aromatic amine), and pyrrole. Pyridine is the strongest base, while pyrrole is by far the weakest (*193*).

6-5-3b. Guanine (II). The structure of guanine is fundamentally similar to that of adenine, the principal differences being that adenine's amine group is on position 2 in guanine instead of position 6; the latter is now occupied by a carbonyl oxygen. A consequence of these differences is that N1 has a hydrogen in guanine; accordingly, the calculated EP of guanine does not show a negative region near N1 (*39*). There is still a strong negative potential in the vicinity of N3, similar to that in adenine, but the principal negative region is now a very large one encompassing both the oxygen and also N7 [Fig. 6-4(a)]. There is a distinct minimum associated with each of these two atoms, the one belonging to N7 being considerably the stronger of the two. This is consistent with the experimental observation that N7 is the favored site for the protonation and alkylation of guanine (*99*, *164*, *165*), although some alkylation does occur at the oxygen (*100*, *176*).

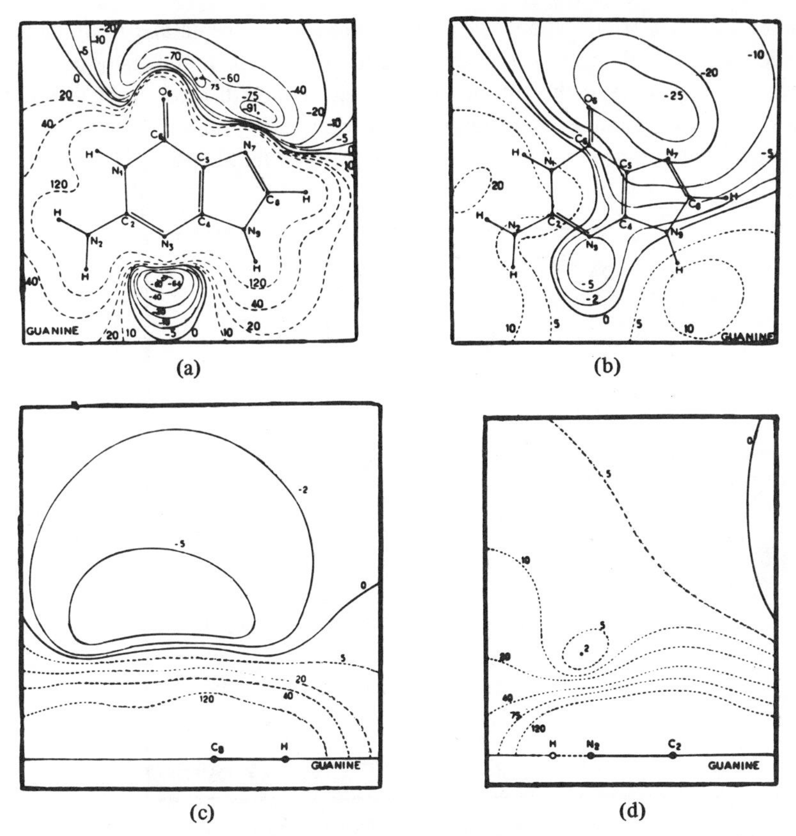

Fig. 6-4. Electrostatic potential of guanine, kcal/mole. (a) Molecular plane. (b) Plane parallel to molecule, at distance of 2 Å. (c) Plane perpendicular to molecule, passing through C8—H bond. (d) Plane perpendicular to molecule, passing through positions C2 and N2. (Reproduced from Ref. 39, courtesy Springer-Verlag.)

A surprising feature of Fig. 6-4 is the absence of any negative potential above the amine group of guanine. There does occur a minimum, but it has a positive value. On the other hand, guanine has a significant negative region above carbon C8, which adenine does not.

Since the wavefunctions for guanine and adenine are in terms of the same basis set, it should be valid to compare their EPs. The comparison shows that the N7 minimum in guanine is more negative than any in adenine, which is in agreement with the fact that guanine is more easily alkylated than any of the other nucleic acid bases (*166*). The more negative character of guanine near C8 is relevant to the experimental finding that the carcinogen *N*-acetoxy-2-acetylaminofluorene binds via electrophilic attack to this position in guanine but almost not at all in adenine (*147, 172, 187*).

6-5-3c. Cytosine (III). The wavefunctions used for cytosine and thymine have smaller basis sets than those for adenine and guanine (*35*), and therefore quantitative comparisons of the EPs in Figs. 6-3 and 6-4 with those in Figs. 6-5 and 6-6 should not be made. Qualitatively, however, they should all be quite compatible with each other, as can be verified by comparing the potentials computed

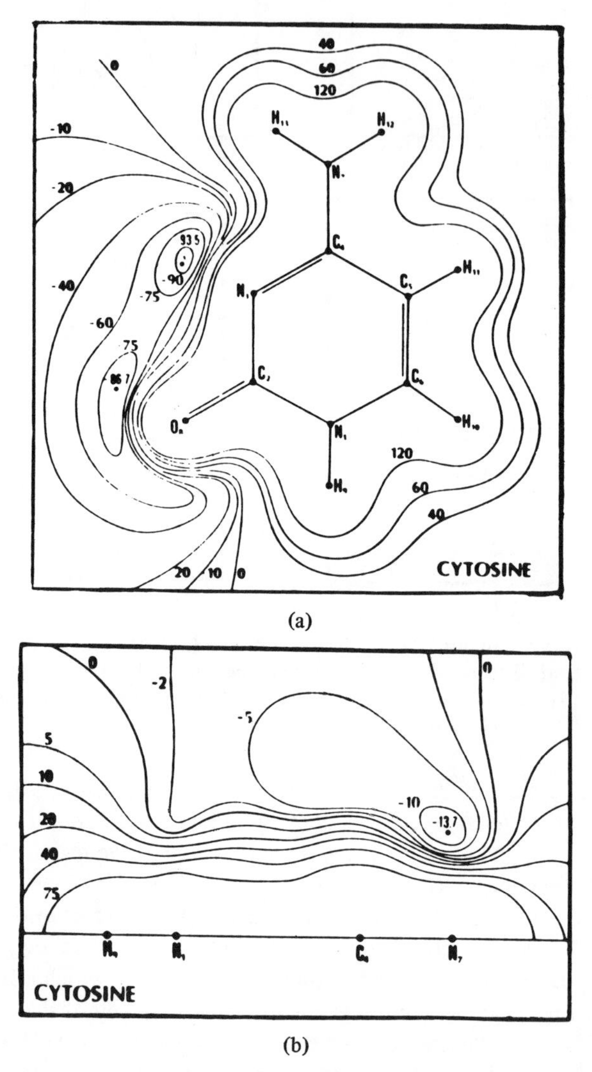

Fig. 6-5. Electrostatic potential of cytosine, kcal/mole. (a) Molecular plane. (b) Plane perpendicular to molecule, passing through positions N1 and C4. (Reproduced from Ref. 35, courtesy Springer-Verlag.)

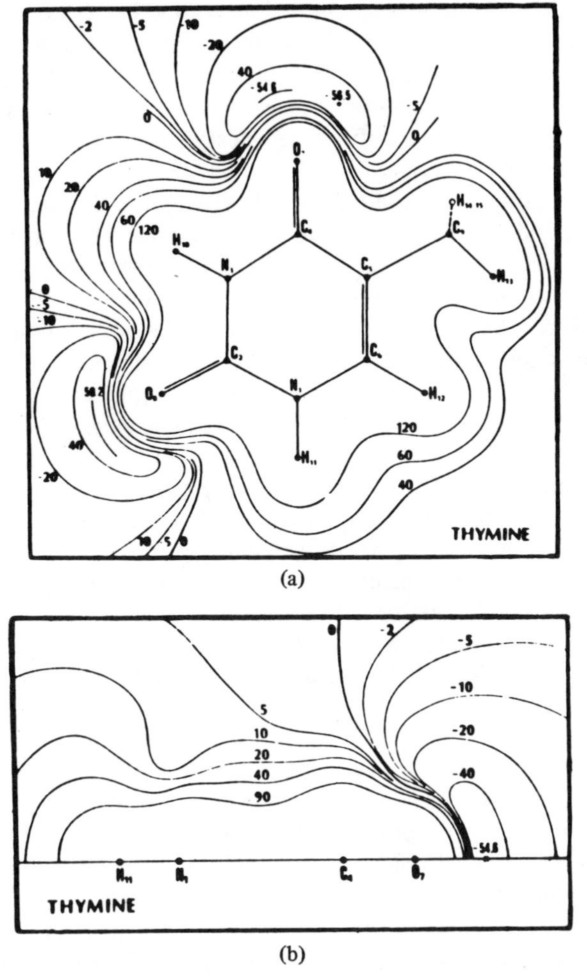

Fig. 6-6. Electrostatic potential of thymine, kcal/mole. (a) Molecular plane. (b) Plane perpendicular to molecule, passing through positions N1 and C4. (Reproduced from Ref. 35, courtesy Springer-Verlag.)

for adenine using wavefunctions having the two different basis sets (*35*, *39*); one of the major differences appears to be that the magnitudes of the minima were found to be somewhat greater using the smaller basis set. Thus the strongest minimum found in cytosine or thymine, -93.5 kcal/mole, would probably be less negative in terms of the adenine-guanine basis set.

Cytosine, like guanine, contains a pyridinelike nitrogen in close proximity to a carbonyl oxygen. As in guanine, these combine to produce a very large region of negative EP, containing two strong minima (Fig. 6-5). There are also two much weaker minima, −13.7 kcal/mole, located above and below the amine nitrogen. The most negative point in cytosine is near the pyridinelike nitrogen, N3, and this atom is indeed the experimentally observed site for both protonation and alkylation (*51*, *140*, *164*, *166*). Cytosine has also been found to complex to Cu(II) through N3, with a significant secondary interaction between the copper ion and the carbonyl oxygen (*157*, *295*, *296*).

Pullman and Armbruster (*260*) have calculated the SCF interaction energies of H^+, CH_3^+, and $C_2H_5^+$ with N3 and O2 of cytosine, and concluded that the preferences of H^+ and $C_2H_5^+$ are for N3 and O2, respectively, while CH_3^+ has no clearcut preference. The fact that most alkylation is found experimentally to take place at N3 was taken to indicate that it does not occur purely by direct electrophilic attack (S_N1 mechanism). These were minimum basis set computations, however, and it is known that the relative magnitudes of interaction energies may change in going from a minimum to a more extended basis set (*95*).

6-5-3d. Thymine (VI). Thymine contains only pyrrolelike nitrogens, which do not have negative potentials (*35*). There are two carbonyl oxygens, however, and each of these has a strong negative region (Fig. 6-6), although not approaching those seen in adenine, guanine, and cytosine. The relatively low basicity of thymine (*166*, *211*) is therefore not surprising.

The preceding discussion shows that the EP is an effective guide to the reactive properties of the nucleic acid bases toward electrophiles, both indicating and ranking the active sites on the molecules.[15] The electronic density distributions alone could not provide such information, as can easily be seen by examining the density maps for adenine, cytosine and thymine (*254*). It is particularly noteworthy that in both guanine and cytosine, one found a nitrogen with a more negative potential than the carbonyl oxygen, in full agreement with experimental observations. This would not have been predicted from the atomic charges obtained by the population analysis method of (6-4); the calculated charge on N3 in cytosine is −0.46, compared to −0.52 for the oxygen (*185*). Indeed, as was pointed out in Section 6-4-2, the EP itself, when computed with a CNDO/2 wavefunction (even if deorthogonalized), fails to show more negative minima by the nitrogens than by the oxygens (*110*, *111*). The potentials corresponding to the adenine–thymine and guanine–cytosine base pairs are discussed in Section 6-5-6.

6-5-4. Five- and Six-Membered Cyclic Molecules

EPs based on SCF wavefunctions have been computed for a large number of cyclic molecules of varying degrees of aromaticity containing zero to four heteroatoms. These all show two basic features: first, there is a large and strong region of negative potential associated with each heteroatom, located symmetrically with respect to the plane of the ring; and second, there are weaker negative regions above and below the ring or some portion of it.

6-5-4a. Benzene and Fluorobenzene. A somewhat better wavefunction was used for benzene (*6*) than for fluorobenzene (*7*), so that the resulting potentials are, from a quantitative standpoint, only roughly comparable. Benzene was found to have very extensive negative regions above and below the ring, each with six minima. These are located near the centers of the C—C bonds, to the inside of the ring, 1.75 Å from the molecular plane. Their values are -14.2 kcal/mole, considerably less than many of the minima associated with heteroatoms.

The substitution of a fluorine atom for a hydrogen produces a region of negative potential that is centered in the molecular plane, on the side of the fluorine that is away from the ring, with a minimum of -23.8 kcal/mole. In addition, there are again extended negative potentials above and below the ring; each of them now has only one minimum, however, located near the *para* carbon, within the ring.[16] Its value was not given by Almlöf et al. (*7*) but it appears to be about -11.5 kcal/mole. They suggest that the presence of this minimum may account for the fact that electrophilic substitution in fluorobenzene occurs almost entirely at the *para* position (*278*).

6-5-4b. Five-Membered Heterocycles. Furan, pyrrole, *N*-methylpyrrole, and thiophene make up an interesting series of molecules having the basic structure VII, where X = 0, NH, NCH_3, and S, respectively. All four molecules undergo

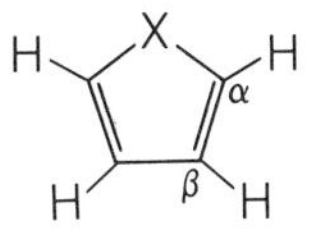

H X H α H β H

VII

electrophilic substitution, which is found experimentally to occur predominantly at the α-positions (*4, 71, 142, 181*). The degree of selectivity varies, however, being greater in furan and thiophene than in the two pyrroles.

EPs have been computed for all four of these molecules, although only for pyrrole (*277*) and thiophene (*104*) are they based on SCF wavefunctions; the potentials for furan and *N*-methylpyrrole were obtained from deorthogonalized CNDO/2 and INDO functions (*235, 237*). The most extensive negative potential found among these molecules is that of thiophene. There is a large negative region to the outside of the sulfur atom which passes through the molecular plane but has minima of about -15 kcal/mole above and below this plane (Fig. 6-7). This negative region continues above and below the ring, and has four more minima, one on each side of each C—C double bond, as shown. These latter have values of -14.0 kcal/mole and are 1.74 Å above the molecular plane; the similarity to benzene is striking, and a comparison does have some validity since the wavefunctions are in terms of similar basis sets. It was suggested (*104*) that the predominance of α-substitution in thiophene may be the combined re-

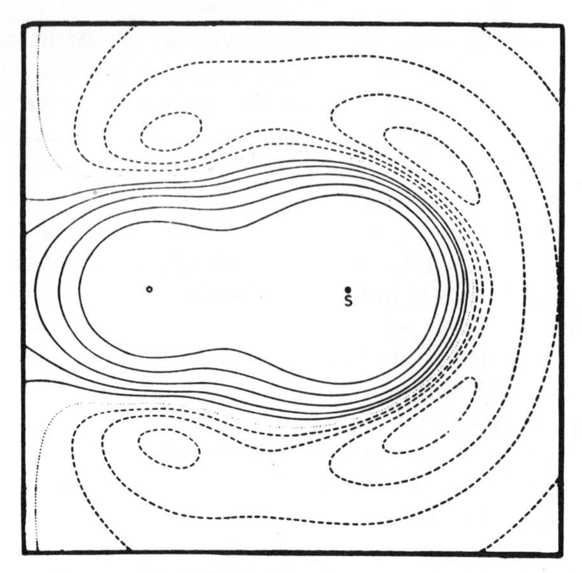

Fig. 6-7. Electrostatic potential of thiophene in plane perpendicular to the molecule through its twofold symmetry axis. The dotted contour corresponds to 0.0 kcal/mole. The solid contours represent 6.3, 12.5, 25.1, 50.2, and 100.4 kcal/mole; the dashed contours correspond to −3.1, −6.3, −9.4, and −12.5 kcal/mole. (Reproduced from Ref. 104, courtesy Springer-Verlag.)

sult of (1) electrophiles approaching through the sulfur negative region and moving over to the α-carbon, and (2) electrophiles attacking via the minima associated with the double bonds and then going preferentially to the α-positions because the highest occupied molecular orbital is concentrated around them.

With regard to the predominance of α-substitution on furan and the pyrroles, an interpretation has been proposed in which is introduced, as a key element, the effect upon the EP of moving (or rotating) out of the molecular plane one or more of the atoms that are attached to the ring (*235, 237*). The reason for bringing in this factor can be seen from Fig. 6-8, which shows the potential computed for pyrrole from an SCF wavefunction (*277*). Consistent with the results obtained for the nucleic acid bases, there is essentially no negative potential associated with the pyrrole nitrogen; there is a definite negative region, but it is in the neighborhood of the β-carbons. There are two minima, but they are near the midpoint of the βC—βC bond, above and below the molecular plane. Fig. 6-8 gives no reason to predict electrophilic preference for the α-carbon; the β-position appears to be the favored one.

Figures 6-3 to 6-8 have brought out the fact that the hydrogen atoms in aromatic systems generally tend to have positive potentials associated with them.

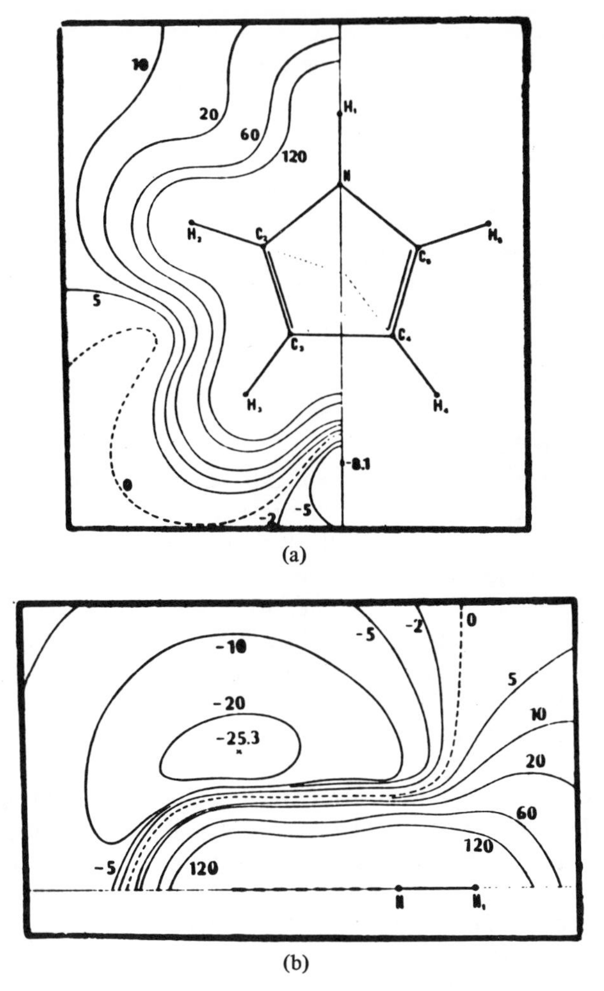

Fig. 6-8. Electrostatic potential of pyrrole, kcal/mole. (a) Molecular plane. (b) Plane perpendicular to molecule through its twofold symmetry axis. (Reproduced from Ref. 277, courtesy Springer-Verlag.)

It has been shown, however, that when a hydrogen that is bonded to a carbon or nitrogen atom of an aromatic ring is rotated out of the plane of the ring, there arises quite a significant region of negative potential near that carbon or nitrogen on the other side of the ring plane (*235*, *237*); this can, of course, serve as a path of approach for an attacking electrophile. It has therefore been proposed that electrophilic attack on furan is preceded by a hydrogen atom moving out of the molecular plane, thereby producing such a negative path to the carbon in question. In the case of furan, the negative potentials resulting from rotating either an α-hydrogen or a β-hydrogen were found to be quite similar, and could not in themselves be used as an explanation for the preponderance of α-substitution. It is necessary, however, to take into account the large, strongly negative region associated with the oxygen in furan. When the α-hydrogen is moved out of the plane, the negative potential that is created becomes simply an extension of the large oxygen region, whereas when the β-hydrogen is rotated, there remains an intervening positive potential between the β-negative region and that of the oxygen. Thus an electrophile which approaches the very attractive oxygen can relatively easily move over to the α-carbon, while a movement to the β-carbon is hindered by a positive barrier. It therefore appears that β-substitution will occur primarily when the initial approach is to the β-carbon, while α-substitution can result from an initial approach to either the α-carbon or the oxygen, and should accordingly occur considerably more frequently.

A similar explanation has been proposed for the α-selectivity of electrophilic substitution on pyrrole and *N*-methylpyrrole (*237*). In these two cases, however, there is the complication that the heteroatom, the pyrrole nitrogen, does not have a strong negative region associated with it, as does the oxygen in furan. It has been postulated, therefore, that in these two molecules the movement of the α-hydrogen out of the molecular plane must occur at a time when the hydrogen or methyl group on the nitrogen is also out of the plane (in the same direction), since the latter action produces a large negative region on the other side of the nitrogen, approximately as strong as that by the furan oxygen. The same reasoning then applies as in the case of furan.

The requirement that the hydrogen or methyl group on the nitrogen be bent out of the ring plane is not an extremely demanding one; the experimentally determined N—H and $N—CH_3$ bending force constants in these molecules are small, significantly less than the C—H bending force constants (*276*). This indicates that the N—H and $N—CH_3$ bonds should be bent out of the molecular plane, at least to some extent, most of the time. Of course they will be in or very near to the plane some fraction of the time, and so the degree of α-selectivity should be less for pyrrole than for furan; it should be yet lower for *N*-methylpyrrole, since the $N—CH_3$ bending force constant is larger than the N—H. These predictions are fully borne out by experimental observations: electrophilic substitution on furan is almost 100% in the α-position, while pyrrole may give as much as 20% of the β-product, and *N*-methylpyrrole even more (*4*, *71*, *142*, *148*, *181*).

It should be noted that these ideas regarding the mode of interaction of electrophiles with furan, pyrrole, and *N*-methylpyrrole are consistent with (but do not necessarily imply) the widely held theory that these processes involve a tetrahedral intermediate (*125*, *148*, *205*, *206*, *267*). With regard to the movement of hydrogens and/or a methyl group out of the molecular plane, which is such a key point in the preceding discussion of furan and the two pyrroles, it seems likely that this movement would be aided by repulsive interaction between the approaching electrophile and the hydrogen or methyl group. The effect upon the EP of moving hydrogens out of a molecular plane has also been investigated for azulene, indole, and benzofuran (*27*) and for toluene (*60*). In each case, the result was the creation of negative potential on the other side of the plane. One can thus explain (*60*) the preference of toluene for protonation in the *para* position; this could not be understood on the basis of the calculated atomic charges.

While the preceding discussion dealt with an apparent similarity in the α-directing mechanisms in furan and pyrrole, an interesting difference in the effects of the two heteroatoms in question, oxygen and nitrogen, was discovered by Chou and Weinstein (*64*). They showed that when a positive point charge is brought near the nitrogen in pyrrole (with no charge transfer being allowed), a greater buildup of electronic charge occurs on the two adjacent α-carbons than on the nitrogen atom. When the point charge is placed near the furan oxygen, however, the greatest increase in electronic density is on the oxygen; similarly, when the point charge is brought near one of the ring carbons in either pyrrole or furan, the main result is charge accumulation on the site being attacked. Thus there appears to be a special property associated with the nitrogen atom. While Chou and Weinstein did not present an analysis of the charge redistribution in terms of sigma and pi contributions, the phenomenon observed here is somewhat reminiscent of Clementi's description of the nitrogen in both pyrrole and ammonia as a charge transformer, which accepts one type of electronic charge (pi or sigma) and donates the other (*69*, *70*).

Another group of five-membered heterocycyles for which EPs have been computed using SCF wavefunctions are imidazole (VIII), pyrazole (IX), oxazole (X), and isoxazole (XI) (*23*, *277*). Each of these contains two heteroatoms, one of which is always a pyridinelike nitrogen, the other either an oxygen or a pyrrolelike nitrogen. As was found for the nucleic acid bases, the most negative potentials in these molecules are associated with the pyridinelike nitrogens and are centered in the molecular plane. The values of these minima correlate with the molecules' experimentally determined solution basicities. There are also strong negative potentials near the oxygens, with minima in the ring planes, but not near the pyrrolelike nitrogens. Finally, each molecule has negative regions above and below the ring; it is interesting that these reach their most negative value in each case in the immediate vicinity of position 4. In general, the calculated EPs of these molecules are in agreement with the available experimental data (*23*).

6-5-4c. Six-Membered Heterocycles. Almlöf et al. (*6*) have made a detailed theoretical study of some of the properties, including EPs, of a series of nitrogen-containing six-membered heterocycles, structures XII–XVII. All of the

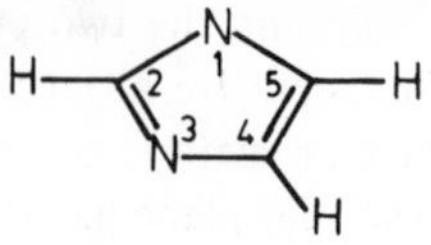

VIII 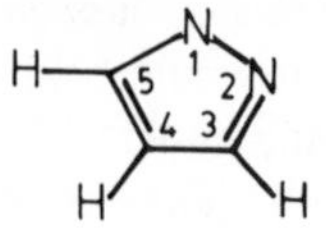 IX 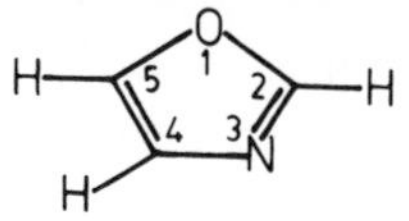 X

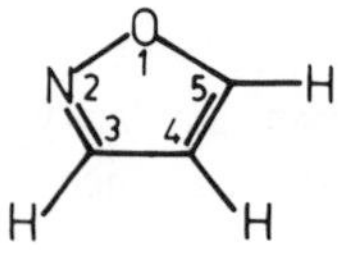

XI 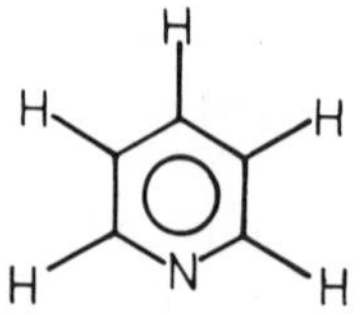 XII 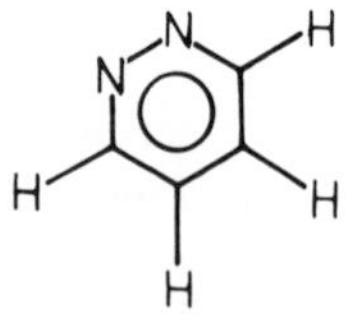 XIII

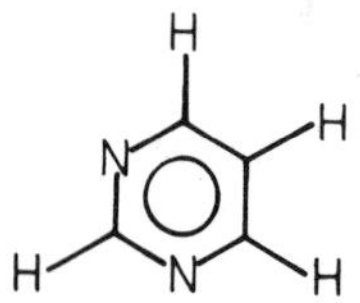

XIV 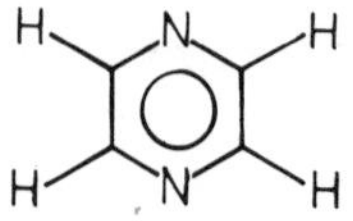 XV 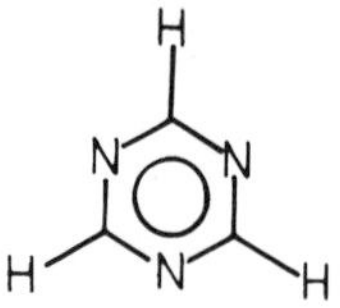 XVI

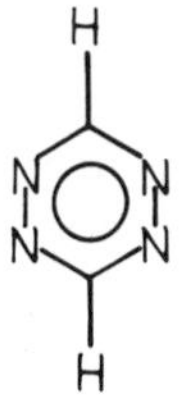

XVII

nitrogens in these molecules are pyridinelike, and accordingly, as has consistently been found to be the case, they all have strong negative regions centered in the molecular planes. The values of the minima, as computed from the respective SCF wavefunctions, ranged from −60.6 kcal/mole for *s*-tetrazine (XVII) to −84.5 kcal/mole for pyridazine (XIII) and −82.7 for pyridine (XII), the magnitude generally becoming greater as the number of nitrogens decreased.[17] Each of these minima is located 1.16 Å from its nitrogen. The potentials elsewhere in the molecular planes are positive. The magnitudes of the nitrogen minima correlate approximately with the observed aqueous solution basicities of the molecules (*6*), except for the pyridazine minimum.

It is interesting that except for pyridine the negative regions associated with the nitrogens are the only ones to be found in these molecules. They do not show negative potentials above and below the ring that can be attributed to the carbon atoms. Almlöf et al. suggested that this inhibits electrophilic attack on these molecules, and indeed they have been found to be relatively unreactive toward electrophiles (*148*). Pyridine does have a rather weak negative region above the β-carbons, and limited electrophilic substitution occurs at these positions.

On the other hand, and not surprisingly, there are much more extensive negative regions in the EP maps of the radical anions of pyrazine (XV), 4-nitropyridine, and the three isomers of dicyanobenzene (*75*, *76*). The potential minima in these systems were found to correspond well with electron spin resonance evidence regarding the ion pairs formed with alkali metal cations.

6-5-5. Multiple Bonds in Noncyclic Systems

Multiple bonds provide a good example of how much more meaningful the EP is as an indicator of reactivity than is the electronic density. The presence of a double or triple bond in a molecule means that there is a relatively high concentration of electronic charge in that particular internuclear region (*231*, *234*, *239*), from which it might be inferred that the region would be attractive toward electrophiles. However, this is not necessarily true because it ignores the effect of the nuclei. The EP associated with a double or triple bond may indeed be negative, but it may also be positive, when the nuclear contribution outweighs the electronic. The potential takes both into account, and thus better reflects what an approaching electrophile sees than does the electronic density alone. Examples of multiple bonds with both positive and negative potentials will be given.

6-5-5a. Ethylene. The only negative regions that have been found around ethylene are two strong ones above and below the molecular plane, their minima being on the axis perpendicular to the C—C bond at its midpoint (*9*). Consistent with these results, recent energy calculations indicate that the most stable form of protonated ethylene is that in which the added hydrogen is situated on this axis (*118*, *318*). Furthermore, there is both theoretical and experimental evidence that the electrophilic addition of Cl_2, Br_2, and I_2 to ethylene proceeds through an analogous halonium ion, $C_2H_4X^+$ (see Ref. 135).

6-5-5b. Acetylene and Some Derivatives. The negative potential associated with acetylene is cylindrically symmetrical about the molecular axis, its minimum being on the line perpendicular to the C—C bond at its midpoint (*68*). Unlike ethylene, the protonated form of acetylene is predicted by energy calculations to have the vinyl structure, in which the added hydrogen is on one of the carbons rather than in a bridging position (*118*). However the computed energy difference between the vinyl and the bridged structures is only 5.7 kcal/mole. Clark and Adams (*68*) suggest that the proton may approach along the perpendicular to the C—C bond, and then move over to one of the carbons.

EPs have also been computed for the fluorine and the methyl derivatives of acetylene, e.g., fluoroacetylene (*68*) and propyne (*307*). The presence of these substituents causes the negative potential associated with the C—C bond to shift toward the other carbon atom.

6-5-5c. CN-Containing Molecules. While both of the isoelectronic molecules acetonitrile, $H_3C—C{\equiv}N$, and propyne, $H_3C—C{\equiv}CH$, contain a formal triple bond, the corresponding EPs are quite different. Although acetonitrile may well have more electronic charge in the triple bond region than does propyne (*239*), its only negative potential is in the nitrogen lone-pair region, in contrast to propyne, in which a negative potential is definitely associated with the triple bond (*307*). These statements are based on INDO wavefunctions for both molecules. It should be noted that the acetonitrile negative potential is very much stronger than that found for propyne.

Similarly, a very negative lone-pair region has been obtained for the end nitrogen in cyanamide, $H_2N—C{\equiv}N$, using an SCF wavefunction (*119*); its minimum is on the N—C—N axis. A much weaker minimum was observed above the amine nitrogen, perpendicular to the NH_2 plane. It is interesting that in isocyanamide, $C{\equiv}N—NH_2$, the lone-pair region of the carbon was found to be essentially as negative as that of the nitrogen in cyanamide (*119*). The geometries used by Hart have the NH_2 group in cyanamide coplanar with the rest of the molecule, whereas in isocyanamide the amine hydrogens are bent out of this plane. As a result, in the latter case there is a very strong negative potential by the amine nitrogen on the other side of the plane from the hydrogens.

There is some uncertainty concerning the diazomethane molecule, $H_2C{=}N{=}N$. Using an SCF wavefunction and computing the EP rigorously, Hart (*119*) obtained a very widely extended negative potential, with a minimum of -143.3 kcal/mole on the molecular axis, in the lone-pair region of the end nitrogen. He also found two other, weaker minima (-56.5 kcal/mole) above and below the carbon atom. It should be mentioned that Hart used bond lengths slightly longer (0.05 Å) than the experimental values. A qualitatively very similar potential map was also computed (*168*) based on a minimum basis SCF function, the integrals again being evaluated rigorously. The EP of diazomethane has also been calculated from two other wavefunctions, using (6-12) to approximate the integrals as discussed in Section 6-4-3. Using an SCF wavefunction having a Gaussian basis set of double-zeta accuracy, there was no indication of any significant negative potential near the end nitrogen; the only minima were above and below the C—N bond, with values of -21.9 kcal/mole (*57*). With an INDO wavefunction and (6-12) there was obtained a negative region around the end nitrogen, but with minima (-13.2 kcal/mole) above and below the molecular plane, rather than in it as Hart had found (*55*). There were also two very slightly stronger minima (-14.9 kcal/mole) near the carbon, above and below the plane. The quantiative and especially the qualitative differences in the potentials resulting from these several computations are rather remarkable. The relative values of the INDO minima may be regarded as indicating the carbon to be the preferred site for proton attack (although the difference of 1.7 kcal/mole is not very significant) and indeed the initial step in the reaction of diazomethane with a protic acid is protonation at the carbon, followed by decomposition to a carbonium ion (*73*). Protonation at the carbon would certainly not have been predicted from Hart's calculated potential. As a possible explanation of this apparent discrepancy, he suggested that the mechanism may be as follows (*119*):

$$H_2CNN \underset{-H^+}{\overset{H^+}{\rightleftharpoons}} H_2CNNH^+ \underset{H^+}{\overset{-H^+}{\rightleftharpoons}} HCNNH \underset{-H^+}{\overset{H^+}{\rightleftharpoons}} HCNNH_2^+$$

$$H_2CNN \underset{-H^+}{\overset{H^+}{\rightleftharpoons}} H_3CNN^+ \longrightarrow CH_3^+ + N_2$$

The reaction supposedly proceeds through H_3CNN^+ because of the irreversible decomposition step.

EP plots have been utilized for investigating the mechanisms of diazomethane reactions with methyllithium (*119*) and with ethylene (*168*). In both instances, potentials representing supposed intermediate stages in the reactions were computed.

6-5-6. Hydrogen Bonding

The nature and characteristics of hydrogen bonds have been analyzed and discussed in great detail (*46*, *141*, *159*, *301*), one purpose being to identify the effects that play significant roles in the formation of these bonds. While the importance of taking into account such factors as charge transfer and exchange has been established, it has also been found that a consideration of only the electrostatic interaction between the unperturbed reactants is often sufficient to permit at least a qualitative description of the hydrogen-bonded system. To quote Morokuma (*138*), who has made extensive molecular orbital studies of hydrogen bonding (*192*, *316*),

> it is interesting to recognize that in an extremely qualitative sense the electrostatic energy E_{es} is often a good indicator of the hydrogen-bond energy E_H with a scaling factor which is less than unity. This conclusion might justify the simple electrostatic model. . . .

The energy E_{es} mentioned above is the electrostatic interaction energy, given by (6-18).

It seems reasonable to anticipate, therefore, that the molecular EP itself (which is in general different from E_{es}) will already provide some useful guidelines concerning hydrogen bonding to the molecule in question. Indeed, for a series of complexes between hydrogen fluoride, serving as a proton donor, and various molecules acting as proton acceptors, Kollman et al. (*160*) found that the SCF-calculated hydrogen bond energies correlated quite well with the EP at a fixed distance from each proton acceptor molecule. There was no good relationship, however, with the calculated charges [via the population analysis equation (6-4)] on the acceptor atoms. Another correlation, but a poorer one, was observed between the potential at a specific distance from the hydrogen for a series of proton donors and the calculated interaction energies with ammonia acting as the proton acceptor. The relationship was again worse when the calculated charges on the hydrogens were used instead of the potentials.

The EP can also serve as a basis for predicting the general directions of hydrogen bonds. This was demonstrated (*169-171*) in theoretical studies of the hydrogen-bonded dimers $(H_2O)_2$, $(H_2S)_2$, and $(HF)_2$. For example, the most negative potential associated with the fluorine atom in HF is not on the H—F axis, but rather [according to Leroy et al. (*171*)] at an angle of about 77° with respect to this axis, to the outside of the fluorine atom. This suggests that when HF acts as a proton acceptor in forming a hydrogen bond, that bond should form an angle of about 77° with the H—F axis, or putting it another way, the H—F—H angle should be 103°. This is in remarkable good agreement with the results of several energy calculations for the $(HF)_2$ system (*88*, *171*) and also with an experimental result of 108° (*93*).

In H_2S, either of the two potential minima associated with the sulfur (Fig. 6-9) serves as a good indicator of the direction of the hydrogen bond in $(H_2S)_2$. In the case of H_2O, however, the two oxygen minima found by Leroy et al. are almost in the molecular plane, whereas both energy calculations (*88*, *171*) and experimental work (*94*) indicate that the hydrogen bond is out of the plane by 52-60°. On the other hand, the H_2O electrostatic potential given by Scrocco and Tomasi (*277*) shows the oxygen minima to be 42° out of the molecular plane, which is much closer to the hydrogen bond angle.

Leroy et al. noted that in each of these three dimers the direction of the hydrogen bond is nearly the same as the direction to the center of charge of a lone-pair localized orbital. This raises the question as to whether the center of charge or the potential minimum has a greater influence upon the hydrogen bond direction. This question has been addressed in the case of glycine, for which it was inferred that the EP has the greater effect (*8*).

Although Leroy et al. treated only the systems $(H_2O)_2$, $(H_2S)_2$, and $(HF)_2$, a recent compilation of energy calculations for all possible pair combinations between HF, H_2O, and NH_3 provides further evidence that the positions of the EP

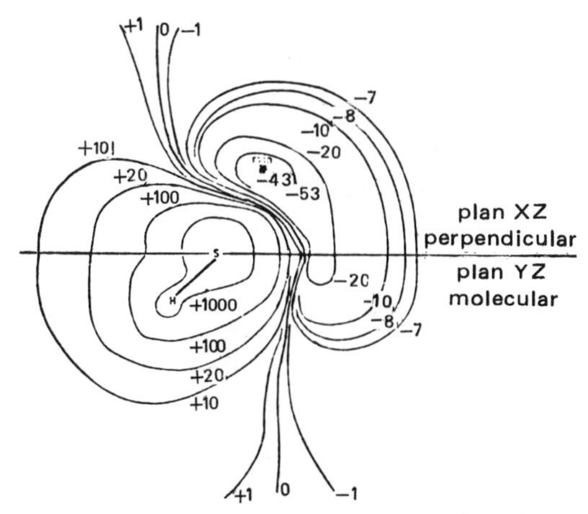

Fig. 6-9. Electrostatic potential of hydrogen sulfide. Bottom half of figure represents molecular plane; top half represents plane perpendicular to molecule through its twofold symmetry axis. The contours represent the following potential values, in kcal/mole, starting with the one encompassing the H—S bonds: +627, +62.7, +12.5, +6.3, +0.63, 0.0, −0.63, −4.4, −5.0, −6.3, −12.5, −22.0. The value of the potential minimum, located in the upper portion of the figure, is −27 kcal/mole. It is at an angle of about 61° relative to the horizontal, taking the sulfur atom as the origin. (Reproduced from Ref. 170, courtesy Sociétés Chimiques Belges.)

minima do permit reasonable estimates of hydrogen bond directions (*160*).[18] It was pointed out that this holds true also for systems in which π electrons act as the proton acceptors; in both C_2H_2 and C_2H_4 the most negative potentials are along the perpendicular bisectors of the C—C bonds (see Section 6-5-5), and it is apparently in the neighborhood of this bisector that hydrogen bonds do form (*87*, *160*). In the cases of H_2CO and HCN, however, both of which also contain π electrons, the most stable hydrogen bonding has been found to occur through the oxygen and nitrogen lone-pair regions (*86*, *87*). This also would have been predicted from the positions of EP minima (*109*, *307*).

Of course, the EP should not be expected to do more than give a rough idea as to the direction of a hydrogen bond. This is immediately evident from the fact that hydrogen bonds with various proton donors and a given acceptor have somewhat different directional properties, even though the location of the EP minimum for the proton acceptor alone remains at the same position (*139*). Nevertheless, it is appropriate, in the context of the preceding discussion, to quote a conclusion reached by Perahia et al. (*216*) in a study of the hydration of the dimethylphosphate anion (DMP^-):

> These results show that . . . the disposition of the electrostatic potential minima around the anionic oxygens has a notable bearing on the positions and energies of hydration sites, and point to the fact that the electrostatic component of the interaction energy between water and DMP^- is an important one.

The phenomenon of hydrogen bonding was examined from another point of view by computing EPs for a number of hydrogen-bonded complexes: $H_2CO \cdot HF$ and $NH_2CHO \cdot HF$ (*274*); $H_2O \cdot HF$, $H_3COH \cdot HF$ and $(CH_3)_2O \cdot HF$ (*1*); and pyridine$\cdot H_2O$, $H_2O \cdot H_2O$, and $NH_3 \cdot xH_2O$, $x = 3, 4, 5$ (*261*). One of the purposes was to determine, in each instance, the effect of the hydrogen bond upon the negative potential and minimum (or minima) previously associated with the proton-accepting atom; in this manner, some insight could be obtained into the likelihood of additional hydrogen bonding interactions with the same proton-acceptor atom. An interesting feature of Pullman and Berthod's (*261*) investigation was their demonstration that quite a satisfactory qualitative picture of the EP of $H_2O \cdot H_2O$ system could be obtained by simply superposing the potentials of two free water molecules, in the appropriate relative geometry. There were, of course, deviations from the more accurate potential computed directly for the $H_2O \cdot H_2O$ complex since polarization, charge transfer, and exchange effects were neglected, but the results were sufficiently encouraging that the same approach was subsequently applied to two very important hydrogen-bonded systems: the adenine–thymine (A–T) and guanine–cytosine (G–C) base pairs of DNA (*219*). Four strongly negative regions were found for each pair; in order of decreasing attractiveness to an approaching electrophile, they were N3(A) > N7(A) > O2(T) > O4(T) in the adenine–thymine pair and N7(G) > N3(G) ≈ O6(G) > O2(C) in the guanine–cytosine combination (see structures II, III, V, and VI). N7 of guanine was the most attractive of all of these sites, in either base pair.

6-5-7. Interactions of Drugs with Cellular Constituents

There have been some interesting utilizations of molecular EPs in studying various aspects of the interactions between certain drugs and appropriate cellular constituents (receptors). Many of these drugs have in common the feature of possessing a tertiary nitrogen. At physiological pH values, most of these nitrogens are protonated; indeed, it is believed that an important factor in the overall drug–receptor interaction is the attraction between the cationic quaternary nitrogen group and an anionic site on the receptor, perhaps a phosphate group of adenosine triphosphate (ATP). There is accordingly an element of ambiguity in calculating the EP as to whether it should be for the neutral molecule or the protonated form. The difference between the two has been demonstrated in the case of 5-hydroxytryptamine (XVIII),

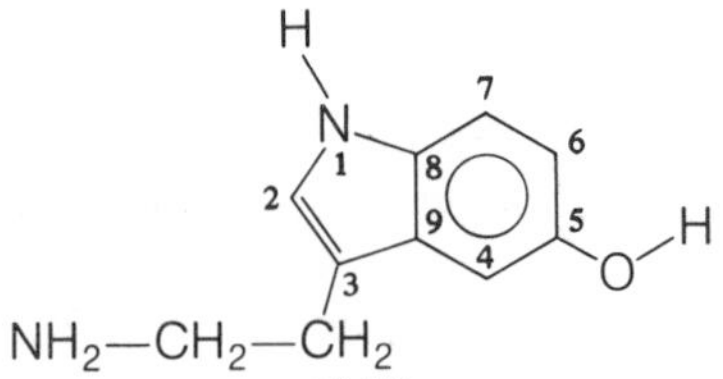

NH_2—CH_2—CH_2

XVIII

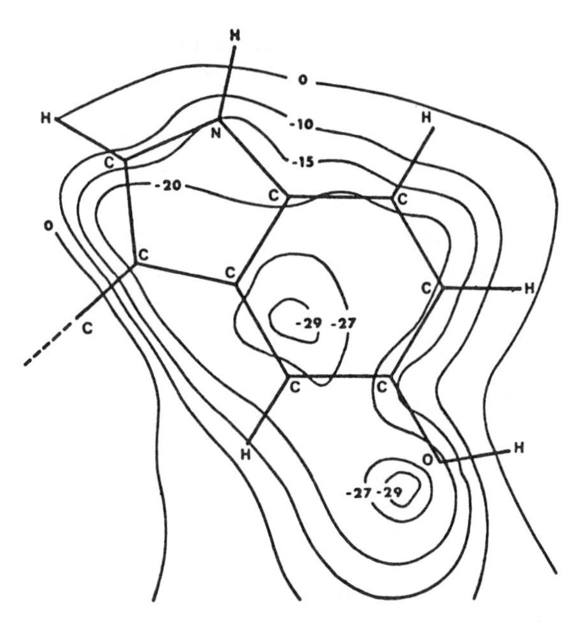

Fig. 6-10. Electrostatic potential in plane 1.6 Å above indole portion of neutral 5-hydroxytryptamine (see structure XVIII). Values are in kcal/mole. (Reproduced from Ref. 309, courtesy John Wiley & Sons, Inc.)

using an SCF wavefunction (*309*, *310*). The neutral molecule has a very large negative region above and below the two rings (the indole portion) with minima near the hydroxyl oxygen and near C4 (Fig. 6-10). When the molecule is protonated on the amine group, these negative regions completely disappear and the potential is positive everywhere above and below the ring. (It is least positive where the oxygen minimum had been). Upon placing a negative ion (OH^-) near the quaternary nitrogen, however, the potential returns to a pattern that is qualitatively very similar to that of the neutral molecule; there are minima in almost the same locations. Calculations for phenethylamine (XIX)

C_6H_5—CH_2—CH_2—NH_2

XIX

also show the protonated form to have positive potentials everywhere, but in this case the introduction of a negative ion (H^-) did not produce as striking a reversal to the neutral molecule potential as for 5-hydroxytryptamine (*182*, *223*). In the work that will be discussed three different approaches have been used to try to take account of the protonation of the drug molecule: (1) the EP was computed for the protonated form; (2) the potential was computed for the protonated form but with a negative ion placed near the quaternary nitrogen to simulate the effect of the anionic receptor site; (3) the potential was computed for the neutral drug molecule, invoking the previously mentioned 5-hydroxytryptamine results to argue that the neutral-molecule potential is qualitatively similar to that of the protonated form in the presence of the anionic site on the receptor.

One of the earliest applications of the EP to the study of drug–receptor interactions (*214*, *307*) investigated the interference, by certain derivatives of 1-cyclohexylpiperidine (XX), in the physiological activity of acetylcholine (XXI).[19]

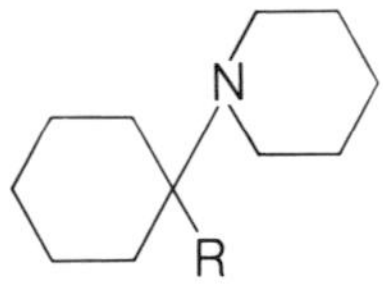

XX

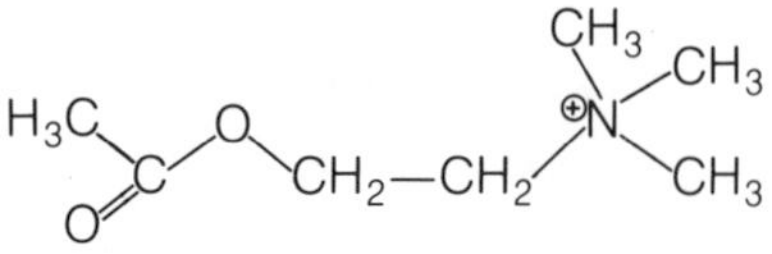

XXI

It was proposed that this occurs because of a basic similarity between the EPs of acetylcholine and the 1-cyclohexylpiperidine derivatives in question, which permits the latter to compete with the acetylcholine for receptor sites. Weinstein et al. (*307*) consider a concentrated negative region that was found near the linking oxygen atom of acetylcholine to be of key importance. They suggested that the interfering 1-cyclohexylpiperidine derivatives are those in which the R portion contains a properly located negative potential, such that if the molecule were superposed on acetylcholine, these two negative regions would coincide. Similar reasoning is the basis for a proposed explanation for the acetylcholine-like behavior of 3-acetoxyquinuclidine and the less similar behavior of its *N*-methyl derivative (*308*).

Two more molecules that interfere with the action of acetylcholine are atropine and scopolamine. Using closely related model systems, Weinstein et al. (*313*) showed that there was again an essential similarity between their EPs and that of acetylcholine. The same general approach has been used to interpret the activities and modes of action of morphine and several of its derivatives (*174*). EP maps for the protonated molecules (in the presence of anions) were computed and compared, and were then used as the basis for discussing the possible manner in which each one interacts with a given receptor (see also Ref. 49).

The acetylcholine and the morphine calculations all involved INDO molecular wavefunctions. Recently, however, electrostatic potentials computed from SCF wavefunctions were used to investigate the origins of the differing pharmacological activities of a group of hydroxy derivatives of tryptamine (*309*). Some results for one of these molecules, 5-hydroxytryptamine, have already been discussed. Each of the four hydroxytryptamines studied had four minima associated with its indole portion, two by the hydroxyl oxygen and two being somewhere above and below the six-membered ring (Fig. 6-10). The relative positions of the two minima on each side of the indole plane are different for each molecule. It was proposed that the range of pharmacological activities shown by these molecules can be attributed in part to the differences in the compatibilities of their EPs with that of the receptor.

Another interesting example of the application of EP calculations to the elucidation of drug interactions is the investigation (*310*) of one aspect of the mode of action of histamine (XXII).

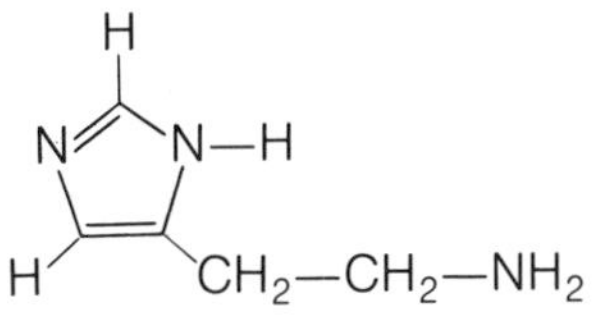

$$\text{(imidazole ring: H, N, N—H, H)}\ CH_2—CH_2—NH_2$$

XXII

It was demonstrated how the relative nucleophilic characters of the two ring nitrogens change in the progression from the neutral molecule to the protonated (at the amine nitrogen) form to the neutralized (by OH^-) form. On the basis of these findings, Weinstein et al. (*310*) proposed a mechanism for the interaction of protonated histamine with a receptor. It is noteworthy that a consideration only of the population-analysis atomic charges calculated using (6-4) would have produced incorrect conclusions regarding the nucleophilic characters of the nitrogens.

Detailed analyses have also been carried out of the potentials associated with phenethylamine (XIX), mentioned earlier, and several related molecules (*182*, *221–223*, *225*). It was proposed in each instance that the potential associated with the phenyl ring and its substituents, which included hydroxyl, nitro, and methyl groups, plays a major role in determining the molecule's biological activity.

6-5-8. Polycyclic Aromatic Hydrocarbons

The reactive properties of polycyclic aromatic hydrocarbons are of great interest because of the widespread occurrence of these compounds in the environment and because some of them are very strong carcinogens. It is widely believed that a key step in the carcinogenic action of such molecules is their interaction with some electrophile, whether in forming a reactive intermediate (see, e.g., Ref. 315) or in acting directly upon cellular constituents (*11*). The knowledge of the hydrocarbons' EPs could therefore help to elucidate the mechanism of the carcinogenic action, and the basis for the wide variations in activity even among hydrocarbons with quite similar structures.

We have computed the potentials for a number of polynuclear aromatic hydrocarbons, using deorthogonalized CNDO/2 wavefunctions (*240, 241*). The potentials were evaluated rigorously, using six-term Gaussian expansions for the STOs (Section 6-4-1). As a test, the EP was computed for naphthalene (XXIII)

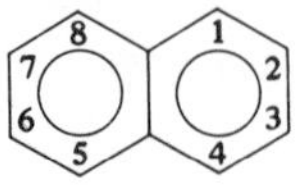

XXIII

using a small Gaussian basis SCF wavefunction and compared to the corresponding CNDO/2 results. The major difference observed was that the SCF potential was negative essentially everywhere above and below the aromatic rings (see Fig. 6-1), with minima inside the C1—C2 (and equivalent) bonds, while the CNDO/2 potential in the same regions was negative only outside of the rings, with minima outside the C1—C2 (and equivalent) bonds. This difference is precisely what would be anticipated on the basis of previous experience with CNDO/2 potentials above aromatic rings (Section 6-4-2). There is also a difference in magnitudes; the SCF potential reaches values as low as -10 kcal/mole, whereas the CNDO/2 does not become more negative than about -1.6 kcal/mole. However, both potentials do predict the same channel for electrophilic attack, a path perpendicular to the C1—C2 (and equivalent) bonds. The hydrocarbons that are of interest as carcinogens are considerably larger than naphthalene, and computation of SCF potentials for them is not presently feasible. However, the above analysis of naphthalene suggests that CNDO/2 potentials can yield useful qualitative information concerning the carcinogenic hydrocarbons, in that they will indicate the relative susceptibilities of the various molecular regions to electrophilic attack.

We have calculated CNDO/2 potentials for phenanthrene, pyrene, chrysene, benz[*a*]anthracene, dibenz[*a*,*h*]anthracene, 7,12-dimethylbenz[*a*]anthracene, 3-methylcholanthrene, and benzo[*a*]pyrene. Each was found to have several negative regions (Figs. 6-11, 6-12); one of these is invariably the so-called *K*-region, which is indicated below for some typical cases.

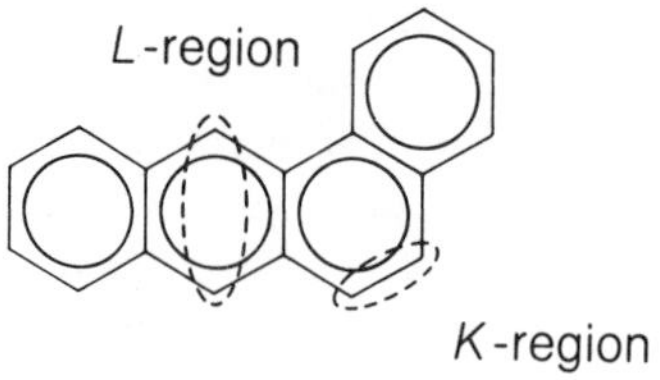

L-region

K-region

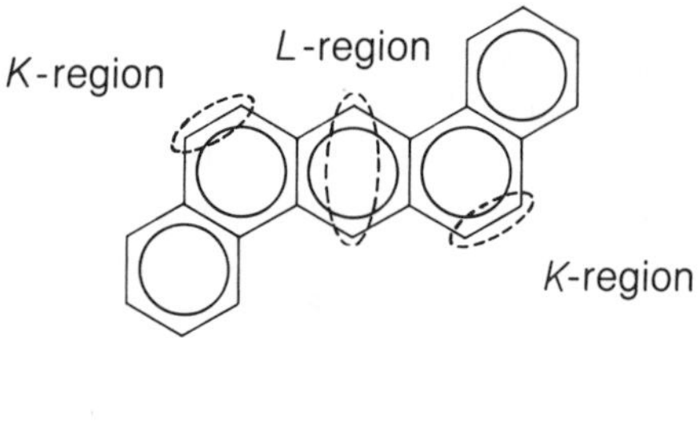

K-region

L-region

K-region

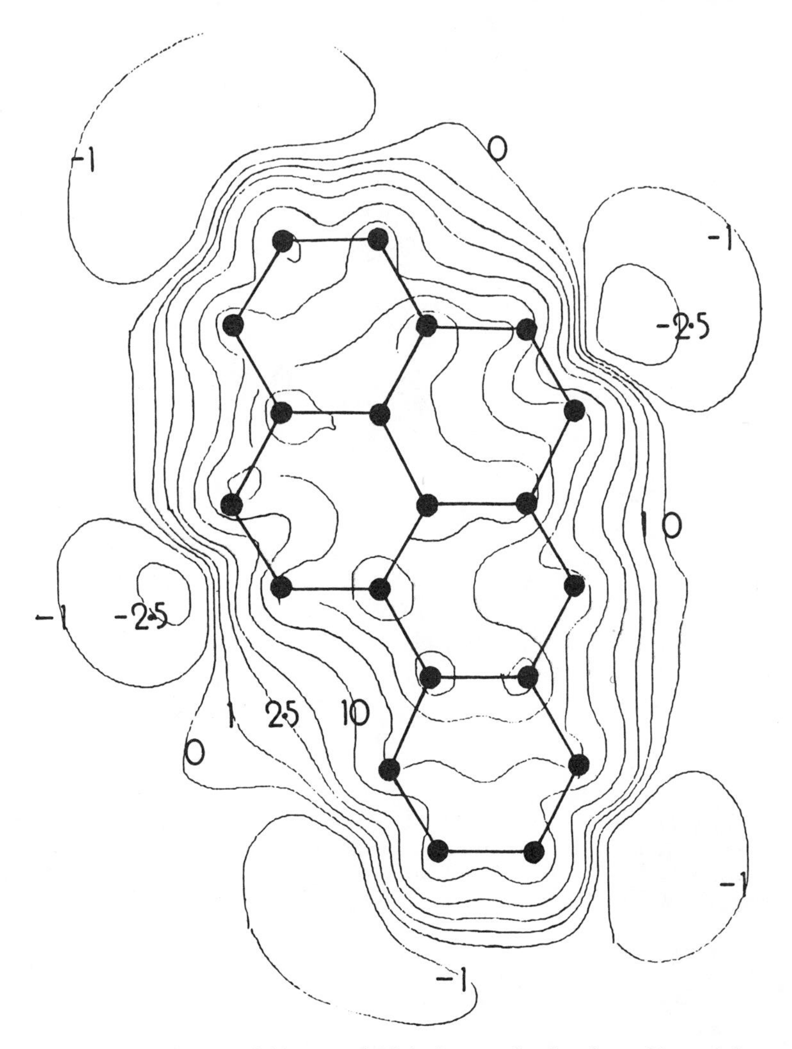

Fig. 6-11. Electrostatic potential in plane 1.32 Å above molecular plane of benzo[*a*]pyrene. Values are in kcal/mole. (Reproduced from Ref. 241, courtesy John Wiley & Sons, Inc.)

A very prominent and long-lasting theory of hydrocarbon carcinogenesis is based upon the concept that the carcinogenicity is related to a high degree of electronic charge localization in the *K*-region accompanied by an inactive or blocked *L*-region (*253*).

An interesting speculation (*240*) is that a key element in the carcinogenic activity of these hydrocarbons may be the presence of two or more suitably located regions of significant negative potential, one of which might be the *K*-

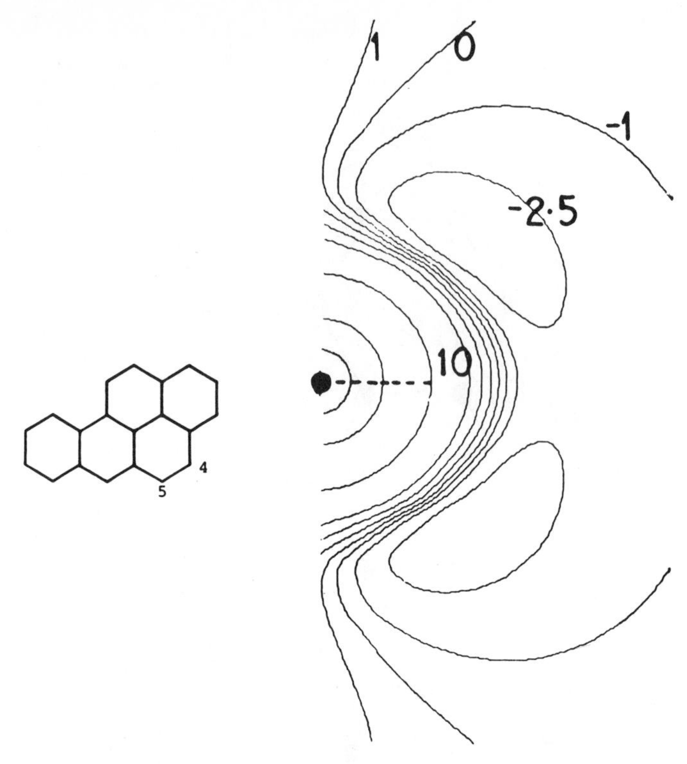

Fig. 6-12. Electrostatic potential in plane perpendicular to C4—C5 bond of benzo[*a*] pyrene at its midpoint (see structure above). Black circle indicates projections of carbons 4 and 5 upon plane of figure; dashed line indicates molecular plane. The C4—C5 bond is one of the *K*-regions of benzo[*a*] pyrene. Values are in kcal/mole. (Reproduced from Ref. 241, courtesy John Wiley & Sons, Inc.)

region. Reasoning along these lines makes it possible to offer an explanation for the fact that 5-methylchrysene (XXIV)

XXIV

is a very potent carcinogen, comparable to benzo[*a*] pyrene, whereas chrysene itself and all of its other monomethyl derivatives are at most weak carcinogens (*241*).

One possible function of such specifically located negative regions is related to the idea, proposed many years ago, that the hydrocarbons intercalate between the base pairs of DNA (*43*, *115*, *136*). Some support for this concept is provided by the remarkable similarity in size and shape of the carcinogenic hydrocarbons and the base pairs (*74*). If such intercalation does occur, then the role of the hydrocarbon negative regions might be to correlate with those associated with the nitrogen and oxygen atoms of the base pairs (Sections 6-5-3 and 6-5-6). In addition, or perhaps alternatively, one or more of the hydrocarbon negative regions might serve as the site for a metabolic process leading to some active intermediate; we have already found some indications of a relationship between negative potential and metabolic reactivity (*240*). There is much recent evidence that hydrocarbon carcinogens interact with DNA in the form of metabolically produced epoxides (see, e.g., Refs. 11, 81, 84, 146, 309, 310).

6-5-9. Excited States

There have been relatively few EP calculations for excited states of molecules. Where such results are available, however, it is interesting to note how the potential changes relative to the ground state. This will be discussed for two specific cases.

6-5-9a. Thymine. SCF-CI wavefunctions for the lowest singlet and triplet excited states of thymine (*285*) have been used to compute the corresponding EPs (*38*). These states may be regarded as arising from $\pi \rightarrow \pi^*$ and $\sigma \rightarrow \pi^*$ transitions; each transition can lead to both a singlet and a triplet state. In the $\pi \rightarrow \pi^*$ excitations, the electron involved remains in essentially the same region of the molecule, that being the neighborhood of the C5—C6 double bond (structure VI), while the predominant contribution to the $\sigma \rightarrow \pi^*$ transition comes from an oxygen lone-pair electron moving to the C5—C6 double-bond region (*285*). It will be seen that the potential maps reflect these differences in the nature of the transitions.

The main features of the EP of the ground state of thymine (Fig. 6-6) are the strong negative regions associated with the two carbonyl oxygens, both of which have minima in the ring plane. The two excited states resulting from the $\pi \rightarrow \pi^*$ transition retain roughly the same potential distribution, apparently reflecting the fact that the rearrangement of electronic density has been relatively small. The principal effect upon the potential is a significant weakening of the negative region near oxygen O2, especially in the triplet state; in the latter, however, the negative region of oxygen O4 is somewhat strengthened.

The $\sigma \rightarrow \pi^*$ excitation produces a much greater change in the EP relative to the ground state, although it is very nearly the same for both the singlet and triplet states. The negative potential associated with O2 is completely eliminated; it is replaced by a new negative region having a minimum near the midpoint of the

C5—C6 double bond (to the outside of the ring) and extending well over toward oxygen O4, which has itself acquired a very much more negative lone-pair region. These changes are quite consistent with an O2 lone-pair electron moving over to the vicinity of the C5—C6 double bond. As Bonaccorsi et al. (*38*) pointed out, these results make it reasonable to surmise that the $\sigma \rightarrow \pi^*$ excitation considerably affects the solvation of the molecule.

These authors also examined the potentials that would be obtained for the excited states if these were described simply by single-determinant wavefunctions, the excitations being taken into account by transferring electrons appropriately from occupied to virtual molecular orbitals. In this rather crude approach, singlet and triplet states have the same electronic density distributions and hence the same EPs. The potential obtained in this manner for the $\pi \rightarrow \pi^*$ excited state was found to be roughly intermediate between those computed for the singlet and triplet states using CI wavefunctions; it was therefore a fair approximation to each of the latter. However, the single-determinant potential for the $\sigma \rightarrow \pi^*$ excited state was very poor, in that it showed essentially no negative region by oxygen O4, completely contrary to what the more accurate treatment indicates, as discussed above. Thus the single-determinant approach may be of some value as an approximation, but only when there is reason to believe that the rearrangement of electronic density accompanying the excitation is rather small.

6-5-9b. Formaldehyde. The EPs of formaldehyde in its ground state (Section 6-4-2), and in its first singlet and triplet excited states have been computed using CI wavefunctions (*85*). Unfortunately, the same geometry was used for the excited states as for the ground state, whereas in reality both of the former have the oxygen atom bent out of the CH_2 plane, unlike the planar ground state (*126*, *153*).

Using initially single-determinant treatments, in which electrons were simply transferred from occupied to virtual molecular orbitals (Section 6-5-9a), the strong negative potential associated with the oxygen in the ground state was found to be completely eliminated by the single orbital excitation to either the $^3A_1$ or the $^1A_1$ state. Instead, significant negative potentials developed above and below the CH_2 portion of the molecule. When CI was included, however, a very marked change occurred; the negative CH_2 regions essentially disappeared, although a positive minimum did remain in the case of the $^3A_1$ state and a considerable negative potential reappeared by the oxygen (but somewhat weaker than in the ground state). The oxygen minima became more negative as more configurations were added, but gradually leveled off and reached stable values of about -39.0 kcal/mole ($^3A_1$) and -41 kcal/mole ($^1A_1$). The latter state proved to be more difficult to treat; for example, the convergence of the minimum values was more erratic than for the ground or $^3A_1$ state. Also, whereas basis sets of different quality produced quite similar results for the latter two, there was a significant difference in the case of the $^1A_1$ state.

The failure to use the true excited-state geometries in the above study is particularly regrettable in view of the markedly different results obtained by Caballol et al. (*56*) with a clearly less accurate procedure but using the correct geometries. Their potential maps, computed with INDO wavefunctions and the potential integrals approximated by means of (6-12), showed negative regions above and below the CH_2 group but none by the oxygen. This is exactly the opposite of what the most accurate CI treatment produced (although for a significantly different geometry); it is, however, qualitatively similar to the results of the single-determinant calculation. What makes the situation particularly interesting is the fact that the potentials computed by Caballol et al. (*56*) are consistent with the discussion of the observed reactive properties of alkanones that has been given by Turro et al. (*298*). It would certainly be very important to carry out CI calculations for these excited states using their true geometries, in order to determine where the negative regions really are.

6-5-10. Indirect Applications to Reactivity

In addition to the direct role that the molecular EP can play in elucidating chemical reactivity, it can also be very useful indirectly by permitting greater insight into certain factors and concepts which can in turn help us to achieve a better understanding of molecular reactivity. Several such indirect applications will now be discussed.

6-5-10a. Lone Pairs. The concept of an electronic lone pair is certainly a well-established and useful one. While its origins are semiempirical, new theoretical support has emerged in recent years as localized-orbital transformations of MO wavefunctions have been found to produce orbitals closely corresponding to the notion of a lone pair (*306*). To a large extent, the EP is quite consistent with the lone-pair concept, sometimes more so than is the electronic density. For example, the oxygen atom in H_2O is normally said to have two lone pairs. These are represented very clearly in the electrostatic potential map by two distinct minima in the plane bisecting the H—O—H angle (*109*, *277*); the electronic density map, however, shows no trace of them (*16*). The discussion in Sections 6-5-3 and 6-5-4 has included frequent mention of potential minima that can be related to lone pairs, such as the strong negative regions and minima associated with pyridinelike nitrogens in heterocyclic rings. There are similar negative potentials and minima in the lone-pair regions of tertiary nitrogens having non-coplanar bonds, such as ammonia (*277*), and various three-membered heterocycles (*31*, *34*): $(CH_2)_2NH$, $CH_2(NH)_2$, and CH_2ONH. When the three bonds are coplanar, however, then there are to be anticipated much weaker negative regions (if any) above and below the nitrogen; these have already been pointed out in the cases of adenine, guanine, and cytosine (Section 6-5-3). This marked dependence of the potential upon the geometry can be demonstrated by means of the results computed with SCF wavefunctions for the H_2NO radical (*96*), even though the effect observed in this instance is for just one electron rather than a pair. When this molecule is planar, there are minima of -17.9 kcal/mole above and below the nitrogen. When the hydrogens are bent 20° out of the plane, a minimum of -34.5 kcal/mole develops on the other side of the nitrogen, and it reaches -56.7 kcal/mole when the hydrogens are bent by 60°.

Some atoms, such as the oxygen in water (see above), are commonly depicted, at least in simple terms, as having more than one lone pair. In many instances, the calculated EP is fully consistent with such a description. Thus there are two distinct minima associated with the sulfur atom in H_2S (*170*); in thiirane, $(CH_2)_2S$ (*31*); and in thiophene (*104*); they are situated symmetrically on both sides of the molecular plane, as anticipated. A rather remarkable case is that of fluorine in the molecule FNO; there were found to be three relatively weak potential minima distributed around the fluorine atom (*30*). There are at least two minima associated with the fluorine in formyl fluoride, HCOF (*61*), but only one in fluorobenzene (*7*).

The situation with respect to oxygen is particularly interesting and important, in view of the atom's widespread occurrence and usually key role. By elementary reasoning, the oxygen atom is usually expected to have two lone pairs, and indeed, in many molecules the EP map shows two distinctly separate minima in the oxygen lone-pair region. In this category are, for example, H_2O and O_3 (*1, 277*); $(CH_2)_2O$ (*31*); H_2CO (*109*); $(H_3C)_2O$ and H_3COH (*1, 190*); CH_2ONH (*34*); H_2NO (*96*); and FNO (*30*). The two minima may not be equivalent, of course, because of the effect of the remainder of the molecule. Thus in oxaziridine, CH_2ONH, the oxygen minima on the two sides of the three-membered ring plane have values of -44.9 and -31.9 kcal/mole. This is because there are two hydrogens on one side of the ring plane and only one on the other.

There are quite a few interesting cases, however, in which oxygen atoms do not show two distinct potential minima. In some molecules, instead of two separate points, one finds an extended minimum partially encircling the oxygen atom, e.g., in formamide (Section 6-5-2, Fig. 6-2). It can also be seen in thymine, especially by oxygen O8 (Fig. 6-6), and in formyl fluoride, HCOF (*61*). An extreme example is the dimethylphosphate anion. The two oxygens which are not bonded to methyl groups and which apparently carry much of the excess negative charge each have in their lone-pair regions a complete ring of minima, along which the potential is always within 5% of its most negative value (*216*). Consistent with this absence of any significant directional preference, the dimethylphosphate anion has been found to have a variety of energetically equivalent hydration possibilities (*256*).

There are also molecules in which only a single potential minimum is associated with an oxygen atom. Both oxazole (X) and isoxazole (XI) fit into this category (*23*, *277*); there is just one minimum near each of their oxygens, that being in the plane of the ring. The same is true of cytosine (III). The EP (Fig. 6-5) shows one minimum in the vicinity of the oxygen, in the molecular plane, at an angle of 55° with respect to the extension of the C—O bond (*35*). Bonaccorsi et al. (*35*) pointed out that a localized orbital transformation of the cytosine wavefunction produced two lone-pair orbitals, one on each side of the C—O bond extension. They suggested that the absence of two potential minima is due to the effects of the neighbors of the carbonyl group. The very strong negative potential of the pyridinelike nitrogen enhances that of the oxygen on one side, while the positive hydrogen diminishes it on the other; the result, according to the authors, is a single oxygen minimum, located on the N3 side of the C—O bond extension.

6-5-10b. Bent Bonds. An interesting feature is shown by the EPs computed from SCF wavefunctions for a series of three-membered ring molecules (*30*, *31*, *277*). It was found that there is a significant negative potential, including a minimum, associated with the C—C single bonds in the molecules C_3H_6, $(CH_2)_2NH$, $(CH_2)_2O$, $(CH_2)_2S$, and $(CH)_2CH_2$. This potential is to the outside of the ring, its minimum being in the ring plane (Fig. 6-13). Such minima have not been

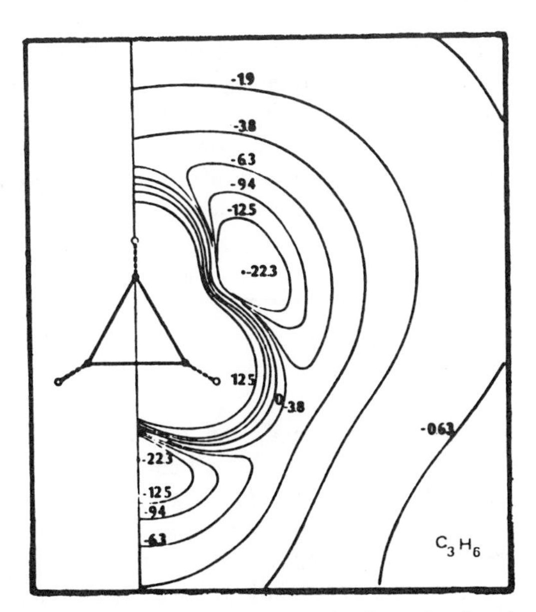

Fig. 6-13. Electrostatic potential of cyclopropane in the molecular plane. Values are in kcal/mole. (Reproduced from Ref. 277, courtesy Springer-Verlag.)

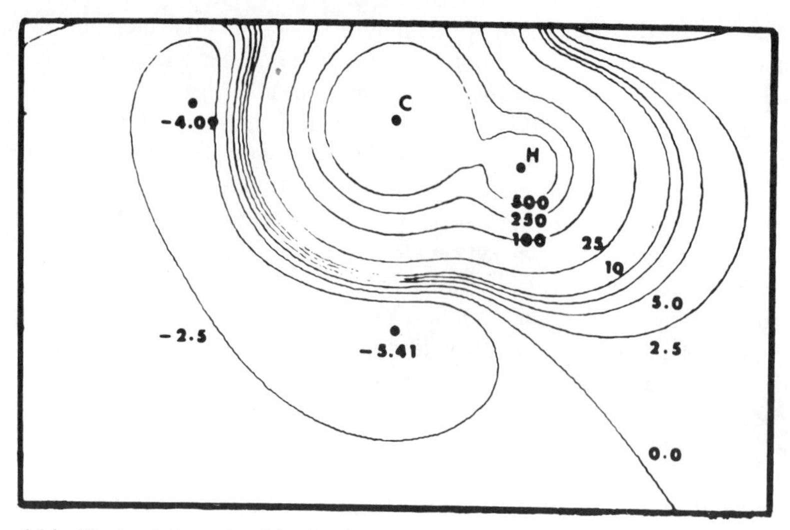

Fig. 6-14. Electrostatic potential of ethane in plane containing both carbons and two hydrogens. Only lower half of plane is shown. Values are in kcal/mole. (Reproduced from Ref. 238, courtesy The North-Holland Publishing Co.)

detected (*238*) near the C—C single bonds in noncyclic molecules, such as ethane (Fig. 6-14).

The localized orbitals corresponding to these C—C bonds are also predominantly outside the ring (*30*, *31*); furthermore, density difference plots obtained theoretically for cyclopropane (*290*) and experimentally for *cis*-1,2,3-tricyanocyclopropane (*120*) show buildups of electronic charge which are centered close to the charge centroids of the localized orbitals. These electronic densities, as well as the calculated EPs,[20] are consistent with the bent-bond concept that has been proposed as a model for the C—C bonds in these molecules (*72*, *78*, *79*, *150*). The fundamental similarity between these negative potentials and that corresponding to the C—C double bond in ethylene (Section 6-5-5) helps to explain the noticeable degree of π character and the somewhat olefinlike properties of some of these molecules, particularly cyclopropane (*63*, *78*, *270*).

6-5-10c. Methyl and Methylene Groups. An intriguing and rather unexpected negative potential has been found to be associated with the methyl groups in a number of molecules (*238*). Its minimum is situated within the cone defined by the three C—H bonds, but to the outside of the C—H portion of the cone. Such negative regions have been observed in CH_4 and C_2H_6 using SCF wavefunctions, and in C_3H_8, 7,12-dimethylbenz[*a*] anthracene, and 3-methylcholanthrene using deorthogonalized CNDO/2 functions. In all cases, the integrals in the expression for the potential, (6-10), were evaluated rigorously. No negative regions were detected near the methyl ends of CH_3CN and CH_3COOH; it may be that the effect is diminished by the presence of a strongly electron-attracting group attached to the methyl carbon.[21] In this connection, it is important to note the first known instance of a negative potential associated with a methyl group in a nonhydrocarbon molecule, acetamide (*207, 208*).

These negative methyl potentials are rather weak, but they are definitely there. In methane there is a minimum of −1.3 kcal/mole at 2.07 Å from the carbon along each threefold symmetry axis; in ethane it is −5.4 kcal/mole at 1.72 Å from each carbon along the molecular axis. It is interesting to note (*238*) that there are also secondary minima of −4.1 kcal/mole in ethane in the cones defined by two hydrogens and a CH_3 group (Fig. 6-14).

The discovery of these negative regions came as something of a surprise, because a striking feature of the EPs previously computed had been the extensive positive regions that often surrounded much of the molecule, and which were attributed to the effect of the peripheral hydrogen atoms (*35*). Indeed, the charges calculated by the population analysis procedure for the hydrogen atoms in methane and ethane, using the same SCF wavefunctions as were utilized for the potentials, are all positive (*212, 227*).

A further important point is brought out by the EP computed for propane, which reveals a negative region and minimum associated with the CH_2 group, in addition to those by the methyls (*238*). The CH_2 minimum is in the H—C—H plane, on the line bisecting the H—C—H angle; it is weaker than the CH_3 minima in propane (−3.3 versus −5.8 kcal/mole, using a CNDO/2 wavefunction). It is notable that such CH_2 minima have not been observed in ethylene (Section 6-5-5) or in cyclopropane (Fig. 6-13).

It appears, therefore, that both CH_3 and CH_2 groups can, in certain types of molecules, give rise to negative EPs. This conclusion is consistent with NMR evidence that the protons in methyl and acyclic methylene groups in saturated-hydrocarbon environments are generally more shielded than are those in other common organic situations (*92, 281*).

It has been suggested that the negative regions of CH_3 and CH_2 groups in polynuclear aromatic hydrocarbons may play significant roles in their metabolic and carcinogenic activities (*240*). Among other possible consequences of these negative potentials, a particularly intriguing one is the idea that they might permit some unanticipated interactions, such as perhaps a kind of weak form of hydrogen bonding, in which the CH_3 or CH_2 group would serve as the proton acceptor. This is purely speculation at present, but some support for it might be provided by the water solubilities of the alkanes. These might be expected to decrease with increasing molecular weight. Instead, the solubility in water increases in going from ethane to propane to butane (*131*). This suggests that some other factor is involved, in addition to size and that for this sequence of molecules it is dominant. That other factor might be the number of CH_3 and

CH_2 groups in each molecule and the resulting opportunities for attractive interactions with water molecules.

In this connection, it is of great interest to note the results of a detailed study of the methane-water interaction (*293*). For a single H_2O molecule in the proximity of a CH_4 molecule, it was found that whether the former was in one of the negative CH_3 or CH_2 regions, its preferred orientation was with one or both of the hydrogens pointing toward the carbon–and therefore toward the region of negative EP.

6-6. RELATIONSHIPS TO INTERACTION ENERGIES

6-6-1. General Interaction Energies

As was discussed in Section 6-3, the EP is a property of an individual molecule in a specific state; it indicates the electrical effects produced in the molecule's environment by its nuclei and electrons. The potential cannot, in any rigorous sense, describe the overall interaction of the given molecule with some other entity, because it is a static rather than a dynamic property; it does not reflect the changes which occur in both reactants as they begin to interact with each other. In general, it must be anticipated that there will be changes both in electronic density distributions (charge transfer and/or polarization) and in geometries. Even when a molecule is interacting with just a proton, in which case the EP of the former is rigorously equal to the first-order interaction energy, there still may well be very significant second-order effects involving geometric and/or electron-density changes in the molecule.

It follows therefore that if the EP is to serve as an effective indicator of the reactive sites on a molecule, then these other factors must either be negligible in the early stages of the interaction (while the site is being determined) or their effects must approximately cancel. The successful applications of the EP that were described in Section 6-5 are evidence that one or the other of these requirements is often satisfied. There have also been a number of theoretical studies in which the roles played by such factors as electronic density rearrangements and changes in geometry were analyzed in detail. The results of some of these investigations will now be summarized.

A thorough analysis of the effects of polarization, charge transfer, and geometrical changes has been made by Pullman (*249, 251*) taking the protonation of formamide as an example. This was done in three stages. First, an SCF wavefunction and energy were computed for the system consisting of the formamide molecule plus a proton, with no basis functions on the proton; its only effect, therefore, could be through the term representing it in the Hamiltonian. In this manner polarization of the formamide electronic density distribution could be taken into account. These calculations were carried out with the proton placed at various points along several different lines of approach to the molecule. The formamide was kept in its ground-state geometry. In the second stage, the whole process was repeated with basis functions on the proton, thereby permitting charge transfer as well as polarization to occur. Finally, in the third stage, some variation of the formamide geometry was allowed.

Taking both polarization and charge transfer into account, the most favored path of approach was in the molecular plane and toward the lone-pair region of the oxygen atom, as would be predicted from the EP of formamide (Section 6-5-2). Consistent with the extended minimum that the potential map (Fig. 6-2) shows in the oxygen lone-pair region, there is a range of about 90° within which the exact angle of approach makes relatively little difference, as far as the SCF total energy of interaction is concerned. As anticipated, the approach of the proton causes a shifting of electronic charge toward the position being attacked, especially when the approach is in the molecular plane. A significant dipole moment is induced; it has a magnitude of better than 0.7 D when the proton is 3.18 Å from the oxygen, in the plane of the molecule. The effect of the polarization is to favor the interaction with the proton; at each point along its path of approach, even if it is one of the least-favored paths, the EP computed with the polarized wavefunction is more negative than with the ground-state function.

It is interesting that the extent of charge transfer is greater for out-of-plane attack than in-plane. Indeed, when this investigation was initially carried out, using CNDO wavefunctions (*186*), the inclusion of charge transfer actually caused attack on nitrogen to be more favored, at a distance of 2 Å, than on oxygen. At 1 Å, however, the oxygen was the preferred site.

Since it appeared likely that bending the NH_2 hydrogens away from an attacking proton would help its approach, the C—N—H angles and the N—H and C—N distances were allowed to vary. This did indeed stabilize the *N*-protonated form considerably (*249*). However, optimizing the C—N and C—O bond lengths in the *O*-protonated structure also lowered its energy somewhat, enough so that it remained the more stable form, by 12 kcal/mole.

Thus, in the case of formamide, a detailed analysis of the effects of electronic rearrangements and geometry changes shows that these do not change the conclusion that was reached on the basis of the EP alone, that the oxygen is the preferred site for protonation (Section 6-5-2). The net result of the electronic density shifts was to reinforce the electrostatic prediction, while the geometry factor opposed it; their overall effect, in this instance at least, was to leave the original conclusion unchanged.

A less extensive study (*60*), but covering some of the same points as the preceding, dealt with the protonation of toluene. First the *ortho*-, *meta*-, and *para*-hydrogens were moved successively out of the plane of the ring, and in each case, the EP near that same carbon but on the other side of the ring plane was computed using CNDO wavefunctions. As has been discussed earlier for other molecules (see Section 6-5-4b), there was always found a negative channel leading to the carbon; that leading to the *para* position was the one that an approaching proton would find most attractive. Then they successively placed a proton in each of these channels and recomputed the wavefunctions, allowing no charge transfer to the proton. The EPs were then recalculated, without including contributions from the proton; they reflected, therefore, only the effect of the polarization of the toluene in each instance. The potentials in the channels of approach were now very much more negative than before (the minima changed from about −10 to about −80 kcal/mole) but the conclusion that the most favorable approach for an electrophile is to the *para* position remained unchanged, in agreement with experimental observations.

It is relevant to mention at this point the calculations (*273, 275*) dealing with the interactions of H_2O and H_2CO with the Li^+ ion. The total SCF stabilization energy, ΔE, was partitioned into electrostatic, exchange, and electron delocalization (polarization plus charge transfer) contributions. For both $H_2O \cdot\cdot Li^+$ and $H_2CO \cdot\cdot Li^+$, it was observed that the electrostatic term followed ΔE relatively closely, essentially all the way to the equilibrium H_2O–Li^+ and H_2CO–Li^+ separations. The exchange and delocalization terms, on the other hand, are much smaller in magnitude and opposite in sign, the former being positive. To some extent, therefore, they cancel, leaving the electrostatic contribution as the dominant factor. These findings are reminiscent, at least in general, qualitative terms, of the situation in hydrogen bonding (Section 6-5-6).

Another approach to the question of how meaningful is the EP as a guide to the interaction pattern was followed by Ghio and Tomasi (*105*), using SCF wavefunctions and taking a group of seven three-membered ring molecules as their test cases. For each molecule, they compared the values of the potential minima with the total SCF energies of interaction when a proton is placed at the positions of each of the minima. As was mentioned at the beginning of this section, the EP at any point is rigorously equal to the first-order energy of interaction with a proton placed at that point. Thus the difference between the potential minimum and the SCF interaction energy for the proton at the same point represents the second- and higher-order contributions: polarization, charge transfer, and exchange. (The molecules were kept at their ground-state geometries.) Although the potential minimum was invariably found to be considerably smaller in magnitude than the SCF interaction energy, indicating the importance of the higher-order terms, there was a good linear correlation between them (correlation coefficient = 0.985). This means that the net effect of the other terms, while large, remained relatively constant through this series of molecules; the potential alone is therefore sufficient to predict the ordering of these protonation energies. Furthermore, the SCF interaction energies, when computed for points along the electrostatically favored lines of approach, predicted the most stable proton positions to be quite close to the locations of the potential minima.[22]

Ghio and Tomasi (*105*) repeated the above SCF energy calculations without basis functions on the proton, which meant that electrostatic and polarization effects were being taken into account, but charge transfer and exchange neglected. The results were intermediate in magnitude between the potential minima and the total SCF protonation energies, but they were still far from being reasonable approximations to the latter. However, these SCF electrostatic plus polarization energies showed a very good linear correlation with the potential minima and about the same degree of success in predicting the position of the proton.

The investigation described in the preceding two paragraphs has also been carried out, for the same molecules, using deorthogonalized CNDO/2 wavefunctions to compute the potentials and interaction energies (*220*). The results were, qualitatively, very much the same as above. The CNDO/2 potential minima, while much smaller in magnitude than the calculated interaction energies, show a good linear correlation with the CNDO/2 protonation energies and, what is especially important, an even better one with the SCF protonation energies; the correlation coefficients are 0.959 and 0.983, respectively.

Unfortunately, the EP is not always as successful a guide to chemical reactivity as the discussion up to this point may seem to indicate. This is strikingly demonstrated by the variation in the proton affinities of ammonia and its methylated derivatives as the number of methyl groups increases. Experimental measurements have shown that the gas-phase proton affinity increases as successive methyl groups are added in going from ammonia to trimethylamine (*13*). It would, of course, be desirable for the potential minima in the nitrogen lone-pair regions to become more negative in the same direction, however much they differ in magnitude from the measured values. Instead, the potentials as computed with SCF wavefunctions show the opposite trend; the most negative minimum is that of NH_3, and it becomes steadily less negative as methyl groups replace hydrogens (*41*, *257*). This situation has been analyzed in detail in two separate studies (*257*, *299*, *300*). The calculated interaction energy of each amine with a proton was broken down into various contributions, such as electrostatic, polarization, and charge transfer.[23] In each of these investigations, the magnitudes of both the polarization and the charge-transfer energies were observed to increase with the degree of methylation, thus counteracting the trend in the electrostatic energies (equal to the potential minima mentioned above). The result was that the calculated proton affinities did increase in the same direction as the experimental ones. However, Pullman and Brochen (*257*) found that the charge-transfer contribution was responsible for this trend, whereas Umeyama and Morokuma's calculations (*299*, *300*) showed the polarization energy, although much smaller in magnitude than the charge-transfer term, to be the key. The important point is that in this series of molecules, polarization and charge transfer together–rather than the electrostatic factor–determined the trend.[24] The same conclusion was reached in regard to the proton affinities of the series H_2O, H_3COH, and $(H_3C)_2O$ (*299*).

It is interesting, therefore, that the experimentally determined affinities of the methylamines for the Li^+ ion decrease with increasing methylation (*264*), as their EPs would predict. Pullman and Brochen attribute this to the great numerical predominance of the electrostatic term in the total interaction energy (see also Ref. 275); thus, for $NH_3 + Li^+$, it is 87% of their calculated total energy of interaction, while for $NH_3 + H^+$ it is 43%. The binding energies of Li^+ to 22 different molecules with basic functional groups have also been calculated (*127*).

In a detailed study of the polarization contribution to the interaction of a molecule with a point charge, Bonaccorsi et al. (*41*), computed the polarization energy as a sum of bond and lone-pair terms for various positions of the point charge. When this was added to the EP, the interaction energy of a proton with a group of nitrogen-containing molecules (N_2, HCN, NH_3, and the alkylamines) increased in magnitude by 16–40 kcal/mole, and the equilibrium $N{-}H^+$ distances were found to be somewhat shorter.

6-6-2. Electrostatic Interaction Energies

As has been discussed, the electrostatic contribution is only one of several terms that comprise the total energy of interaction between any two systems, and which may be of quite significant magnitudes. However, it can happen that these other terms may partially cancel or their sum may remain fairly constant in a series of related systems, so that the electrostatic contribution becomes dominant (Section 6-6-1). Thus, in the case of two molecules combining to form a complex, it might be that by following the changes in just the electrostatic interaction energy as the two molecules assume different orientations relative to one another, the most stable geometry of the complex could be determined. This would presumably be most likely when the molecules are polar, and when the complex is a relatively weak one, so that polarization, charge transfer, etc. can be expected to play minor roles.

One such type of interaction to which this approach has been applied is hydration. Bonaccorsi et al. (*33*) have analyzed the mode of interaction of one water molecule with another one, and then with each of several three-membered ring molecules: oxirane, $(CH_2)_2O$; aziridine, $(CH_2)_2NH$; oxaziridine, CH_2ONH; and cyclopropene, $(CH)_2CH_2$ (*2*).

The H_2O-H_2O system was used as a test case. The electrostatic interaction energy was computed by a modified form of (6-19). V_A was the EP of one H_2O molecule, obtained rigorously from an SCF wavefunction; ρ_B, however, was approximated by a point charge representation of the electronic density of the other H_2O molecule, using point charges (Section 6-4-5) established by requiring them to reproduce as closely as possible the previously mentioned SCF EP and also the SCF dipole moment. With this approximation, (6-19) becomes

$$E_{es} = \sum_k V_A(r_k)\,Q_k + \sum_j Z_j V_A(R_j), \qquad (6\text{-}20)$$

where r_k is the location of point charge Q_k. Equation (6-20) was used to compute the electrostatic interaction energies corresponding to various orientations of one H_2O molecule with respect to the other. The results were most encouraging in that they closely paralleled the SCF total interaction energies (obtained holding the geometries of the individual H_2O molecules fixed). The various relative minima in the interaction energy, as well as the absolute minimum, came at very nearly the same orientations, and even their magnitudes were quite similar; the discrepancies were less than 1 kcal/mole in values of about 6 kcal/mole.[25]

The same approach was applied to the systems $(CH_2)_2O \cdot H_2O$, $(CH_2)_2NH \cdot H_2O$, $CH_2ONH \cdot H_2O$, and $(CH)_2CH_2 \cdot H_2O$, still using the point-charge model for H_2O (*2*). In the first three of these, the interaction can be regarded as ordinary hydrogen bonding. In agreement with the discussion in Section 6-5-6, the most stable directions of the hydrogen bonds were found to be approximately those indicated by the locations of the primary minima in the molecular EPs. In the case of cyclopropene, this involves an H_2O hydrogen pointing toward the midpoint of the C—C double bond. An interesting feature of these four systems is that a stable (although weak) hydrate was detected in which the H_2O is associated with the C—C single bonds. As pointed out in Section 6-5-10, these "banana" bonds have negative potentials. The H_2O molecule is predicted by these calculations to align itself with one hydrogen pointing toward the midpoint of the C—C single bond. This evidence of hydrogen bonding to a C—C single banana bond gives some plausibility to the speculation that some such interaction may be possible with those CH_3 and CH_2 groups that have negative potentials (Section 6-5-10).

Equation (6-20) and the same point-charge representation for H_2O have further been used to investigate the hydration of formamide, $H_2N \cdot CHO$ (*3*), and *N*-methylacetamide, $H_3CNH \cdot COCH_3$ (*255*). For the former, the SCF total interaction energies were also calculated, keeping the individual H_2O and $H_2N \cdot CHO$ geometries fixed. For the most part, these two sets of results for the hydration of formamide showed essentially the same good agreement in predicting the most stable orientations and the corresponding interaction energies as before in the case of the water dimer, $(H_2O)_2$ (*33*).

While discussing hydrated systems it is worth digressing briefly to note, in one instance at least, how the EP of an individual water molecule is affected by the central ion and by the other waters. Using minimum-basis-set SCF wavefunctions, Berthod and Pullman (*25*) showed, for the dimethylphosphate anion with six or seven waters of hydration, that the negative minima by the oxygen atoms of the outer water molecules change from -70.8 kcal/mole in a single free H_2O molecule to -147.2 kcal/mole in the hydrated system (in the $H_2O \cdot H_2O$ dimer, the value was -86.2 kcal/mole). These effects, which were attributed both to polarization and to some charge transfer to the water, should temper the optimism expressed in Section 6-5-6 concerning the meaningfulness of superimposing the potentials of the components of a complex. Some of the inner water molecules were so oriented that their negative regions reinforced each other, and the minimum went as low as -176.8 kcal/mole.

Returning now to the previous discussion, it was pointed out in Section 6-4-4 that a convenient way to compute the electrostatic interaction energy is to express the potential V_A in (6-19) as a multipole expansion. This simplified the calculations, but the cost is a decrease in the accuracy of V_A unless many terms are included in the expansion. This technique has been tested for the $(CH_2)_2NH \cdot H_2O$ complex (*37*). Using a single-center expansion for V_A, it was found that octopole terms had to be included in order to reproduce at least qualitatively the results obtained with a more accurate V_A. There have been a number of other computations of electrostatic interaction energies using multipole expansions and a modified form of (6-19) in which E_{es} is expressed as a sum of interactions between various multipoles. This approach has especially been applied to interactions (including hydration) involving biologically active molecules, such as the DNA bases (*248*, *262*, *265*, *266*, *288*). It was also used to calculate the electrostatic contribution to the lattice energy of crystalline formic acid (*284*). Using an expansion through the octopole terms, the electrostatic energy was found to be relatively large in magnitude and to vary significantly with the orientation of the molecules in the crystal lattice. It was also observed to be a driving force for distortion of the molecules upon crystallization. When this electrostatic energy was computed using a point-charge (monopole) approximation [Mulliken population analysis, (6-4)], the results were much smaller in magnitude but showed a similar orientation dependence.[26]

6-6-3. Proton Affinities

It has already been pointed out that the molecular EP at a point is rigorously equal to the first-order term in the energy of interaction of the molecule with a proton placed at that point. This does not mean, however, that the EP will provide a good estimate of the proton affinity; on the contrary, their magnitudes often differ by a factor of 2 or more. For example, the experimentally measured proton affinities of H_2O and NH_3 are 169 and 202 kcal/mole, respectively (*149*); their most negative electrostatic potentials, on the other hand, are only -79 and -117 kcal/mole (*30*). Note, however, that while these magnitudes are certainly far apart, the difference in the measured proton affinities (33 kcal/mole) is similar to the difference in the potential minima (38 kcal/mole). This indicates that although the second- and higher-order contributions to the proton affinity, such as those due to polarization, charge transfer, etc., are very significant, their net effect is nearly the same for water and ammonia (see, e.g., Ref. 132).

The same situation has been found in a series of three-membered ring molecules (*105*). In a study that has already been discussed, Ghio and Tomasi found a good linear relationship (correlation coefficient = 0.985) between the calculated SCF proton affinities of these molecules (for fixed geometries) and the magnitudes of their potential minima. Although the latter quantities were in all instances less than half the former, the existence of this correlation means that the sum of the second- and higher-order energy terms is approximately the same for all of these molecules. If this were generally true for groups of related molecules, then the EP would be a useful means for predicting trends in proton affinities and even for estimating their magnitudes, through correlations such as the above. Unfortunately, at least two series of molecules are already known in which the magnitudes of the potential minima and the proton affinities inincrease in opposite directions (*257, 299, 300*); the first series consists of NH_3, $(CH_3)NH_2$, $(CH_3)_2NH$, and $(CH_3)_3N$, while the second is made up of H_2O, CH_3OH and $(CH_3)_2O$ (see Section 6-6-1, and Refs. 15, 47, and 50).

It does appear, however, that the location of the EP minimum may be a good indication of the equilibrium position of the proton in the protonated system. Thus the most negative potential in the neighborhood of the ethylene molecule was found to be only about -20 kcal/mole, whereas its measured proton affinity is 159 kcal/mole (*9*). However, when a proton was placed at the position of the potential minimum and the SCF total energy computed (without optimizing the geometry) to calculate the proton affinity, the result was 149 kcal/mole, quite close to the experimental value. Equally good results were obtained in this manner for aziridine and oxirane (*58*). Similarly, the EP minimum for H_2CO^- (*96*) indicates a proton position very close to that obtained in a recent ab initio SCF computation for H_2COH, in which the bond lengths and bond angles were optimized (*114*). This result is of particular significance because H_2CO^- is an open-shell system; most computations of EPs have been for closed-shell systems. Ghio and Tomasi (*105*) showed, for their series of three-membered ring molecules, that the equilibrium bond distance for an added proton as determined by an SCF total energy calculation is close (within about 0.08 Å) to that predicted by the EP.

In the present context, it is interesting to note the formal similarity between the EP approximation (6-8) to the proton affinity, which can be written in the form

$$PA(r) = -\sum_A \frac{Z_A}{|R_A - r|} + \sum_{i=1}^{N} \int \frac{\psi_A^*(r_1, r_2, \ldots, r_N)\, \psi_A(r_1, r_2, \ldots, r_N)\, dr_1\, dr_2 \cdots dr_N}{|r_i - r|} \tag{6-21a}$$

and the expression that results from the integral Hellmann–Feynman theorem (*152, 213*; see Chapter 4):

$$\mathrm{PA}(r) = -\sum_{\mathrm{A}} \frac{Z_{\mathrm{A}}}{|R_{\mathrm{A}} - r|} + \frac{\sum_{i=1}^{N} \int \psi^*_{\mathrm{AH}^+}(r_1, \ldots, r_N) \dfrac{1}{|r_i - r|} \psi_{\mathrm{A}}(r_1, \ldots, r_N)\, dr_1 \cdots dr_N}{\int \psi^*_{\mathrm{AH}^+}(r_1, \ldots, r_N)\, \psi_{\mathrm{A}}(r_1, \ldots, r_N)\, dr_1 \cdots dr_N}. \quad (6\text{-}21\mathrm{b})$$

$\mathrm{PA}(r)$ is the proton affinity for a proton located at position r. ψ_{AH^+} and ψ_{A} are the true wavefunctions of the protonated and unprotonated species, respectively. For the true wavefunctions, (6-21a) is of first-order accuracy while (6-21b) is exact (except for the assumption that the internuclear distances within A are not changed in going to AH^+). The striking similarity between the two equations can easily be seen; simply replacing the first ψ_{A} in (6-21a) with ψ_{AH^+} yields (6-21b). This single change essentially takes account of the effects, such as polarization and charge transfer, that are not included in the first-order relationship.

We have used (6-21b) in conjunction with minimum-basis SCF wavefunctions to estimate the proton affinities of several small molecules (*243*), and have obtained better values than would result from taking the difference of the computed energies corresponding to those wavefunctions, $E_{\mathrm{A}} - E_{\mathrm{AH}^+}$. For CO, for instance, (6-21b) predicted a proton affinity of 157 kcal/mole, while the energy difference was 175 kcal/mole. The experimental value is 139 kcal/mole (*149*). In these approximate calculations, an important factor is the overlap between ψ_{A} and ψ_{AH^+}; the best estimate is obtained when this overlap is maximized (for a related discussion, see Ref. 179). An important application of (6-21b) should be in connection with wavefunctions computed by semiempirical methods, such as the extended-Hückel, which do not even yield directly a total energy, so that the quantity $E_{\mathrm{A}} - E_{\mathrm{AH}^+}$ cannot even be calculated.

6-6-4. Perturbation-Theory Expansions

As has already been discussed (Sections 6-3 and 6-6-1), there is, in general, no direct relationship between the EP of a molecule M and its total energy of interaction with some other entity. First, the potential takes no account of the nature of the other entity and how it may change as the interaction proceeds, and second, the potential of the undistorted ground-state molecule M does not reflect changes, due to such effects as polarization, charge transfer, etc., that occur in M itself. If the interaction is with a point charge (such as a proton), then the first problem does not arise, since the only relevant property of a point charge is the magnitude of its charge, which is unaffected by the interaction and which can easily be incorporated into any energy calculation. But even with a point charge, such factors as polarization and charge transfer can have very significant effects upon molecule M and upon the total interaction energy.

What is rigorously exact is that the EP of molecule M at any point r is directly related to the first-order term in a perturbation-theory expansion of the total interaction energy of M with a point charge located at that position (Section 6-3). If the value of the point charge is $\pm Q$, and if the wavefunction for molecule M is in terms of molecular orbitals, then this first-order energy is [invoking (6-8)]

$$E_1(r) = \pm QV(r) = \sum_{\mathrm{A}} \frac{\pm QZ_{\mathrm{A}}}{|R_{\mathrm{A}} - r|} - \sum_i N_i \int \frac{Q\psi_i(r')\,\psi_i(r')\,dr'}{|r' - r|}, \tag{6-22}$$

where Z_{A} is the charge on nucleus A and N_i is the number of electrons in molecular orbital ψ_i.

If one is seeking a reasonable approximation to the total interaction energy ΔE of molecule M with point charge $\pm Q$, and if the first-order term E_1 is known, then a possible next step is to compute the second-order term in the expansion, E_2:

$$\Delta E = E_1 + E_2 + \cdots \tag{6-23}$$

By applying uncoupled perturbation theory within the framework of the self-consistent-field method (*21*, *289*), an expression for the second-order energy term can be obtained:

$$E_2(r) = 2 \sum_j^{\mathrm{occ}} \sum_k^{\mathrm{unocc}} \frac{\left| \left\langle \psi_k(r') \left| \frac{\pm Q}{|r' - r|} \right| \psi_j(r') \right\rangle \right|^2}{\epsilon_j - \epsilon_k}. \tag{6-24}$$

The summations are over the doubly occupied MOs ψ_j and the unoccupied MOs ψ_k of the unperturbed molecule M. ϵ_j and ϵ_k are the respective orbital energies. Equation (6-24) is applicable only to molecules having all MOs either doubly occupied or unoccupied.

The second-order energy term E_2 reflects at least a portion of the polarization of the electronic distribution of A that results from the proximity of the point charge $\pm Q$. Note that the same types of integrals are involved in computing E_2 as E_1.

There has been a limited number of investigations of the effects of including terms beyond E_1 in the perturbation-theory expansion of ΔE [see (6-23)]; most of this work has been carried out by Weinstein and co-workers. In a detailed study of the protonation of adenine (V) and 7-methyladenine, Bartlett and Weinstein (*21*) calculated E_1 and E_2 for a proton situated, in turn, near each of the possible nitrogen bonding sites. The effect of E_2 in these instances was not found to be beneficial. Experimentally, the favored protonation sites on adenine and 7-methyladenine have been found to be the N1 and N3 positions, respectively (Section 6-5-3). Bartlett and Weinstein computed the EP for both molecules, using INDO wavefunctions. These correctly predicted protonation at N1 in adenine, where the potential minimum was -117.8 kcal/mole (versus -114.5 at N7 and -109.5 at N3); in 7-methyladenine, however, the most negative potential was near N9, -133.9 kcal/mole, compared with -130.9 at N3 and -113.9 at N1. For protonation, these potentials are equal to E_1 in (6-23), since $Q = +1$. When E_2 was calculated and added to E_1, N1 was no longer predicted to be the favored site on adenine; ΔE was more negative for N7, -164.5 versus -155.5 kcal/mole. For 7-methyladenine, the inclusion of E_2 simply increased the margin between N9 and N3: $\Delta E(\text{N9}) = -171.6$, $\Delta E(\text{N3}) = -164.6$ kcal/mole. These results may reflect a weakness in the computational procedure, or they may indicate that more terms are needed in the E expansions in these two cases to describe more adequately the polarization of the molecules.

In some instances, in contrast, the E_2 term has played a key role in relating theoretical and experimental results. Van der Neut (*302*), studying the protonation of the propadiene radical anion, found that the EP alone favored an approach to the center of one of the C—C bonds; when the second-order term was taken into account, however, the middle carbon became the indicated site of interaction. This is consistent with experimental evidence cited by Van der Neut. In computing E_2, it was necessary to use a modified form of (6-24) because of the singly occupied MO of the propadiene anion. Iterative extended-Hückel wavefunctions were used in these calculations.

Bertran and co-workers (*27*, *28*) presented general forms of (6-22) and (6-24) for the case of one molecule or ion interacting with another, removing the requirement that one of them be a point charge. They investigated some interactions of several aromatic molecules and ions with two electrophiles (H^+ and NO_2^+) and one nucleophile (OH^-). Some important approximations were made, however, in evaluating E_1 and E_2; for instance, a point-charge approach, using calculated atomic charges, was used to calculate one- and two-center integrals.

As part of their study of the pharmacological activities of some hydroxytryptamines (Section 6-5-7), Weinstein et al. (*309*, *310*) utilized (6-24) and SCF wavefunctions to compute and plot E_2 in the neighborhoods of these molecules. Figure 6-15 shows a map of E_2 in a plane parallel to the indole portion of 5-hydroxytryptamine; it is the plane in which the EP of this molecule is shown in

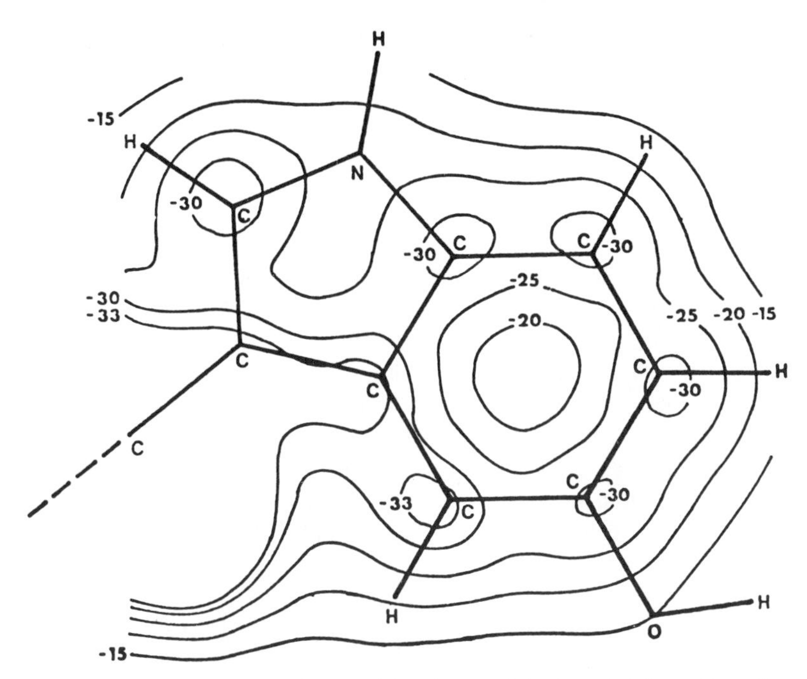

Fig. 6-15. E_2 (polarization term) plotted in plane 1·6 Å above indole portion of neutral 5-hydroxytryptamine (see structure XVIII). Values are in kcal/mole. (Reproduced from Ref. 309, courtesy John Wiley & Sons, Inc.)

Fig. 6-10. The value of E_2 at any point on this plot indicates the polarization interaction with the molecule of a +1 point charge situated at that position. It is interesting to note, with respect to the indole portion of the molecule, that this polarization map emphasizes interactions at the ring carbons, whereas the EP map (Fig. 6-10) tends more to emphasize the regions between the carbons. This is reminiscent of the results for the propadiene anion, mentioned above.

Thus far, this section has dealt primarily with the manner in which the interaction of a molecule with a point charge is affected by the accompanying polarization of the molecule. However it is also very important to consider how the interaction of a molecule with one point charge affects its reactive properties toward another one that approaches it at some other site. As Bartlett and Weinstein (*21*) pointed out, most large molecules, especially those of biological interest, have several reactive sites, and it is quite possible that the activity of each such site is affected by what occurs at the others. For instance, both experimental and theoretical studies indicate that electrophilic attack on adenine is most likely at either the N1 or the N3 positions (see structure V and the discussion in Section 6-5-3). Yet, once N1 is protonated, the favored site for a second proton is not N3 but rather N7!

Bartlett and Weinstein applied a double-perturbation approach to this problem, representing the total interaction energy by

$$\Delta E = E_{1,0} + E_{0,1} + E_{1,1} + E_{2,0} + E_{0,2} + \cdots + \frac{QQ'}{|r - r'|}. \tag{6-25}$$

$E_{1,0}$ and $E_{2,0}$ are the first- and second-order terms in the interaction energy of the molecule with a single point charge Q located at r; they can therefore be calculated with (6-22) and (6-24). Analogously, $E_{0,1}$ and $E_{0,2}$ are the first- and second-order terms corresponding to the interaction of the molecule with a single charge Q', situated at r', and can also be obtained with (6-22) and (6-24). These four terms take no account whatsoever of the fact that there are two point charges. The key term is $E_{1,1}$; it represents the effect that the presence of one point charge has upon the interaction of the molecule with the other. Bartlett and Weinstein presented a derivation of $E_{1,1}$ within the framework of the SCF method which, in terms of the uncoupled approximation, led to the result:

$$E_{1,1} = 4 \sum_{j}^{\text{occ}} \sum_{k}^{\text{unocc}} \frac{\left\langle \psi_j(r'') \left| \frac{Q}{|r'' - r|} \right| \psi_k(r'') \right\rangle \left\langle \psi_k(r'') \left| \frac{Q'}{|r'' - r'|} \right| \psi_j(r'') \right\rangle}{\epsilon_j - \epsilon_k}. \tag{6-26}$$

The quantities in (6-26) are defined exactly as they were for (6-24). Note that the integrations in (6-26) are over r''.

Bartlett and Weinstein used an INDO wavefunction to compute $E_{1,1}$ for two protons at all three possible combinations of the N1, N3, and N7 sites on adenine (V). In all instances, $E_{1,1}$ was found to be positive, indicating that a proton at any one of these sites decreases the reactivity of either of the other two toward a second proton. Of the three possible pairs of sites for diprotonation, the calculated ΔE values from (6-25) predicted the N1–N7 pair to be preferred, in agreement with experimental results (Section 6-5-3).

In the case of 7-methyladenine, in contrast, the two combinations of sites that were considered (N1–N3 and N3–N9) produced essentially the same ΔE values, −225.4 and −225.2 kcal/mole, respectively, so that no preference could be inferred. If, however, the $E_{2,0}$ and $E_{0,2}$ terms were not included in (6-25), then the N3–N9 site combination was favored, as is indeed observed experimentally (Section 6-5-3). It is noteworthy that while $E_{1,1}$ for the N1–N3 case was again positive, it was negative for the N3–N9 site pair. This means that protonation of either N3 or N9 makes the other one more susceptible to a second proton.

Bartlett and Weinstein interpret this in terms of the proximity of the two sites; protonation at one causes a shifting of electronic charge into that general region, which aids protonation at the other.

An interesting point that is brought out by Bartlett and Weinstein is the possibility of using $E_{1,1}$ to partially correct the molecular EP for the effects of monoprotonation of the molecule. They showed that the effect of a proton at site $\mathbf{r}'$ upon the adenine EP at site $\mathbf{r}$ could be taken into account reasonably well by adding the appropriate $E_{1,1}$ and $QQ'/|\mathbf{r} - \mathbf{r}'|$ terms to the EP of unprotonated adenine at site $\mathbf{r}$. This was confirmed by computing a wavefunction and EP directly for the $\mathbf{r}'$-protonated form of adenine. Of course $E_{1,1}$ and $QQ'/|\mathbf{r} - \mathbf{r}'|$ are both energy quantities and can be added to EP only if the point charge at $\mathbf{r}$ is taken to have a value of +1. Bartlett and Weinstein showed that this corrected EP properly predicted the site of attack of the second proton for both adenine and 7-methyladenine. Thus $E_{1,1}$, together with the QQ' term, permits an estimate of how a molecule's EP at any particular point is affected by a perturbing point charge.

In an important extension of this concept of a corrected EP, Weinstein (*305*) studied the effect that the interaction of adenine with a water molecule may have upon the adenine potential. The water molecule was represented by the set of thirteen point charges (Section 6-4-5) that had been developed by Bonaccorsi et al. (*33*). For three different locations of the water molecule, all of which were above the adenine plane, the corrected adenine potential was computed, using a deorthogonalized INDO wavefunction, at the positions of the potential minima near the atoms N1, N3, and N7. The method used was exactly as described above, except that instead of one perturbing point charge, there were thirteen. The second integral in (6-26), therefore, involved a summation over thirteen terms. These corrected potentials were tested by being compared to those obtained using a wavefunction for the whole adenine-H_2O complex. The agreement was good; the same trends were observed, and the difference between the values was never more than 3%. The effect of the perturbing water molecule upon the potential was invariably small, 1-7 kcal/mole. As Weinstein pointed out, however, if there are several sites on a molecule that differ relatively little in reactivity, the perturbing effect of a polar molecule such as water could conceivably alter their order of activity. The technique being discussed might therefore play an important role in such situations.

More recently, Chang et al. (*61*) have presented a generalized treatment of the multiple reactive sites problem. It involves calculating the interaction energy of the perturbed molecule with a point charge, but using an expanded representation of ΔE containing more terms than (6-25). The effect of the perturbing molecule is now taken into account more fully and more accurately. As an example, using SCF wavefunctions, they considered the interaction of formyl fluoride, HCOF, with a proton while the former was being perturbed by a hydrogen fluoride molecule. Again, comparison with SCF results for the whole HCOF·HF complex showed generally good agreement; what is particularly significant is that the added terms in the ΔE expansion invariably improved this agreement.

Of course, even this extended representation of ΔE does not exactly reproduce the SCF results for the whole complex. In order to learn whether the remaining difference is due to the truncation of the perturbation expansion or to its neglect of overlap and exchange effects between the molecules forming the complex, Weinstein et al. (*312*) repeated the $HCOF-HF-H^+$ study, using an interaction-field-modified-Hamiltonian (IFMH) method that was designed to be equivalent to an infinite perturbation expansion. They found the IFMH results to be, in general, very similar to those obtained from the truncated perturbation expansion, and concluded that these differ from the SCF results for HCOF·HF complex largely because of the neglect of overlap and exchange contributions.

Chang and Weinstein (*62*) have now expanded the multiple-reactive-sites perturbation technique to include the effect of two perturbing molecules upon the EP of a third one. They treated the case of one or two water molecules interacting with adenine (V). Looking at the three most nucleophilic sites in adenine (N1, N3, and N7), they found that monohydration at either N1 or N7 leads to less negative EPs at the other two sites (making them less susceptible to protonation), whereas monohydration at N3 results in more negative potentials near the other sites, making them more reactive toward a proton. When two water molecules were allowed to interact with the adenine at any two of the three sites being considered (N1, N3, and N7), the third became less attractive to a proton except when it was N7, which became more attractive. In general, the EPs obtained from these perturbation calculations were in good agreement with those computed from an SCF treatment of the whole monohydrated or dihydrated adenine complex. In one instance, Chang and Weinstein made an estimate of the overlap contribution that is neglected by the perturbation approach; it was found to be negligible at the adenine-water separation being used (N · · · O distance $\geqslant$2.85 Å).

6-7. ELECTROSTATIC FORCES

6-7-1. Force Fields in Molecules

It was pointed out in Section 6-3 that the negative gradient of the EP at a point in the space around a molecule is directly related to the electrostatic force that is exerted by the molecule's nuclei and electrons upon a point charge placed at the point. The direction of the gradient vector gives the direction of the force. Thus, for a point charge $\pm Q$ situated at r, this force is

$$F(r) = -\nabla[\pm QV(r)]. \tag{6-27}$$

Inserting $V(r)$ as given by (6-8), (6-27) becomes

$$F(r) = \sum_{A} \frac{\pm QZ_A(r - R_A)}{|r - R_A|^3} - \int \frac{\pm Q\rho(r')(r - r')\,dr'}{|r - r'|^3}. \tag{6-28}$$

By calculating $F(r)$ at various points, it is possible to obtain a representation of the electrostatic force field in the neighborhood of the molecule showing lines of force that indicate the path that would be followed by the charge $\pm Q$ if it were acted upon only by the electrostatic effect of the undistorted molecular charge distribution.

Such a force field should be a valuable guide to the route that would be followed by an electrophile approaching the molecule. However, there has been very little application of (6-27) to the study of chemical reactivity. One interesting example has been provided by Dovesi et al. (*90*) who used a multipole expansion of $V(r)$ to obtain the force field in the plane of the pyrimidine molecule (Fig. 6-16). The arrows indicate lines of force; these would direct an approaching electrophile away from the hydrogens and toward the nitrogen atoms.

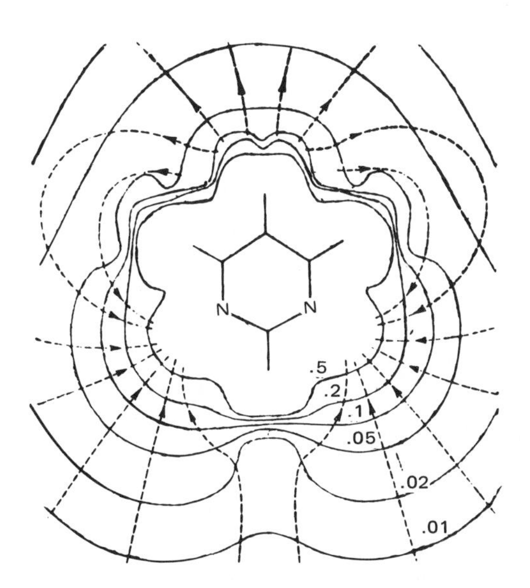

Fig. 6-16. Electrostatic force field in the molecular plane of pyrimidine. Force contours (solid curves) indicate force exerted upon +1 point charge; values are in eV/au (1.0 au = 0.529 Å). Lines of force (dashed arrows) show direction of force exerted upon positive point charge. (Reproduced from Ref. 90, courtesy The Chemical Society, London.)

6-7-2. Forces on Nuclei

According to the Hellmann-Feynman (H-F) theorem (*98*, *124*), the force exerted upon any nucleus A in a molecule by the electrons and the other nuclei can be obtained by simply applying (6-28), with $Q = Z_A$:

$$F_A(R_A) = \sum_{B \neq A} \frac{Z_A Z_B (R_A - R_B)}{|R_A - R_B|^3} - \int \frac{Z_A \rho(r')(R_A - r')\, dr'}{|R_A - r'|^3}. \qquad (6\text{-}29)$$

When the nuclei are at their equilibrium positions, the force on each one is zero.

A consideration of the forces acting upon nuclei can yield some interesting insights into bonding phenomena. For instance, Feynman (*98*) suggested in his original paper on the H-F theorem that the van der Waals forces between two neutral atoms arise from the polarization of each atom's electronic charge distribution toward the other and the subsequent "attraction of each nucleus for the distorted charge distribution of its *own* electrons. . . ." This conjecture was later confirmed (*130*) by a perturbation calculation of the long-range interaction of two hydrogen atoms, which showed that the electronic density distribution of each atom is indeed polarized toward the other.[27] They also obtained an expression for the force on either nucleus as a function of R^{-7}, from which they calculated the leading (R^{-6}) term in the interaction energy with an accuracy of four significant figures. This was a direct confirmation of the H-F theorem (see Chapter 7).

Bader and Chandra (*17*) have studied the variation in the force on each nucleus as two hydrogen atoms approach each other and form the H_2 molecule. They used the optimized-valence-configuration H_2 wavefunction (*83*) to compute the forces. Their calculated force, on either nucleus, is plotted against the internuclear distance in Fig. 2-4, which also shows the force curve obtained from an experimental potential energy curve. The agreement is very good. Since the H_2 wavefunction used by Bader and Chandra was constructed from STOs centered on nuclei A and B, the integral term in (6-29), which represents the force exerted upon nucleus A by the electrons, is composed of terms involving all the products of these atomic orbitals taken two at a time. Bader and Chandra partitioned this total electronic contribution to the force on nucleus A into three portions (see also Sections 2-3 and 2-4): (1) an atomic force, composed of those terms that involve only atomic orbitals centered on A; (2) an overlap force, arising from products of atomic orbitals on A and B; and (3) a screening force, due to those terms that involve only atomic orbitals centered on nucleus B. At larger internuclear distances, between 2.5 and 4.0 Å, the overlap force is very small while the screening force is practically equal in magnitude but opposite in direction to the repulsive force between the nuclei, so that they essentially cancel. Thus the net force on each nucleus is its atomic force, arising from the polarized nature of its own electronic density distribution and having the effect of pulling that nucleus toward the other. This is in full agreement with Feynman's idea, and with the conclusions of Hirschfelder and Eliason (*130*) mentioned earlier.

In an interesting series of papers, Nakatsuji and co-workers have applied force concepts to the analysis of various molecular properties (*151*, *196-203*). Nakatsuji also partitions the electronic force on a nucleus, in a manner somewhat similar to that of Bader and Chandra. First, there is an atomic dipole force F_{AD}, defined in exactly the same way as their atomic force. Second, Nakatsuji introduces an exchange force F_{EC}, which is taken to represent "the attraction between nucleus A and the electron distribution piled up in the region between nuclei A and B through electron exchange. . . ." Finally, there is a gross charge force F_{GC}, corresponding to the interaction of nucleus A with the gross charge on atom B. The repulsion between nuclei A and B is added into this last term (see Section 3-3 for details).

As an example of an application of this force partitioning, the interaction of two methyl radicals to form an ethane molecule has been analyzed in detail (*199*). First each radical was fixed in a planar geometry and the two were brought together along an axis connecting the two carbons and perpendicular to each CH_3 plane. The two methyl radicals were taken to have a staggered relative orientation. It was found that at all the carbon-carbon distances considered the total force on each carbon nucleus was repulsive, tending to pull it away from the other carbon. Only the term F_{EC} corresponding to exchange between the carbons was attractive, and it was insufficient to drive the reaction toward the formation of a C—C bond. The total force on each hydrogen nucleus was also repulsive, pulling it away from the other CH_3 radical. The same calculation was subsequently repeated without restricting the methyl radicals to planarity; at each stage of their approach, the hydrogens were bent back (away from the other methyl) sufficiently to make the total force on each hydrogen nucleus equal to zero. Now, the total force on each carbon nucleus was consistently attractive, pulling it toward the other carbon. The magnitude of this force, which constitutes a portion of the driving force of the reaction, naturally diminished as the nuclei approached their equilibrium separation. A major reason for this change in the total force on each carbon nucleus is that the bending back of the hydrogens produces a polarization of each carbon's electronic charge toward the other carbon, resulting in an attractive atomic dipole force F_{AD}. Another important factor which affects F_{GC} is the decrease in the repulsive interaction between the carbon nucleus on one methyl and the hydrogen nuclei on the other when the latter are bent back. Thus, a force analysis such as the preceding can indicate how the driving force of a reaction is affected by various factors, such as changes in geometry.[28]

A somewhat similar approach (*300*) analyzed the interactions of some molecules to form donor-acceptor complexes. These authors decomposed the total force of attraction bringing two molecules together into contributions reflecting electrostatic effects, polarization, exchange, etc., in order to gain insight into the factors responsible for bringing the molecules together in each case (see also Ref. 143).[28]

A rather unusual but very promising application of the force concept is the gradient method for determining the structures of the reactants, transition states, and products of reactions (*161*, *162*, *184*). This approach is based on the fact that the gradient vector of the potential energy of a reacting system is zero at those points on the energy surface that correspond to the initial reactants, the final products and any transition states. The procedure, therefore, is to compute ∇E for various configurations of the reacting system and seek those for which $\nabla E = 0$. Since the direction of the vector ∇E is the direction of steepest descent of E (which is intrinsically negative), then by following the direction indicated by each computed ∇E the system is necessarily moving toward the nearest minimum in E. Thus the minima can be located much more efficiently than if the entire potential energy surface, or even a large portion of it, had to be computed. Furthermore, it is easier to locate transition states with the gradient, since they correspond to zero values, whereas they are only saddle points (not minima) on the energy surface. ∇E is of course a force, and the magnitude of the gradient vector can be regarded as a sum of internal forces within the reacting system that are operating to bring it to an equilibrium or transition-state geometry.

Komornicki and McIver (*161*) have used the above procedure, only slightly modified, to study several reactions, using semiempirical wavefunctions. More recently the method was applied to the HNC → HCN isomerization (*137*, *162*), using SCF wavefunctions. Geometries were obtained for both HNC and HCN, as well as for the transition state, which was found to be a three-membered ring. The process was predicted to be exothermic by 19.3 kcal, with an activation barrier of 55.4 kcal, the latter being in good agreement with the result of a configuration-interaction calculation, which yielded a value of 49.5 kcal (*215*).

6-8. SUMMARY

Our primary aim in this chapter has been to convey some feeling for the effectiveness of the electrostatic potential (EP) as a guide to the reactive properties of molecules. We have attempted to present a balanced picture, showing both successes and failures. The EP is a real physical property, with certain definite physical meanings (Section 6-3). Considerable caution should be exercised in trying to broaden its interpretation. For instance, it should not in general be anticipated that the potential will serve as a direct measure of interaction energies. The EP can, however, reveal the regions on a molecule that are most susceptible to electrophilic attack, and it can give some indication of their relative reactivities toward electrophiles. This should make it especially useful in investigations that involve large molecules, which may have several competing reactive sites. In summary, therefore, we have found that, on the whole, the EP can indeed be a very effective approach to the study of molecular reactivity.

Notes

1. This procedure has been formulated as a solution to the Poisson equation (*286*; see also Ref. 191).

2. In this chapter, the term *self-consistent-field* will be applied to a wavefunction only if it is also ab initio.

3. There is no obvious pattern; the better wavefunction sometimes has a more negative minimum, other times a less negative one.

4. One hartree (h) = 27.21 electron-volts (eV) = 627 kcal/mole.

5. In the case of HNO_2, going from a minimum-basis-set to a somewhat better one changed the relative ordering (although not the locations) of three of the four minima associated with the molecule. However, these had initially been within 6 kcal/mole of each other (*82*, *95*).

6. A possible intermediate level has been examined by Osman and Weinstein (*209*), who computed the EPs for several indole-related systems using wavefunctions that had been computed by a pseudopotential method developed by Ewig et al. (*97*). This procedure, in which the atomic cores are replaced by effective potentials, required considerably less computer time and core than the corresponding minimum-basis-set ab initio SCF calculation and produced potential maps that are very similar in their general features to the ab initio results.

7. In the CNDO/2 method, as well as in other semiempirical procedures that consider only the valence electrons explicitly, the core electrons are regarded as part of an effective nucleus having a positive charge $Z' = Z - N_{core}$, where N_{core} is the number of core electrons. Thus, Z' must replace Z in (6-10).

8. In reality, of course, the CNDO/2 basis orbitals are not orthogonal; the overlaps between them are simply neglected in computing the CNDO/2 wavefunction. The suggestion that the latter may be regarded as composed of orthogonalized atomic orbitals was made by Pople and Segal (*245*).

9. A comparison of potentials obtained with deorthogonalized and nondeorthogonalized INDO wavefunctions was given by Weinstein et al. (*308*).

10. For n STOs, each written as a sum of m Gaussian functions, the total number of integrals for one point is $n(n+1)(m^2)/2$.

11. For the ensuing discussion, the total net charge, $\Sigma_A Z_A - N$, will be called the *monopole strength*.

12. Various localization procedures have been reviewed (*306*).

13. For instance, Swissler and Rein (*294*) report that the calculation of all H_2O and CH_4 multipole moment components through octopole, both single- and many-center, required less than 10 seconds on the CDC 6400 computer; calculating the iterative extended-Hückel wavefunctions that were used took 4.5 seconds. Bonaccorsi et al. (*37*) found that the time required to determine $V(r)$

at one point in the vicinity of the pyridine molecule, using an SCF wavefunction and an IBM 360/67 computer, was 2.7 seconds with (6-10) and rigorous evaluation of all integrals, and only 0.01 seconds with a multipole expansion carried through the octopole terms.

14. Adenine exists in two tautomeric forms, in one of which N7 has a hydrogen and in the other, N9. The equilibrium constant between these tautomers, $K = [N7H]/[N9H]$, has been estimated to be 0.28 at 20°C (*91*).

15. It must always be kept in mind, however, that the order of preference depends to some extent on the electrophile, as well as on the EP of the substrate. For example, the binding preferences of an Na^+ ion to the nucleic acid bases (*218*), do not exactly follow the pattern that would be predicted from a consideration of the potentials alone.

16. Basically similar potentials are associated with the phenyl rings in phenol, C_6H_5OH, and anisole, $C_6H_5OCH_3$ (*223*).

17. The fact that the pyridazine minimum is more negative than that of pyridine may reflect, in the former molecule, an enhancement of each nitrogen's negative potential by that of its neighbor.

18. For more examples, see Section 6-6-2.

19. Note that XXI contains a quaternary nitrogen even without protonation. Its potential was calculated with an anion (NH_2^- or OH^-) near this nitrogen, representing the anionic site on the biological receptor.

20. It should be mentioned that Catalan and Yanez (*58*), using larger basis sets than those of Bonaccorsi et al., did not find a negative potential associated with the C—C bond in oxirane, $(CH_2)_2O$, although the positive potential did reach a minimum value there. They did find a negative C—C potential in aziridine, $(CH_2)_2NH$.

21. Negative methyl potentials have also been obtained for propene, with an INDO wavefunction (H. Weinstein, private communication) and for the singlet and triplet $\pi \rightarrow \pi^*$ excited states of thymine (Section 6-5-9). In addition, and particularly interesting, the group potential of CH_3 (Section 6-4-6) has a negative region analogous to that described above (*40*).

22. A similar success has been reported recently for the interaction of the radical anion H_2CO^- with H^+ (*96*).

23. Each of these three energies was negative.

24. It is well known that the aqueous-solution basicities of the methylamines vary in a markedly different manner from the gas-phase proton affinities. This has been discussed in detail and analyzed in terms of electrostatic effects by Aue et al. (*14*).

25. In these calculations, the distance between the oxygen atoms was kept at 2.74 Å and only the relative orientations of the molecules were varied. The method could not be used to determine the optimum O—O separation; E_{es} simply became monotonically more negative as this separation decreased.

26. Recently, Bertaut (*22*) has discussed the use of series expansions in Fourier space for calculating the EPs in crystal lattices.

27. Many elementary textbooks incorrectly attribute van der Waals forces to the formation of instantaneous atomic dipoles and the subsequent interaction of the positive portion of one with the negative part of another.

28. The reaction $NH_2 + H \rightarrow NH_3$ has recently been studied in a similar manner (*203*).

References

1. Adamowicz, L. and Sadlej, J., 1977, *Adv. Mol. Relax. Interact. Processes*, **10**, 283.
2. Alagona, G., Cimiraglia, R., Scrocco, E., and Tomasi, J., 1972, *Theoret. Chim. Acta* (Berlin), **25**, 103.
3. ——, Pullman, A., Scrocco, E., and Tomasi, J., 1973, *Int. J. Peptide Protein Res.*, **5**, 251.
4. Albert, A., 1968, *Heterocyclic Chemistry*, 2nd ed. (Oxford University Press, New York).
5. Almlöf, J., Johansen, H., Roos, B., and Wahlgren, U., 1972, in USIP Report 72-16, University of Stockholm, Institute of Physics.
6. ——, Johansen, H., Roos, B., and Wahlgren, U., 1973, *J. Electron Spectrosc. Related Phen.*, **2**, 51.
7. ——, Henriksson-Enflo, A., Kowalewski, J., and Sundbom, M., 1973, *Chem. Phys. Lett.*, **21**, 560.
8. ——, Kvick, A., and Thomas, J. O., 1973, *J. Chem. Phys.*, **59**, 3901.
9. —— and Støgard, A., 1974, *Chem. Phys. Lett.*, **29**, 418.
10. ——, Haselbach, E., Jachimowiez, F., and Kowalewski, J., 1975, *Helv. Chim. Acta*, **58**, 2403.
11. Arcos, J. C. and Argus, M. F., 1974, in *Chemical Induction of Cancer*, vol. IIA (Academic Press, New York), p. 135.
12. Armbruster, A. M. and Pullman, A., 1974, *FEBS Lett.*, **49**, 18.
13. Arnett, E. M., 1973, *Acc. Chem. Res.*, **6**, 404.
14. Aue, D. H., Webb, H. M., and Bowers, M. T., 1976, *J. Am. Chem. Soc.*, **98**, 318.
15. ——, Webb, H. M., and Bowers, M. T., 1972, *J. Am. Chem. Soc.*, **94**, 4726.
16. Aung, S., Pitzer, R. M., and Chan, S. I., 1968, *J. Chem. Phys.*, **49**, 2071.
17. Bader, R. F. W. and Chandra, A. K., 1968, *Can. J. Chem.*, **46**, 953.
18. ——, Beddall, P. M., and Cade, P. E., 1971, *J. Am. Chem. Soc.*, **93**, 3095.
19. Barnett, M. P. and Coulson, C. A., 1951, *Phil. Trans. Roy. Soc.* (London), **243**, 221.
20. ——, 1963, in *Methods in Computational Physics*, vol. 2, eds. B. Alder, S. Fernbach, and M. Rotenberg (Academic Press, New York), p. 95.
21. Bartlett, R. J. and Weinstein, H., 1975, *Chem. Phys. Lett.*, **30**, 441.
22. Bertaut, E. P., 1978, *J. Phys. Chem. Solids*, **39**, 97.
23. Berthier, G., Bonaccorsi, R., Scrocco, E., and Tomasi, J., 1972, *Theoret. Chim. Acta* (Berlin), **26**, 101.
24. Berthod, H. and Pullman, A., 1975, *Chem. Phys. Lett.*, **32**, 233.
25. —— and Pullman, A., 1977, *Chem. Phys. Lett.*, **46**, 249.
26. —— and Pullman, A., 1978, *Theoret. Chim. Acta* (Berlin), **47**, 59.
27. Bertran, J., Silla, E., Carbo, R., and Martin, M., 1975, *Chem. Phys. Lett.*, **31**, 267.

28. ——, Silla, E., and Fernandez-Alonso, J. I., 1975, *Tetrahedron*, **31**, 1093.
29. Bishop, D. M., 1967, in *Advances in Quantum Chemistry*, vol. 3, ed. P.-O. Löwdin (Academic Press, New York), p. 25.
30. Bonaccorsi, R., Scrocco, E., and Tomasi, J., 1970, in Theoretical Section Progress Report 1969/1970, Laboratorio di Chimica Quantistica ed Energetica Molecolare del C.N.R. ed Istituto di Chimica-Fisica dell'Università di Pisa, Pisa, Italy, p. 35.
31. ——, Scrocco, E., and Tomasi, J., 1970, *J. Chem. Phys.*, **52**, 5270.
32. ——, Petrongolo, C., Scrocco, E., and Tomasi, J., 1970, in *Quantum Aspects of Heterocyclic Compounds in Chemistry and Biochemistry*, eds. E. D. Bergmann and B. Pullman (Academic Press, New York), p. 181.
33. ——, Petrongolo, C., Scrocco, E., and Tomasi, J., 1971, *Theoret. Chim. Acta* (Berlin), **20**, 331.
34. ——, Scrocco, E., and Tomasi, J., 1971, *Theoret. Chim. Acta* (Berlin), **21**, 17.
35. ——, Pullman, A., Scrocco, E., and Tomasi, J., 1972, *Theoret. Chim. Acta* (Berlin), **24**, 51.
36. ——, Pullman, A., Scrocco, E., and Tomasi, J., 1972, *Chem. Phys. Lett.*, **12**, 622.
37. ——, Cimiraglia, R., Scrocco, E., and Tomasi, J., 1974, *Theoret. Chim. Acta* (Berlin), **33**, 97.
38. ——, Scrocco, E., and Tomasi, J., 1974, in *Chemical and Biochemical Reactivity*, eds. E. D. Bergmann and B. Pullman (D. Reidel, Dordrecht, Holland), p. 387.
39. ——, Scrocco, E., Tomasi, J., and Pullman, A., 1975, *Theoret. Chim. Acta* (Berlin), **36**, 339.
40. ——, Scrocco, E., and Tomasi, J., 1976, *J. Am. Chem. Soc.*, **98**, 4049.
41. ——, Scrocco, E., and Tomasi, J., 1976, *Theoret. Chim. Acta* (Berlin), **43**, 63.
42. ——, Scrocco, E., and Tomasi, J., 1977, *J. Am. Chem. Soc.*, **99**, 4546.
43. Boyland, E., 1964, *Brit. Med. Bull.*, **20**, 121.
44. Boys, S. F., 1950, *Proc. Roy. Soc.* (London), **A200**, 542.
45. ——, 1960, *Proc. Roy. Soc.* (London), **A258**, 402.
46. Bratož, S., 1967, in *Advances in Quantum Chemistry*, vol. 3, ed. P.-O. Löwdin (Academic Press, New York), p. 209.
47. Brauman, J. I., Riveros, J. M., and Blair, L. K., 1971, *J. Am. Chem. Soc.*, **93**, 3914.
48. Breon, T-L., Petersen, H., and Paruta, A. N., 1978, *J. Pharmaceut. Sci.*, **67**, 67.
49. ——, Petersen, H., and Paruta, A. N., 1978, *J. Pharmaceut. Sci.*, **67**, 73.
50. Briggs, J. P., Yamdagmi, R., and Kebarle, P., 1972, *J. Am. Chem. Soc.*, **94**, 5128.
51. Brookes, P. and Lawley, P. D., 1962, *J. Chem. Soc.*, 1348.
52. Brown, R. D., 1964, in *Molecular Orbitals in Chemistry, Physics, and Biology*, eds. P.-O. Löwdin and B. Pullman (Academic Press, New York), p. 485.

53. Bryan, R. F. and Tomita, K., 1962, *Acta Cryst.*, **15**, 1179.
54. Caballol, R., Gallifa, R., Martin, M., and Carbo, R., 1974, *Chem. Phys. Lett.*, **25**, 89.
55. ——, Carbo, R., and Martin, M., 1974, *Chem. Phys. Lett.*, **28**, 422.
56. ——, Gallifa, R., Martin, M., and Carbo, R., 1976, in *Proceedings of the Conference on Excited States of Biological Molecules, Lisbon 1974*, (John Wiley & Sons, New York), p. 66.
57. Carbo, R. and Martin, M., 1975, *Int. J. Quantum Chem.*, **9**, 193.
58. Catalan, J. and Yanez, M., 1978, *J. Am. Chem. Soc.*, **100**, 1398.
59. Cavalieri, L. F. and Rosenberg, B. H., 1957, *J. Am. Chem. Soc.*, **79**, 5352.
60. Chalvet, O., Decoret, C., and Royer, J., 1976, *Tetrahedron*, **32**, 2927.
61. Chang, S. Y., Weinstein, H., and Chou, D., 1976, *Chem. Phys. Lett.*, **42**, 145.
62. —— and Weinstein, H., 1978, *Int. J. Quantum Chem.*, **14**, 801.
63. Charton, M., 1970, in *Chemistry of the Alkenes*, vol. 3, ed. J. Zabicky (Interscience, New York), Chapter 10.
64. Chou, D. and Weinstein, H., 1978, *Tetrahedron*, **34**, 275.
65. Christensen, J. J., Rytting, J. H., and Izatt, R. M., 1970, *Biochem.*, **9**, 4907.
66. Christensen, D. H., Kortzeborn, R. N., Bak, B., and Led, J. J., 1970, *J. Chem. Phys.*, **53**, 3912.
67. Cimiraglia, R. and Tomasi, J., 1977, *J. Am. Chem. Soc.*, **99**, 1135.
68. Clark, D. T. and Adams, D. B., 1973, *Tetrahedron*, **29**, 1887.
69. Clementi, E., Clementi, H., and Davis, D. R., 1967, *J. Chem. Phys.*, **46**, 4725.
70. ——, 1967, *J. Chem. Phys.*, **47**, 2323.
71. Cooksey, A. R., Morgan, K. J., and Morrey, D. P., 1970, *Tetrahedron*, **26**, 5101.
72. Coulson, C. A. and Moffitt, W. E., 1949, *Phil. Mag.*, **40**, 1.
73. Cowell, J. W. and Ledwith, A., 1970, *Quart. Rev. Chem. Soc.*, **24**, 119.
74. Craig, M. and Isenberg, I., 1970, *Biopolymers*, **9**, 689.
75. Cremaschi, P., Gamba, A., and Simonetta, M., 1975, *Theoret. Chim. Acta* (Berlin), **40**, 303.
76. ——, Gamba, A., Morosi, G., Oliva, C., and Simonetta, M., 1975, *J. Chem. Soc. Faraday II*, **71**, 1829.
77. ——, Gamba, A., Morosi, G., and Simonetta, M., 1976, *Theoret. Chim. Acta* (Berlin), **41**, 177.
78. Cromwell, N. H. and Graff, M. A., 1952, *J. Org. Chem.*, **17**, 414.
79. Cunningham, G. L., Jr., Boyd, A. W., Myers, R. J., Gwinn, W. D., and Le Van, W. I., 1951, *J. Chem. Phys.*, **19**, 676.
80. Cusachs, L. C. and Politzer, P., 1968, *Chem. Phys. Lett.*, **1**, 529.
81. Daly, J. W., Jerina, D. M., and Witkop, B., 1972, *Experientia*, **28**, 1129.
82. Dargelos, A., El Ouadi, S., Liotard, D., Chaillet, M., and Elguero, J., 1977, *Chem. Phys. Lett.*, **51**, 545.
83. Das, G. and Wahl, A. C., 1966, *J. Chem. Phys.*, **44**, 87.
84. Daudel, P., Duquesne, M. D., Vigny, P., Grover, P. L., and Sims, P., 1975, *FEBS Lett.*, **57**, 250.

85. Daudel, R., LeRouzo, H., Cimiraglia, R., and Tomasi, J., 1978, *Int. J. Quantum. Chem.*, **13**, 537.
86. Del Bene, J. E. and Marchese, F. T., 1973, *J. Chem. Phys.*, **58**, 926.
87. ——, 1974, *Chem. Phys. Lett.*, **24**, 203.
88. Dill, J. D., Allen, L. C., Topp, W. C., and Pople, J. A., 1975, *J. Am. Chem. Soc.*, **97**, 7220.
89. Doggett, G., 1969, *J. Chem. Soc.*, A, 229.
90. Dovesi, R., Pisani, C., Ricca, F., and Roetti, C., 1974, *J. Chem. Soc. Faraday II*, **70**, 1381.
91. Dreyfus, M., Dodin, G., Bensaude, O., and Dubois, J. E., 1975, *J. Am. Chem. Soc.*, **97**, 2369.
92. Dyer, J. R., 1965, *Applications of Absorption Spectroscopy of Organic Compounds* (Prentice-Hall, Englewood Cliffs, N.J.), p. 86.
93. Dyke, T. R., Howard, B. J., and Klemperer, W., 1972, *J. Chem. Phys.*, **56**, 2442.
94. ——, Mack, K. M., and Muenter, J. S., 1977, *J. Chem. Phys.*, **66**, 498.
95. Edwards, W. D. and Weinstein, H., 1978, *Chem. Phys. Lett.*, **56**, 582.
96. Ellinger, Y., Subra, R., Berthier, G., and Tomasi, J., 1975, *J. Phys. Chem.*, **79**, 2440.
97. Ewig, C. S., Osman, R., and Van Wazer, J. R., 1977, *J. Chem. Phys.*, **66**, 3557.
98. Feynman, R. P., 1939, *Phys. Rev.*, **56**, 340.
99. Fiskin, A. M. and Beer, M., 1965, *Biochem.*, **4**, 1289.
100. Friedman, O. M., Mahapatra, G. N., and Stevenson, R., 1963, *Biochim. Biophys. Acta*, **68**, 144.
101. Fukui, K., 1964, in *Molecular Orbitals in Chemistry, Physics, and Biology*, eds. P.-O. Löwdin and B. Pullman (Academic Press, New York), p. 513.
102. ——, 1965, in *Modern Quantum Chemistry*, Istanbul Lectures, Part I, ed. O. Sinanoglu (Academic Press, New York), p. 49.
103. ——, 1971, *Acc. Chem. Res.*, **4**, 57.
104. Gelius, U., Roos, B., and Siegbahn, P., 1972, *Theoret. Chim. Acta* (Berlin), **27**, 171.
105. Ghio, C. and Tomasi, J., 1973, *Theoret. Chim. Acta* (Berlin), **30**, 151.
106. ——, Scrocco, E., and Tomasi, J., 1976, in *Environmental Effects of Molecular Structure and Properties*, ed. B. Pullman (D. Reidel, Dordrecht, Holland), p. 329.
107. Giessner-Prettre, C. and Pullman, A., 1968, *Theoret. Chim. Acta* (Berlin), **11**, 159.
108. —— and Pullman, A., 1971, *C. R. Acad. Sci. Paris*, **272C**, 750.
109. —— and Pullman, A., 1972, *Theoret. Chim. Acta* (Berlin), **25**, 83.
110. —— and Pullman, A., 1974, *Theoret. Chim. Acta* (Berlin), **33**, 91.
111. —— and Pullman, A., 1975, *Theoret. Chim. Acta* (Berlin), **37**, 335.
112. Goldblum, A. and Pullman, B., 1978, *Theoret. Chim. Acta* (Berlin), **47**, 345.
113. Greenwood, H. H. and McWeeny, R., 1966, in *Advances in Physical Organic Chemistry*, vol. 4, ed. V. Gold (Academic Press, New York), p. 73.

114. Ha, T. K., 1975, *Chem. Phys. Lett.*, **30**, 379.
115. Haddow, A., 1957, in *Proceedings of the Second Canadian Cancer Research Conference*, ed. R. W. Begg (Academic Press, New York), p. 361.
116. Hall, G. G., 1961, *Phil. Mag.*, **6**, 249.
117. ——, 1973, *Chem. Phys. Lett.*, **20**, 501.
118. Hariharan, P. C., Lathan, W. A., and Pople, J. A., 1972, *Chem. Phys. Lett.*, **14**, 385.
119. Hart, B. T., 1973, *Austral. J. Chem.*, **26**, 461.
120. Hartman, A. and Hirshfeld, F. L., 1966, *Acta Cryst.*, **20**, 80.
121. Hayes, D. M. and Kollman, P. A., 1976, *J. Am. Chem. Soc.*, **98**, 3335.
122. —— and Kollman, P. A., 1976, *J. Am. Chem. Soc.*, **98**, 7811.
123. Hehre, W. J., Stewart, R. F., and Pople, J. A., 1969, *J. Chem. Phys.*, **51**, 2657.
124. Hellmann, H., 1937, *Einführung in die Quantenchemie* (Franz Deuticke, Leipzig).
125. Hermann, R. B., 1968, *Int. J. Quantum Chem.*, **2**, 165.
126. Herzberg, G., 1966, *Electronic Spectra of Polyatomic Molecules* (D. Van Nostrand, Princeton, N.J.).
127. Hinton, J. F., Beeler, A., Harpool, D., Griggs, R. W., and Pullman, A., 1977, *Chem. Phys. Lett.*, **47**, 411.
128. Hirshfeld, F. L., 1977, *Theoret. Chim. Acta* (Berlin), **44**, 129.
129. Hirschfelder, J. O., Curtiss, C. F., and Bird, R. B., 1954, *Molecular Theory of Gases and Liquids* (John Wiley & Sons, New York), p. 839.
130. —— and Eliason, M. A., 1967, *J. Chem. Phys.*, **47**, 1164.
131. Hodgman, C. D., ed., 1961, *Handbook of Chemistry and Physics*, 43rd ed. (Chemical Rubber Publishing Co., Cleveland, Ohio).
132. Hopkinson, A. C., Holbrook, N. K., Yates, K., and Csizmadia, I. G., 1968, *J. Chem. Phys.*, **49**, 3596.
133. —— and Csizmadia, I. G., 1973, *Can. J. Chem.*, **51**, 1432.
134. —— and Csizmadia, I. G., 1974, *Theoret. Chim. Acta* (Berlin), **34**, 93.
135. ——, Lien, M. H., Yates, K., and Csizmadia, I. G., 1977, *Theoret. Chim. Acta* (Berlin), **44**, 385.
136. Huggins, Ch. and Yang, N. C., 1962, *Science*, **137**, 257.
137. Ishida, K., Morokuma, K., and Komornicki, A., 1977, *J. Chem. Phys.*, **66**, 2153.
138. Iwata, S. and Morokuma, K., 1973, *J. Am. Chem. Soc.*, **95**, 7563.
139. Janda, K. C., Steed, J. M., Novick, S. E., and Klemperer, W., 1977, *J. Chem. Phys.*, **67**, 5162.
140. Jardetzky, O., Pappas, P., and Wade, N. G., 1963, *J. Am. Chem. Soc.*, **85**, 1957.
141. Joesten, M. D. and Schaad, L. J., 1974, *Hydrogen Bonding* (Marcel Dekker, New York).
142. Jones, A. R., 1970, in *Advances in Heterocyclic Chemistry*, vol. 11, eds. A. R. Katritzky and A. J. Boulton (Academic Press, New York), p. 383.
143. Joshi, B. D. and Morokuma, K., 1977, *J. Chem. Phys.*, **67**, 4880.
144. Jug, K., 1969, *Theoret. Chim. Acta* (Berlin), **14**, 91.

145. Julg, A., 1975, in *Topics in Current Chemistry*, vol. 58, (Springer-Verlag, Berlin).
146. Kapitulnik, J., Levin, W., Conney, A. H., Yagi, H., and Jerina, D. M., 1977, *Nature*, **266**, 378.
147. Kapuler, A. M. and Michelson, A. M., 1971, *Biochim. Biophys. Acta*, **232**, 436.
148. Katritzky, A. R. and Lagowski, J. M., 1968, *The Principles of Heterocyclic Chemistry* (Academic Press, New York).
149. Kebarle, P., 1977, *Ann. Rev. Phys. Chem.*, **28**, 445.
150. Kilpatrick, J. E. and Spitzer, R., 1946, *J. Chem. Phys.*, **14**, 463.
151. Koga, T. and Nakatsuji, H., 1976, *Theoret. Chim. Acta* (Berlin), **41**, 119.
152. Kim, H. J. and Parr, R. G., 1964, *J. Chem. Phys.*, **41**, 2892.
153. Kirby, G. H. and Miller, K., 1971, *J. Mol. Struc.*, **8**, 373.
154. Kistenmacher, T. J. and Shigematsu, T., 1974, *Acta Cryst.*, **B30**, 166.
155. —— and Shigematsu, T., 1974, *Acta Cryst.*, **B30**, 1528.
156. —— and Shigematsu, T., 1975, *Acta Cryst.*, **B31**, 211.
157. ——, Shigematsu, T., and Weinstein, H., 1975, *J. Mol. Struc.*, **25**, 125.
158. Klopman, G., 1968, *J. Am. Chem. Soc.*, **90**, 223.
159. Kollman, P. A. and Allen, L. C., 1972, *Chem. Rev.*, **72**, 283.
160. ——, McKelvey, J., Johansson, A., and Rothenberg, S., 1975, *J. Am. Chem. Soc.*, **97**, 955.
161. Komornicki, A. and McIver, J. W., Jr., 1974, *J. Am. Chem. Soc.*, **96**, 5798.
162. ——, Ishida, K., and Morokuma, K., 1977, *Chem. Phys. Lett.*, **45**, 595.
163. Koutecky, J., Zahradnik, R., and Cizek, J., 1961, *Trans. Faraday Soc.*, **57**, 169.
164. Lawley, P. D., 1957, *Biochim. Biophys. Acta*, **26**, 450.
165. ——, 1957, *Proc. Chem. Soc.*, 290.
166. ——, 1966, *Progr. Nucleic Acid Res. Mol. Biol.*, **5**, 89.
167. Leroy, G. and Louterman-Leloup, G., 1975, *J. Mol. Struc.*, **28**, 33.
168. —— and Sana, M., 1975, *Tetrahedron*, **31**, 2091.
169. ——, Louterman-Leloup, G., and Ruelle, P., 1976, *Bul. Soc. Chim. Belg.*, **85**, 205.
170. ——, Louterman-Leloup, G., and Ruelle, P., 1976, *Bul. Soc. Chim. Belg.*, **85**, 219.
171. ——, Louterman-Leloup, G., and Ruelle, P., 1976, *Bul. Soc. Chim. Belg.*, **85**, 229.
172. Levine, A. F., Fink, L. M., Weinstein, I. B., and Grunberger, D., 1974, *Cancer Research*, **34**, 319.
173. Levy, M., Stevens, W. J., Shull, H., and Hagstrom, S., 1974, *J. Chem. Phys.*, **61**, 1844.
174. Loew, G. H., Berkowitz, D., Weinstein, H., and Srebrenik, S., 1974, in *Molecules and Quantum Pharmacology*, eds. E. D. Bergmann and B. Pullman (D. Reidel, Dordrecht, Holland), p. 335.
175. ——, Weinstein, H., and Berkowitz, D., 1976, in *Environmental Effects on Molecular Structure and Properties*, ed. B. Pullman (D. Reidel, Dordrecht, Holland), p. 239.

176. Loveless, A., 1969, *Nature*, **223**, 206.
177. Löwdin, P.-O., 1950, *J. Chem. Phys.*, **18**, 365.
178. ——, 1953, *J. Chem. Phys.*, **21**, 374.
179. Lowe, J. P. and Mazziotti, A., 1968, *J. Chem. Phys.*, **48**, 877.
180. Margenau, H. and Murphy, G. M., 1956, *The Mathematics of Physics and Chemistry*, 2nd ed. (D. Van Nostrand, Princeton, N.J.), p. 98.
181. Marino, G., 1971, in *Advances in Heterocyclic Chemistry*, vol. 13, eds. A. R. Katritzky and A. J. Boulton (Academic Press, New York), p. 235.
182. Martin, M., Carbo, R., Petrongolo, C., and Tomasi, J., 1975, *J. Am. Chem. Soc.*, **97**, 1338.
183. McGlynn, S. P., Vanquickenborne, L. G., Kinoshita, M., and Carroll, D. G., 1972, *Introduction to Applied Quantum Chemistry* (Holt, Rinehart and Winston, Inc., New York).
184. McIver, J. W., Jr. and Komornicki, A., 1971, *Chem. Phys. Lett.*, **10**, 303.
185. Mely, B. and Pullman, A., 1969, *Theoret. Chim. Acta* (Berlin), **13**, 278.
186. —— and Pullman, A., 1972, *C. R. Acad. Sci. Paris*, **274C**, 1371.
187. Miller, J. A., 1970, *Cancer Research*, **30**, 559.
188. Mo, O. and Yanez, M., 1978, *Theoret. Chim. Acta* (Berlin), **47**, 263.
189. Møller, C. and Plesset, M. S., 1934, *Phys. Rev.*, **46**, 618.
190. Momany, F. A., 1978, *J. Phys. Chem.*, **82**, 592.
191. Morgan, J. van W., 1977, *J. Phys.*, **C10**, 1181.
192. Morokuma, K., 1971, *J. Chem. Phys.*, **55**, 1236.
193. Morrison, R. T. and Boyd, R. N., 1973, *Organic Chemistry*, Chapter 31, (Allyn and Bacon, Inc., Boston, Mass).
194. Mulliken, R. S., 1949, *J. Chim. Phys.*, **46**, 675.
195. ——, 1955, *J. Chem. Phys.*, **23**, 1833.
196. Nakatsuji, H., 1973, *J. Am. Chem. Soc.*, **95**, 345.
197. ——, 1973, *J. Am. Chem. Soc.*, **95**, 354.
198. ——, 1973, *J. Am. Chem. Soc.*, **95**, 2084.
199. ——, Kuwata, T., and Yoshida, A., 1973, *J. Am. Chem. Soc.*, **95**, 6894.
200. ——, 1974, *J. Am. Chem. Soc.*, **96**, 30.
201. —— and Koga, T., 1974, *J. Am. Chem. Soc.*, **96**, 6000.
202. ——, Koga, T., and Yonezawa, T., 1978, *J. Am. Chem. Soc.*, **100**, 1029.
203. ——, Matsuda, K., and Yonezawa, T., 1978, *Chem. Phys. Lett.*, **54**, 347.
204. Oie, T., Maggiora, G. M., and Christoffersen, R. E., 1976, *Int. J. Quantum Chem., Quantum Biol. Symp. No. 3*, ed. P.-O. Löwdin (John Wiley & Sons, New York), p. 119.
205. Olah, G. A., Schlosberg, R. H., Kelley, D. P., and Mateescu, G. D., 1970, *J. Am. Chem. Soc.*, **92**, 2546.
206. ——, Schlosberg, R. H., Porter, R. D., Mo, Y. K., Kelley, D. P., and Mateescu, G. D., 1972, *J. Am. Chem. Soc.*, **94**, 2034.
207. Orita, Y. and Pullman, A., 1977, *Theoret. Chim. Acta* (Berlin), **46**, 251.
208. —— and Pullman, A., 1977, *Theoret. Chim. Acta* (Berlin), **45**, 257.
209. Osman, R. and Weinstein, H., 1977, *Chem. Phys. Lett.*, **49**, 69.
210. Pack, G. R., Wang, H., and Rein, R., 1972, *Chem. Phys. Lett.*, **17**, 381.
211. Pal, B. C., 1962, *Biochem.*, **1**, 558.
212. Palke, W. E. and Lipscomb, W. N., 1966, *J. Am. Chem. Soc.*, **88**, 2384.

213. Parr, R. G., 1964, *J. Chem. Phys.*, **40**, 3726.
214. Paster, Z., Maayani, S., Weinstein, H., and Sokolovsky, M., 1974, *Eur. J. Pharmacol.*, **25**, 270.
215. Pearson, Schaefer, H. F., III, and Wahlgren, U., 1975, *J. Chem. Phys.*, **62**, 350.
216. Perahia, D., Pullman, A., and Berthod, H., 1975, *Theoret. Chim. Acta* (Berlin), **40**, 47.
217. ——, Pullman, A., and Pullman, B., 1976, *Theoret. Chim. Acta* (Berlin), **42**, 23.
218. ——, Pullman, A., and Pullman, B., 1977, *Theoret. Chim. Acta* (Berlin), **43**, 207.
219. —— and Pullman, A., 1978, *Theoret. Chim. Acta* (Berlin), **48**, 263.
220. Petrongolo, C. and Tomasi, J., 1973, *Chem. Phys. Lett.*, **20**, 201.
221. —— and Tomasi, J., 1974, in *Chemical and Biochemical Reactivity*, eds. E. D. Bergmann and B. Pullman (D. Reidel, Dordrecht, Holland), p. 513.
222. ——, Tomasi, J., Macchia, B., and Macchia, F., 1974, *J. Med. Chem.*, **17**, 501.
223. —— and Tomasi, J., 1975, *Int. J. Quantum Chem., Quantum Biol. Symp. No. 2*, ed. P.-O. Löwdin (John Wiley and Sons, New York), p. 181.
224. —— and Tomasi, J., 1976, private communication.
225. ——, Macchia, B., Macchia, F., and Martinelli, A., 1977, *J. Med. Chem.*, **20**, 1645.
226. ——, Preston, H. J. T., and Kaufman, J. J., 1978, *Int. J. Quantum Chem.*, **13**, 457.
227. Pitzer, R. M., 1967, *J. Chem. Phys.*, **46**, 4871.
228. Politzer, P., 1966, *J. Phys. Chem.*, **70**, 1174.
229. —— and Cusachs, L. C., 1968, *Chem. Phys. Lett.*, **2**, 1.
230. —— and Harris, R. R., 1970, *J. Am. Chem. Soc.*, **92**, 6451.
231. —— and Harris, R. R., 1971, *Tetrahedron*, **27**, 1567.
232. —— and Mulliken, R. S., 1971, *J. Chem. Phys.*, **55**, 5135.
233. ——, Smith, R. K., and Kasten, S. D., 1972, *Chem. Phys. Lett.*, **15**, 226.
234. —— and Reggio, P. H., 1972, *J. Am. Chem. Soc.*, **94**, 8308.
235. ——, Donnelly, R. A., and Daiker, K. C., 1973, *J. C. S. Chem. Comm.*, 617.
236. —— and Parr, R. G., 1974, *J. Chem. Phys.*, **61**, 4258.
237. —— and Weinstein, H., 1975, *Tetrahedron*, **31**, 915.
238. —— and Daiker, K. C., 1975, *Chem. Phys. Lett.*, **34**, 294.
239. —— and Kasten, S. D., 1976, *J. Phys. Chem.*, **80**, 283.
240. ——, Daiker, K. C., and Donnelly, R. A., 1976, *Cancer Lett.*, **2**, 17.
241. —— and Daiker, K. C., 1977, *Int. J. Quantum Chem., Quantum Biol. Symp. No. 4*, ed. P.-O. Löwdin (John Wiley & Sons, New York), p. 317.
242. ——, 1977, in *Homoatomic Rings, Chains, and Macromolecules of Main Group Elements*, ed. A. L. Rheingold (Elsevier, Amsterdam), p. 95.
243. —— and Daiker, K. C., 1978, *J. Chem. Phys.*, **68**, 5289.
244. Pollak, M. and Rein, R., 1967, *J. Chem. Phys.*, **47**, 2045.

245. Pople, J. A. and Segal, G. A., 1965, *J. Chem. Phys.*, **43**, S136.
246. —— and Beveridge, D. L., 1970, *Approximate Molecular Orbital Theory* (McGraw-Hill, Inc., New York).
247. —— and Seeger, R., 1975, *J. Chem. Phys.*, **62**, 4566.
248. Port, G. N. J. and Pullman, A., 1973, *FEBS Lett.*, **31**, 70.
249. Pullman, A., 1973, *Chem. Phys. Lett.*, **20**, 29.
250. ——, 1974, in *Chemical Carcinogenesis*, Part A, eds. P. O. P. Ts'o and J. A. Dipaolo (Marcel Dekker, New York), p. 375.
251. ——, 1974, in *Chemical and Biochemical Reactivity*, eds. E. D. Bergmann and B. Pullman (D. Reidel, Dordrecht, Holland), p. 1.
252. ——, 1974, *Int. J. Quantum Chem., Quantum Biol. Symp. No. 1*, ed. P.-O. Löwdin (John Wiley & Sons, New York), p. 33.
253. —— and Pullman, B., 1955, "New Developments" in *Advances in Cancer Research*, vol. 3, eds. J. P. Greenstein and A. Haddow (Academic Press, New York), p. 117.
254. ——, Dreyfus, M., and Mely, B., 1970, *Theoret. Chim. Acta* (Berlin), **17**, 85.
255. ——, Alagona, G., and Tomasi, J., 1974, *Theoret. Chim. Acta* (Berlin), **33**, 87.
256. ——, Berthod, H., and Gresh, N., 1975, *Chem. Phys. Lett.*, **33**, 11.
257. —— and Brochen, P., 1975, *Chem. Phys. Lett.*, **34**, 7.
258. —— and Berthod, H., 1976, *Chem. Phys. Lett.*, **41**, 205.
259. —— and Berthod, H., 1977, *Int. J. Quantum Chem., Quantum Biol. Symp. No. 4*, ed. P.-O. Löwdin (John Wiley & Sons, New York), p. 327.
260. —— and Armbruster, A. M., 1977, *Theoret. Chim. Acta* (Berlin), **45**, 249.
261. —— and Berthod, H., 1978, *Theoret. Chim. Acta* (Berlin), **48**, 269.
262. —— and Perahia, D., 1978, *Theoret. Chim. Acta* (Berlin), **48**, 29.
263. Pullman, B., Goldblum, A., and Berthod, H., 1977, *Biochem. Biophys. Res. Comm.*, **77**, 1166.
264. Regis, A. and Corset, J., 1973, *Can. J. Chem.*, **51**, 3577.
265. Rein, R., 1973, in *Advances in Quantum Chemistry*, vol. 7, ed. P.-O. Löwdin (Academic Press, New York), p. 335.
266. ——, Rabinowitz, J. R., and Swissler, T. J., 1972, *J. Theoret. Biol.*, **34**, 215.
267. Ridd, J. H., 1971, in *Physical Methods in Heterocyclic Chemistry*, vol. 4, ed. A. R. Katritzky (Academic Press, New York), p. 55.
268. Robb, M. A. and Csizmadia, I. G., 1968, *Theoret. Chim. Acta* (Berlin), **10**, 269.
269. Roothaan, C. C. J., 1951, *J. Chem. Phys.*, **19**, 1445.
270. Rosowsky, A., 1964, in *Heterocyclic Compounds with Three- and Four-Membered Rings*, Part One, ed. A. Weissberger (Interscience, New York), p. 1.
271. Ros, P. and Schuit, G. C. A., 1966, *Theoret. Chim. Acta* (Berlin), **4**, 1.
272. Rouse, R. A., 1976, *Theoret. Chim. Acta* (Berlin), **41**, 149.
273. Russegger, P. and Schuster, P., 1973, *Chem. Phys. Lett.*, **19**, 254.
274. Sadlej, J., 1977, *Rocz. Chem.*, **51**, 1013.

275. Schuster, P., Marius, W., Pullman, A., and Berthod, H., 1975, *Theoret. Chim. Acta* (Berlin), **40**, 323.
276. Scott, D. W., 1971, *J. Mol. Spectrosc.*, **37**, 77.
277. Scrocco, E. and Tomasi, J., 1973, in *Topics in Current Chemistry, New Concepts II*, No. 42 (Springer-Verlag, Berlin), p. 95.
277a. —— and Tomasi, J., 1978, in *Advances in Quantum Chemistry*, vol. 11, ed. P.-O. Löwdin (Academic Press, New York), p. 115.
278. Sheppard, W. A. and Sharts, C. M., 1969, *Organic Fluorine Chemistry* (W. A. Benjamin, New York).
279. Shillady, D. D., Billingsley, F. P., II, and Bloor, J. E., 1971, *Theoret. Chim. Acta* (Berlin), **21**, 1.
280. Silla, E., Scrocco E., and Tomasi, J., 1975, *Theoret. Chim. Acta* (Berlin), **40**, 343.
281. Silverstein, R. M. and Bassler, G. C., 1967, *Spectrometric Identification of Organic Compounds*, 2nd ed. (John Wiley & Sons, New York).
282. Silverstone, H. J., 1968, *J. Chem. Phys.*, **48**, 4106.
283. Singer, K., 1960, *Proc. Roy. Soc.* (London), **A258**, 412.
284. Smit, P. H., Derissen, J. L., and Van Duijneveldt, F. B., 1977, *J. Chem. Phys.*, **67**, 274.
285. Snyder, L. C., Shulman, R. G., and Neumann, D. B., 1970, *J. Chem. Phys.*, **53**, 256.
286. Srebrenik, S., Weinstein, H., and Pauncz, R., 1973, *Chem. Phys. Lett.*, **20**, 419.
287. ——, Pauncz, R., and Weinstein, H., 1975, *Chem. Phys. Lett.*, **32**, 420.
288. Stamatiadou, M. N., Swissler, T. J., Rabinowitz, J. R., and Rein, R., 1972, *Biopolymers*, **11**, 1217.
289. Stevens, R. M., Pitzer, R. M., and Lipscomb, W. N., 1963, *J. Chem. Phys.*, **38**, 550.
290. ——, Switkes, E., Laws, E. A., and Lipscomb, W. N., 1971, *J. Am. Chem. Soc.*, **93**, 2603.
291. Stout, E. W., Jr., and Politzer, P., 1968, *Theoret. Chim. Acta* (Berlin), **12**, 379.
292. Sung, S. S., Chalvet, O., and Daudel, R., 1960, *J. Chim. Phys.*, **57**, 31.
293. Swaminathan, S., Harrison, S. W., and Beveridge, D. L., 1978, *J. Am. Chem. Soc.*, **100**, 5705.
294. Swissler, T. J. and Rein, R., 1972, *Chem. Phys. Lett.*, **15**, 617.
295. Szalda, D. J., Marzilli, L. G., and Kistenmacher, T. J., 1975, *Inorg. Chem.*, **14**, 2076.
296. ——, Marzilli, L. G., and Kistenmacher, T. J., 1975, *Biochem. Biophys. Res. Comm.*, **63**, 601.
297. Tait, A. D. and Hall, G. G., 1973, *Theoret. Chim. Acta* (Berlin), **31**, 311.
298. Turro, N. J., Dalton, J. C., Dawes, K., Farrington, G., Hautala, R., Morton, D., Niemczyk, M., and Schore, N., 1972, *Acc. Chem. Res.*, **5**, 92.
299. Umeyama, H. and Morokuma, K., 1976, *J. Am. Chem. Soc.*, **98**, 4400.
300. —— and Morokuma, K., 1976, *J. Am. Chem. Soc.*, **98**, 7208.
301. —— and Morokuma, K., 1977, *J. Am. Chem. Soc.*, **99**, 1316.

302. Van der Neut, R. N., 1975, *Tetrahedron*, **31**, 2547.
303. Van der Helm, D., 1973, *J. Cryst. Mol. Struct.*, **3**, 249.
304. Weinstein, H., 1974, *Int. J. Quantum Chem., Quantum Chem. Symp. No. 8*, ed. P.-O. Löwdin (John Wiley & Sons, New York), p. 123.
305. ——, 1975, *Int. J. Quantum Chem., Quantum Biol. Symp. No. 2*, ed. P.-O. Löwdin (John Wiley & Sons, New York), p. 59.
306. ——, Pauncz, R., and Cohen, M., 1971, in *Advances in Atomic and Molecular Physics*, vol. 7, eds. D. R. Bates and I. Esterman (Academic Press, New York), p. 97.
307. ——, Maayani, S., Srebrenik, S., Cohen, S., and Sokolovsky, M., 1973, *Mol. Pharmacol.*, **9**, 820.
308. ——, Maayani, S., Srebrenik, S., Cohen, S., and Sokolovsky, M., 1975, *Mol. Pharmacol.*, **11**, 671.
309. ——, Chou, D., Kang, S., Johnson, C. L., and Green, J. P., 1976, *Int. J. Quantum Chem., Quantum Biol. Symp. No. 3*, ed. P.-O. Löwdin (John Wiley & Sons, New York), p. 135.
310. ——, Chou, D., Johnson, C. L., Kang, S., and Green, J. P., 1976, *Mol. Pharmacol.*, **12**, 738.
311. —— and Osman, R., 1977, *Int. J. Quantum Chem., Quantum Biol. Symp. No. 4*, ed. P.-O. Löwdin (John Wiley & Sons, New York), p. 253.
312. ——, Eilers, J. E., and Chang, S. Y., 1977, *Chem. Phys. Lett.*, **51**, 534.
313. ——, Srebrenik, S., Maayani, S., and Sokolovsky, M., 1977, *J. Theoret. Biol.*, **64**, 295.
314. ——, Osman, R., Edwards, W. D., and Green, J. P., 1978, *Int. J. Quantum Chem., Quantum Biol. Symp. No. 5*, ed. P.-O. Löwdin (John Wiley & Sons, New York), p. 449.
315. Weinstein, I. B., Jeffrey, A. M., Jannette, K. W., Blobstein, S. H., Harvey, R. G., Harris, C., Autrup, H., Kasai, H., and Nakanishi, K., 1976, *Science*, **193**, 592.
316. Yamabe, S. and Morokuma, K., 1975, *J. Am. Chem. Soc.*, **97**, 4458.
317. Yoshimine, M. and McLean, A. D., 1967, *Int. J. Quantum Chem., Quantum Chem. Symp. No. 1*, ed. P.-O. Löwdin (John Wiley & Sons, New York), p. 313.
318. Zurawski, B., Ahlrichs, R., and Kutzelnigg, W., 1973, *Chem. Phys. Lett.*, **21**, 309.

7 Miscellaneous Applications of the Hellmann–Feynman Theorem

B. M. Deb
Department of Chemistry
Indian Institute of Technology
Bombay, India

Contents

7-1. INTRODUCTION AND OVERVIEW

This chapter is a medley of assorted gems and loose threads. In this we discuss diverse uses of the general and electrostatic Hellmann–Feynman (H–F) theorems, in order to further illustrate their wide applicability to atomic and molecular physics, solid state physics, statistical thermodynamics, etc. Because of space limitations, our discussion of individual works in this section will be quite brief.

7-1-1. A Survey of Some Results

In order to compute the force on a nucleus in a polyatomic molecule, using either ab initio or approximate LCAO–MO wavefunctions, one must evaluate one-, two-, and three-center one-electron force integrals (see Sections 2-4-1 and 3-3-2 for the significance of these integrals). The one- and two-center integrals may be evaluated through the formulas listed by earlier workers (*7, 11, 27*). By using expansion techniques (*4, 11, 26*), the three-center integrals can also be readily evaluated either analytically or seminumerically; the master formulas for such integrals involving upto $2s$ and $2p$ STOs are listed by Deb (*32*). Alternatively, one may employ the Neumann expansion (*32, 67*) for $1/r_k$, where the force operator $f_k = -\text{grad}_k(1/r_k)$. It turns out that invoking a Mulliken-type approximation can diminish the values of three-center force integrals by as much as 40%.

In spite of accurate evaluation of three-center integrals, the sad fact is that unless the wavefunctions employed are close to the Hartree–Fock limit, e.g., those employed by Bader et al. for diatomic molecules (Chapter 2), they result in quite unrealistic forces, especially on the heavier nuclei in a molecule. This occurs because the H–F forces are extremely sensitive to small errors in the wavefunction (Section 1-8), especially near the nuclei of interest, so that an inadequate description of inner-shell polarization results in large errors. But this drawback can be turned into an advantage, by employing it as a sensitive test of the accuracy of a calculated wavefunction. The departure of the equilibrium force on a nucleus from zero value is a measure of the "badness" of a wavefunction. An example of this approach was provided by Kern and Karplus (*59*) who examined several single-determinant wavefunctions for HF. From their work and that of Coulson and Deb (*29*) it was reasonably certain that ab initio minimum-basis-set single-determinant wavefunctions will not satisfy the condition of electrostatic equilibrium unless this condition is preimposed as a constraint in the variational procedure. In particular, the virtual orbitals (see Refs. 56 and 89) obtained in the closed-shell ($2n$ electrons) Hartree-Fock-Roothaan procedure result in peculiar forces. Further, Hartree-Fock AOs lead to more realistic charge densities than with STOs.[1]

Constrained variational procedures, employing force on a nucleus as a constraint, have also been carried out (*22*, *23*, *66*). This idea of using the experimental value of an observable as a constraint in the variational procedure for improving the quality of the wavefunction by sacrificing only a small amount of energy was suggested by Mukherji and Karplus (*71*). Loeb and Rasiel (*66*) also used two constraints for LiH, one being the force on Li nucleus[2] and the other being either the experimental dipole moment[3] or satisfaction of the virial theorem. However, in view of the accompanying increase in total energy, it is doubtful whether the use of more than one constraint would really be helpful. The presence of low-lying excited states is a sufficient, but not necessary, condition to ensure that the energy sacrifice remains small compared to the gain in expectation values of other observables. Loeb and Rasiel showed that, for LiH, using the force on Li nucleus in a single-constraint procedure resulted in only a very small increase in energy, while the expectation values of several operators were improved. Bender and Rothenberg (*14*), however, questioned the numerical accuracy of these calculations. In this context, it may be remembered that, like other one-electron properties, the force calculated using Hartree-Fock wavefunctions is correct to second order in the correlation error.

Anderson and Parr (*2*) have derived a Poisson-type equation for nuclear motion in diatomic molecules,

$$\nabla_A^2 E = 4\pi Z_A \rho_B(A), \tag{7-1}$$

where Z_A is the nuclear charge of atom A, E is the total energy, and ρ_B is that part of electron density which follows perfectly the motion of nucleus B. For both homo- and heteronuclear diatomics, the calculated quadratic, cubic, and quartic force constants agree well with experimental values.

By means of the Thomas-Reiche-Kuhn sum rule, Cohen and Drake (*25*) verified that, through first order in perturbation theory, the nucleus of an atom subjected to a uniform electric field is acted upon by a force equal to the net charge of the atom times the external field vector. In turn, Young (*108*) has proved that if a molecule is subjected to an external static electromagnetic field, the force and torque felt by the electrons are transmitted to the nuclei. To these one adds the force and torque, respectively, exerted *directly* on the nuclei by the external field, in order to obtain the *total* effect on the nuclei. Analogous results are valid for oscillating electromagnetic fields (see also Ref. 94a).

Using the general H-F theorem and assuming that only a Fermi contact mechanism is responsible for electron-coupled interactions between nuclear spins in a molecule, Pople et al. (*83, 84*) have carried out finite perturbation calculations for isotropic nuclear spin-spin coupling constants. The parameters chosen are nuclear magnetic moments. The results reproduce most experimental trends regarding coupling constants involving carbon and hydrogen, whereas coupling constants involving fluorine are less well reproduced, longer-range couplings to fluorine turning out poorly.

Epstein (*38*; see also Chapter 1) has discussed how one can relate the electric and magnetic dipole moment operators for a system to the first derivatives of the Hamiltonian with respect to electric and magnetic fields. Also, the electric polarizability and magnetic shielding tensors are related to first and second hyperpolarizabilities or hypersusceptibilities (see Refs. 3 and 30) and involve third derivatives (at zero field) of the energy with respect to electric or magnetic field components. There has been little attempt to calculate polarizabilities and hyperpolarizabilities of molecules in terms of expectation values of second and higher derivatives of the Hamiltonian mainly because

$$\left\langle \psi \left| \frac{\partial^n H}{\partial \lambda^n} \right| \psi \right\rangle \neq \frac{\partial^n E}{\partial \lambda^n}, \quad n \geqslant 2. \tag{7-2}$$

Equality in (7-2) is satisfied only for a very special class of wavefunctions (*33*).

McKinley (*69*) has derived nonrelativistic and relativistic classical analogs of the H-F theorem by considering the invariance of classical action over the whole class of classical cyclic orbits. Using the classical nonrelativistic relation

$$E(\lambda) = \frac{p^2(\lambda)}{2m} + V(\lambda) \tag{7-3}$$

the following result was obtained:

$$\frac{\partial E(\lambda)}{\partial \lambda} = \left[\frac{\partial V(\lambda)}{\partial \lambda}\right]_{\text{av}}, \tag{7-4}$$

where "av" refers to time average. The mass dependence of energy may be written as

$$m\left(\frac{\partial E}{\partial m}\right) = -\left(\frac{p^2}{2m}\right)_{\text{av}} = -(T)_{\text{av}}, \tag{7-5}$$

where T is the kinetic energy. For a classical relativistic particle,

$$E(\lambda) = [p^2(\lambda)\, c^2 + m^2 c^4]^{1/2} + V(\lambda) \tag{7-6}$$

and one again obtains

$$\frac{\partial E(\lambda)}{\partial \lambda} = \left[\frac{\partial V(\lambda)}{\partial \lambda}\right]_{\text{av}}$$

$$m\left(\frac{\partial E}{\partial m}\right) = mc^2 \left[\left(1 - \frac{v^2}{c^2}\right)^{1/2}\right]_{\text{av}}. \tag{7-7}$$

If one uses the Dirac Hamiltonian for the particle,

$$H_D = c\boldsymbol{\alpha} \cdot \boldsymbol{p} + mc^2\beta + V(\lambda), \tag{7-8}$$

then the quantum mechanical analog of (7-7) is

$$m\left(\frac{\partial E}{\partial m}\right) = mc^2\langle\psi|\beta|\psi\rangle. \tag{7-9}$$

In addition, McKinley (*69*) derived the following relations:

classical:

$$c\left(\frac{\partial E}{\partial c}\right) = (\boldsymbol{v} \cdot \boldsymbol{p})_{\text{av}} + 2mc^2 \left[\left(1 - \frac{v^2}{c^2}\right)^{1/2}\right]_{\text{av}}$$

$$= (\boldsymbol{r} \cdot \nabla V)_{\text{av}} + 2mc^2 \left[\left(1 - \frac{v^2}{c^2}\right)^{1/2}\right]_{\text{av}}, \tag{7-10}$$

since classical virial theorem says

$$(\boldsymbol{v} \cdot \boldsymbol{p})_{\text{av}} = (\boldsymbol{r} \cdot \nabla V)_{\text{av}}; \tag{7-11}$$

quantum mechanical:

$$c\left(\frac{\partial E}{\partial c}\right) = \langle\psi|c\boldsymbol{\alpha}\cdot\boldsymbol{p}|\psi\rangle + 2mc^2\langle\psi|\beta|\psi\rangle$$

$$= \langle\psi|\boldsymbol{r}\cdot\nabla V|\psi\rangle + 2mc^2\langle\psi|\beta|\psi\rangle, \tag{7-12}$$

since the quantum virial theorem for a Dirac particle says

$$\langle\psi|c\boldsymbol{\alpha}\cdot\boldsymbol{p}|\psi\rangle = \langle\psi|\boldsymbol{r}\cdot\nabla V|\psi\rangle. \tag{7-13}$$

Epstein and Epstein (*37*) discuss the H–F theorem with radial quantum numbers.

There also exist statistical thermodynamic analogs of the H-F theorem and curvature theorem (*33*, *93*). These relations, derived by Golden (*46*), are, respectively,

$$\frac{\partial F_M}{\partial \lambda_j} = \mathrm{Tr}\left(\frac{\partial H}{\partial \lambda_j}\right) \rho\{M\}, \quad \text{for all } j \tag{7-14}$$

and

$$\sum_{j,k} \chi_j \chi_k^* \frac{\partial^2 F_M}{\partial \lambda_j \partial \lambda_k} \leqslant \sum_{j,k} \chi_j \chi_k^* \,\mathrm{Tr}\left(\frac{\partial^2 H}{\partial \lambda_j \partial \lambda_k}\right) \rho\{M\}. \tag{7-15}$$

In (7-14) and (7-15) the Hamiltonian H is a function of a finite number of real parameters, $\lambda_1, \lambda_2, \ldots$, etc.; F_M is the Helmholtz function defined in terms of the partition function; $\rho\{M\}$ is a distribution operator defined in terms of the partition function and a projection-diagonal Hamiltonain $H\{M\}$ (*46*), and the χ's are finite arbitrary numbers. From (7-14) and (7-15) the quantum mechanical H-F and curvature theorems follow for a single parameter in the limit $T \to 0$ if the ground state of $H\{M\}$ is nondegenerate; in case of degeneracy, the quantum mechanical theorems correspond to a summation over the degenerate states. Both (7-14) and (7-15) are applicable to systems of bosons or fermions, or suitable mixtures of them. Golden and Guttmann (*47*) have discussed quantum statistical cases involving a nonlinear dependence of the Hamiltonian on a single parameter, while Okubo (*78*) deals with cases involving a linear dependence of the Hamiltonian on several parameters.

In the H-F theorem the variable time was introduced explicitly by Kerner (*60*). By assuming that the nuclei move classically while the electrons move quantum mechanically, he obtained the equation

$$M \frac{d^2 X}{dt^2} = -\frac{\partial V_{nn}}{\partial X} - \left\langle \psi(t) \left| \frac{\partial V_{ne}}{\partial X} \right| \psi(t) \right\rangle, \tag{7-16}$$

where M is the mass of a nucleus and X its coordinate. Although (7-16) and Kerner's iteration scheme to tackle it seem to have little practical value, this dynamic Feynman's theorem led Clinton (*24*) to propose the following interpretation of the dynamic Jahn-Teller effect: consider a molecule having three identical nuclei in a D_{3h} configuration whose electronic state is doubly degenerate (for such a system, see Ref. 85). There exists a doubly degenerate normal mode Q such that $(\partial E/\partial Q)$ is not zero in the D_{3h} configuration. Under certain conditions, the time-dependent perturbed wavefunction can be interpreted as a rotating vector in the plane of the molecule, the rotation having been caused by infinitesimal perturbations due to the coupling of nuclear and electronic motions.

An interesting application of the H–F theorem results when the parameter involved is identified with a scale factor for nuclear charge (*40*, *45*). In this case one can examine changes in the total energy and MO energies, the nature of electron reorganization, etc., during molecular "transmutations" of the type $CO \rightarrow N_2$. This involves scaling up the carbon nuclear charge and scaling down the oxygen nuclear charge by unity. The H–F theorem is also useful in the calculation of dipole moment derivatives (*16*, *17*). These derivatives are necessary for explaining the intensities in vibrational spectra, e.g., via the polar tensor formalism (*77*, *80*). The application of the theorem to a pseudo-eigenvalue problem (*94*) was essential for a direct calculation of single-particle density matrix elements through third order. Harris and Heller (*49*) resorted to the H–F theorem for testing the internal consistency of a proposed theory. For semi-empirical MO theories employing adjustable parameters the theorem can provide useful insights into the role each parameter plays in such a theory. The H–F theorem has been used to prove a reciprocity relation in the context of a density-functional theory of chemisorption on metal surfaces (*107*), while Blount (*15*) discusses the applicability of the theorem under periodic boundary conditions. Wannier and Meissner (*100*) examine the possibility that the electrostatic H–F theorem might not hold in its assumed form in the case of crystalline solids and define a stress tensor (see Section 7-3) for solids as a generalization of the H–F theorem. Swenson and Danforth (*98*) have applied a set of hypervirial theorems and the H–F theorem to a general anharmonic oscillator. The exact energy and expectation values of powers of the position coordinate are expanded in a power series of the anharmonic coupling constant. These authors have shown that use of the above theorems enables one to express each term in these expansions solely in terms of the unperturbed energy and known constants. One can thus eliminate the usual tedious calculations of sums over intermediate states of products of matrix elements which arise in nth order Rayleigh–Schrödinger perturbation theory. Further, Freed (*42*) has shown that the evaluation of force constants by the differentiation of Hartree–Fock potential energy curves is formally equivalent to their calculation at a single internuclear distance by the use of the fully coupled Hartree–Fock perturbation calculation of the second-order-type expression involved (the relaxation contribution, see Chapter 5). In the process, he has provided diagrammatic representations of certain terms involving either an operator $\partial A/\partial\lambda$ or the matrix elements of such an operator. Another simplification yielded by the H–F theorem in calculating integrals over transcorrelated wavefunctions has been discussed by Hurley (*55*), who indicates how such $3N$-dimensional integrals can be reduced to, at worst, (pseudo-) six-dimensional integrals. Nakatsuji (*74*) shows that in the H–F and integral H–F theorems terms arising due to the quantum potential defined by

Bohm (see, e.g., Ref. 13), indicating the presence of a quantum force as distinct from a classical force, are hidden. This occurs because in the final forms of these two theorems the quantum potential terms vanish by integration over all space, thereby permitting a classical interpretation of these theorems. It is interesting to note that the same Bohm potential makes its appearance in the fluid-dynamical interpretation of quantum mechanics (see also Section 7-3).

We shall now discuss certain other applications of H-F theorem and forces in molecules in somewhat greater detail.

7-1-2. Expectation Values Involving the Operator $\partial H/\partial\lambda$

In many second-order perturbation calculations one comes across infinite sums like

$$S \equiv \sum_{n \neq m} \frac{\langle m|A|n\rangle \left\langle n \left| \frac{\partial \Omega}{\partial \lambda} \right| m \right\rangle}{E_n - E_m} \tag{7-17a}$$

$$S^* \equiv \sum_{n \neq m} \frac{\left\langle m \left| \frac{\partial \Omega}{\partial \lambda} \right| n \right\rangle \langle n|A|m\rangle}{E_n - E_m}, \tag{7-17b}$$

where A is a linear operator, $\{|n\rangle\}$ is a complete set of orthonormal eigenvectors with eigenvalues $\{E_n\}$ of the Hermitian linear operator Ω (e.g., H). $S = \pm S^*$ depending on whether A is Hermitian or anti-Hermitian, respectively. Starting from the relation

$$\left\langle n \left| \frac{\partial \Omega}{\partial \lambda} \right| m \right\rangle = (E_m - E_n) \left\langle n \left| \frac{\partial m}{\partial \lambda} \right. \right\rangle \tag{7-18}$$

Salem (*86*, *87*) proves the sum rule

$$S + S^* = \left\langle m \left| \frac{\partial A}{\partial \lambda} \right| m \right\rangle - \frac{\partial}{\partial \lambda} \langle m|A|m\rangle. \tag{7-19}$$

Equation (7-19) has been applied to finding the continuum contribution in quantities like S in an H-F treatment of a spherical atom (e.g., H or He) in a uniform external field (*88*). Such a continuum contribution can sometimes outweigh that from discrete states.

By replacing Ω in (7-18) with the Born-Oppenheimer Hamiltonian, one obtains the following general relation between two potential energy curves $E_m(R)$

and $E_n(R)$ for a diatomic molecule:

$$E_m(R) = E_n(R) + \frac{\left\langle n \left| \frac{\partial H}{\partial \lambda} \right| m \right\rangle}{\left\langle n \left| \frac{\partial m}{\partial \lambda} \right. \right\rangle}, \tag{7-20}$$

where $\langle n | \partial m/\partial\lambda \rangle \neq 0$. Taking λ as R, the internuclear distance, $\langle n | \partial H/\partial\lambda | m \rangle$ becomes a coupling term between the electronic and nuclear motions (*19*). One can also apply (7-18) to relate molecular constants like equilibrium geometry, dissociation energy, harmonic vibration frequency, force constant, anharmonicity constant, etc. between two electronic states (*20*). Chen (*21*) has further discussed situations where scale factors might be chosen as parameters. Pandres (*79*) derives the result

$$i\hbar\lambda \frac{\partial E}{\partial \lambda} \langle \psi | \psi \rangle = \langle \psi | [W, H] | \psi \rangle, \tag{7-21}$$

where λ is a scale factor and W is a Hermitian operator involving coordinates and momenta. Thus, if λ is varied to optimize E, the corresponding wavefunction satisfies the hypervirial theorem. Pandres has also discussed a tensor form of (7-21).

7-1-3. Proton Affinities and Inner-Shell Binding Energies (ESCA): Nuclear Charge as a Parameter

In recent years proton affinities and ESCA binding energies have been interpreted in terms of the electrostatic potential (see Chapter 6) at a nucleus. These may be defined as energy differences between isoelectronic molecules whose Hamiltonians differ only in the magnitude of the central nuclear charge Z, namely, $(E_{\text{ion}} - E_0)$ where E_{ion} is the energy of the protonated or ionized state. Assuming that the change of state does not cause a significant change in molecular geometry,[4] one can write

$$E_{\text{ion}} - E_0 \simeq E(Z+1) - E(Z). \tag{7-22}$$

Using the H-F and curvature theorems, Davis and Rabalais (*31*) derived the result[5]

$$E(Z+1) - E(Z) \simeq -V_p(Z) - \tfrac{1}{2}[V_p(Z+1) - V_p(Z)], \tag{7-23}$$

where $V_p(Z)$ is the potential energy of an electron at nucleus p of charge Z. Equation (7-23) may be used to calculate core-level chemical shifts (ESCA) and correctly predicts relative proton affinities for carbonyl compounds, alcohols, ethers, and amines; the absolute proton affinities, however, are either somewhat overestimated or underestimated. Also, alkyl substitution increases the proton affinities of neutral species through a delocalization of positive charge. This approach reveals that, within a homologous series of molecules, a linear correlation exists between proton affinities and inner-shell binding energies.

7-1-4. Variation in the Binding Energy of Biexcitons with Electron-to-Hole Mass Ratio

Consider a molecule (biexciton) in which two excitons with the same hole masses are bound together. Taking $\sigma = m_e/m_h$, the Hamiltonian for the system may be written as (*1*)

$$H = -\frac{1}{2(1+\sigma)}[\Delta_{e_1} + \Delta_{e_2} + \sigma(\Delta_{h_1} + \Delta_{h_2})] + \frac{1}{r_{e_1 e_2}} + \frac{1}{r_{h_1 h_2}} - \frac{1}{r_{e_1 h_1}} - \frac{1}{r_{e_1 h_2}} - \frac{1}{r_{e_2 h_1}} - \frac{1}{r_{e_2 h_2}}. \quad (7\text{-}24)$$

The biexciton is bound if $W = 1 + E < 0$, where E is the lowest-energy eigenvalue. The symmetry of the Hamiltonian implies (*101*)

$$\left(\frac{\partial W}{\partial \sigma}\right)_{\sigma=1} = 0, \quad (7\text{-}25)$$

which was not satisfied by previous calculations (*90, 91*). Further, (7-24) yields

$$\frac{\partial^2 H}{\partial \sigma^2} = -\frac{2}{1+\sigma}\frac{\partial H}{\partial \sigma}, \quad (7\text{-}26)$$

which, together with (7-25), implies

$$\frac{\partial W}{\partial \sigma} \geqslant 0, \quad 0 \leqslant \sigma \leqslant 1 \quad (7\text{-}27)$$

and

$$\frac{\partial^2 W}{\partial \sigma^2} \leqslant 0. \quad (7\text{-}28)$$

Inequalities (7-27) and (7-28) indicate that W is a monotonic function of σ, and is convex upwards (*1, 92*). Using similar arguments, Bednarek and Adamoski

(*12*) obtained an estimate of the binding energy of a biheteroexciton with different hole masses.

7-2. INTERMOLECULAR FORCES

In his original work Feynman (*41*, Part I) suggested that the long-range attraction between two atoms in their ground states originates from the following effect. Because of the presence of the atom B the electron density in atom A is distorted and shifted toward B, the extent of polarization being proportional to R^{-7}, where R is the internuclear distance. A similar distortion occurs with the electron density of B, so that the centroid of the electron density in each atom lies between the nuclei of A and B. Thus, each nucleus is pulled toward the other by virtue of the attractive force exerted by its own distorted electron density (see Fig. 2-3, $R = 8.0$ au). This conjecture of Feynman was verified much later, first by Hirschfelder and Eliason (*51*), and then by Bader and Chandra (*6*).

The problem in verification, as usual, was the requirement that one must have a reliable one-electron density function in order to compute satisfactory forces via the electrostatic H–F theorem. Thus, even for the long-range interaction of two ground-state hydrogen atoms, Frost (*44*) was unable to find an approximate wavefunction which would satisfactorily reproduce the dipole–dipole part (R^{-7} dependent) of the intermolecular force. Further, since electron correlation plays an important role in intermolecular forces (*52*), it is clear that one should really employ one-electron densities of greater than Hartree–Fock accuracy in order to account for intermolecular forces over long, intermediate, and short ranges. Consequently, the application of the electrostatic H–F theorem has been restricted to very small systems, the most detailed study being on the intermolecular force between two hydrogen atoms (for a detailed account of intermolecular forces, see ref. 68).

7-2-1. Long-Range Forces Between Two Ground-State Hydrogen Atoms

The long-range interaction energy between two hydrogen atoms in the ground state may be written as

$$U(R) = C_6 R^{-6} + C_8 R^{-8} + \cdots . \tag{7-29}$$

The force on the proton A in the z direction (i.e., toward the proton B) is

$$\begin{aligned} F_{Az} &= \frac{\partial U}{\partial R} = -6C_6 R^{-7} - 8C_8 R^{-9} - \cdots \\ &= -R^{-2} + \int \frac{z_{A1}}{r_{A1}^3} \rho_1 \, d\tau_1 , \end{aligned} \tag{7-30}$$

where ρ_1 is the total one-electron density. Hirschfelder and Eliason (*51*) have been able to calculate accurately the leading term $-6C_6R^{-7}$ in the above expression, and in the process found out the origin of the van der Waals attraction between the two atoms. Their perturbation approach is briefly described here, following the account by Hirschfelder and Meath (*52*).

Assuming that electron 1 is associated with the isolated atom A and electron 2 is associated with the isolated atom B, the interaction or perturbation potential is given by

$$V = -r_{A2}^{-1} - r_{B1}^{-1} + r_{12}^{-1} + R^{-1}. \tag{7-31}$$

V may be expressed as a multipole expansion in powers of R^{-1},

$$V = \sum_{n=1}^{\infty} V_n/R^n, \tag{7-32}$$

where

$$V_n = \sum_{l_A=0}^{n-1} \sum_{m=-l_<}^{l_<} \frac{(-1)^{l_B}(n-1)!\, Q_{l_A}^m(1)\, Q_{l_B}^{-m}(2)}{[(l_A - m)!\,(l_A + m)!\,(l_B - m)!\,(l_B + m)!]^{1/2}} \tag{7-33}$$

with $l_<$ as the lesser of l_A and $l_B = n - l_A - 1$, and the Q_l^m are the irreducible tensor components of the electrostatic multipole operators for the interacting atoms (see Ref. 52).

Now, the Heitler–London wavefunction for the H_2 molecule may be written as

$$\psi = \frac{1}{\sqrt{2}} [\phi_1(A1, B2) \pm \phi_2(B1, A2)], \tag{7-34}$$

where $\phi_1(A1, B2)$ means that electron 1 is associated with atom A and electron 2 with atom B. At large R, the overlap integral

$$\int \phi_1 \phi_2 \, d\tau_2$$

may be neglected and ρ_1 may be partitioned as

$$\rho_1 = \rho_1(A) + \rho_1(B), \tag{7-35}$$

where

$$\rho_1(\mathrm{A}) = \int \phi_1^* \phi_1 \, d\tau_2$$

$$\rho_1(\mathrm{B}) = \int \phi_2^* \phi_2 \, d\tau_2 .$$

One can expand the function ϕ_1 in powers of R^{-1}:

$$\phi_1 = N[\psi_0 + R^{-3}\psi_3^{(1)} + R^{-4}\psi_4^{(1)} + R^{-5}\psi_5^{(1)} + R^{-6}\{\psi_6^{(1)} + \psi_3^{(2)}\} + R^{-7}\{\psi_7^{(1)} + \psi_{3,4}^{(1,1)}\} + \cdots], \quad (7\text{-}36)$$

where N is a normalization constant; ψ_0 is equal to $A_0(1)B_0(2)$, where A_0 is the $1s$ hydrogenic orbital on atom A; $\psi_n^{(1)}$ are the first-order perturbed wavefunctions corresponding to the perturbation potential V_n; $\psi_3^{(2)}$ is the second-order perturbed function corresponding to V_3; and $\psi_{3,4}^{(1,1)}$ is a mixed second-order perturbed function, first-order separately with respect to V_3 and V_4. It now remains to examine separately the contributions of $\rho_1(\mathrm{A})$ and $\rho_1(\mathrm{B})$ to $F_{\mathrm{A}z}$. Let us consider $\rho_1(\mathrm{A})$ first.

Using the orthonormality of spherical harmonics it is possible to argue that

$$\int \psi_0^* \psi_n^{(1)} \, d\tau_2 = 0. \quad (7\text{-}37)$$

Also, one can express $\psi_{3,4}^{(1,1)}$ as

$$\psi_{3,4}^{(1,1)} = \chi(\mathrm{A}1)B_0(2) + \Lambda(\mathrm{A}1, \mathrm{B}2), \quad (7\text{-}38)$$

where χ is a function to be determined and Λ contains all those parts of $\psi_{3,4}^{(1,1)}$ that correspond to the excitation of electron 2. Hence

$$\int \psi_0^* \Lambda \, d\tau_2 = 0. \quad (7\text{-}39)$$

One can now write $\rho_1(\mathrm{A})$, up to terms of the order of R^{-7}, as

$$\rho_1(\mathrm{A}) = N^2 [A_0^*(1)A_0(1) + R^{-6}\{\rho_{33}^{(1)}(\mathrm{A}) + \rho_{03}^{(2)}(\mathrm{A})\} + R^{-7}\{\rho_{34}^{(1)}(\mathrm{A}) + \rho_{34}^{(2)}(\mathrm{A})\}], \quad (7\text{-}40)$$

where

$$\rho_{33}^{(1)}(\mathrm{A}) = \int \psi_3^{*(1)} \psi_3^{(1)} \, d\tau_2$$

$$\rho_{03}^{(2)}(\mathrm{A}) = \int [\psi_0^* \psi_3^{(2)} + \psi_3^{*(2)} \psi_0] \, d\tau_2$$

$$\rho_{34}^{(1)}(\mathrm{A}) = \int [\psi_3^{*(1)} \psi_4^{(1)} + \psi_4^{*(1)} \psi_3^{(1)}] \, d\tau_2$$

$$\rho_{34}^{(2)}(\mathrm{A}) = A_0^*(1)\, \chi(\mathrm{A}1) + \chi^*(\mathrm{A}1)\, A_0(1). \qquad (7\text{-}41)$$

As it happens, $\rho_{33}^{(1)}(\mathrm{A})$ and $\rho_{03}^{(2)}(\mathrm{A})$ are symmetric functions of $z_{\mathrm{A}1}$, and therefore they will not contribute to the force integral

$$\int \frac{z_{\mathrm{A}1}}{r_{\mathrm{A}1}^3} \rho_1 \, d\tau_1,$$

since $-\infty \leqslant z_{\mathrm{A}1} \leqslant +\infty$.

Next, we consider the contribution of $\rho_1(\mathrm{B})$ to $F_{\mathrm{A}z}$. It is possible to write

$$\rho_1(B) = B_0^*(1)\, B_0(1) + O(R^{-6}). \qquad (7\text{-}42)$$

One can readily show that at large separations

$$\int \frac{z_{\mathrm{A}1}}{r_{\mathrm{A}1}^3} B_0^*(1)\, B_0(1)\, d\tau_1 \simeq R^{-2}; \qquad (7\text{-}43)$$

therefore, this cancels the nuclear-nuclear repulsion force $-R^{-2}$. Terms of the order of R^{-6} also do not contribute to the leading term of the London dispersion force, for the following reason: $z_{\mathrm{A}1}/r_{\mathrm{A}1}^3$ can be expanded about the center B in a power series of R^{-1}:

$$\frac{z_{\mathrm{A}1}}{r_{\mathrm{A}1}^3} = R^{-2} - 2R^{-3} z_{\mathrm{B}1} + \cdots \qquad (7\text{-}44)$$

Thus, $O(R^{-6})$ contributes terms of the order of R^{-8}.

Summarizing, the leading term in F_{Az} varies as R^{-7}, where

$$F_{Az} = -6C_6R^{-7} - \cdots$$

$$= R^{-7}\left[\int \frac{z_{A1}}{r_{A1}^3}\rho_{34}^{(1)}(A)\,d\tau_2 + \int \frac{z_{A1}}{r_{A1}^3}\rho_{34}^{(2)}(A)\,d\tau_2\right] + \cdots \quad (7\text{-}45)$$

The perturbed densities $\rho_{34}^{(1)}(A)$ and $\rho_{34}^{(2)}(A)$ have been evaluated, and Hirschfelder and Eliason showed that

$$-6C_6R^{-7} = R^{-7}(2.717 + 36.284)$$

$$= 39.001R^{-7} \text{ au.} \quad (7\text{-}46)$$

The exact value (not corrected for mass polarization) is 38.994 R^{-7} au. Thus, these authors were able to obtain a very accurate estimate of the dipole-dipole part of the London dispersion force. This work also explains why Frost's (*44*) earlier work was unable to reproduce the dipole-dipole force. In order to do so, one requires the R^{-3}, R^{-4}, and R^{-7} parts of the approximate wavefunction. A wavefunction that is correct up to R^{-4} gives only about 7% of the dipole-dipole force, although it yields a good estimate for the corresponding energy term[6] C_6R^{-6} as part of the expectation value of the Hamiltonian. Further, the dipole-dipole force (R^{-7} term) originates from the R^{-7} components of $\rho_1(A)$, i.e., $\rho_{34}^{(1)}(A)$ and $\rho_{34}^{(2)}(A)$. The centroids of these two components are somewhat shifted from A toward B. Thus, the atom A appears as an atomic dipole whose nucleus is dragged toward B by its own electron density. The same is true for the atom B, thereby verifying Feynman's early conjecture. It is also clear that the R^{-6} component of $\rho_1(A)$, resulting from the R^{-3} part of the wavefunction, cannot be responsible for either the dipole-dipole or the dipole-quadrupole force, because this component forms a quadrupolar charge distribution centered on the proton A. This will result in only the quadrupole-quadrupole force (R^{-11} term). As pointed out before, Bader and Chandra (*6*) have also demonstrated pictorially that the origin of London dispersion forces between two hydrogen atoms lies in the simultaneous inward polarization of the two electronic densities as the atoms approach each other (Fig. 2-3, R = 8.0 au).

7-2-2. Long-Range Forces Between a Ground-State Hydrogen Atom and a Proton or an Excited Hydrogen Atom

According to the model proposed by Nakatsuji and Koga (*75*), based on Nakatsuji's electrostatic force theory (ESF, see Ref. 73 and Chapter 3), the long-range interaction between two atoms originates from the atomic dipole (AD) and the extended gross charge (EGC) forces. Feynman's explanation in terms of atomic dipoles thus appears as a special case involving the long-range interaction between two neutral S-state atoms for which only the AD force is responsible. Another special case arises when one of the interacting atoms is just a bare nucleus. The long-range forces then may be regarded as originating only from the EGC force. This may be seen from the following brief account of Nakatsuji and Koga's work.

Consider two atoms A and B separated by a long distance R. Neglecting electron exchange, the μth set of electrons is associated with isolated atom A and the νth set of electrons with isolated atom B. The Hamiltonian of the system may be split into zero-order and interaction Hamiltonians,

$$H = H^{(0)} + H^{(1)}, \tag{7-47a}$$

where

$$\begin{aligned} H^{(0)} &= \sum_{\mu}\left(-\frac{1}{2}\Delta_{\mu} - \frac{Z_{\mathrm{A}}}{r_{\mathrm{A}\mu}}\right) + \sum_{\mu>\mu'}\frac{1}{r_{\mu\mu'}} \\ &\quad + \sum_{\nu}\left(-\frac{1}{2}\Delta_{\nu} - \frac{Z_{\mathrm{B}}}{r_{\mathrm{B}\nu}}\right) + \sum_{\nu>\nu'}\frac{1}{r_{\nu\nu'}} \end{aligned} \tag{7-47b}$$

$$H^{(1)} = -\sum_{\nu}\frac{Z_{\mathrm{A}}}{r_{\mathrm{A}\nu}} - \sum_{\mu}\frac{Z_{\mathrm{B}}}{r_{\mathrm{B}\mu}} + \sum_{\mu,\nu}\frac{1}{r_{\mu\nu}} + \frac{Z_{\mathrm{A}}Z_{\mathrm{B}}}{R}, \tag{7-48}$$

and all other quantities have their usual significance. Similarly, one can write the total force operator for A, in the z-direction, as

$$\mathbb{F}_{\mathrm{A}} = \mathbb{F}_{\mathrm{A}}^{(0)} + \mathbb{F}_{\mathrm{A}}^{(1)}, \tag{7-49}$$

where

$$\mathbb{F}_{\mathrm{A}}^{(0)} = -\frac{\partial H^{(0)}}{\partial R_{\mathrm{A}}} = Z_{\mathrm{A}}\sum_{\mu} f_{\mathrm{A}\mu}^{(0)}$$

$$\mathbb{F}_{\mathrm{A}}^{(1)} = -\frac{\partial H^{(1)}}{\partial R_{\mathrm{A}}} = Z_{\mathrm{A}}\left(\sum_{\nu} f_{\mathrm{A}\nu}^{(1)} - \frac{Z_{\mathrm{B}}}{R^2}\right)$$

$$f_{\mathrm{A}\mu}^{(0)} = \frac{z_{\mathrm{A}\mu}}{r_{\mathrm{A}\mu}^3};\quad f_{\mathrm{A}\nu}^{(1)} = \frac{z_{\mathrm{A}\nu}}{r_{\mathrm{A}\nu}^3}. \tag{7-50}$$

The total force on the nucleus of A is given by a sum of EC (exchange charge), AD, and EGC forces. At large R, the EC force varies as e^{-2R} and can be neglected in comparison with terms varying as powers of R^{-1}. Thus, neglecting electron exchange, Nakatsuji and Koga obtain the expressions for the force on A in the z-direction (toward B) as

$$F_A = F_A(\mathrm{AD}) + F_A(\mathrm{EGC})$$

$$F_A(\mathrm{AD}) = \langle \psi | \mathbb{F}_A^{(0)} | \psi \rangle = Z_A \int \rho_1(\mathrm{A}) f_{A1}^{(0)} \, d\tau_1$$

$$F_A(\mathrm{EGC}) = \langle \psi | \mathbb{F}_A^{(1)} | \psi \rangle = Z_A \left\{ \int \rho_1(\mathrm{B}) f_{A1}^{(1)} \, d\tau_1 - \frac{Z_B}{R^2} \right\} \tag{7-51}$$

$$\rho_1 = \rho_1(\mathrm{A}) + \rho_1(\mathrm{B}), \tag{7-52}$$

where ψ is the exact wavefunction of the perturbed system. The EGC force is due to the electrostatic interaction of the nucleus A with the electron density and the nucleus of the interacting atom B. In a similar perturbative way, one can write for the nth order correction to the interaction force

$$F_A^{(n)} = F_A^{(n)}(\mathrm{AD}) + F_A^{(n)}(\mathrm{EGC}), \tag{7-53}$$

where

$$F_A^{(n)}(\mathrm{AD}) = Z_A \int \rho_1^{(n)}(\mathrm{A}) f_{A1}^{(0)} \, d\tau_1$$

$$F_A^{(n)}(\mathrm{EGC}) = Z_A \left\{ \int \rho_1^{(n-1)}(\mathrm{B}) f_{A1}^{(1)} \, d\tau_1 - \delta_{n1} \frac{Z_B}{R^2} \right\}, \tag{7-54}$$

where $\rho_1^{(n)}$ is the nth order correction to the density. Thus, the calculation of EGC force requires a wavefunction correct to $(n - 1)$th order, unlike the AD force, which requires a wavefunction correct to nth order. If $\rho_1(\mathrm{B})$ is spherically symmetric then for large R one can show that for a neutral atom B

$$\int \rho_1(\mathrm{B}) \frac{z_{A1}}{r_{A1}^3} \, d\tau_1 = \frac{Z_B}{R^2}. \tag{7-55}$$

Thus, for two neutral atoms in S states the EGC force vanishes. Similarly, one can argue that if the atom A is a bare nucleus, the AD force, $F_A^{(n)}(\mathrm{AD})$, vanishes for all n with no-resonance wavefunctions. For resonance wavefunctions the total interaction force is partitioned between the AD and EGC effects.

Consider now the long-range resonance force between the hydrogen atoms, one in the 1s state and the other in a 2p (σ or π) state. The zero-order wavefunction in the resonance form is taken as

(a) $1s2p_\sigma$ *states:*

$$|0\rangle = \frac{1}{\sqrt{2}}\,[1s(\mathrm{A1})\,2p_\sigma(\mathrm{B2}) \pm 2p_\sigma(\mathrm{A1})\,1s(\mathrm{B2})] \tag{7-56}$$

(b) $1s2p_\pi$ *states:*

$$|0\rangle = \frac{1}{\sqrt{2}}\,[1s(\mathrm{A1})\,2p_\pi(\mathrm{B2}) \pm 2p_\pi(\mathrm{A1})\,1s(\mathrm{B2})]. \tag{7-57}$$

In (a) the plus sign refers to ${}^1\Sigma_g^+$ and ${}^3\Sigma_u^+$ states, and the minus sign refers to ${}^1\Sigma_u^+$ and ${}^3\Sigma_g^+$ states, whereas in (b) the plus sign corresponds to ${}^1\Pi_u$, ${}^3\Pi_g$ and the minus sign to ${}^1\Pi_g$ and ${}^3\Pi_u$ states. Using a 220-term expansion for the first-order wavefunction (in resonance form), Nakatsuji and Koga have calculated the leading terms (R^{-4} dependent) of the interaction. Their results (Table 7-1) show excellent agreement with the corresponding energy calculations. The EGC force is attractive for $1s2p_\sigma$ states because the $2p_\sigma$ orbital on B can shield the nucleus A adequately from the nuclear repulsion by B. In the $1s2p_\pi$ states such shielding would obviously be quite inadequate. On the other hand, the AD force is repulsive for $1s2p_\sigma$ states because the centroid of the distorted electron density on A lies in the outward (away from B) direc-

Table 7-1. Coefficients (in au) of the Leading R^{-4}-Dependent Terms of the Resonance Force between H(1*s*) and H(2*p*).[a]

Force	State			
	$1s2p_\sigma$		$1s2p_\pi$	
	${}^1\Sigma_g^+, {}^3\Sigma_u^+$	${}^1\Sigma_u^+, {}^3\Sigma_g^+$	${}^1\Pi_u, {}^3\Pi_g$	${}^1\Pi_g, {}^3\Pi_u$
$F_\mathrm{A}^{(1)}$(AD)	-21.329568	-14.670434	+7.335217	+10.664784
$F_\mathrm{A}^{(1)}$(EGC)	+18.0	+18.0	-9.0	-9.0
$F_\mathrm{A}^{(1)}$(total)	-3.329568	+3.329566	-1.664783	+1.664784
$\partial E^{(1)}/\partial R$	-3.329574	+3.329574	-1.664787	+1.664787

[a] A positive sign indicates attraction and the negative sign repulsion; $E^{(1)}$ is the first-order interaction energy.

tion. For the $1s2p_\pi$ states the charge density on A is polarized inward, resulting in an attractive AD force.

7-2-3. Intermolecular Forces in Other Systems

Chandra and Sebastian (*18*) have studied the interaction between the He^+ ion and the H atom, which leads to the formation of the lowest $^3\Sigma$ and $^1\Sigma$ excited states of the HeH^+ ion.[7] The equilibrium internuclear distances in the triplet and singlet states are 4.5 a_0 and 5.65 a_0, respectively. The molecular one-electron density (the subscript 1 is omitted for convenience) is partitioned as

$$\rho = \rho_{\mathrm{H}} + \rho_{\mathrm{He}} + \Delta\rho, \tag{7-58}$$

and the behavior of f_{A}^{ρ} is studied as a function of R, where f_{A}^{ρ} is the charge equivalent (see Section 2-4-1) of the electronic force along the bond on the nucleus A by the charge density ρ,

$$f_{\mathrm{A}}^{\rho} = R^2 \int \rho(r) \frac{z_{\mathrm{A}1}}{r_{\mathrm{A}1}^3} \, d\tau_1 . \tag{7-59}$$

Figure 7-1 depicts the variations of the forces on the proton and He nucleus, employing Michels's (*70*) wavefunctions. For both the states

$$\lim_{R \to \infty} f_{\mathrm{H}}^{\Delta\rho} = 1,$$

indicating that at very large separations the H density shields the proton from the He nucleus and there is no net force on the proton. As the H atom approaches the He^+ ion, $f_{\mathrm{H}}^{\Delta\rho} > 1$, because the H density is increasingly polarized toward the He nucleus along the bond, causing an attractive force on the proton. However, $f_{\mathrm{H}}^{\rho_{\mathrm{He}}}$ maintains its limiting value of unity until $R \simeq 2.0\ a_0$, when the proton begins to penetrate the undeformed electron cloud of He^+. On the other hand, $f_{\mathrm{He}}^{\Delta\rho}$ remains approximately zero until $R \simeq 5.5\ a_0$ ($^1\Sigma$ state) or 4.5 a_0 ($^3\Sigma$ state), indicating that the He^+ density is only slightly polarized. For lesser values of R, $f_{\mathrm{He}}^{\Delta\rho}$ becomes negative, indicating that the He^+ density is back-polarized, causing an effectively repulsive force on the He nucleus. An added repulsive effect comes from the descreening of the proton due to the penetration of the He nucleus into the H electron cloud (Fig. 7-1). Therefore, the net force on the He nucleus is never strongly attractive for both the states. In fact, neither of the two nuclei is descreened to a marked extent by the penetration of the other nucleus, and the intermolecular attraction is due to the inward polarization of the H density induced by the He^+ ion (ion-induced-dipole at-

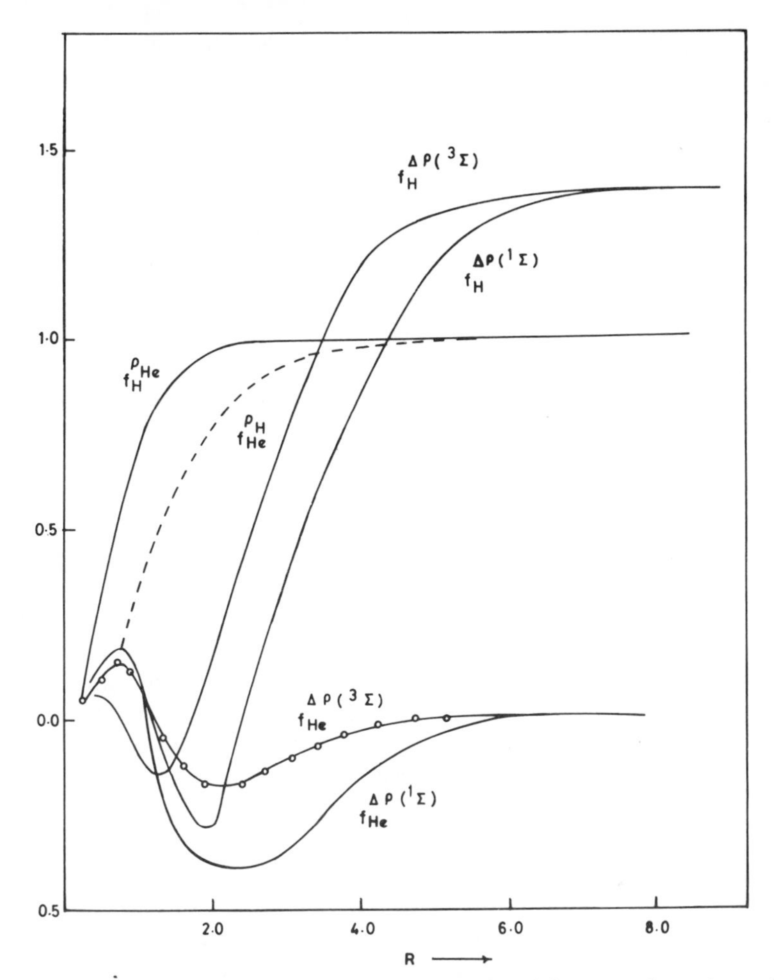

Fig. 7-1. Variation of the charge equivalents of partitioned forces on the proton and He nucleus with internuclear distance, for the $^1\Sigma$ and $^3\Sigma$ excited states of the HeH^+ ion [see (7-58) and (7-59)]. All quantities are in au. (Reproduced from Ref. 18, courtesy Taylor and Francis Ltd., London.)

traction). The density in the immediate vicinity of He^+ is back-polarized to counterbalance the attraction caused by the polarized H density on the He nucleus.

The intermolecular forces in somewhat larger systems have also been studied. Thus, Nakatsuji et al. (*76*) have employed semiempirical extended-Hückel (*53*) wavefunctions to study the intermediate-range interaction between two initially planar methyl radicals to form the ethane molecule. The interaction results in a bending force on the protons, partly because of orbital preceding of the C—H bond electron cloud (see Section 3-5-5). The resulting bending of each methyl radical changes the direction of a carbon atomic dipole so that it becomes oriented toward the other carbon atom. This AD force on a carbon nucleus is a significant driving force, especially in the early stages of the reaction. The EGC force also favors the occurrence of the reaction. However, the most important part of the driving force is the EC (C—C) force, which represents the attraction of a carbon nucleus to the electron density accumulated between the two carbon nuclei. This EC force changes little because of bending. If, on the other hand, planarity of the methyl radicals is forced on the system at all stages of the interaction, the reaction will not proceed (see Section 3-5-6).

Bader (5) has extended Platt's (*81*) model for determining the force constants in diatomic hydrides in order to study hydrogen-bonded systems as well as the properties of transition states in proton transfer reactions. For H-bonded systems Bader's force calculations show good agreement with the approximately linear relationship observed experimentally between the shift in A—H stretching frequency and the overall length of the hydrogen bond. For proton transfer reactions Bader could predict the geometry of the transition state and its perpendicular bending frequencies.

The electrostatic H-F theorem has also been applied to other systems. For example, Makin (*67*) had tried to examine (i) the three-body forces in a system of three ground-state hydrogen atoms with parallel spins, and three ground-state helium atoms, the atoms in each system being either in an equilateral triangular or a linear symmetric array; (ii) the interaction between a ground-state helium atom and a ground-state hydrogen molecule; (iii) the interaction between two ground-state hydrogen molecules. In case (i), with an equilateral triangular array, the force on a nucleus was generally repulsive except at small internuclear distances, when the force became attractive. For the helium system, such attractive forces were operative at internuclear distances $\leqslant 3.5\ a_0$, indicating a polarization of the charge cloud into the triangle. In the linear symmetric configuration the force on a terminal nucleus was generally attractive toward the middle atom. For case (ii), the hydrogen basis AOs require some p character to be incorporated, so that the force on a helium nucleus is always repulsive as the helium atom approaches the hydrogen molecule along the perpendicular bisector of the H—H length. In case only 1s orbitals are taken for the hydrogen atoms, the force on the He nucleus is at first attractive and then becomes repulsive at a distance of $\sim 6.0\ a_0$ from the protons. Case (iii) was apparently not pursued to its conclusion.

7-2-4. Coordinate Dependence of Long-Range Forces

Our discussion of intermolecular forces so far has been restricted to forces in laboratory-fixed (L) coordinate systems in which all the nuclei and electrons of the interacting system are measured independently from a common origin. While this choice of coordinate system has certain conceptual advantages, it does raise a somewhat paradoxical situation, as pointed out by Steiner (*97*). For example, in calculating the leading term in the long-range force between two ground-state hydrogen atoms the function ψ_7, a part of the second-order wavefunction, provides about 93% of C_6. Thus, energetically one requires ψ_0 and ψ_3 (first-order wavefunction) to calculate the leading term, while the force treatment additionally requires ψ_4 and ψ_7, i.e., the second-order wavefunction (*51*).

This apparent perturbation-theoretic paradox between the force and the energy approaches disappears when one looks at the interaction force in an alternative coordinate system, the so-called relative (R) system in which the nuclei and electrons of each interacting subsystem are measured from an origin fixed in each subsystem.[8] The latter can be defined unambiguously only when electron exchange between the subsystems can be neglected, i.e., only for long-range interactions. Thus, the L-coordinate system has wider applicability than the R-coordinate system. However, the R-coordinate system has certain advantages over the L-coordinate system (*61*) from a perturbation-theoretic viewpoint. These are: (1) The nth-order force in the R-coordinate system is equal to the derivative of the nth order perturbation energy, and (2) the wavefunctions and their derivatives up to order n determine the force to order $2n + 1$ (see also Ref. 106). Thus, as far as long-range forces are concerned, the force approach is not numerically inferior to the energy approach (cf. Ref. 88). Steiner (*97*) has derived some interesting relations between forces in the two coordinate systems.

Koga and Nakatsuji (*61*) call the force in the R-system force on whole particles (i.e., all the nuclei and electrons in the interacting subsystem), while the force in the L-system is designated force on a nucleus. The force on whole particles comes entirely from the perturbation Hamiltonian, in contrast to the force on a nucleus, where the unperturbed Hamiltonian is also involved. For exact wavefunctions, the force on whole particles reduces to the sum of the forces acting on all the nuclei of the subsystem. These ideas have been illustrated with three systems: $H(1s) + H^+$, $H(1s) + H(1s)$, and $NH_3 + H^+$. In the second system, for example, the calculation of the leading term of the interaction force requires the first-order wavefunction, as in the energetic treatment. Koga and Nakatsuji (*62*) have also examined the R-coordinate representation in relation to the virial and hypervirial theorems.[9] They find that (1) in the R-coordinates, the kinetic energy operator is separated into intra- and intermolecular parts, enabling one to partition the virial and hypervirial theorems into intra- and intermolecular theorems; (2) under the Born–Oppenheimer approximation, the electronic Hamiltonian in R-coordinates is the same as that in L-coordinates, so that the intramolecular H–F and virial theorems are identical in the two coordinate systems; (3) use of the partitioned virial and hypervirial theorems reduces the accuracy of the wavefunction required in perturbation calculations by at least one order.

The foregoing account of intermolecular forces makes it clear that the application of the H–F theorem provides new and significant insights into these very important phenomena. These gains have resulted partly because the electrostatic H–F theorem employs the electron density in the three-dimensional space. However, it is also clear that this conceptual simplification is achieved at some cost. Thus, the force treatments of intermolecular forces do not reveal clearly the effects due to electron spin, nor do they show up directly the effects of electron correlation, although a sufficiently accurate single-particle density would contain these effects in an implicit way.

7-3. INTERNAL STRESSES IN MOLECULES: A GENERALIZATION OF THE FORCE CONCEPT

Based on a suggestion by Feynman (*41*), Deb and Bamzai (*34, 35*) have presented a stress formalism as an extension of the force concept: Associated with each point in the 3-dimensional space of a molecule one can define a stress tensor S_ν^μ in terms of internal fields arising from the nuclear and electronic distributions ρ_N and ρ, respectively. The index ν refers to the direction of the local force and μ refers to the direction of the positive outward normal to the area on which the local force acts. The single-particle density $\rho(r)$ is sufficient to construct S_ν^μ, whose components have the same form as Maxwell's stress tensor for classical electromagnetic fields.[10]

For many-electron systems the covariant derivative of S_ν^μ yields the local electrostatic force density

$$S_{\nu,\mu}^\mu = \rho_N(E_\nu + F_\nu) + \rho(E_\nu + F_\nu), \tag{7-60}$$

where E_ν and F_ν are field components arising due to ρ_N and ρ, respectively, satisfying respective Poisson's equations. One can prove that

$$S_\nu^\mu = \frac{1}{4\pi}\left[G^\mu G_\nu - \frac{1}{2}\delta_\nu^\mu G \cdot G\right] \tag{7-61}$$

satisfies (7-60), where

$$G = E + F = -\nabla V_{\text{mol}}, \tag{7-62}$$

V_{mol} being the molecular electrostatic potential (see Chapter 6).

A simple model for molecular interaction, e.g., chemical binding, can be constructed (*35*) starting from the stress tensors $S_\nu^\mu(\text{I}), S_\nu^\mu(\text{II})$ as those for the initial undistorted systems to form $S^{\mu}_{\nu}(\mathrm{I} + \mathrm{II})$, i.e., the stress tensor for the final system after a charge migration $\Delta\rho$ has taken place. Thus, one obtains

$$S^{\mu}_{\nu}(\mathrm{I}+\mathrm{II}) = S^{\mu}_{\nu}(\mathrm{I}) + S^{\mu}_{\nu}(\mathrm{II}) + S^{\mu}_{\nu}(\mathrm{I}, \mathrm{II}) + S^{\mu}_{\nu}(\text{interaction}), \tag{7-63}$$

where $S^{\mu}_{\nu}(\mathrm{I}, \mathrm{II})$ is a zero-order interaction term persisting even at very large distances and

$$\begin{aligned} S^{\mu}_{\nu}(\text{interaction}) = {} & \frac{1}{4\pi}\,[G(\mathrm{I})^{\mu}J_{\nu} + J^{\mu}G(\mathrm{I})_{\nu} - \delta^{\mu}_{\nu}G(\mathrm{I})\cdot J] \\ & + \frac{1}{4\pi}\,[G(\mathrm{II})^{\mu}J_{\nu} + J^{\mu}G(\mathrm{II})_{\nu} - \delta^{\mu}_{\nu}G(\mathrm{II})\cdot J] \\ & + \frac{1}{4\pi}\left[J^{\mu}J_{\nu} - \frac{1}{2}\,\delta^{\mu}_{\nu}J\cdot J\right], \end{aligned} \tag{7-64}$$

J being related to $\Delta\rho$ by the Poisson equation

$$J^{\mu}_{,\mu} = 4\pi\Delta\rho. \tag{7-65}$$

Therefore, the interaction between systems I and II can be studied by monitoring the local changes in total stress, electrostatic force density, and interaction stress at certain specified points of interest. Using this approach, Deb and Bamzai (*35*) have studied chemical binding in the H_2 molecule and obtained certain interesting features. For example, for a point in the antibinding region (see Chapter 2), the total stress [see (7-63)], interaction stress [see (7-64)], and third interaction stress [last term in (7-64)] were studied for various R (Fig. 7-2). Two of these pass through minima at an R value close to $R_{\mathrm{eq}} = 1.40165$ au. These detailed local variations can be rationalized by means of classical electrostatic arguments, and they indicate that chemical binding in H_2 occurs because of the variation of electrostatic pressure from point to point in such a manner as to cause the vanishing of either the total electrostatic force density or the difference force density or both at certain points on the internuclear axis. Similar calculations have also been carried out for the H_2^+ molecule (*10*). Thus, the stress formalism seems to be a promising alternative approach to the study of molecular behavior, by virtue of its information content, retention of classical concepts, and inclusion of the H-F viewpoint as a special case. It is likely that fresh insights into transferability[11] and reactivity will be obtained by studies on other diatomics.

While the above work is an electrostatic viewpoint Deb and Ghosh (*36*) have derived a comprehensive stress tensor for a many-electron polyatomic system,

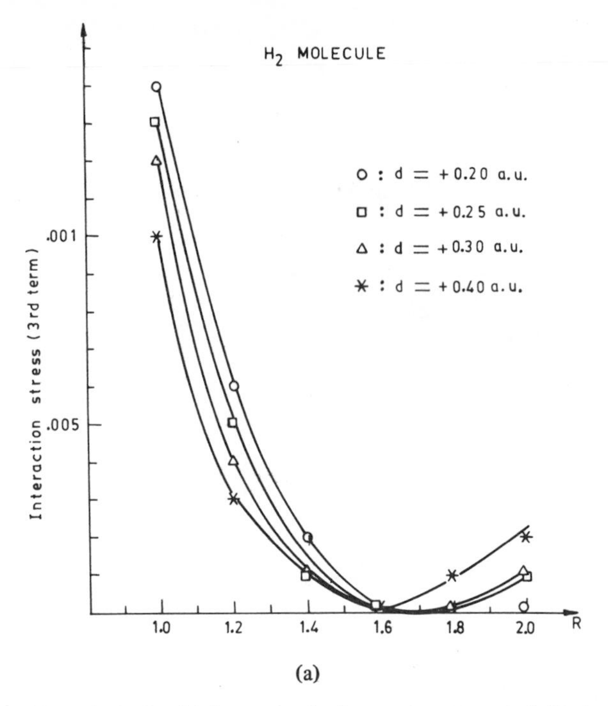

(a)

Fig. 7-2. Variations of (a) the third term in the interaction stress, and (b) the total interaction stress and the total stress (p. 413) as functions of the internuclear distance R in the H_2 molecule; d is the distance (in au) of the point considered on the internuclear axis, in the antibinding region, from the nearest proton. In (b) curves 1 to 4 represent total stress; in au (n_1, n_2, n_3) represent stress value (n_1) at origin 0, R value (n_2) at origin 0, and 1 div = n_3 stress, for curves 1 to 8: 1 (24.30, 0.8, 0.02); 2 (4.60, 0.8, 0.005); 3 (9.80, 0.8, 0.01); 4 (3.36, 0.8, 0.005); 5 (0, 0.8, 0.02); 6 (0, 0.8, 0.05); 7 (0, 0.8, 0.02); 8 (0, 0.8, 0.02).

consisting of contributions from kinetic energy, Coulomb energy, and exchange-correlation energy, starting from the local density-functional theory (*48, 54, 63, 109*). The kinetic energy part is derived from the Bohm potential (*13, 110*) for each electron. For stationary densities, the covariant derivative of this stress tensor, i.e., the local force density, vanishes. This prevents a spontaneous collapse of electronic charge onto the nuclei and a spontaneous oozing out of electronic charge from the system (see also Ref. 65). The vanishing of the local force density also leads to the Euler or Navier–Stokes equation in fluid dynamics[12] for such static charge densities and can be employed, in principle, for testing the local accuracy of calculated (approximate) electron densities. It might also indicate a route toward a possible deterministic equation for the single-particle density $\rho(r)$ consistent with the work of Balázs (*9*), who showed that any molecular $\rho(r)$ which is expressed as a local functional of the electrostatic potential alone (e.g., the Thomas–Fermi density) will not lead to chemical binding, but

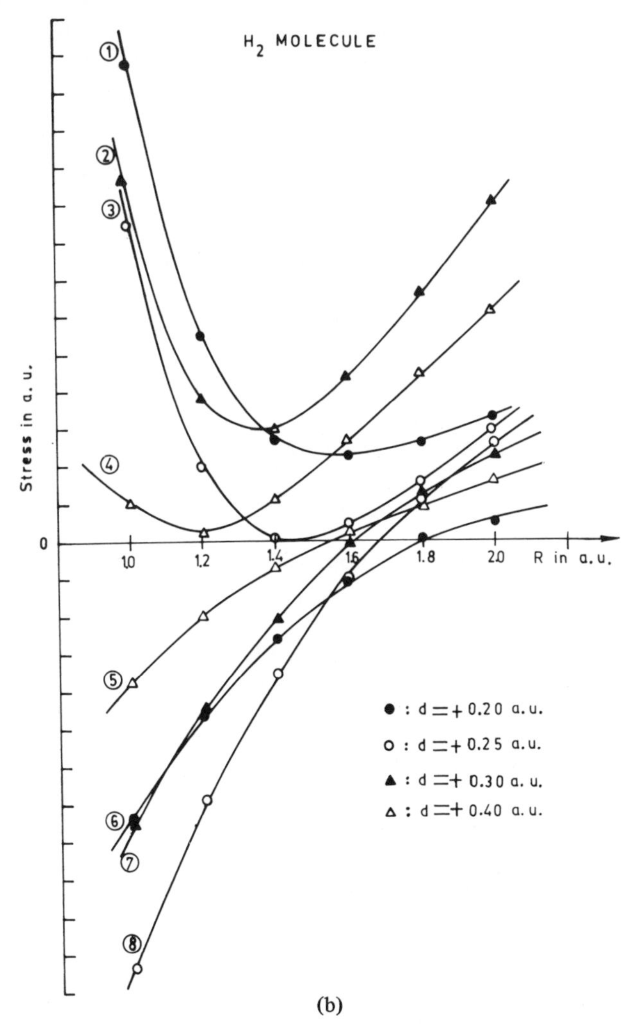

(b)

Fig. 7-2. *(Continued)*

incorporating the potential gradient in the functional will give rise to binding. Quantum stress tensors, incorporating only the kinetic energy, have also been defined by Wong and co-workers (*104, 105, 113*) Takabayashi (*99*), Pauli (*111*), Epstein (*39*), and Rosen (*112*); see also Ref. 109a.

7-4. SUMMARY

In spite of accurate evaluation of 3-center force integrals, single-determinant LCAO-MO wavefunctions of polyatomic molecules generally yield unsatisfactory forces unless the wavefunctions are close to the Hartree–Fock limit. Since forces sharply reflect the quality of a wavefunction, these may be employed as variational constraints for improving a wavefunction. Through the use of H–F and curvature theorems, one can obtain a Poisson-type equation for nuclear motions; correlate molecular constants between two electronic states, proton affinities and inner-shell binding energies (ESCA), nuclear spin coupling constants, binding energy of biexcitons, etc. These theorems have analogs in statistical thermodynamics and classical mechanics (non-relativistic and relativistic). The H–F theorem yields new and significant physical insights into intermolecular forces. If one chooses the relative coordinate system, then the force treatment of long-range intermolecular forces is numerically as good as the corresponding energetic treatment. The chapter concludes with a brief discussion of internal stresses in molecules, a promising new concept capable of yielding fresh insights into molecular structure and reactivity.

Notes

1. The Jahn–Teller force calculations (Section 3-4-3) of Coulson and Deb (*28*) demonstrated that $3d$ STOs are inadequate for molecular calculations.
2. The force on the proton could not be included in the double constraint procedure because the corresponding value of the Lagrange multiplier was too large.
3. Note that Bader and Jones (*8*) have also used known values of forces and dipole moments to obtain wavefunctions for a few molecules in a nonvariational method. However, in this method the sacrifice in energy must be considerable.
4. This is a drastic assumption and generally is not true (see Chapter 3).
5. Compare with the works of Wilson (*103*), Frost (*43*), and Politzer and Parr (*82*). See also Chapters 1 and 2.
6. This is an apparent paradox which can be resolved by a different choice of the coordinate system (see Section 7-2-4).
7. For an early energetic study of the intermolecular forces between excited atoms, see Ref. 72. For a density-functional formulation, see Ref. 110a.
8. For a discussion of the coordinate dependence of general and electrostatic H–F theorems, see Chapter 1.
9. For a review of these theorems, see Ref. 102.
10. For an introduction to tensors, see Ref. 95.
11. The integral of (7-60) is a characteristic property of a virial fragment. See Ref. 96 and Section 2-7-2.
12. The fluid-dynamical interpretation of Schrödinger equation has recently received considerable attention (see, e.g., Refs. 50, 57, 58, and 105). The stress tensor appears in fluid-dynamical equations of motion (see, e.g., Ref. 64).

References

1. Adamowski, J., Bednarek, S., and Suffczynski, M., 1971, *Solid State Comm.*, **9**, 2037.

2. Anderson, A. B. and Parr, R. G., 1970, *J. Chem. Phys.*, **53**, 3375.
3. Atkins, P. W., 1970, *Molecular Quantum Mechanics*, Part III (Oxford University Press, New York and London), p. 398.
4. Bader, R. F. W., 1962, *Can. J. Chem.*, **40**, 2140.
5. ——, 1964, *Can. J. Chem.*, **42**, 1822.
6. —— and Chandra, A. K., 1968, *Can. J. Chem.*, **46**, 953.
7. —— and Jones, G. A., 1961, *Can. J. Chem.*, **39**, 1253.
8. —— and Jones, G. A., 1963, *Can. J. Chem.*, **41**, 586, 2251, 2791.
9. Balázs, N. L., 1967, *Phys. Rev.*, **156**, 42.
10. Bamzai, A. S. and Deb, B. M., 1981, forthcoming.
11. Barnett, M. P., 1963, *Methods in Computational Physics*, vol. 2, eds. B. Alder et al. (Academic Press, New York), p. 95.
12. Bednarek, S. and Adamowski, J., 1972, *Phys. Lett.*, **41A**, 347.
13. Belinfante, F. J., 1973, *Survey of Hidden Variables Theory* (Pergamon Press, Oxford).
14. Bender, C. F. and Rothenberg, S., 1971, *J. Chem. Phys.*, **55**, 2000.
15. Blount, E. L., 1971, *Phys. Rev.*, **B3**, 3554.
16. Chandra, A. K., 1968, *Mol. Phys.*, **14**, 577.
17. ——, 1971, *Mol. Phys.*, **21**, 365.
18. —— and Sebastian, K. L., 1976, *Mol. Phys.*, **31**, 1489.
19. Chen, J. C. Y., 1963, *J. Chem. Phys.*, **38**, 283.
20. ——, 1963, *J. Chem. Phys.*, **38**, 832.
21. ——, 1963, *J. Chem. Phys.*, **39**, 3167.
22. Chong, D. P. and Byers Brown, W., 1966, *J. Chem. Phys.*, **45**, 392.
23. —— and Rasiel, Y., 1966, *J. Chem. Phys.*, **44**, 1819.
24. Clinton, W. L., 1960, *J. Chem. Phys.*, **32**, 626.
25. Cohen, M. and Drake, G. W. F., 1967, *Proc. Phys. Soc.*, **92**, 23.
26. Coulson, C. A., 1937, *Proc. Cambridge Phil. Soc.*, **33**, 104.
27. ——, 1942, *Proc. Cambridge Phil. Soc.*, **38**, 210.
28. —— and Deb, B. M., 1969, *Mol. Phys.*, **16**, 545.
29. —— and Deb, B. M., 1971, *Int. J. Quantum Chem.*, **5**, 411.
30. Davies, D. W., 1967, *The Theory of the Electric and Magnetic Properties of Molecules* (John Wiley & Sons, New York), Chapter 4.
31. Davis, D. W. and Rabalais, J. W., 1974, *J. Am. Chem. Soc.*, **96**, 5305.
32. Deb, B. M., 1971, *Proc. Indian Nat. Sci. Acad.*, **37A**, 349.
33. ——, 1972, *Chem. Phys. Lett.*, **17**, 78.
34. —— and Bamzai, A. S., 1978, *Mol. Phys.*, **35**, 1349.
35. —— and Bamzai, A. S., 1979, *Mol. Phys.*, **38**, 2069.
36. —— and Ghosh, S. K., 1979, *J. Phys. B: Atom. Mol. Phys.*, **12**, 3857.
37. Epstein, J. H. and Epstein, S. T., 1962, *Am. J. Phys.*, **30**, 266.
38. Epstein, S. T., 1974, *The Variation Method in Quantum Chemistry* (Academic Press, New York), pp. 89, 247.
39. ——, 1975, *J. Chem. Phys.*, **63**, 3573.
40. ——, Hurley, A. C., Wyatt, R. E., and Parr, R. G., 1967, *J. Chem. Phys.*, **47**, 1275.
41. Feynman, R. P., 1939, B.S. Dissertation, M.I.T., unpublished.
42. Freed, K. F., 1970, *J. Chem. Phys.*, **52**, 253.

43. Frost, A. A., 1962, *J. Chem. Phys.*, **37**, 1147.
44. ——, 1966, University of Wisconsin Theoretical Chemistry Report, WIC-TCI-204.
45. Garcia-Sucre, M., 1976, *J. Chem. Phys.*, **65**, 280.
46. Golden, S., 1975, *Chem. Phys. Lett.*, **31**, 195.
47. —— and Guttmann, C., 1965, *J. Chem. Phys.*, **43**, 1894.
48. Harris, J. and Jones, R. O., 1978, *J. Chem. Phys.*, **68**, 1190.
49. Harris, R. A. and Heller, D. F., 1975, *J. Chem. Phys.*, **62**, 3601.
50. Heller, D. F. and Hirschfelder, J. O., 1977, *J. Chem. Phys.*, **66**, 1929.
51. Hirschfelder, J. O. and Eliason, M. A., 1967, *J. Chem. Phys.*, **47**, 1164.
52. —— and Meath, W. A., 1967, *Adv. Chem. Phys.*, **12**, 3.
53. Hoffmann, R., 1963, *J. Chem. Phys.*, **39**, 1397.
54. Hohenberg, P. and Kohn, W., 1964, *Phys. Rev.*, **136**, B864.
55. Hurley, A. C., 1976, *Electron Correlation in Small Molecules* (Academic Press, London), p. 253.
56. Huzinaga, S., McWilliams, D., and Cantu, A. A., 1973, in *Advances in Quantum Chemistry*, vol. 7, ed. P.-O. Löwdin (Academic Press, New York), p. 187.
57. Janossy, L., 1976, *Found. Phys.*, **6**, 341.
58. Kan, K. K. and Griffin, J. J., 1977, *Phys. Rev.*, **C15**, 1126.
59. Kern, C. W. and Karplus, M., 1964, *J. Chem. Phys.*, **40**, 1374.
60. Kerner, E. H., 1959, *Phys. Rev. Lett.*, **2**, 152.
61. Koga, T. and Nakatsuji, H., 1976, *Theoret. Chim. Acta* (Berlin), **41**, 119.
62. —— and Nakatsuji, H., 1976, *Chem. Phys.*, **16**, 189.
63. Kohn, W. and Sham, L. J., 1965, *Phys. Rev.*, **140A**, 1133.
64. Landau, L. and Lifshitz, E. M., 1959, *Fluid Mechanics* (Pergamon Press, London).
65. Lieb, E. H., 1976, *Rev. Mod. Phys.*, **48**, 553.
66. Loeb, R. J. and Rasiel, Y., 1970, *J. Chem. Phys.*, **52**, 4995.
67. Makin, D., 1967–70, Progress Report, Wave Mechanics and Quantum Theory Group, University of Oxford, Session 1967–68, pp. 61, 64; Session 1968–69, pp. 40, 43; Session 1969–70, pp. 44, 46 (unpublished).
68. Margenau, H. and Kestner, N., 1967, *Theory of Intermolecular Forces* (Pergamon Press, New York).
69. McKinley, W. A., 1971, *Am. J. Phys.*, **39**, 905.
70. Michels, H. H., 1966, *J. Chem. Phys.*, **44**, 3834.
71. Mukherji, A. and Karplus, M., 1963, *J. Chem. Phys.*, **38**, 44.
72. Mulliken, R. S., 1960, *Phys. Rev.*, **120**, 1674.
73. Nakatsuji, H., 1973, *J. Am. Chem. Soc.*, **95**, 345, 354.
74. ——, 1977, *J. Chem. Phys.*, **67**, 1312.
75. —— and Koga, T., 1974, *J. Am. Chem. Soc.*, **96**, 6000.
76. ——, Kuwata, T., and Yoshida, A., 1973, *J. Am. Chem. Soc.*, **95**, 6894.
77. Newton, J. H. and Person, W. B., 1976, *J. Chem. Phys.*, **64**, 3036.
78. Okubo, S., 1971, *J. Math. Phys.*, **12**, 1123.
79. Pandres, D., 1963, *Phys. Rev.*, **131**, 886.
80. Person, W. B. and Newton, J. H., 1974, *J. Chem. Phys.*, **61**, 1040.

81. Platt, J. R., 1950, *J. Chem. Phys.*, **18**, 932.
82. Politzer, P. and Parr, R. G., 1974, *J. Chem. Phys.*, **61**, 4258.
83. Pople, J. A., McIver, J. W., and Ostlund, N. S., 1968, *J. Chem. Phys.*, **49**, 2960.
84. ——, McIver, J. W., and Ostlund, N. S., 1968, *J. Chem. Phys.*, **49**, 2965.
85. Porter, R. N., Stevens, R. M., and Karplus, M., 1968, *J. Chem. Phys.*, **49**, 5163.
86. Salem, L., 1962, *Phys. Rev.*, **125**, 1788.
87. ——, 1963, *Ann. Phys.*, **8**, 169.
88. —— and Wilson, E. B., 1962, *J. Chem. Phys.*, **36**, 3421.
89. Schaefer, H. F., 1972, *Electronic Structure of Atoms and Molecules: A Survey of Rigorous Quantum Mechanical Results* (Addison-Wesley, Reading, Mass.), pp. 305–307.
90. Sharma, R. R., 1968, *Phys. Rev.*, **170**, 770.
91. ——, 1968, *Phys. Rev.*, **171**, 36.
92. Shimamura, I., 1976, *Chem. Phys. Lett.*, **39**, 285.
93. Silverman, J. N. and van Leuven, J. C., 1970, *Chem. Phys. Lett.*, **7**, 37.
94. Simons, J., 1973, *J. Chem. Phys.*, **59**, 2436.
94a. Sorbello, R. S. and Dasgupta, B. B., 1980, *Phys. Rev.*, **B21**, 2196.
95. Spiegel, M. R., 1959, *Theory and Problems of Vector Analysis and Introduction to Tensor Analysis* (Schaum Publishing Company, New York).
96. Srebrenik, S. and Bader, R. F. W., 1975, *J. Chem. Phys.*, **63**, 3945.
97. Steiner, E., 1973, *J. Chem. Phys.*, **59**, 2427.
98. Swenson, R. J. and Danforth, S. H., 1972, *J. Chem. Phys.*, **57**, 1734.
99. Takabayashi, T., 1952, *Prog. Theor. Phys.* (Japan), **8**, 143.
100. Wannier, G. H. and Meissner, G., 1971, *Phys. Rev.*, **B3**, 1240.
101. Wehner, R. K., 1969, *Solid State Comm.*, **7**, 457.
102. Weislinger, E. and Olivier, G., 1974, *Int. J. Quantum Chem. Symp.*, **8**, 389.
103. Wilson, E. B., 1962, *J. Chem. Phys.*, **36**, 2232.
104. Wong, C. Y., 1976, *J. Math. Phys.*, **17**, 1008.
105. ——, Maruhn, J. A., and Welton, T. A., 1975, *Nucl. Phys.*, **A253**, 469.
106. Yaris, R., 1963, *J. Chem. Phys.*, **39**, 863.
107. Ying, S. C., Smith, J. R., and Kohn, W., 1975, *Phys. Rev.*, **B11**, 1483.
108. Young, R. H., 1969, *Mol. Phys.*, **16**, 509.
109. Zunger, A. and Freeman, A. J., 1977, *Phys. Rev.*, **B15**, 4716.

Supplementary References

109a. Bartolotti, L. J. and Parr, R. G., 1980, *J. Chem. Phys.*, **72**, 1593.
110. Bohm, D., 1952, *Phys. Rev.*, **85**, 166, 180.
110a. Clugston, M. J., 1978, *Adv. Phys.*, **27**, 893.
111. Pauli, W., 1958, *Encyclopedia of Physics*, vol. 5, part 1, ed. S. Flügge (Springer-Verlag, Berlin), pp. 23–28.
112. Rosen, N., 1974, *Nuovo Cim.*, **B19**, 90.
113. Wong, C. Y., 1976, *Phys. Lett.*, **B63**, 395.

8

Calculation of First- and Second-Order Reduced Density Matrices

A. J. Coleman

Department of Mathematics and Statistics
Queen's University
Kingston, Canada

Contents

8-1. INTRODUCTION

In his summary of the Boulder Conference on Molecular Quantum Mechanics in June 1959, Charles Coulson made the following remarks (*15*):

> It has frequently been pointed out that a conventional many-electron wavefunction tells us more than we need to know. . . . There is an instinctive feeling that matters such as electron correlation should show up in the two-particle density matrix. . . but we still do not know the conditions that must be satisfied by the density matrix. Until these conditions have been elucidated, it is going to be very difficult to make much progress along these lines.

In private conversation, Professor Coulson stated that these ideas had occurred to him several years before 1960 and that he had devoted some months to a fruitless effort to find the constraints on the 2-matrix which would ensure that it actually represented an N-electron system. He therefore took an intense interest in the effort, which was in progress during the latter years of his life, to solve the N-representability problem.

Although Husimi (*30*) had pointed out much earlier that the exact energy of an atom or molecule could be expressed in terms of the 2-matrix, it was not until Löwdin's (*37*, *38*) basic papers appeared that the current interest in the analysis of the 1-matrix and the 2-matrix in terms of their eigenfunctions–the so-called natural spin orbitals and natural spin geminals–got underway. Shull demonstrated the value of such an analysis for the hydrogen molecule wavefunction in an important paper (*47*) which concluded with the following intriguing assertion:

> The very fact that so much of chemical behavior can be described empirically quite satisfyingly in terms of "bonds," which are two-electron concepts, suggests that the next invariants one should look for occur in the reduction of polyelectron functions into optimum two-electron parts. Each of these two-electron functions[1] would then be the wave mechanical correspondent of the chemist's "bond," and would in turn be subject to a natural spin orbital analysis similar to that of this paper. An obvious place to look for the desired characteristic invariants associated with the "geminals" is the second-order density matrix, since the elements of the matrix are two-electron functions of fundamental significance.

Since 1960 the idea of natural spin orbital has become a commonplace for many chemists and physicists. They should agree with the view expressed by D. W. Smith in his article in Ref. 11: (i) for any approximate wavefunction the natural orbitals and their occupation numbers should be exhibited in order to facilitate a meaningful invariant comparison with other proposed functions for the same system, and (ii) expanding the wavefunction in terms of natural orbitals frequently suggests that many configurations can be dropped without sensible loss of accuracy thus simplifying calculations. However, Shull's proposal that chemists should search for "bonds" in a geminal analysis of chemical systems has not been seriously pursued hitherto, possibly because geminals are much more difficult to work with than orbitals. Hopefully, there will be fruitful research in the direction to which Shull pointed in the coming years. This appears as an exciting unexplored area for the discovery of new concepts which could be of great value to chemists.

The observation of Coulson and others that most of the properties of an N-electron system which are of interest to chemists and physicists (for example, the exact energy of a state) can be expressed in terms of the 2-matrix, immediately gives rise to the hope that we might calculate the 2-matrix directly without first obtaining the wavefunction. If a basis of r spin orbitals is used, a complete multiconfiguration calculation requires the determination of $\binom{r}{N}$ coefficients, whereas the 1-matrix is determined by $\binom{r+1}{2}$ coefficients and the 2-matrix by $\frac{1}{8}r(r-1)(r^2-r+2)$. For large N, the resulting economy would be considerable.

We might attempt to find the 2-matrix D^2, by a variational or by a perturbation approach. If we pursued the first of these we would attempt to choose D^2 in such a manner as to minimize the energy and thus obtain the 2-matrix of the ground state. In the process of varying D^2 it would be essential to impose on it the constraints to which Coulson alluded, i.e., N-representability. While necessary and sufficient conditions for N-representability of the 1-matrix have been given in relatively convenient form (*8*), such is not the case for the 2-matrix. In the present chapter we shall summarize what is known about N-representability conditions for the 2-matrix and the problems of implementing them.

Serious attempts to calculate the 1-matrix or the 2-matrix directly by perturbation methods are relatively recent. Most of them are inspired by the Green's-function (GF) methods developed in the quantum theory of fields. The common methods involve various rules for truncating an infinite series—frequently nonconvergent—and seldom take cognizance of the requirement that the Green's functions and their associated 1- or 2-matrix be N-representable. Thus the bulk of the approximations developed for Green's functions hitherto are somewhat dubious to say the least. However, recently a number of chemists have used GF methods with some success to obtain natural orbitals for a few systems involving a small number of electrons ($N \leqslant 4$).

Of course, in this general context the most striking result in recent years was the Hohenberg-Kohn Theorem (*28*) which asserted that the energy of a nondegenerate ground state of an atom or molecule is determined by the electron density $\rho(r)$ alone. This implies that the electron density by itself carries enough information to determine the ground state of a system. When first announced, it was a startling result. However, it follows from Kato's cusp condition (*31,52*) that for a nucleus of charge Z at the origin

$$\lim_{r \to 0} (\ln \tilde{\rho}(r))' = -2Z$$

[$\tilde{\rho}$ is the spherically averaged $\rho(r)$]. This condition may easily be verified for hydrogenlike atoms, for example. Also these cusps at the nuclei are the only singularities of $\rho(r)$. Thus $\rho(r)$ contains a record of the position and strength of the nuclear charges, whereas $\int \rho(r)\, dv = N$, the number of electrons. But this is enough information to specify the molecule and hence its ground state. Thus the energy of the ground state is a functional of the electron density function: $\rho \to E(\rho)$. Several authors have attempted to determine this functional, but it is known with some assurance only for a uniform, high-density electron gas (see Refs. 25, 55-58 for discussions on this problem and other related problems).

8-2. BASIC PROPERTIES OF REDUCED DENSITY MATRICES

In this section we recall the definition of reduced density matrices and, without proof, recount some of their basic properties. We shall discuss only fermion systems for which the wavefunction is antisymmetric though much of the theory applies with only minor modification to bosons (see Ref. 54 for a review).

If a system of N indistinguishable fermions–electrons, for example–is in a pure state this will be described in the Schrödinger representation by a function ψ of N particles $1, 2, \ldots, N$.

$$\psi: (1, 2, \ldots, N) \to \psi(1, 2, \ldots, N), \tag{8-1}$$

where ψ is antisymmetric with respect to the interchange of particles. Thus, for example,

$$\psi(1, 2, 3, \ldots, N) = -\psi(2, 1, 3, \ldots, N) = \psi(2, 3, 1, 4, \ldots, N). \tag{8-2}$$

That is, the wavefunction merely changes sign when the particles undergo an odd permutation and is unchanged by an even permutation. Associated with any fixed N-particle function ψ there is a linear operator P_ψ called the projector onto ψ defined as follows. If f is an arbitrary N-particle function, then

$$P_\psi f = \langle \psi | f \rangle \, \psi, \tag{8-3}$$

where $\langle \psi | f \rangle$ is the scalar product defined, as usual, by

$$\langle \psi | f \rangle = \int \psi^*(1, \ldots, N) f(1, \ldots, N)\, d(1, \ldots, N). \tag{8-4}$$

In the familiar Dirac bracket notation,

$$P_\psi = |\psi\rangle \langle \psi| \tag{8-5}$$

so that

$$P_\psi |f\rangle = \langle \psi | f \rangle | \psi \rangle. \tag{8-6}$$

However, it is illuminating to regard P_ψ as an integral operator, so we recall the following ideas. Suppose $h(x)$ and $k(x, y)$ are functions of one and two variables, respectively, then to h we can associate another function of one variable which we denote by Kh and is such that

$$Kh(x) = \int k(x,y)\, h(y)\, dy, \tag{8-7}$$

where integration is over the domain of h. Since the operation which sends the function h into the function Kh is defined by (8-7), it is called an integral operator, is denoted by K, and has kernel $k(x, y)$. The one- and two-particle Green's functions are the kernels of such integral operators. From (8-3) and (8-4) we see that P_ψ is an integral operator on the antisymmetric functions of N-particles and that it has kernel

$$\psi(1,2,\ldots,N)\, \psi^*(1',2',\ldots,N'). \tag{8-8}$$

The $2N$-particle function (8-8) is the N-particle density matrix associated with the state ψ. The use of the term *matrix* in this context is perhaps unfortunate but seems foisted on us by history. It is justified by regarding the integration in (8-7) as summation, with x the row and y the column index. For an N-particle operator with kernel such as (8-8), $(1,2,\ldots,N)$ is the "row" and $(1',2',\ldots,N')$ the "column" index. By contracting or tracing on $N - 2$ particles in (8-8) we obtain a function of four particles

$$D^2(1,2;1',2') = \int \psi(1,2,3,\ldots,N)\, \psi^*(1',2',3,\ldots,N)\, d(3,\ldots,N), \tag{8-9}$$

which is the second-order reduced matrix or, briefly, the 2-matrix of the state ψ. Clearly $D^2(1,2;1',2')$ could be regarded as the kernel of a 2-particle operator D^2 such that if $f(1,2)$ is any 2-particle function, then $D^2 f$, where

$$D^2 f(1,2) = \int D^2(1,2;1',2')\, f(1',2')\, d(1',2'), \tag{8-10}$$

is the new function obtained from f by action on it by means of the integral operator D^2 with kernel defined in (8-9).

Similarly, if we contract (8-8) on $N - 1$ particles we obtain

$$D^1(1;1') = \int \psi(1,2,\ldots,N)\,\psi^*(1',2,\ldots,N)\,d(2,\ldots,N) \qquad (8\text{-}11)$$

$$= \int D^2(1,2;1',2)\,d(2), \qquad (8\text{-}12)$$

which is the first-order reduced density matrix, or briefly, the 1-matrix of the pure state ψ. Assuming, as we shall throughout this chapter, that ψ is normalized, we have

$$\int D^2(1,2;1,2)\,d(1,2) = \int D^1(1;1)\,d(1) = 1. \qquad (8\text{-}13)$$

By making use of the antisymmetry of ψ it is easy to show that the expectation, for the state ψ, of any 1-particle observable is determined by D^1, and of any 2-particle observable by D^2. Indeed, suppose A is a 1-particle observable which is represented by an operator

$$A = A(1) + A(2) + \cdots + A(N), \qquad (8\text{-}14)$$

which is symmetric in the N particles, then the expectation value of A for the state ψ is

$$\langle\psi|A|\psi\rangle = N\int [A(1)\,D^1(1;1')]_{1=1'}\,d(1). \qquad (8\text{-}15)$$

For example, if A were the kinetic energy of N electrons of an atom

$$A = -\frac{\hbar^2}{2m}\sum_i \Delta_i \qquad (8\text{-}16)$$

where Δ_i is the Laplacian acting on the ith electron, and the integrand on the right-hand side of (8-15) is

$$[A(1)\,D^1(1;1')]_{1=1'} = -\frac{\hbar^2}{2m}\left[\left(\frac{\partial^2}{\partial x_1^2} + \frac{\partial^2}{\partial y_1^2} + \frac{\partial^2}{\partial z_1^2}\right) D^1(1;1')\right]_{1=1'},$$

indicating that after differentiating with respect to the unprimed variables, the primed and unprimed variables are identified before the integration is effected.

When the operator $A(i)$ merely denotes multiplication of functions, e.g., if

$$A(i) = -\frac{Ze^2}{r_i}, \tag{8-17}$$

the potential energy between a nucleus at the origin with charge Ze and an electron, then $[A(1)D^1(1;1')]_{1=1'}$ is merely $A(1)D^1(1;1)$.

Similarly the expectation value of a two-particle operator

$$B = \sum_{i<j} B(i,j) \tag{8-18}$$

is given by

$$\langle\psi|B|\psi\rangle = \binom{N}{2} \int [B(1,2)D^2(1,2;1',2')]_{\substack{1=1'\\2=2'}} d(1,2). \tag{8-19}$$

In particular, if the Hamiltonian

$$H = \sum_i H(i) + \sum_{i<j} H(i,j), \tag{8-20}$$

then the exact energy of the state ψ is given by

$$E = \frac{N}{2} \int [K^2(1,2;1',2')D^2(1,2;1',2')]_{\substack{1=1'\\2=2'}} d(1,2) \tag{8-21}$$

$$= \frac{N}{2} \operatorname{tr}(K^2 D^2), \tag{8-22}$$

where the reduced Hamiltonian is given by

$$K^2 = H(1) + H(2) + (N-1)H(1,2). \tag{8-23}$$

As mentioned in Section 8-1, reduced density matrices are of interest chiefly because they enter into formulas such as (8-15), (8-19), and (8-22), showing that in order to study those properties of a single state which are of interest to physicists and chemists, we need know only functions of 2 or 4 particles rather than the wavefunction of N particles. It is therefore a problem of great importance, to which a satisfactory solution has not yet been achieved, to find direct methods of obtaining $D^1(1;1')$ and $D^2(1,2;1',2')$ for a given state without first calculating $\psi(1,2,\ldots,N)$. At first one might think that to obtain D^2 for the ground state of a system it would merely be necessary to minimize the expression (8-22) by varying D^2 subject to the normalization (8-13). However, we easily see that, in addition to (8-13), D^2 must satisfy the conditions

$$D^2(1,2;1',2') = D^{2*}(1',2';1,2) \tag{8-24}$$

$$D^2(1,2;1',2') = -D^2(2,1;1',2') = -D^2(1,2;2',1') \tag{8-25}$$

and, further

$$D^2 \geqslant 0, \tag{8-26}$$

i.e., D^2 acting as an operator on the space of all antisymmetric functions of 2 variables has nonnegative eigenvalues. Condition (8-24) on the 2-matrix is equivalent to the condition that D^2 be a hermitian operator, but (8-26) is even stronger. However, the conditions (8-24), (8-25), and (8-26) are not the real difficulty, which is somewhat subtle. The variation of D^2 in (8-22) must be restricted by the condition that $D^2(1,2;1',2')$ can be represented as in (8-9) by an integral involving the wavefunction which is antisymmetric in N particles. The antisymmetry is essential, since to forget it would be tantamount to forgetting that electrons obey Fermi statistics! The number N is essential in defining the atom or molecule under consideration. The problem of obtaining intrinsic criteria by which to recognize when a function of four particles can be expressed in the form (8-9) was named[2] the N-representability problem (*8*), but it had been noticed, perhaps in less precisely formulated terms, by others including Coulson before 1960. A simple solution of the N-representability problem would imply that the N-body problem is simple. This seems unlikely, but even a partial solution could shed important light on correlation phenomena and many other matters of concern to physicists and chemists.

However, most research on the N-representability problem has dealt with the problem of ensemble N-representability, rather than with the problem as formulated in (8-9) which is better described as pure N-representability since in (8-9) $D^2(1,2;1',2')$ is represented by a pure state.

Von Neumann (*42*) showed that in order to discuss mixtures or ensembles of various states of a system, it is convenient to specify the ensemble by means of a positive operator on the space of antisymmetric N-particle functions. If this operator, which is called the N-particle or von Neumann density operator, is denoted by D^N, then D^N is a positive operator of trace unity such that the expectation value of an observable such as the energy H is given by $\operatorname{tr}(HD^N)$. In the particular case that the system is in a pure state ψ, $D^N = P_\psi$ and $\operatorname{tr}(HD^N) = \operatorname{tr}(HP_\psi) = \langle\psi|H|\psi\rangle$, which agrees with the familiar formula for a pure state.

However, if the system is an ensemble of pure states ψ_i weighted by w_i, where $w_i > 0$ and $\Sigma w_i = 1$, then

$$D^N = \sum w_i P_{\psi_i}. \tag{8-27}$$

The N-matrix associated with this operator is

$$D^N(1,2,\ldots,N;1',2',\ldots,N') = \sum w_i \psi_i(1,2,\ldots,N)\,\psi_i^*(1',2',\ldots,N'). \tag{8-28}$$

We shall denote the set of all D^N of the form (8-28) by γ_N (or γ_N^N). It is a matter of capital importance that γ_N is a convex set.

Recall that a subset C of a linear space is convex if together with any two points of C all the points in the straight segment with these two points as end points are also contained in C. Thus, in the plane, circles, triangles, and squares are convex. A fact of great importance in this chapter is that a convex set is determined by its extreme points. For a plane convex polygon, the vertices are extreme points. For a circle or an ellipse each point of the circumference is extreme. In general, a point of a convex set is extreme if it is not an interior point of any straight-line segment completely contained in the set.

Just as we contracted (8-8) to obtain (8-9) so we could contract (8-28) and obtain

$$D^2(1,2;1',2') = \sum w_i D^2(\psi_i | 1,2;1',2'). \tag{8-29}$$

When there exist $w_i > 0$, $\Sigma w_i = 1$, and antisymmetric normalized N-particle functions ψ_i such that (8-29) obtains, we say that D^2 and $D^2(1,2;1',2')$ are ensemble N-representable.

An even more general situation is possible. For example, the grand canonical ensemble in statistical mechanics is a weighted sum of states of different particle numbers. Indeed, one of von Neumann's motivations in introducing the density matrix notation was to lay a proper mathematical foundation for statistical mechanics. The corresponding reduced density matrices are referred to merely as representable. However, hitherto there has been little discussion of reduced density matrices in statistical mechanics. The chief interest of chemists is in pure states so pure N-representability is the problem of central interest. Currently, however, it seems totally intractable (except for 2-representability of the 1-matrix!) and so research activity has been directed largely to ensemble N-representability or representability as a more tractable problem which may throw some light on pure N-representability (for certain pure-representable 1-matrices, see Ref. 3). A major advantage of the set of representable or the set of ensemble N-representable 2-matrices is that they are convex, and the considerable body of tools developed by mathematicians to study convexity may be brought to bear upon them. There is the additional consideration that, in the variational method of seeking D^2 for the ground state, if one varies over ensemble N-representable D^2 the solution obtained will automatically be pure N-representable.

In attempting to approximate D^1 or D^2 for any specific system it is usual to operate within the possibilities of a chosen set of basis orbitals in terms of which D^1 and D^2 will be described by matrices with complex elements. Let $\{\phi_i\}$ be a complete orthonormal set of one-particle functions (spin orbitals) and suppose that

$$\psi(1,2,\ldots,N) = \sum_i c_{i_1 i_2 \ldots i_N}\, \phi_{i_1}(1)\, \phi_{i_2}(2) \cdots \phi_{i_N}(N) \tag{8-30}$$

$$D^1(1;1') = \sum_{i,j} d^1_{ij}\, \phi_i(1)\, \phi_j^*(1') \tag{8-31}$$

and

$$D^2(1,2;1',2') = \sum_{i,j,k,l} d^2_{ijkl}\, \phi_i(1)\, \phi_j(2)\, \phi_k^*(1')\, \phi_l^*(2'). \tag{8-32}$$

Then if D^1 and D^2 are pure N-representable by ψ,

$$d^1_{ij} = \sum_{i_2 \ldots i_N} c_{i i_2 \ldots i_N}\, c^*_{j i_2 \ldots i_N} \tag{8-33}$$

and

$$d^2_{ijkl} = \sum_{i_3 \ldots i_N} c_{i j i_3 \ldots i_N}\, c^*_{k l i_3 \ldots i_N}. \tag{8-34}$$

The antisymmetry of ψ in the particle variables is, by (8-30), equivalent to the antisymmetry of the tensor $c_{i_1 i_2 \ldots i_N}$ with respect to the interchange of indices. In this context the N-representability problem can be formulated as follows: given d^1_{ij} and d^2_{ijkl}, provide criteria to recognize whether or not there is an antisymmetric $c_{i_1 i_2 \ldots i_N}$ satisfying equations (8-33) and (8-34). This formulation enables us to recognize immediately that N-representability is invariant with respect to unitary transformations of the basis set $\{\phi_i\}$. This statement can be interpreted in two ways: (i) Suppose that in (8-31) the ϕ_i are replaced by $\tilde{\phi}_i$ but d^1_{ij} are left unchanged. Then D^1 will change to $\tilde{D}^1$, but because of (8-33) $\tilde{D}^1$ will be pure N-representable if and only if D^1 is. (ii) Express $\tilde{D}^1$ in terms of

$\{\phi_i\}$ so the coefficients d^1_{ij} will be replaced by $\tilde{d}^1_{ij}$ obtained by action of the unitary group on its indices. The new matrix $\tilde{d}^1_{ij}$ is pure N-representable if and only if d^1_{ij} is. It is a short step from the preceding to the following conclusion (8):

Theorem: Each of the sets γ^1_N and γ^2_N of ensemble N-representable 1- and 2-matrices is partitioned into orbits under the action of the unitary group $U(r)$ of transformations of the orthonormal spin-orbital basis set.

Many questions about density matrices and the closely related Green's functions are best discussed in the notation of second quantization. The present is a convenient context in which to make this connection. Associated with our complete orthonormal basis $\{\phi_i\}$ there is a set of annihilators $\{a_i\}$ and creators $\{a_i^+\}$ which act on Fock space and satisfy the commutation relations

$$a_i a_j^+ + a_j^+ a_i = \delta_{ij} \tag{8-35}$$

$$a_i^+ a_j^+ + a_j^+ a_i^+ = a_i a_j + a_j a_i = 0. \tag{8-36}$$

Fock space is a Hilbert space whose elements can represent the state of a system with any fixed number of particles or coherent ensembles of such states. It contains a unique element representing the vacuum which we denote by $|0\rangle$ and which has the property $a_i|0\rangle = 0$ for all i. All elements of Fock space can be obtained from the vacuum state by repeated action of the creation operators a_i^+ and linear combination. Thus $\chi_i = a_i^+|0\rangle$ is the Fock-space vector for the spin orbital which we denoted in Schrödinger representation by ϕ_i. By means of the mixed symbol $\Phi(1) = \Sigma_i \phi_i(1)\, a_i$, we may pass from the one-particle state χ_i in Fock space to the corresponding state in configuration space by evaluating

$$\begin{aligned}\langle 0|\Phi(1)|\chi_i\rangle &= \sum_j \phi_j(1)\,\langle 0|a_j a_i^+|0\rangle \\ &= \sum_j \phi_j(1)\,\langle 0|a_j a_i^+ + a_i^+ a_j|0\rangle \\ &= \sum_j \phi_j(1)\,\delta_{ij}\,\langle 0|0\rangle = \phi_i(1).\end{aligned}$$

Similarly, $a_i^+ a_j^+|0\rangle$ corresponds to the normalized Slater function

$$\frac{1}{\sqrt{2}}(\phi_i(1)\,\phi_j(2) - \phi_i(2)\,\phi_j(1)) = \frac{1}{\sqrt{2}}\langle 0|\Phi(2)\,\Phi(1)\,a_i^+ a_j^+|0\rangle.$$

In general,

$$(N!)^{-1/2}\langle 0|\Phi(N)\cdots\Phi(1)\,a_{i_1}^{+}a_{i_2}^{+}\cdots a_{i_N}^{+}|0\rangle \tag{8-37}$$

is the normalized Slater function formed from the spin orbitals $\phi_{i_1}, \phi_{i_2}, \ldots, \phi_{i_N}$. Further, if

$$|\psi\rangle = \sum_{i_1<i_2\ldots<i_N} c_{i_1 i_2 \ldots i_N}\, a_{i_1}^{+}a_{i_2}^{+}\cdots a_{i_N}^{+}|0\rangle$$

then $\langle 0|\Phi(N)\,\Phi(N-1)\cdots\Phi(1)|\psi\rangle$, up to normalization, is identical to ψ of (8-30). Further,

$$\langle\psi|\Phi^{+}(1')\,\Phi(1)|\psi\rangle = ND^{1}(1;1') = \rho_1(1;1') \tag{8-38}$$

$$\langle\psi|a_j^{+}a_i|\psi\rangle = Nd_{ij}^{1} \tag{8-39}$$

$$\begin{aligned}\langle\psi|\Phi^{+}(2')\,\Phi^{+}(1')\,\Phi(1)\,\Phi(2)|\psi\rangle &= N(N-1)\,D^{2}(1,2;1',2')\\ &= \rho_2(1,2;1',2')\end{aligned} \tag{8-40}$$

$$\langle\psi|a_l^{+}a_k^{+}a_i a_j|\psi\rangle = N(N-1)\,d_{ijkl}^{2}. \tag{8-41}$$

From (8-39) and (8-41) we can immediately derive important necessary conditions for pure N-representability.

From the commutation relations it follows that $N_i = a_i^{+}a_i$ is an operator with the property $N_i^2 = N_i = N_i^{+}$. Thus N_i is an orthogonal projector and will have the property that for any $\psi, \langle\psi|N_i\psi\rangle = \langle N_i\psi|N_i\psi\rangle$ is a nonnegative real number less than or equal to 1. For $i \neq j, a_j^{+}a_i^{+}a_i a_j = a_j^{+}a_j a_i^{+}a_i = N_j N_i$ which is also a projector. Taking account of normalization properly one easily obtains the following important result:

Theorem: If ϕ_i is a normalized spin orbital and $[\phi_i\,\phi_j]$ is a normalized Slater determinant, then

$$0 \leqslant N\langle\phi_i|D^{1}\,\phi_i\rangle \leqslant 1 \tag{8-42}$$

$$0 \leqslant \binom{N}{2}\langle[\phi_i\,\phi_j]\,|D^{2}\,[\phi_i\,\phi_j]\,\rangle \leqslant 1. \tag{8-43}$$

These restrictions on the diagonal elements of the matrices representing D^1 and D^2 in a Slater basis are frequently referred to as the Pauli conditions on the reduced density matrices. They are necessary but not sufficient conditions for N-representability.

Since N-representability is invariant with respect to a unitary transformation of the spin-orbital basis set, and since the one-particle operator D^1 can be diagonalized by means of a unitary transformation, it follows that N-representability of D^1 can be characterized by conditions on the eigenvalues λ_i^1 of D^1 alone. Condition (8-42) implies that for pure N-representability the eigenvalues of D^1 are not greater than N^{-1}. However, the following important result (*8*) is true:

Theorem: A necessary and sufficient condition for D^1 to be ensemble N-representable is that its eigenvalues be nonnegative and less than or equal to N^{-1}.

The possibility of such a simple, relatively complete, and satisfactory characterization of ensemble N-representability of the 1-matrix is due to the fact that in this case the eigenvalues λ_i^1 of D^1 constitute a complete set of invariants under the action of $U(r)$. However, for reduced density matrices of order higher than the first, a complete set of invariants under the action of $U(r)$ is more complicated and contains many more elements, but some progress can be made by considering the eigenfunctions of D^2, which we shall relate to the definitive paper of Erhard Schmidt on integral equations with nonsymmetric kernels (*46*), in which he showed that if a complex-valued function f of two variables is expanded in the form

$$f(x,y) = \sum c_{ij} g_i(x)\, h_j(y), \tag{8-44}$$

in which $\{g_i\}$ is an orthonormal set, then the most efficient (in the sense of least squares) set of $g_i(x)$ are the eigenfunctions of an integral operator D_x with kernel

$$D_x(x;x') = \int f(x,y) f^*(x',y)\, dy. \tag{8-45}$$

By most efficient we mean the following. Suppose we truncate the series (8-44) at $i = n$ and set $\Sigma c_{ij} h_j(y) = c_i k_i(y)$. Then we ask for what choice of $\{g_i\}$ will

$$\int \left| f(x,y) - \sum_1^n c_i g_i(x)\, k_i(y) \right|^2 dx\, dy \tag{8-46}$$

be a minimum? Schmidt showed that the best expansion in this sense is

$$f(x,y) = \sum c_i g_i(x)\, k_i(y), \tag{8-47}$$

where g_i is an eigenfunction of D_x with eigenvalue $|c_i|^2$, and k_i is an eigenfunction of

$$D_y(y;y') = \int f(x,y) f^*(x,y')\, dx, \tag{8-48}$$

also with eigenvalue $|c_i|^2$. If in Schmidt's argument we replace x by $x_1, x_2, \ldots, x_p$ and y by $x_{p+1}, \ldots, x_N$ we find that the most efficient least-squares approximation to ψ is of the form

$$\psi(x_1, x_2, \ldots, x_N) = \sum c_i \phi_i^p(x_1, \ldots, x_p)\, \phi_i^q(x_{p+1}, \ldots, x_N),$$

where $p + q = N$. Here ϕ_i^p is an eigenfunction of

$$D^p(x_1, x_2, \ldots, x_p; x_1', \ldots, x_p')$$

$$= \int \psi(x_1, \ldots, x_N)\, \psi^*(x_1', \ldots, x_p', x_{p+1}, \ldots, x_N)\, d(x_{p+1}, \ldots, x_N) \tag{8-49}$$

with eigenvalue $|c_i|^2$, and ϕ_i^q is an eigenfunction of

$$D^q(x_{p+1}, \ldots, x_N; x_{p+1}', \ldots, x_N')$$

$$= \int \psi(x_1, \ldots, x_N)\, \psi^*(x_1, \ldots, x_p, x_{p+1}', \ldots, x_N')\, d(x_1, \ldots, x_p) \tag{8-50}$$

with the same eigenvalue $|c_i|^2$. It follows that the eigenfunctions of D^1 are particularly well suited for expanding ψ and therefore they were called *natural spin orbitals* (*37*). The eigenfunctions of D^2 have similar virtues and have been called *natural geminals.* Anyone with a penchant for abbreviation will not resist the usage *norbs* and *nags*. In the next section we report numerical results which illustrate the advantage of using natural orbitals in a multiconfiguration approximation to wavefunctions.

To conclude this section we return to the basic formula (8-22) for the exact energy of a system of N electrons. If g_i are the nags of D^2 with corresponding eigenvalue λ_i^2 then

$$D^2 = \sum \lambda_i^2 |g_i\rangle\langle g_i| \tag{8-51}$$

and from (8-22),

$$E = \frac{N}{2} \sum \lambda_i^2 \langle g_i | K^2 g_i \rangle. \tag{8-52}$$

This means that the exact total energy of the state is a sum of contributions associated with the natural geminals g_i. This observation gives added force to the suggestion of Shull quoted in Section 8-1. In fact, hitherto there has been relatively little research effort devoted to obtaining explicitly the natural geminals of simple molecules or studying the electron density in the natural geminals. This would appear to us as an essential first step in exploring Shull's suggestion and may well prove an important new research area.

The relation of formulas (8-51) and (8-52) to superconductivity is particularly interesting and well understood. Yang was the first to associate "large" eigenvalues of D^2 to superconductivity (*53*). It is a trivial matter to see that the 2-matrix of an N-particle Slater determinant has $\binom{N}{2}$ nonzero eigenvalue $\lambda_i^2 = \binom{N}{2}^{-1}$ and that the nags are Slater geminals, that is, 2×2 determinants. We already noticed, (8-43), that if g_i is a Slater determinant, $\lambda_i^2 \leqslant \binom{N}{2}^{-1}$. Further it has been proved (*12*; see also Ref. 13) that D^2 has at least $\binom{N}{2}$ nags so that the average value of λ_i^2 is not greater than $\binom{N}{2}^{-1}$. For some time it was believed that $\binom{N}{2}^{-1}$ was an upper bound for λ_i^2. Indeed an erroneous proof of this conjecture was published. However, it was finally shown that λ_i^2 is less than, but can become arbitrarily close to, N^{-1} (*6–10, 45, 53*). Since K^2 involves a factor $(N - 1)$, when $\lambda_i^2 \to N^{-1}$ the contribution to E in (8-52) associated with the geminal g_i is proportional to N. A geminal describes correlation so that associated with "large" λ_i^2 there is an infinite-range correlation. This fact was first proved by Yang for superconductivity by an argument which was later simplified (*2*). The preceding argument is more general and covers both superconductivity and the many varieties of magnetic ordering. Indeed, with minor modification the argument can include superfluidity (*14*).

8-3. ANALYZING WAVEFUNCTIONS

The information of interest to a chemist about the electronic state of an atom or molecule is encoded in the 2-matrix. Therefore in comparing various approximate wavefunctions for the same state it is sufficient to compare the 1-matrices and 2-matrices derived from them. This can be accomplished most easily by displaying the eigenfunctions and corresponding eigenvalues of the reduced density matrix, since these have an intrinsic significance independent of the original basis sets used in calculating the various wavefunctions. The earliest examples of such natural-orbital analysis of wavefunctions revealed that most of the information in the original multiconfiguration wavefunction could have been contained in a function of much fewer configurations using as a basis the most highly occupied natural orbitals.

This fact was first noticed by Löwdin and Shull (*38*) and developed in their paper on the helium atom (*48*). They took Laguerre functions as their basic spin orbitals, using as many as 21 and including configurations up to $(6s)^2$. The best energy achieved with this basis set was -2.878970 au. However, using as a basis set 1, 3, and 6 natural orbitals they obtained -2.585000, -2.876022, and -2.878962, respectively. This result of Löwdin and Shull illustrates vividly the advantage of employing norbs as a basis—an advantage which derives from Schmidt's observation that the natural p-functions provide the most efficient basis for approximating a function in the sense of least squares (*8*).

In a subsequent paper, Shull (*47*) applied these ideas to a study of the hydrogen molecule and concluded that "approximate natural spin orbital occupation numbers are nearly invariant under a wide variety of choices of basis functions and therefore are particularly suitable for comparison of different approximate functions and for discussion of their respective properties." This work of Shull on H_2 was extended by Davidson and Jones (*16*) who obtained the norbs of the 50-term ground-state wavefunction of H_2 obtained by Kolos and Roothaan. Using only *four* norbs they obtained a function which had an overlap of 0.999667 with the original function and which gave 90% of the correlation energy. In subsequent work, Davidson (*17*) studied Kinoshita's 39-term function for He and noted, however, that even with many spin orbitals the cusp at $r_{12} = 0$ is not well approximated. The type of natural analysis described above for two-electron systems was extended to four- and six-electron systems (*1*, *51*).

The preceding results suggest that any published approximate wavefunction should be accompanied by an analysis of the 1-matrix, certainly to the extent of reporting the occupancy of the norbs. There has been relatively little work so far published on the natural geminals (nags) of the 2-matrix. The first was due to Smith and Fogel (*50*), who analyzed the 2-matrix of Watson's 37-configuration wavefunction for Be, normalized to have trace 6. They found occupancy of 6 nags at levels 1.00084 to 0.916971; 12 at level 0.026999; and the remainder at less than 0.0007. A Slater wavefunction would give six nags all with occupancy 1. At one time it had been conjectured that the occupancy of a nag could not exceed that of a Slater function. This was disproved subsequently, so it was of some interest that indeed, the Slater value was exceeded by Watson's Be wavefunction. The same situation occurs in the ^1S state of carbon, which has a "supergeminal" with occupation number approximately 1.25 (*4*). However, the fact that occupancy of a nag can achieve $(N/2)$ is the basic condition permitting the existence of superconductivity and superfluidity. Smith and Fogel studied the change in D^2 in passing from HF to a separated pair to a multiconfiguration wavefunction and concluded that one might interpret the effect of outer-shell correlation in Be as causing a "forced delocalization of the natural geminals and a splitting of the singlet–triplet degeneracy which would otherwise be present in the geminal occupation numbers."

We referred in Section 8-1 to Shull's assertion that the natural geminals might prove of capital significance for a deeper understanding of the chemical bond. Unhappily, the geminal analysis of wavefunctions begun by Darwin Smith and his students has had few imitators. It is most desirable that geminal analyses of a wide variety of atoms and molecules be published in order for us to develop an intuitive feeling and a sense of familiarity with geminals analogous to that with orbitals which grew out of the widespread experimentation with various kinds of orbitals between 1930 and 1950. We need studies of the variation of the nags and their occupancy over several series of homologous atoms and molecules, and of their differences between strikingly different systems.

8-4. *N*-REPRESENTABILITY

In previous sections we have made several allusions to the N-representability of the reduced density matrices, which essentially is the requirement that electrons satisfy Fermi statistics. N-representability is therefore rather crucial. Conditions for N-representability play the role of constraints if we attempt to find D^2 directly for the ground state by a variational approach. In the next section we report on some attempts to do this. In the present section, we attempt to summarize current knowledge about conditions for ensemble N-representability. However, because of space limitations we shall be forced to omit most proofs and also some results which require too extensive prefatory explanations.

To introduce the notion of the *dual* or *polar* of a convex cone, consider the situation depicted in Fig. 8-1. Let $\mathcal{V}$ be the set of vectors emanating from O and pointing in directions between and including OA and OB. If $\boldsymbol{v}_1$ and $\boldsymbol{v}_2$ belong to $\mathcal{V}$, then their weighted sum $\boldsymbol{v} = m_1\boldsymbol{v}_1 + m_2\boldsymbol{v}_2$, with m_1 and m_2 any nonnegative real number, is a vector in $\mathcal{V}$. Indeed, $\boldsymbol{v}$ would point in a direction between $\boldsymbol{v}_1$ and $\boldsymbol{v}_2$. Therefore, the set $\mathcal{V}$ is closed under the operation of taking weighted sums (with nonnegative weights) of its members. $\mathcal{V}$ is said to be a convex cone. Furthermore, as defined, $\mathcal{V}$ is closed–i.e., the limit of any convergent sequence of vectors in $\mathcal{V}$ is also in $\mathcal{V}$. We could have defined a cone $\mathcal{W}$, say to consist of all vectors in directions between but not including $\boldsymbol{OA}$ and $\boldsymbol{OB}$. $\mathcal{W}$ is also a convex cone but it is not closed. Indeed, any vector $\boldsymbol{v}$ in $\mathcal{W}$ is surrounded by a set of vectors, a neighborhood of $\boldsymbol{v}$, all of which are also in $\mathcal{W}$. Thus $\mathcal{W}$ is an open cone. The closure of $\mathcal{W}$ is $\mathcal{V}$. If we had included $\boldsymbol{OA}$ but not $\boldsymbol{OB}$, we would have a cone which is neither open nor closed. The intersection of the line l in the diagram with $\mathcal{V}$ is a segment $[MN]$ which is called a base for $\mathcal{V}$, since $\mathcal{V}$ can be obtained by drawing all possible vectors through O and the various points of $[MN]$. $\mathcal{V}$ being closed and convex implies that $[MN]$ is closed and convex. M and N are extreme points of $[MN]$. Similarly $\boldsymbol{OA}$ and $\boldsymbol{OB}$ are called ex-

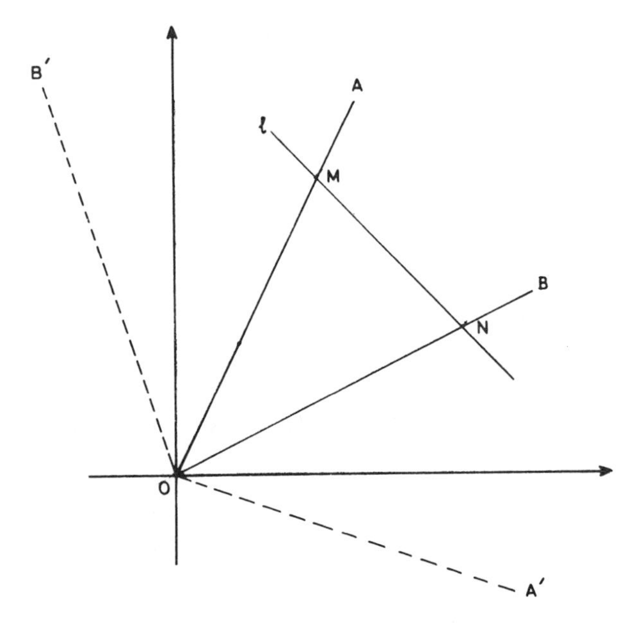

Fig. 8-1. The notion of a convex cone.

treme rays of $\mathcal{O}$. It is an important fact that a compact convex set is determined by its extreme points. Also, a closed convex cone which has a compact base is determined by its extreme rays–i.e., the cone is the smallest closed convex set containing the extreme rays.

Now consider the set of all vectors $\boldsymbol{b}$ which have nonnegative scalar product with all the vectors in $\mathcal{O}$. Let $\tilde{\mathcal{O}} = \{\boldsymbol{b} \,|\, \boldsymbol{v} \in \mathcal{O} \Rightarrow \boldsymbol{b} \cdot \boldsymbol{v} \geqslant 0\}$. Clearly, $\tilde{\mathcal{O}}$ is the closed convex cone with extreme rays OA' and OB', where $OA' \perp OA$ and $OB' \perp OB$. Indeed, it is trivial to see that if $\boldsymbol{b}_1$ and $\boldsymbol{b}_2$ belong to $\tilde{\mathcal{O}}$, then for $m_1, m_2 \geqslant 0$, $(m_1\boldsymbol{b}_1 + m_2\boldsymbol{b}_2) \cdot \boldsymbol{v} = m_1\boldsymbol{b}_1 \cdot \boldsymbol{v} + m_2\boldsymbol{b}_2 \cdot \boldsymbol{v} \geqslant 0$ for all $\boldsymbol{v} \in \mathcal{O}$. Therefore, $m_1\boldsymbol{b}_1 + m_2\boldsymbol{b}_2 \in \tilde{\mathcal{O}}$. $\tilde{\mathcal{O}}$ is called the dual or polar cone to $\mathcal{O}$. If we now ask for the polar cone of $\tilde{\mathcal{O}}$, namely $\tilde{\tilde{\mathcal{O}}} = \{\boldsymbol{c} \,|\, \boldsymbol{b} \in \tilde{\mathcal{O}} \Rightarrow \boldsymbol{c} \cdot \boldsymbol{b} \geqslant 0\}$, we easily see that $\tilde{\tilde{\mathcal{O}}} = \mathcal{O}$. This illustrates the famous Bipolar Theorem which motivates much of the research on ensemble N-representability: *The polar of the polar of a convex cone is the closure of the original cone.* Thus a closed convex cone is determined by its polar. Each element $\boldsymbol{b}$ of the polar defines a condition, namely $\boldsymbol{b} \cdot \boldsymbol{v} \geqslant 0$, which must be satisfied by all vectors $\boldsymbol{v}$ of the original cone. One easily sees that it will be sufficient to know the conditions defined by the extreme rays of the polar. If they are satisfied, then all other conditions will be automatically guaranteed.

The set of pth order reduced matrices which are ensemble N-representable form a closed convex cone and the subset of trace 1 constitute a base of this cone which we denote by Υ_N^p. We denote by L_N^p the contraction map which integrates on $n - p = q$ variables. If D^N is a normalized N-particle, or von Neumann, density matrix, then $D^p = L_N^p D^N$ is an ensemble N-representable pth order reduced density matrix, or a p-matrix for short. If Υ^N denotes the set of all normalized positive operators on the space of antisymmetric N-particle wave functions, then $\Upsilon_N^p = L_N^p \Upsilon^N$.

We denote the polar of Υ_N^p by $\tilde{\Upsilon}_N^p$. This will be a convex cone the elements of which are not necessarily normalized. (Clearly $\boldsymbol{b} \cdot \boldsymbol{v} \geqslant 0 \Rightarrow 3\boldsymbol{b} \cdot \boldsymbol{v} \geqslant 0$.) The kernel of $\boldsymbol{b} \in \tilde{\Upsilon}_N^p$ is defined as $\{D^p | \boldsymbol{b} \cdot D^p = 0, D^p \in \Upsilon_N^p\}$. An element of $\tilde{\Upsilon}_N^p$ is interior if and only if its kernel is empty. If $\boldsymbol{b} \in \partial \tilde{\Upsilon}_N^p$, the surface of $\tilde{\Upsilon}_N^p$, but $\boldsymbol{b}$ is not extreme, then for some $\boldsymbol{b}_1$ and $\boldsymbol{b}_2, \boldsymbol{b} = m_1 \boldsymbol{b}_1 + m_2 \boldsymbol{b}_2$ with $m_1 > 0$, $m_2 > 0, \boldsymbol{b}_1 \in \tilde{\Upsilon}_N^p$. But $\boldsymbol{b} \cdot \boldsymbol{v} = 0 \Rightarrow \boldsymbol{b}_1 \cdot \boldsymbol{v} = \boldsymbol{b}_2 \cdot \boldsymbol{v} = 0$. Therefore the kernel of $\boldsymbol{b}$ is contained in the kernel of $\boldsymbol{b}_1$ and the kernel of $\boldsymbol{b}_2$, and both $\boldsymbol{b}_1$ and $\boldsymbol{b}_2 \in \partial \tilde{\Upsilon}_N^p$.

Suppose B^p is an operator acting on the particles $1, 2, \ldots, p$; f is a function of particles $1, 2, \ldots, N$; A_N is the antisymmetrizer acting on particles $1, 2, \ldots, N$; and I^q is the identity operator on the particles $p + 1, \ldots, N$. Then, the tensor product $B^p \otimes I^q$ denotes the operation which sends f into $B^p f$, whereas $B^p \wedge I^q = A_n B^p \otimes I^q A_N$ is an operation which first antisymmetrizes N particles, then acts with $B^p \otimes I^q$ and finally antisymmetrizes. Thus $B^p \wedge I^q$ is an operation which is linear in B^p and sends antisymmetric functions of N-particles into antisymmetric functions of N-particles. Thus if we define the lifting operator Γ_p^N by

$$\Gamma_p^N B^p = B^p \wedge I^q \tag{8-53}$$

we see that the lifting operator will send an operator B^p which acts on antisymmetric p-particle functions into one which acts on antisymmetric N-particle functions. It is a key observation (*32*) that

$$\mathrm{tr}\,(B^p \wedge I^q \cdot D^N) = \mathrm{tr}\,(\Gamma_p^N B^p)(D^N) = \mathrm{tr}\,(B^p L_N^p(D^N)). \tag{8-54}$$

That is, with respect to the trace scalar product, Γ_p^N and L_N^p are hermitian conjugate operators. Furthermore, we have the following key result:

Theorem:

$$B^p \in \tilde{\Upsilon}_N^p \Longleftrightarrow B^p \wedge I^q \geqslant 0. \tag{8-55}$$

That is, the polar of Υ_N^p consists precisely of those operators on p-space which when lifted to N-space give rise to a positive semidefinite operator.

Since in (8-54) D^N can be any positive operator, it follows that a necessary and sufficient condition that $B^p \in \tilde{\Upsilon}_N^p$ is that $B^p \wedge I^q$ be a positive semidefinite operator on $\wedge^N \mathcal{H}$. Since $I^p \wedge I^q = I^N$, it is clear that $kI^p \in \tilde{\Upsilon}_N^p$ for $k > 0$. If, as we always assume, the dim $(\mathcal{H}) = r$ is finite, the dimension of $\wedge^N \mathcal{H}$ is $\binom{r}{N} = s$ and $B^p \wedge I^q$ is represented by an $s \times s$ matrix. Let

$$\Delta(B^p) = |B^p \wedge I^q| \tag{8-56}$$

denote the determinant of $B^p \wedge I^q$. Δ will be a homogeneous polynomial of degree s in the coefficients of the matrix representing B^p with respect to any fixed basis of $\wedge^p \mathcal{H}$. For $\gamma > 0$, the hypersurface of dimension $s - 1$ on which $\Delta(B^p) = \gamma$ consists of several distinct connected portions which can be classified by the number of negative eigenvalues of $B^p \wedge I^p$. Since γ is equal to the product of these eigenvalues, the number of negative eigenvalues is even for $\gamma > 0$. The different portions of this hypersurface are separated by the hypersurface on which $\Delta(B^p) = 0$. There is one portion, on which all the eigenvalues of $B^p \wedge I^p$ are positive and this, by the preceding result, is interior to $\tilde{\Upsilon}_N^p$. Let us denote this particular surface by Σ_γ. It is not difficult to show that Σ_γ for $\gamma > 0$ is a convex hypersurface. Furthermore it is smooth, which means that at each point it possesses a unique tangent hyperplane. As $\gamma \to 0$, $\Sigma_\gamma \to \partial \tilde{\Upsilon}_N^p$, the surface of $\tilde{\Upsilon}_N^p$.

Since for $\gamma > 0$, Σ_γ is smooth, at any point B^p on the surface Σ_γ the gradient of Δ, which we shall denote by $\partial\Delta/\partial B^p$, is well defined. It points in a direction perpendicular to the tangent plane to Σ_γ at B^p. From Fig. 8-1 we see that the polar of the convex cone $\mathcal{C}$ determined by $\boldsymbol{OA}$ and $\boldsymbol{OB}$ is determined by the vectors $\boldsymbol{OB}'$ and $\boldsymbol{OA}'$ which are perpendicular to the boundary of $\mathcal{C}$. In an analogous manner, the polar of the convex set with boundary Σ_γ is determined by the set

$$\tilde{\Sigma}_\gamma = \left\{ \frac{\partial \Delta}{\partial B^p} \middle| B^p \in \Sigma_\gamma \right\}. \tag{8-57}$$

Indeed, as γ varies over all positive real numbers, Σ_γ sweeps out the whole of the interior of the cone with base $\tilde{\Upsilon}_N^p$. By considering the limit at γ approaches 0 through positive values we are led to the following basic result:

Theorem: The cone based on Υ_N^p, i.e., the set of unnormalized N-representable reduced p-matrices, is the closure of

$$\left\{ \frac{\partial \Delta}{\partial B^p} \middle| B^p \wedge I^q > 0 \right\}, \tag{8-58}$$

where $p + q = N$. The condition $B^p \wedge I^q > 0$, which means that $\Gamma_p^N B^p$ is a strictly positive definite operator on $\wedge^N \mathcal{H}$, is equivalent to requiring that B^p belong to the interior of the polar cone, $\tilde{\Upsilon}_N^p$, of Υ_N^p.

The boundary $\partial \tilde{\Upsilon}_N^p$ of $\tilde{\Upsilon}_N^p$ is a convex hypersurface and is a connected portion of the complicated hypersurface Σ_0 which divides the set of hermitian operators on $\wedge^p \mathcal{H}$ into 2^s regions indexed by the signs of the eigenvalues of $B^p \wedge I^q$, $(p + q = N)$. We conjecture that these regions are grouped into $s + 1$ connected sets indexed by the number of negative eigenvalues of $B^p \wedge I^q$, from 0 to s. In any case, we are interested only in $\tilde{\Upsilon}_N^p$ in which $B^p \wedge I^q$ has no negative eigenvalues.

When presented in the preceding general form, the geometry of this situation may appear rather complicated. It might be helpful to consider the following example, which illustrates the main ideas involved. Consider the set of positive semidefinite 2×2 matrices,

$$A = \begin{pmatrix} a & c \\ c & b \end{pmatrix} \geqslant 0. \tag{8-59}$$

Let $f(A) = \det A = ab - c^2$. The gradient of f is given by

$$\frac{\partial f}{\partial a} = b, \quad \frac{\partial f}{\partial b} = a, \quad \frac{\partial f}{\partial c} = -2c.$$

We associate to f the matrix

$$\begin{pmatrix} b & -c \\ -c & a \end{pmatrix}, \tag{8-60}$$

which we denote by $\partial f/\partial A$, so that

$$A \frac{\partial f}{\partial A} = f(A)\, I. \tag{8-61}$$

Thus $\partial f/\partial A$ is essentially the inverse of A. It is sometimes called the adjugate of A.

It is known that the cone of positive matrices is self-dual. Indeed, as A of (8-59) ranges over the positive semidefinite matrices, the matrix (8-60) also ranges over the positive semidefinite matrices.

For $\gamma \geqslant 0$, let Σ_γ denote the surface

$$f(A) = \gamma \tag{8-62}$$

regarded as a subset of the real space $\mathcal{L}$ of all hermitian 2×2 matrices. We easily see that Σ_0 is a right circular cone which divides $\mathcal{L}$ into three regions in which A has 0, 1, or 2 negative eigenvalues. On the surface Σ_0, A has one zero eigenvalue except at O, where it has two. The dimension of the set of A with 1 or 2 zero eigenvalues is 2 or 0, respectively. Notice that requiring that one more eigenvalue be zero reduces the dimension by 2 rather than by 1. This explains why Σ_0 divides $\mathcal{L}$ into only $3 = 2 + 1$, rather than $2^s = 4$ connected regions as might have been expected. For $\gamma > 0$, Σ_γ is a hyperboloid of two sheets which is asymptotic to the cone Σ_0 at infinity.

The preceding theorem can be rephrased as follows, for $p = 2$: *As B^2 varies over $\tilde{\Upsilon}_N^2$, $\partial\Delta/\partial B^2$ varies over the cone with base Υ_N^2.*

We have thus achieved a parametrization of the set Υ_N^2 of N-representable 2-matrices by means of the 2-matrices belonging to $\tilde{\Upsilon}_N^2$. Furthermore, we can characterize $\tilde{\Upsilon}_N^2$ as the connected set of hermitian operators which contains I^2 and in which $\Delta(B^2)$ is always nonnegative. We are thus led to the following strategy for obtaining the ground-state energy of an N-electron system:

i. Choose a suitable basis of r spin orbitals. (For this step the chemist's experience and knowhow will be crucial.)
ii. Take the $\binom{r}{2}^2$ elements of the matrix B^2 as parameters.
iii. Calculate $\Delta(B^2)$ in terms of the coefficients of B^2.
iv. Calculate $\partial\Delta/\partial B^2$, the gradient of Δ for general B^2.
v. Starting from B^2 proportional to I^2 vary B^2 subject to the following conditions:
   a. $\Delta(B^2) \geqslant 0$;
   b. $\mathrm{tr}\,(\partial\Delta/\partial B^2) = 1$ so as to minimize $\mathrm{tr}\,(K^2(\partial\Delta/\partial B^2))$, where K^2 is the reduced Hamiltonian.

Since Δ is a homogeneous polynomial of degree $\binom{r}{N}$ in the coefficients of B^2, for large r and N calculating with it will pose grave difficulties. However, by using computer languages designed to manipulate algebraic expressions, it should be possible to implement the above strategy for small values of r and N.

For $p = 1$, B^1 can be diagonalized by means of a transformation in $U(r)$. When this is done, $B^1 \wedge I^{N-1}$ admits a simple explicit representation. To obtain this, suppose the eigenvalues of B^1 are b_i, $1 \leqslant i \leqslant r$, and that K runs over the $\binom{r}{N}$ sets of N integers between 1 and r. Defining

$$b_K = \sum_{i \in K} b_i, \tag{8-63}$$

then

$$\Delta(B^1) = \prod_K b_K. \tag{8-64}$$

Similarly, if B^2 is diagonal in a Slater basis with eigenvalues b_{ij}, if we define

$$\beta_K = \sum b_{ij}, \tag{8-65}$$

where the summation is over distinct pairs (i, j) in K, then

$$\Delta(B^2) = \prod_K \beta_K. \tag{8-66}$$

However, if B^2 cannot be diagonalized by means of a basis of Slater geminals, then, in general, $\Delta(B^2)$ will be rather complicated. In this case, if we interpret the b_{ij} as the diagonal terms of B^2 referred to a Slater basis, then the terms of the right-hand side of (8-66) will occur in the full expansion and it might be possible to make some progress by retaining only particular "important" terms in $\Delta(B^2)$ involving off-diagonal coefficients of B^2. This is an avenue which, to the author's knowledge, has not been explored so far.

8-5. SOME NECESSARY CONDITIONS FOR *N*-REPRESENTABILITY

The strategy proposed in the previous section for imposing N-representability on the 2-matrix is mathematically complete and correct but is quite impractical for large r, since it involves a polynomial of degree $\binom{r}{N}$. Therefore a great deal of effort has gone into the search for alternative methods of assuring N-representability of D^2, if not exactly then at least approximately. We are encouraged in this search by the fact that the N-representability of the 1-matrix can be expressed in a very simple manner.

Indeed, in the case of the 1-matrix the theorem of Section 8-2 ensures that the Pauli conditions alone are necessary *and* sufficient for N-representability. It may easily be seen that these conditions can be expressed by

$$0 \leqslant N\langle\phi|D^1|\phi\rangle \leqslant 1 \tag{8-67}$$

for all normalized spin orbitals ϕ. Alternatively, these conditions are equivalent to

$$D^1 \geqslant 0 \text{ and } I^1 - ND^1 \geqslant 0, \tag{8-68}$$

which assert that two $r \times r$ matrices are positive semidefinite. In (8-67) and (8-68) we assume that D^1 is normalized so that $D^0 = \operatorname{tr} D^1 = 1$. The condition that an unnormalized D^1 is N-representable takes the form

$$D^1 \geqslant 0 \text{ and } D^0 I^1 - ND^1 \geqslant 0. \tag{8-69}$$

This form has the advantage of explicitly displaying the fact that the second condition is linear in D^1 and shows that the set of 1-matrices satisfying the conditions (8-69) is convex.

There are a number of known necessary conditions satisfied by an N-representable D^2 which are analogous to (8-69) in requiring that certain matrices linear in D^2 be positive semidefinite. Four such conditions have been given considerable attention. They are frequently referred to as the D, Q, B, and G conditions.

The *D-condition* asserts

$$D^2 \geqslant 0. \tag{8-70}$$

That is, D^2 is a positive semidefinite operator acting on the Hilbert space of all antisymmetric two-particle functions, or geminals, g. The D-condition is strictly equivalent to assuming that for all geminals g, the projectors P_g belong to $\tilde{\Upsilon}_N^2$, the polar cone of the N-representable 2-matrices. That is,

$$P_g = |g\rangle\langle g| \in \tilde{\Upsilon}_N^2. \tag{8-71}$$

The *Q-condition* asserts

$$Q(D^2) = D^0 I^2 - 2ND^1 \wedge I^1 + \binom{N}{2} D^2 \geqslant 0, \tag{8-72}$$

and is equivalent to

$$q(g) = I^2 - 2ND^1(g) \wedge I^1 + \binom{N}{2} P_g \in \tilde{\Upsilon}_N^2 \tag{8-73}$$

for all g. Here we assume that the geminal g is normalized, and $D^1(g)$ is its 1-matrix. $Q(D^2)$ is often called the hole-matrix because, if $D^2 = L_N^2(P_\Psi)$, where Ψ is an N-fermion function, then $Q(D^2)$ is proportional to the matrix which, in second-quantization notation, is given by

$$\langle \Psi | a_i a_j a_k^\dagger a_l^\dagger | \Psi \rangle.$$

This can be compared with D^2, which is proportional to $\langle \Psi | a_i^\dagger a_j^\dagger a_k a_l | \Psi \rangle$.

The *B-condition* (*12, 13*) asserts that

$$B(D^2) = D^0 I^2 - (N-2) D^1 \wedge I^1 - (N-1) D^2 \geqslant 0 \tag{8-74}$$

and is equivalent to

$$b(g) = I^2 - (N-2) D^1(g) \wedge I^1 - (N-1) P_g \in \tilde{\Upsilon}_N^2 \tag{8-75}$$

for all g.

When $D^1(g)$ is proportional to I^1, $b(g)$ becomes essentially the Hamiltonian of the simplest form of the BCS theory of superconductivity. The ground state of $b(g)$ is the so-called antisymmetrized geminal product (AGP) wavefunction formed from one geminal g. It is also called the projected BCS wavefunction and the pairing wavefunction. The B-condition does not seem to have a perspicuous form in second-quantization notation. It is related to an irreducible representation of the symmetric group on $N+2$ objects (*12*). The operator $b(g)$ in (8-75) is of interest because $\Gamma_2^N b(g)$ can be used as a model Hamiltonian which involves interaction in an essential manner and for which the ground state is known explicitly. For various choices of g the corresponding ground states are quite different, ranging from Hartree-Fock to BCS.

The *G-condition* (*24*) asserts that for arbitrary complex numbers c^{ij}

$$c^{*ij} c^{kl} [\delta_{ik} D^1_{lj} - (N-1) D^2_{lijk}] \geqslant 0, \tag{8-76}$$

which is equivalent to asserting that

$$G_{ijkl} = \delta_{ik} D^1_{lj} - (N-1) D^2_{lijk} \tag{8-77}$$

are the components of a positive semidefinite operator on the tensor product $\mathcal{H} \otimes \mathcal{H}$. We are, of course, assuming the Einstein summation on repeated subscripts and superscripts. If α denotes the one-particle operator represented by the matrix a_s^r such that $\alpha\phi_s = a_s^r \phi_r$, where $\{\phi_r\}$ is an orthonormal basis of $\mathcal{H}$, then denote by $\bar{\alpha} \wedge \alpha$ the operator on $\wedge^2 \mathcal{H}$ obtained by antisymmetrizing $\bar{a}_k^i a_l^j$ in the pairs i, j, and k, l, and let γ_k^i be the one-particle operator with matrix $\bar{a}_k^s a_s^i$. Then the G-condition on D^2 is equivalent to

$$\gamma \wedge I^1 - (N-1) \bar{\alpha} \wedge \alpha \in \tilde{\Upsilon}_N^2 \tag{8-78}$$

for all α (*32*).

The G-condition has an intuitively clearer presentation in second-quantization notation. Let $A = c^{ij} a_i^+ a_j$ be an arbitrary one-particle operator; then for D^2 derived from an N-particle wavefunction Ψ, the condition (8-76) is equivalent to

$$\langle \Psi | A^+ A | \Psi \rangle \geqslant 0 \tag{8-79}$$

for all A. Erdahl and Rosina have shown that the G-condition implies the B-condition (*21*).

A fifth necessary condition discovered only recently (*33*) appears to be closely related to the G-condition, though the relation is not fully understood. This so-called *C-condition* is given by

$$C(D^2) = (r - N + 2) D^1 \wedge I^1 - (N-1) D^2 \geqslant 0 \tag{8-80}$$

and is equivalent to

$$c(g) = (r - N + 2) D^1(g) \wedge I^1 - (n - 1) P_g \in \tilde{\Upsilon}_N^2 \tag{8-81}$$

for all geminals g.

Kummer et al. (*33*) define a two-dimensional set of necessary conditions generated by operator homomorphisms of Υ^2 into Υ^2. They show that this two-dimensional set is convex, with four extreme points corresponding precisely to the D, Q, B, and C conditions. We refer the reader to their paper and to that of Kummer and Absar (*34*) for complete details.

All five conditions are linear in D^2. They are necessary conditions for N-representability. Thus the set of D^2 which satisfy such conditions is a convex set $\mathcal{C}$ which contains Υ_N^2. The energy E is linear in D^2. Thus the minimum of E on $\mathcal{C}$ will be a lower bound to the ground-state energy.

The first calculations employing this approach were quite encouraging. Fusco (*22*) made use of the D, Q, and G conditions to obtain a value of -14.61425 au for the ground-state energy of Be atom. This may be compared with the Hartree-Fock (HF) value of -14.57294 and R. E. Watson's IC value of -14.61414.

Mihailovic and Rosina (*40*) similarly made calculations for various model nuclear energies with the following results (in MeV).

	Variational	IC
^{15}O	-21.69	-20.14
^{16}O	-36.60	-35.00
^{20}Ne	-43.10	-41.61
^{28}Si	-172.45	-153.18

In each case, the variational calculation gave *lower* bounds to the exact model IC values, as was to be expected, but they seemed not to be very sharp for larger N.

For additional numerical results and for details about the theory and methods of calculation by which these results were attained, the reader may consult the literature (*23*).

8-6. RELATION TO GREEN'S-FUNCTION METHODS

Öhrn has recently surveyed the application of Green's functions (GF) to atomic and molecular structure (*43*). His clearly written and comprehensive article contains an extensive bibliography. We therefore refrain from any attempt to develop GF methods, since it would take more space than is available and the author cannot hope to improve on Öhrn's treatment.

A reduced density matrix is an initial value of a particular GF of the same order. Conversely, we may describe a GF as a time-dependent reduced density matrix. It follows that the N-representability conditions on the RDM are also conditions on the GF. It is only in recent years that the importance of N-representability has been appreciated by the GF devotees. The GFs of various orders satisfy a hierarchy of coupled equations up to an equation of order N. If they are all satisfied then N-representability is automatic. However, in practice the hierarchy of equations is truncated at order one or two. But, since necessary and sufficient conditions for N-representability for second-order reduced density matrices are not known, it is impossible to verify that any particular ansatz for truncation ensures N-representability. This fact was already implicit in Hone's (*29*) discussion of the apparent contradiction between results of Anderson and Morel, and those of Gorkov and Galitskii.

In the literature of GFs there has been extensive study of their mathematical properties by means of Fourier representation, complex function theory, and perturbation theory. Most of the latter is of practical value only in homogeneous systems. However, in recent years there have been some encouraging attempts to adopt methods originally developed for quantum field theory to atomic and molecular systems. For example, Doll and Reinhardt (*20*) obtained surprisingly good values for the electron density, natural orbitals, ionization potential, and total ground-state energy for He and Be by means of a second-order perturbation expression for the first-order GF. Their implicit starting point is a HF wavefunction, although the wavefunction never appears explicitly in their calculation. For further information and many more references, the interested reader may consult Refs. 36 and 43.

Corresponding to the hierarchy of equations for GFs there is a hierarchy of coupled equations satisfied by reduced density matrices of increasing order (*5*). It has been shown by Nakatsuji (*41*) that this hierarchy can be truncated at order $p = 4$ in a manner which could give exact results once the N-representability problem has been solved.

8-7. SUMMARY AND CONCLUSION

We began this chapter by quoting Charles Coulson and Harrison Shull to the effect that most—probably all—of the information about the state of a physical system which is of practical significance is encoded in the second-order reduced density operator (RDO) of the state and the associated first-order RDO. We have outlined some of the efforts of the past twenty years which were inspired by this insight. To this end we recalled the definition of the RDOs associated with the state of a system of N identical particles. The eigenstates of these operators and their associated eigenvalues are intrinsically related to the state and determine the RDO. Thus we defined the natural spin orbitals and natural spin geminals of the first- and second-order RDO.

The chapter was largely devoted to a review of efforts to solve the ensemble N-representability problem. It was necessary to recall basic mathematical ideas about convex cones and their polar cones. The set of ensemble N-representable RDOs is convex, and each polar element gives rise to a linear equation which is a necessary condition for an operator to be N-representable. The most important known conditions were described, namely the D, Q, B, G, and C conditions. The numerical results which have been obtained by means of these conditions were briefly discussed.

Because of lack of space, our account has necessarily been fragmentary. To tell the full story would require a large treatise. Significant omissions include the following points:

(i) Dirac (*19*) introduced the 1-matrix for a single Slater function and used it in developing his refinement of the Thomas–Fermi theory. He showed that the self-consistent-field method of Hartree–Fock is essentially a procedure for finding $\rho_1 = ND^1$, subject to the strong N-representability sufficiency condition that the wavefunction is a single Slater determinant.

(ii) Harriman et al. (*26, 27*) obtained formulas for the 1- and 2-matrix of a spin-projected single determinant. As is well known, for the ground state such wavefunctions frequently account for a large portion of the correlation energy—up to 90%.

(iii) Simons and Harriman (*49*) introduced the concept of approximate N-representability and studied how optimizing N-representability interacts with minimizing the energy.

(iv) Ruskai (*44*), extending ideas of the present author expounded at Sanibel Island in 1964, developed an algorithm for deciding on the pure N-representability of a given 2-matrix. Unfortunately, the algorithm seems of little practical value. At any stage it could show that a given 2-matrix is not N-representable, but it does not finally decide on N-representability until arriving at the corresponding wavefunction.

(v) At the conferences at Queen's University on RDOs in 1967 and 1974, Rosina outlined his ideas for exploring the excited states which can be obtained as one-particle excitations from the ground state. He proposed obtaining the 2-matrix of the ground state by a variational approach and using it as the point of departure for obtaining many low-lying states (*11, 21*). The interested reader will also find much material which we have been unable to include in this brief chapter in Refs. 17, 35, and 39.

It will be apparent that the line of thought initiated by Dirac, Husimi, Coulson, and Löwdin is still in full train and that there are many avenues of significant research to be pursued (see, e.g., Ref. 18), including (a) the numerical exploration of the norbs and nags of a variety of simple chemical systems; (b) the discovery of necessary N-representability conditions which are more stringent than those hitherto obtained; and (c) the development of new numerical optimization techniques when the feasible region is a convex set of matrices M of very large degree defined by conditions which impose positive definiteness on matrices which are linear in M.

Notes

1. At this point Shull has a footnote in which he proposes, for the first time we believe, the term *geminal* for a two-electron function.

2. Coleman first noticed the problem while studying Frenkel's treatment of second quantization in 1952 when he was a member of the Summer Research Institute of the Canadian Mathematical Congress. He gleefully announced that he had reduced the N-body problem to the $2\frac{1}{2}$-body problem! His announcement was a trifle premature, as subsequent events have shown.

References

1. Barnett, G. P., Linderberg, J., and Shull, H., 1965, *J. Chem. Phys.*, **43**, S80.
2. Bloch, F., 1965, *Phys. Rev.*, **A137**, 787.
3. Borland, R. E. and Dennis, K., 1972, *J. Phys. B: Atom. Mol. Phys.*, **5**, 7.
4. Brown, R. E. and Smith, V., 1974, Queen's Papers on Pure and Applied Mathematics, No. 40, pp. 75–84.
5. Cohen, L. and Frishberg, C., 1976, *Phys. Rev.*, **A13**, 927.
6. Coleman, A. J., 1959, *Can. Math. Bull.*, **2**, 209.
7. ——, 1961, *Can. Math. Bull.*, **4**, 209.
8. ——, 1963, *Rev. Mod. Phys.*, **35**, 668.
9. ——, 1965, *J. Math. Phys.*, **6**, 1425.
10. ——, 1967, *Can. J. Phys.*, **45**, 1271.
11. —— and Erdahl, R. M., 1968, Queen's Papers on Pure and Applied Mathematics, No. 11, Reduced density matrices with applications to physical and chemical systems.
12. ——, 1972, *J. Math. Phys.*, **13**, 214.
13. ——, 1973, *Rep. Math. Phys.*, **4**, 113.
14. ——, 1975, A general theory for super phenomena in physics, in *Quantum Statistics and the Many-Body Problem*, eds. S. B. Trickey, W. P. Kirk, and J. W. Duffy (Plenum Press, New York), p. 239.
15. Coulson, C. A., 1960, *Rev. Mod. Phys.*, **32**, 175.
16. Davidson, E. R. and Jones, L. L., 1962, *J. Chem. Phys.*, **32**, 2966.
17. ——, 1976, *Reduced Density Matrices in Quantum Chemistry* (Academic Press, New York).

18. Dennis, K., 1977, The *N*-representability problem and some related atomic calculations, Dissertation, University of London.
19. Dirac, P. A. M., 1930, *Proc. Cambridge Phil. Soc.*, **26**, 376.
20. Doll, J. D. and Reinhardt, W. P., 1972, *J. Chem. Phys.*, **57**, 1169.
21. Erdahl, R. M., 1974, Queen's Papers on Pure and Applied Mathematics, No. 40, Reduced density operators with applications to physical and chemical systems–II.
22. Fusco, M. A., 1974, Density matrices and atomic structure problem, Dissertation, University of California (Davis).
23. Garrod, C., Mihailovic, M. V., and Rosina, M., 1975, *J. Math. Phys.*, **16**, 868.
24. —— and Percus, J. K., 1964, *J. Math. Phys.*, **5**, 1763.
25. Gilbert, T. L., 1975, *Phys. Rev.*, **B12**, 2111.
26. Hardisson, A. and Harriman, J. E., 1967, *J. Chem. Phys.*, **46**, 3639.
27. Harriman, J. E., 1964, *J. Chem. Phys.*, **40**, 2827.
28. Hohenberg, P. and Kohn, W., 1964, *Phys. Rev.*, **136**, B864.
29. Hone, D., 1962, *Phys. Rev. Lett.*, **8**, 369.
30. Husimi, K., 1940, *Proc. Phys. Math. Soc. Jap.*, **22**, 264.
31. Kato, T., 1957, *Comm. Pure Appl. Math.*, **10**, 151.
32. Kummer, H., 1967, *J. Math. Phys.*, **8**, 2063.
33. ——, Absar, I., and Coleman, A. J., 1977, *J. Math. Phys.*, **18**, 329.
34. —— and Absar, I., 1977, *J. Math. Phys.*, **18**, 335.
35. Krachko, E. C. and Krugliak, Yu A., 1975, in *Physics of Molecules*, vol. 1 (in Russian, published by the Ukrainian Academy of Science), p. 1.
36. Linderberg, J. and Öhrn, Y., 1973, *Propagators in Quantum Chemistry* (Academic Press, New York).
37. Löwdin, P.-O., 1955, *Phys. Rev.*, **97**, 1474, 1490, 1510.
38. —— and Shull, H., 1955, *Phys. Rev.*, **101**, 1730.
39. Mestechkin, M. M., 1977, *The Density Matrix Method in the Theory of Molecules* (in Russian), Naukova Dumka, Kiev.
40. Mihailovic, M. V. and Rosina, M., 1975, *Nucl. Phys.*, **A237**, 229.
41. Nakatsuji, H., 1976, *Phys. Rev.*, **A14**, 41.
42. Neumann, J. von, 1930, *Mathematischen Grundlagen der Quantum Mechanik* (Springer-Verlag, Berlin).
43. Öhrn, Y., 1976, Proceedings of the Second World Congress on Quantum Molecular Science, New Orleans.
44. Ruskai, B., 1972, *Phys. Rev.*, **A5**, 1336.
45. Sasaki, F., 1965, *Phys. Rev.*, **138**, B1338.
46. Schmidt, E., 1907, *Math. Ann.*, **63**, 433.
47. Shull, H., 1959, *J. Chem. Phys.*, **30**, 1405.
48. —— and Löwdin, P.-O., 1959, *J. Chem. Phys.*, **30**, 617.
49. Simons, J. and Harriman, J. E., 1970, *Phys. Rev.*, **A2**, 1034.
50. Smith, D. W. and Fogel, S. J., 1965, *J. Chem. Phys.*, **40**, 591.
51. Smith, V. H., 1967, *Theoret. Chim. Acta* (Berlin), **7**, 245.
52. Steiner, E., 1961, *J. Chem. Phys.*, **39**, 2365.
53. Yang, C. N., 1962, *Rev. Mod. Phys.*, **34**, 694.

Supplementary References

54. Coleman, A. J., 1978, *Int. J. Quantum Chem.*, **13**, 67.
55. Gunnarsson, O. and Jones, R. O., 1980, *Phys. Scrip.*, **21**, 394.
56. Lawes, G. P. and March, N. H., 1980, *Phys. Scrip.*, **21**, 402.
57. Parr, R. G., Gadre, S. R., and Bartolotti, L. J., 1979, *Proc. Nat. Acad. Sci. USA*, **76**, 2522.
58. Rajagopal, A. K., 1980, in *Advances in Chemical Physics*, vol. 41, eds. I. Prigogine and S. A. Rice (Wiley-Interscience, New York), p. 59.

9
Calculation of Forces by Non-Hellmann–Feynman Methods

P. Pulay

Department of General and Inorganic Chemistry
Eötvös Loránd University
H-1088 Budapest
Hungary.

Contents

9-1. INTRODUCTION

In spite of its great theoretical significance (see, e.g., Ref. 50), the Hellmann-Feynman (H-F) theorem has been of surprisingly little value for practical calculations, e.g., evaluation of molecular geometries and force constants. The present chapter deals with direct analytical calculation of the gradient of total molecular energy with respect to nuclear coordinates,

$$\partial/\partial X_a \langle \psi | H | \psi \rangle = \langle \psi | H | \psi \rangle^a, \tag{9-1}$$

where the superscript a denotes differentiation with respect to the nuclear coordinate X_a. The exact force ($-\langle \psi | H | \psi \rangle^a$) and the H-F force ($-\langle \psi | H^a | \psi \rangle$) are equal for the exact wavefunction and for certain fully optimized wavefunctions, e.g., Hartree-Fock functions, but they differ significantly for all practical wavefunctions. In this chapter we shall discuss the calculation and the use of exact forces and show that their use is much superior (although much more expensive) to the H-F forces.

In principle, the derivatives of molecular energy surface can be determined from a pointwise calculation of the energy for a number of geometry parameters, followed by a numerical differentiation procedure which usually involves curve fitting. For several years the present author has stressed two significant advantages over this pointwise procedure of a direct analytic calculation of the first energy derivative (followed by numerical differentiation if higher derivatives are needed):

1. It greatly enhances information about the energy surface obtainable from a single wavefunction. Since there are $3N - 6$ independent forces ($N \equiv$ number of nuclei) versus a single energy value, the gradient calculation is equivalent to $3N - 6$ separate energy calculations. This advantage becomes more significant from triatomics onwards, making possible, e.g., complete geom-

etry optimization in polyatomic molecules, which is hardly practical using the pointwise method.

2. High numerical accuracy is easily attainable. In contrast, the pointwise method is usually plagued by numerical difficulties associated with taking small differences of large numbers. Direct calculation of second and higher derivatives would increase their numerical accuracy even more, but this does not seem feasible at present.

9-2. CALCULATION OF THE ENERGY GRADIENT

We shall show that the evaluation of energy gradient from a single wavefunction is possible under certain circumstances, and derive the necessary conditions for this.[1] We specify a one-electron basis set $\{\chi_m(\boldsymbol{p})\}$ by the general functional form (fixed) of each basis function and by a set of parameters $\boldsymbol{p}$ which typically includes orbital exponents and the positions of the (centers of the) basis functions. The main peculiarity of the perturbation caused by nuclear displacements (as opposed to, e.g., weak external fields) is that some of the parameters, the function centers, must depend on nuclear coordinates (Section 9-3). Let us consider a normalized wavefunction $\psi(x, C_k, p_t)$ constructed from the basis functions and characterized by a set of variational parameters $\boldsymbol{C} = \{C_k\}$ which typically includes orbital coefficients and possibly the coefficients of configurations in a multideterminant wavefunction. We assume that the orbitals and/or configurations obey a set of normalization and orthogonality constraints, written as

$$f_m(\boldsymbol{C}, \boldsymbol{S}) = 0, \tag{9-2}$$

where $\boldsymbol{S}$ is the overlap matrix, $S_{ij} = \langle \chi_i | \chi_j \rangle$. For simplicity we also assume that all basis functions and coefficients are real. This is not a serious restriction.

The energy gradient

$$E^a = \langle \psi | H^a | \psi \rangle + 2\langle \psi^a | H | \psi \rangle \tag{9-3}$$

comprises both H–F force (first term) and the "wavefunction force" (79), the latter vanishing for exact wavefunctions. ψ depends on X_a, since both C_k and p_t do so. Then the wavefunction force becomes

$$\langle \psi^a | H | \psi \rangle = -f_{\text{int}} - f_{\text{norm}}, \tag{9-4}$$

$$-f_{\text{int}} = \sum_t (\partial/\partial p_t \langle \psi | H | \psi \rangle)\, p_t^a, \tag{9-4a}$$

$$-f_{\text{norm}} = 2 \sum_k \langle \partial\psi/\partial C_k | H | \psi \rangle\, C_k^a. \tag{9-4b}$$

In practical situations the parameters p_t are simple fixed functions of nuclear coordinates (e.g., basis functions move rigidly with nuclei and their exponents are fixed), and the integral force, (9-4a), is straightforward to evaluate (p_t^a is discussed in Section 9-3). The energy $\langle \psi | H | \psi \rangle$ is a linear combination of integrals over the basis functions, and its derivative with respect to p_t is a linear combination of integral derivatives. Despite its theoretical simplicity, (9-4a) takes up most of the numerical work in evaluating the exact force (Section 9-4-2). In general, the evaluation of the derivatives C_k^a, for calculating the normalization force, (9-4b), is a complex task. Even for a Hartree-Fock wavefunction, a linear system of $n(m - n)$ equations must be solved for this, n being the number of occupied orbitals and m the number of basis functions (*44*). The computational effort is proportional to $n^3(m - n)^3$, thus increasing with the sixth power of molecular size for a fixed m/n. It seems that analytical evaluation of C_k^a is not practical.[2] Thus, it is important that for general SCF wavefunctions the derivatives C_k^a should not be needed for calculating the exact force.

General SCF wavefunctions are characterized by the fact that all parameters entering the wavefunction either (a) have been completely optimized with respect to energy, or (b) have a prescribed simple dependence on nuclear coordinates. In what follows we assume that the linear parameters C_k are optimized and the nonlinear parameters p_t have a simple fixed dependence on X_a. To determine the normalization force, we start with the optimization condition for C_k,

$$\partial/\partial C_k \left[\langle \psi | H | \psi \rangle - \sum_m \lambda_m f_m(C, S) \right] = 0, \tag{9-5}$$

where the orthonormality constraints (9-2), multiplied by Lagrange multipliers λ_m, have been subtracted from energy. The determination of the wavefunction usually furnishes the values of the Lagrange multipliers (e.g., Hartree-Fock orbital energies). However, a formulation will be given later which does not require them.

Since neither H nor S depends on C_k, (9-5) is equivalent to

$$2\langle \partial/\partial C_k \psi | H | \psi \rangle = \sum_m \lambda_m \partial/\partial C_k f_m(C, S). \tag{9-6}$$

Substituting this into (9-4b) gives

$$-f_{\text{norm}} = \sum_k \sum_m \lambda_m [\partial/\partial C_k f_m(C, S)]\, C_k^a. \tag{9-7}$$

Now, the crucial point in deriving a gradient formula is that the unknown quantities C_k^a can be eliminated from (9-7) through the identity

$$0 = f_m(C,S)^a$$
$$= \sum_k [\partial/\partial C_k f_m(C,S)]\, C_k^a + \sum_l [\partial/\partial S_l f_m(C,S)]\, S_l^a, \qquad (9\text{-}8)$$

where the indices of S have been replaced by the single index l. The derivatives S_l^a appear because the basis set depends on nuclear coordinates through the nonlinear parameters p_t. Substituting (9-8) into (9-7) we obtain

$$-f_{\text{norm}} = \sum_m \lambda_m \sum_l [\partial/\partial S_l f_m(C,S)]\, S_l^a. \qquad (9\text{-}9)$$

The term f_{norm} does not arise without explicit orthonormality constraints. All quantities in (9-9) are either known or easily evaluated: λ_m is known from the SCF-like procedure, and $\partial/\partial S_l f_m(C,S)$ and S_l^a are easy to evaluate.

The above formula is valid only for a wavefunction in which all C_k are optimized. Although this is apparently a substantial restriction on the applicability of the force method, one should remember that a wavefunction which has not been totally optimized in the above sense probably does not yield a smooth potential surface (even if its global features are correct). A truncated CI expansion without orbital optimization yields an uneven surface from which derivatives are difficult or impossible to extract (*107*). A Hartree-Fock determinant with the inclusion of *all* single and double substitutions (*66*) yields a substantially smoother surface. Even this wavefunction is not completely optimized in our sense, as the variation of orbitals in the reference determinant cannot be exactly replaced by single substitutions (*35*, *74*). However, the difference is small, involving only triple and higher substitutions. The modified SCEP (self-consistent electron-pair) method (*66*, *67*) leads to a multiconfiguration SCF-type wavefunction and is a very important development, because it permits the application of the gradient method to highly correlated wavefunctions.

Equation (9-9) may be formulated in another way to cover the case when the Lagrange multipliers are unknown, e.g., the SCEP wavefunction. Let a set of numbers $\{D_k\}$ be known which obeys (9-8) if substituted for C_k^a. Multiplying by λ_m and summing over m we obtain

$$\sum_k \left[\sum_m \lambda_m \partial/\partial C_k f_m\right] D_k + \sum_l \sum_m \lambda_m (\partial/\partial S_l f_m)\, S_l^a = 0. \qquad (9\text{-}10)$$

Since according to (9-5) the term in brackets is $\partial/\partial C_k\langle\psi|H|\psi\rangle$,

$$\sum_k [\partial/\partial C_k\langle\psi|H|\psi\rangle]\, D_k = -\sum_{m,l} \lambda_m(\partial/\partial S_l f_m)\, S_l^a, \tag{9-11}$$

i.e., the left-hand side is equal to the correct normalization force. Thus, any D_k can be used to calculate f_{norm}, even if it differs from the true C_k^a, provided it obeys (9-8). It is sometimes easier to find a solution[3] D to (9-8) than to determine the Lagrange multipliers λ_m. The meaning of (9-11) is as follows. For an infinitesimal change δX_a, the orthonormality constraints are violated because the overlap matrix changes. Any change in the coefficients $\delta \boldsymbol{C} = \boldsymbol{D}\delta X_a$ which restores orthonormality leads to the correct normalization force if the wavefunction is completely optimized with respect to $\boldsymbol{C}$.

Summarizing, the total energy derivative can be easily determined analytically for wavefunctions which contain only two kinds of parameters: (1) those which depend in a prescribed simple way on nuclear coordinates, and (2) those which minimize total energy, possibly under some (orthonormality) constraints. For such a wavefunction, the exact force has three components: the H-F force, the integral force, and the normalization force. Determination of the H–F force is simple, both theoretically and numerically, while that of the integral force is straightforward but requires the bulk of computation. The normalization force is usually easy to obtain numerically, but is the most complicated term theoretically.

9-3. BASIS SETS AND FORCES

9-3-1. Dependence of the Basis Set on Nuclear Coordinates

There are three reasonable ways to define the dependence of basis-set parameters on nuclear coordinates (*60*). The form of f_{int} depends critically upon this definition. (i) Define an arbitrary but physically reasonable and simple functional dependence of the p_t on nuclear coordinates and evaluate (9-4a) with this definition of $p_t^a = (\partial p_t/\partial X_a)$. (ii) Hold the parameter p_t constant. In this case the contribution of p_t to both (9-4a) and (9-9) vanishes (the overlap matrix depends only through p_t on nuclear coordinates). (iii) Optimize the parameter p_t. As $\boldsymbol{S}$ in (9-2) depends on p_t, it is not the energy itself but the functional

$$\langle\psi|H|\psi\rangle - \sum_m \lambda_m f_m$$

which is to be minimized,[4] as in the case of C_k. This yields

$$\partial/\partial p_t\langle\psi|H|\psi\rangle = \sum_m \lambda_m \partial f_m/\partial p_t. \tag{9-12}$$

If all parameters p_t are either constant or optimized then, multiplying (9-12) by p_t^a and summing over t, we obtain

$$\sum_t (\partial/\partial p_t \langle\psi|H|\psi\rangle)\, p_t^a = \sum_{m,t} \lambda_m (\partial f_m/\partial p_t)\, p_t^a$$

$$= \sum_m \sum_t \sum_l \lambda_m (\partial f_m/\partial S_l)(\partial S_l/\partial p_t)\, p_t^a$$

$$= \sum_{m,l} \lambda_m (\partial f_m/\partial S_l)\, S_l^a, \qquad (9\text{-}13)$$

where we have used the fact that the constraints f_m depend only through $\boldsymbol{S}$ on p_t. The left-hand side of (9-13) is $-f_{\text{int}}$, while the right-hand side is f_{norm}. This shows that under the above conditions these two forces cancel out and the exact force is equal to the H–F force. Basis sets which contain only parameters obeying conditions (ii) and (iii) are called floating functions (*50*);[5] wavefunctions built from floating functions are termed stable with respect to nuclear coordinates (*46*).

As the calculation of exact forces is particularly simple for stable wavefunctions, it is important to compare floating basis sets to the more customary bases defined by (i). Unfortunately, both methods (ii) and (iii) increase the computational effort out of proportion to the gain due to simplification in the force calculation. Optimization of basis function positions and exponents has been used with success (*43*) for subminimal basis sets, but the extension of this method to larger basis sets does not seem practical. The other approach, the use of fixed basis functions, is applicable only to the few molecules which can be described by one-center basis sets (*69*), and its usefulness is disputed even in these cases. In the general case, it is necessary to include the polarization functions $\partial\chi_r/\partial x_r$, $\partial\chi_r/\partial y_r$, etc., for every basis function χ_r which is significantly populated to describe correctly even an infinitesimal perturbation due to the motion of the nuclei. This increases computational time by about two orders of magnitude.

It is indeed definition (i) which has been adopted intuitively for calculating potential surfaces in most cases. The simplest physically reasonable dependence of the nonlinear parameters on nuclear coordinates is the following (*60*): (a) the exponents of basis functions are constant; (b) the positions of basis functions are linear combinations of nuclear position vectors (in the usual special case this includes functions moving rigidly with a particular nucleus). With fixed orbital coefficients these conditions describe properly the bulk of the change which takes place during distortion of the nuclear framework (Sections 3-5 and 5-5): The atomic part of the electron density follows the nuclei. In order to describe the remaining finer changes by the linear coefficients C_k the basis set must be sufficiently flexible. Restriction (a) requires the presence of functions with different exponents for every important type of basis functions. Restriction (b) can be compensated for by adding polarization functions to the basis set, particularly in the valence region where deviations from atomic behavior are largest (note that basis functions centered on other atoms also offer a certain polarization).

It has been argued (*7, 8, 28*) that reoptimization of orbital exponents at every nuclear conformation is important (note that Ref. 8 incorrectly equates the exact forces to H–F forces). Recent results do not support this view. It may hold for limited one-center expansions where this is the only way for the wavefunction to adapt itself to changing nuclear positions. For extended multicenter basis sets the superiority of extending the basis set, compared to exponent optimization, has been clearly demonstrated (*26*). Exponent optimization for extended basis sets is very time consuming. This usually forces one to carry out the optimization only partially, resulting in an uneven potential surface from which derivatives, especially the higher ones, cannot be evaluated. The results on N_2 (*27*) illustrate this danger. In their high-quality Hartree–Fock calculation the authors obtain a slightly uneven potential curve because of incomplete exponent optimization. The calculated anharmonicity parameter $\omega_e x_e$ varies wildly with the choice of points on the curve, the minimum and maximum values being 2.85 and 3975(!) cm^{-1} versus the experimental 14.2 cm^{-1}. A fixed-exponent $(11s6p2d1f)$ Gaussian basis set yields a Hartree–Fock value (*62*) of 13.2 cm^{-1} at the theoretical energy minimum, in reasonable agreement with experiment.

9-3-2. Hellmann–Feynman versus Exact Forces

As the basis set for an SCF wavefunction is extended, deviations from the H–F theorem tend to diminish. In the limiting case of a complete basis set the H–F theorem should hold because a complete basis set may be independent of nuclear coordinates; this implies that both f_{int} and f_{norm} vanish. The question arises whether it is more advantageous to work with a very good basis set and the H–F force or with a more modest basis and the exact force. Practical experience indicates that the second course is much preferable. This is easily understood by considering that, for a wavefunction correct to order ϵ, the H–F force is correct to order ϵ only, as H^a does not commute with the Hamiltonian H, while the approximate energy is correct to order ϵ^2; this is valid for the exact force also under the rather plausible condition that the quality of the description does not change rapidly with nuclear coordinates (*44*). This means that the exact forces converge quadratically toward the accurate value and the H–F forces only linearly.

Numerical tests (*6, 39, 51, 96, 98*) show that H–F forces are generally very unreliable, even for sophisticated wavefunctions. An example (*97*) helps to explain why H-F forces are so sensitive. Consider a neutral spherical atom in an electric field F. If the polarization of the electron cloud is neglected as a first approximation, then the electronic contribution to the H-F force vanishes and the total H-F force is ZF instead of the correct value, zero. Note that the exact force is correct even with the spherical wavefunction. If the polarization of the electron cloud is taken properly into account, then the electrons exert an H-F force on the nucleus which cancels the ZF term. It is clear from this example that a necessary condition for usable H-F forces is the correct description of the small and for most purposes unimportant polarization of the atomic cores. Suitable polarization functions ensure that the basis set does not carry forces and therefore ensure the validity of the H-F theorem. However, as mentioned above, this is likely to increase the computational effort by two orders of magnitude.

Another difficulty with the H-F forces is that their sum and torque do not necessarily vanish if they are calculated from nonstable wavefunctions.[6] This precludes a meaningful transformation of the forces to internal valence coordinates, which are the relevant physical quantities.

9-4. APPLICATION TO SCF WAVEFUNCTIONS

9-4-1. The Closed-Shell Case

Consider first the simplest and most important SCF wavefunction, the closed-shell Hartree-Fock case. Let C_{ri} denote the orbital coefficients of the doubly occupied MOs,

$$\psi_i = \sum_r C_{ri} \chi_r, \quad i = 1, \ldots, n; r = 1, \ldots, m.$$

The Hartree-Fock energy is

$$E = \sum_{r,s} P_{rs} \langle r|h|s\rangle + \tfrac{1}{2} \sum_{r,s,t,u} (P_{rs} P_{tu} - \tfrac{1}{2} P_{rt} P_{su}) J_{rstu} + \Omega. \quad (9\text{-}14)$$

Here $\langle r|h|s\rangle = \langle \chi_r|h|\chi_s\rangle$, where h is the one-electron part of the Hamiltonian; $J_{rstu} = \langle \chi_r(1)\, \chi_s(1)| 1/r_{12} |\chi_t(2)\, \chi_s(2)\rangle$; Ω is the nuclear repulsion energy; and $\boldsymbol{P} = 2\boldsymbol{C}\boldsymbol{C}^+$ is the density matrix. The H-F force,

$$-f_{\mathrm{HF}} = \sum_{r,s} P_{rs} \langle r|h^a|s\rangle + \Omega^a \quad (9\text{-}15)$$

is simply the expectation value of the field operator at the particular nucleus times the nuclear charge. Its calculation is simple, as h^a is a one-electron operator, i.e., the number of integrals to be evaluated is only $O(m^2)$.

To obtain f_{norm}, write the orthonormality constraints for the orbitals as

$$C^{+}SC - I_n = 0, \tag{9-16}$$

where I_n is the $n \times n$ unit matrix. In the closed-shell case (and generally in all cases when the occupation numbers of the orbitals are equal) the orbitals can be chosen in a way that the matrix of the Lagrange multipliers, corresponding to the n^2 [or, taking into account the symmetry of the matrix, $n(n + 1)/2$] constraints (9-16), becomes diagonal,

$$\lambda_{ij} = 2\epsilon_i \delta_{ij} \tag{9-17}$$

(in order to maintain the usual convention, the half of the Lagrangian is denoted by ϵ_i). Substituting this into the general formula (9-9) we obtain

$$f_{\text{norm}} = \sum_{i,j} 2\epsilon_i \delta_{ij} \left(\sum_{k,l} C_{ki} C_{lj} \right) S^a_{kl} = 2 \sum_{k,l} \left(\sum_i C_{ki} \epsilon_i C_{li} \right) S^a_{kl}$$

$$= \text{Tr}\,(\boldsymbol{R}\boldsymbol{S}^a), \tag{9-18}$$

where the densitylike matrix $\boldsymbol{R}$ is defined as

$$R = 2C\epsilon C^{+}. \tag{9-19}$$

Finally, f_{int} can be written as

$$-f_{\text{int}} = \sum_{r,s} P_{rs} \langle r^a | h | s \rangle + \tfrac{1}{2} \sum_{r,s,t,u} (P_{rs} P_{tu} - \tfrac{1}{2} P_{rt} P_{su}) J_{r^a stu}. \tag{9-20}$$

Here the integrals containing the derivatives of the basis functions are defined as

$$\langle r^a | h | s \rangle = \sum_t \langle \partial \chi_r / \partial p_t | h | \chi_s \rangle \, p^a_t, \tag{9-21a}$$

$$J_{r^a stu} = \sum_t \langle \partial \chi_r(1) / \partial p_t \chi_s(1) | 1/r_{12} | \chi_t(2)\, \chi_u(2) \rangle. \tag{9-21}$$

The sum of (9-15), (9-18), and (9-20) gives the exact force for closed-shell Hartree-Fock wavefunctions. However, for one-center basis sets (*1, 2, 24, 25*) the exact force simplifies to the H-F force, which explains the moderate agreement with experiment in such calculations. These results were generalized (*7*) to include optimized (variable) orbital exponents.[7] Other equivalent expressions for the exact force have been derived (*10, 44, 70, 79*).

Realizing the significance of exact forces for geometry and force constants, the first calculations were performed by Pulay (*79–81*) using the MOLPRO program (*58*). Currently five ab initio gradient programs seem to be in use: MOLPRO, two extensions (*72*, *102*) of Gaussian 70 (*47*), TEXAS (*90*), and an extension (*49*) to POLYATOM (75). Of these only MOLPRO and TEXAS allow the use of d-type functions.

9-4-2. Computational Considerations

The integral force is the major computational task. We write this as

$$-f_{\text{int}} = \sum_{r \geqslant s} P_{rs}(\langle r^a|h|s\rangle + \langle r|h|s^a\rangle)(2 - \delta_{rs})$$

$$+ \tfrac{1}{2} \sum_{\substack{r \geqslant s,\, t \geqslant u \\ rs \geqslant tu}} (P_{rs}P_{tu} - \tfrac{1}{4} P_{rt}P_{su} - \tfrac{1}{4} P_{ru}P_{st})(J_{r^a stu} + J_{rs^a tu}$$

$$+ J_{rst^a u} + J_{rstu^a})(2 - \delta_{rs})(2 - \delta_{tu})(2 - \delta_{rs,tu}). \tag{9-22}$$

We fix the exponents of basis functions and take the position vectors of their centers as (see Section 9-3)

$$\chi_r = \sum_a \mu_{ra} X_a. \tag{9-23}$$

Here μ_{ra}, the coefficient of following for X_a and the basis set positional coordinate r, corresponds to p_t^a in (9-21). If each basis function follows one nucleus rigidly, and the X_a are simple Cartesians, then each sum (9-20) contains only one nonvanishing term. We can avoid multiplication by μ_{ra} by introducing the forces f_r acting on the position vectors r,

$$-f_r = \partial E/\partial x_r = -\text{Tr}\,(R\partial S/\partial x_r) + 2 \sum_s P_{rs}\langle r'|h|s\rangle$$

$$+ 2 \sum_{s,t,u} (P_{rs}P_{tu} - \tfrac{1}{2}P_{rt}P_{su})\, J_{r'stu}, \tag{9-24}$$

where $r' = \partial\chi_r/\partial x_r$. The force on X_a is then

$$f_a = -E^a = f_a^{\text{HF}} + \sum_r f_r \mu_{ra}. \tag{9-25}$$

Note that a large f_r indicates the need for including the polarization function $\partial\chi/\partial x_r$, except for core functions where a small polarization can cause large forces. The computational advantages of the above formulas are as follows: (a) The contributions to f_{int} can be summed up immediately after their calculation, requiring no external storage; (b) One can calculate forces for all nuclear coordinates by going through the integral list only once. This is one of the main advantages of the gradient method, since the amount of information varies as N (number of nuclei) while the computational effort is independent of N. From the vanishing of sum and torque one can determine forces for six (five for linear molecules) nuclear coordinates. However, the saving is significant only for molecules where most basis functions belong to a single atom. Integrals containing the derivatives $\partial\chi/\partial x$ are easily reduced to those over functions having quantum number $l' = l \pm 1$; for both Gaussian and Slater functions the radial part remains essentially unchanged.

Although the evaluation of exact forces is simple in principle, it requires considerable computational effort because all derivatives of two-electron integrals must be evaluated. From a single integral J_{rstu}, twelve derivatives arise from twelve coordinates of four function centers. Although there are relations between the derivatives (e.g., sum of all x components vanish), the ratio $T_{\text{forces}}/T_{\text{int}} \simeq$ 3–6 for different programs. With modern integral evaluation techniques, however, the force calculation may even be cheaper than the calculation of SCF energy. Force calculations offer even more advantages with semiempirical SCF wavefunctions (e.g., of ZDO type) then with ab initio ones (*45, 56, 78, 84, 93*), since integral evaluation is only a minor time consumer in ZDO methods.

9-4-3. Open Shells and Multiconfigurational Wavefunctions

Force calculation for open-shell and multiconfigurational SCF wavefunctions is straightforward in principle, but some computational difficulties arise. For all these methods the energy is (in terms of basis functions)

$$E = \sum_{r,s} P_{rs} \langle r|h|s\rangle + \tfrac{1}{2} \sum_{r,s,t,u} \mathcal{P}_{rs,tu} J_{rstu} + \Omega. \tag{9-26}$$

The actual expressions of the density matrices $\mathbf{P}$ and $\mathcal{P}$ in terms of the coefficients $\mathbf{C}$ depend on the wavefunction and the CI coefficients, but it is no more as simple as in the closed-shell Hartree–Fock case where

$$\mathbf{P} = 2\mathbf{C}\mathbf{C}^{+} \quad \text{and} \quad \mathcal{P}_{rs,tu} = P_{rs}P_{tu} - \tfrac{1}{2} P_{rt}P_{su}. \tag{9-27}$$

This causes a problem in calculating the two-electron part of f_{int} because all one-electron density matrices from which $\mathcal{P}$ is constructed must be stored in the core memory; alternatively, the whole second-order density matrix must be calculated in advance and read in piecewise from external storage.

Another difficulty is that now the off-diagonal Lagrange multipliers must be retained. Depending on the method of calculation, these may or may not be available for the calculation of f_{norm}. If they are available, then

$$f_{\text{norm}} = \text{Tr}\,(\boldsymbol{R}\boldsymbol{S}^a), \tag{9-18}$$

with

$$\mathbf{R} = \mathbf{C}\boldsymbol{\epsilon}\mathbf{C}^{+}, \tag{9-28}$$

still holds but now $\boldsymbol{\epsilon}$, the matrix of the Lagrange multipliers, is no longer diagonal. If $\boldsymbol{\epsilon}$ is not available, the alternative technique of Section 9-3 must be used. The orthonormality constraints can be written in general as

$$\mathbf{C}^{+}\mathbf{S}\mathbf{C} - \mathbf{I} = 0, \tag{9-29}$$

where the columns of $\mathbf{C}$ contain the orbital coefficients. Equation (9-8) now becomes

$$\mathbf{C}^{+}\mathbf{S}\mathbf{D} + \mathbf{D}^{+}\mathbf{S}\mathbf{C} + \mathbf{C}^{+}\mathbf{S}^{a}\mathbf{C} = 0. \tag{9-30}$$

One possible solution to this equation is

$$\mathbf{D} = -\tfrac{1}{2}\mathbf{S}^{-1}\mathbf{S}^{a}\mathbf{C}. \tag{9-31}$$

The other factor in f_{norm}, $\partial E/\partial C_{ri}$, is easily calculated, once the actual form of the density matrices in (9-26) is given.

In a multiconfigurational wavefunction, normalization of the total wavefunction does not follow from orthonormality of the orbitals. In principle the total normalization constraint also contributes to f_{norm}. However, if the configurations are orthonormal by virtue of orbital orthogonality, as usual, then the total normalization condition

$$\sum_k A_k^2 - 1 = 0 \tag{9-32}$$

does not contain the overlap matrix $\boldsymbol{S}$. According to the general formula (9-9), this causes the contribution of (9-32) to vanish.

Little practical advance has been made so far in calculating exact forces for general SCF wavefunctions. MOLPRO (*58*) contains both an unrestricted and a restricted (Roothaan-type open-shell) SCF version. The computer programs mentioned before are limited to the closed-shell case (see also Ref. 36).

9-5. TRANSFORMATION OF CARTESIAN FORCES AND FORCE CONSTANTS TO INTERNAL COORDINATES

Although it is advantageous to calculate the forces in Cartesian coordinates first (Section 9-4), most applications require forces and force constants (FC) in internal valence coordinates, e.g., bond lengths, angles or their linear combinations (*79*). Let $X = (X_1, \ldots, X_{3N})^+$ denote the column vector of Cartesian displacements from a reference configuration, and $\boldsymbol{q} = (q_1, \ldots, q_M)^+$ a chosen set of internal displacement coordinates. Define the forces in internal coordinates as $\phi_i = -(\partial E/\partial q_i)$ at the reference configuration, with all $q_j = 0$ for $i \neq j$. This presupposes that the coordinates $\{q_i\}$ form a complete and nonredundant set. Compare the energy expression for Cartesian and internal coordinates, truncated at the quadratic terms:

$$E = E_0 - f^+X + \tfrac{1}{2} X^+KX = E_0 - \boldsymbol{\phi}^+q + \tfrac{1}{2} q^+Fq, \tag{9-33}$$

where f and $\boldsymbol{\phi}$ are the column vectors of the forces, $\boldsymbol{K}$ and $\boldsymbol{F}$ are the harmonic FC matrices in Cartesian and internal coordinates, respectively. Let us express the internal coordinates by Cartesian ones through quadratic terms:

$$q_i = B_iX + \tfrac{1}{2} X^+C^iX, \tag{9-34}$$

where $\boldsymbol{B}_i$ is the ith row of the matrix which relates $\boldsymbol{q}$ and $\boldsymbol{X}$ to first order (*112*). Substituting (9-34) into the left-hand side of (9-33) and comparing the coefficients of like terms we obtain

$$f = B^+\boldsymbol{\phi} \tag{9-35}$$

$$K = B^+FB - \sum_i \phi_i C^i. \tag{9-36}$$

In order to express the internal forces and FCs, multiply (9-35) from the left by A^+, (9-36) from the left by A^+ and from the right by A where

$$A = mB^+(BmB^+)^{-1} \tag{9-37}$$

and $\boldsymbol{m}$ is any $3N \times 3N$ matrix for which $\boldsymbol{BmB}^+$ is nonsingular. The $N \times M$ matrix $\boldsymbol{A}$ obviously satisfies

$$BA = A^+B^+ = I_M. \tag{9-38}$$

The above procedure gives

$$\boldsymbol{\phi} = A^{+} f \tag{9-39}$$

$$F = A^{+} K A + \sum_i \phi_i A^{+} C^i A. \tag{9-40}$$

By including the six translational and rotational coordinates in the set $\{q\}$ it can be shown that any of the matrices (9-37) yields the same internal forces if the sum and torque of the Cartesian forces vanish. They also yield the same internal FCs if $\boldsymbol{K}$ obeys the translational and rotational invariance conditions. The simplest choice for $\boldsymbol{m}$ is obviously the unit matrix $\boldsymbol{I}_{3N}$.

The second term on the right-hand side of (9-40) is usually neglected because FCs are sought at the equilibrium geometry. However, in quantum chemical calculations it may be preferable to determine the FCs at some other reference geometry, notably at the experimental equilibrium where the theoretical ϕ_i are nonvanishing. In this case the whole formula (9-40) must be used; this differs from the usual transformation formula for FCs by the right-hand sum, indicating that the calculated harmonic normal frequencies are not strictly invariant against anharmonic (nonlinear) coordinate transformations, e.g., they may be different for Cartesian and for internal valence coordinates. Evaluation of FCs at a non-equilibrium reference geometry can be visualized as adding a linear term to the potential function which shifts the minimum of the theoretical energy surface to the reference geometry. However, a function which is linear in Cartesians contains quadratic (and higher) terms in valence coordinates and vice versa. Experience shows that one should regard the FCs in valence coordinates as the relevant physical quantities.

9-6. CALCULATION OF MOLECULAR GEOMETRIES AND REACTION PATHS

9-6-1. Geometries

It is well known that for reasonably smooth functions gradient minimization methods are much more efficient than methods which use only the function values, especially if a sufficiently accurate starting approximation is known. This also applies to the determination of molecular geometries. If we are close enough to equilibrium so that the potential energy can be regarded as quadratic in the nuclear coordinates, and a reasonable approximation $\boldsymbol{F}_0$ to the theoretical FC matrix can be found, then the force relaxation method (*79*) may be used:

$$q_{i+1} = q_i + F_0^{-1} \boldsymbol{\phi}_i, \tag{9-41}$$

where q_i and $\boldsymbol{\phi}_i$ denote the nuclear coordinates and forces in the ith step. The use of internal valence coordinates is almost indispensable in this calculation because only these allow the estimation of $\boldsymbol{F}_0$. Note that all coordinates are optimized simultaneously in this procedure and the final geometry does not depend on $\boldsymbol{F}_0$, which controls only the rate of convergence.

For ordinary molecules, especially for organic ones, finding a sufficiently accurate $\boldsymbol{F}_0$ does not usually present difficulties. For unusual molecules, ions, and transition states the best procedure is probably to determine the FCs for those (usually few) degrees of freedom which might have changed significantly relative to the stable molecule. As shown in Section 9-7, only one additional force calculation is necessary to obtain a whole row of the FC matrix (one-sided distortions suffice for this relatively crude purpose). The information obtained during the iteration (9-41) can be used to improve the FC matrix $\boldsymbol{F}_0$. This is the basis of the variable-metric optimization methods (*40*), which are quadratically convergent in contrast to the linear convergence of the force relaxation method. The following formulation (*56*, *73*) is particularly simple:

$$\boldsymbol{F}_{i+1}^{-1} = \boldsymbol{F}_i^{-1} - z_i z_i^+ / c_i, \tag{9-42}$$

where $z_i = \Delta q_i + \boldsymbol{F}_i^{-1} \Delta\boldsymbol{\phi}_i$, $c_i = \Delta\boldsymbol{\phi}_i^+ z_i$, and Δq_i and $\Delta\boldsymbol{\phi}_i$ are the changes of coordinates and forces, respectively, in the ith step. The idea behind this formula is that $\boldsymbol{F}_{i+1}$ satisfies the equation

$$\Delta q_i = -\boldsymbol{F}_{i+1}^{-1} \Delta\boldsymbol{\phi}_i, \tag{9-43}$$

which is a necessary condition for the true inverse FC matrix. Meyer (*59*) suggested a similar approximation,

$$\boldsymbol{F}_{i+1} = \boldsymbol{F}_i - (\boldsymbol{\phi}_{i+1} \Delta q_i^+ + \Delta q_i \boldsymbol{\phi}_{i+1}^+ - c\Delta q_i \Delta q_i^+)/b, \tag{9-44}$$

with $b = \Delta q_i^+ \Delta q_i$ and $c = \boldsymbol{\phi}_{i+1}^+ \Delta q / b$. Equation (9-41) gives the updated internal coordinates. However, for the next iteration step the Cartesians are usually needed. To first order they are obtained by the transformation

$$\Delta^1 X = A \Delta \boldsymbol{q}, \tag{9-45}$$

where A is the matrix introduced in (9-37). By using (9-45) iteratively, it is possible to perform this transformation exactly; this is a task often needed, e.g., for FC calculations. First determine the internal displacements $\Delta^1 \boldsymbol{q}$ corresponding to the Cartesian displacements $\Delta^1 X$. Because (9-45) is valid only to first order, $\Delta^1 \boldsymbol{q} \neq \Delta \boldsymbol{q}$. Let

Table 9-1. Methyl Group Tilt and Distortion.[a]

	CH_3NH_2		CH_3OH		CH_3NO	
	exp.	calc.	exp.	calc.	exp.	calc.
tilt (degrees)	3.5	3.4	3.3	3.9	2.5	2.2
$r_1 - r_2$ (Å)	0.007	0.008	-0.006	-0.006	–	-0.001

[a]Data taken from Ref. 41. The tilt is considered positive if the symmetry axis of the methyl group passes through the HNH triangle, the OH bond, or the NO bond, respectively; r_1 is the CH bond in the molecular symmetry plane. The conformation of methylamine and methyl alcohol is staggered, that of nitrosomethane is eclipsed.

$$\Delta^2 X = A(\Delta q - \Delta^1 q) + \Delta^1 X \tag{9-46a}$$

$$\Delta^3 X = A(\Delta q - \Delta^2 q) + \Delta^2 X \tag{9-46b}$$

and so on until the difference $\Delta q - \Delta^i q$ becomes negligible. On the basis of (9-41), (9-44), and (9-45) an automatic geometry search routine has been implemented in MOLPRO (*58*). The program TEXAS (*90*) uses (9-41), (9-42), and (9-46).

It is clear that gradient optimization works best for potential surfaces having exactly one energy minimum. In case of multiple minima or highly anharmonic surfaces a pointwise mapping of the surface may be necessary. Also, in systems with very low FCs in certain directions (i.e., when the lowest eigenvalue of the FC matrix is small), small forces on nuclei do not ensure closeness to the energy minimum. Then one should again determine these low FCs, although large corrections to geometry remain unreliable because the surface is no longer quadratic. The type of work which is best suited to the gradient method is, e.g., the tilt and asymmetry of methyl groups in asymmetric environments (*41*). Experimental studies usually neglect these small (but significant) effects. However, the excellent agreement (Table 9-1) with microwave results shows that one should constrain these variables at ab initio values instead of neglecting them.

9-6-2. Saddle Points and Reaction Paths

The force method is especially suited for determining saddle points on energy surfaces, since these points cannot be characterized by such global criteria as energy minima. A method (*57*) based on minimizing the gradient vector, though successful in semiempirical calculations, has certain drawbacks, e.g., slow convergence and appearance of false minima. In an ab initio study of (methyl) isocyanide–cyanide rearrangement the present author found the force relaxation method more advantageous. For a saddle point, usually one of the diagonal elements (corresponding approximately to the reaction coordinate) of the FC matrix is negative, and one then has to determine the FCs for those degrees of freedom which are likely to participate strongly in the reaction coordinate (this also ensures that the FC matrix has only one negative eigenvalue). Note that one may even use scaled CNDO FCs in ab initio geometry calculations (*103*).

For constructing adiabatic reaction paths,[8] one frequently fixes one coordinate (the reaction coordinate) at a series of values, and at each value optimizes all other coordinates. This constrained geometry optimization assumes an infinitely large diagonal FC for the constrained coordinate. Note that reaction paths are not physically defined, and changing the definition of reaction coordinate results in a discontinuity of the reaction path or (if the change occurs at a saddle point) of its derivative. Nevertheless the choice of reaction coordinate is often relatively unambiguous. If the system is much less rigid along the reaction coordinate than in other directions (e.g., for most internal rotations), the reaction path is approximately invariant regarding coordinate choice. One can then separate out (Born–Oppenheimer type) the slow motion along the reaction coordinate. The slow motion takes place on an effective potential curve, which can be approximated by the adiabatic reaction path, to which the zero-point energy of the rigid modes should be added. Examples of such calculations are the rotational barrier in nitrous acid (*108*) and hydrogen peroxide (*33*).

9-7. CALCULATION OF FORCE CONSTANTS

Let the nuclear coordinates $q_1, \ldots, q_M$ characterize a molecule; the coordinates are generally chosen as valence coordinates or their symmetrized linear combinations. The FC $F_{ijk\ldots}$ is defined as

$$F_{ijk\ldots} = \partial\partial\partial \cdots E/\partial q_i \partial q_j \partial q_k \cdots \qquad (9\text{-}47)$$

at a suitably chosen reference geometry. Since FCs are not directly accessible to measurement, many workers prefer to compare theoretical and experimental (spectroscopic) observables instead of FCs. However, because the error in such an observable is determined roughly by the dominant error in the contributing FCs, it is better to compare the latter if reliable experimental values are available.

The FCs (see also Chapter 5) must be evaluated by numerically differentiating the forces. For harmonic FCs the difference quotient may be used:

$$F_{ij} = -\Delta\phi_i/\Delta q_j. \qquad (9\text{-}48)$$

The effect of cubic anharmonicity is eliminated if the two points lie symmetrically with respect to the reference geometry at $q_j^0 \pm \frac{1}{2}\Delta q_j$. There are several points worth noting. (1) From only two force calculations a whole row of the harmonic FC matrix F_{ij}, $i = 1, \ldots, M$, can be calculated. By including forces at the reference geometry, all diagonal and semidiagonal (F_{ijj}) cubic FCs can be obtained. (2) The results are much more accurate than those from energy surfaces. In the past, numerical inaccuracies often prevented the calculation of smaller deformational and coupling force constants. (3) Only $2M$ force evaluations are necessary in the general (asymmetric) case to determine the whole harmonic force field whereas the energy surface method requires at least M^2 evaluations. Force calculations are reduced further by using symmetry, e.g., all 43 harmonic FCs in benzene result from 11 force calculations. As a numerical check, two values of each coupling constant $F_{ij}, i \neq j$ are compared, namely, $\Delta\phi_i/\Delta q_j$ and $\Delta\phi_j/\Delta q_i$. Frequently, one can also omit force calculations for certain unimportant diagonal FCs (notably CH stretches). (4) The distortions $\frac{1}{2}\Delta q_j$ (multiples of 0.01 Å for bond stretchings and of 1° for deformations) are a compromise between numerical errors, which increase with decreasing Δq, and the formula error in (9-48), which increases with increasing Δq. If high numerical accuracy is ensured, then one may use much smaller and one-sided distortions, reducing computations by nearly half.

There are arguments in favor of using either the theoretical or experimental equilibrium as reference geometry. Using the former is wholly consistent; it is somewhat ambiguous to define theoretical FCs at the experimental geometry, although this is not apparent when valence coordinates are used. However, this author shares the view *(106)* that better values are obtained at the experimental geometry. For a diatomic molecule the difference between theoretical and experimental quadratic FCs,

$$F_2^{\text{th}}(R^{\text{th}}) - F_2^{\text{exp}}(R^{\text{exp}}) = F_2^{\text{th}}(R^{\text{exp}}) - F_2^{\text{exp}}(R^{\text{exp}}) + (R^{\text{th}} - R^{\text{exp}})F_3 + \cdots \qquad (9\text{-}49)$$

has two contributions, the first due to incorrect curvature of the theoretical potential curve and the second due to the error in theoretical geometry. Most stretching potential curves are so anharmonic that the second term dominates. Unfortunately, most ab initio calculations do not separate the two terms, creating a wrong impression that the Hartree–Fock FCs are very poor.

The curvature error for near-Hartree–Fock potential curves is less than 14% even for N_2 (worst case, Table 9-2). The calculation of $F_2^{\text{th}}(R^{\text{exp}})$ in effect adds an empirical linear term (Section 9-5), which shifts the theoretical energy minimum to the experimental geometry. This is justified because (1) the domi-

Table 9-2. Harmonic Force Constants of Some Diatomic Molecules.[a]

Molecule	$F_2^{HF}(R^{th})$	$F_2^{HF}(R^{exp})$	$F_2^{cor}(R^{exp})$	$F_2^{HF}+F_2^{cor}$	F_2^{exp}
BH	3.36	3.18	−0.08	3.10	3.05
CH	5.08	4.64	−0.16	4.48	4.48
NH	6.97	6.20	−0.16	6.04	5.97
HF	11.30	9.84	−0.03	9.81	9.66
N_2	31.4	26.14	–	–	22.97

[a]Units are aJ/Å². Abbreviations: HF = Hartree–Fock; cor = correlation contribution; exp = experimental. $F_2(R^{th})$ is the FC evaluated at the theoretical equilibrium geometry. The data for the hydrides have been calculated from the potential curves of Meyer and Rosmus (*65*). For nitrogen an ab initio calculation was performed, using a (9*s*5*p*1*d*) Gaussian basis set. Experimental FCs have been calculated from the spectroscopic constants listed by Mizushima (*68*).

nant term in correlation energy is linear in nuclear coordinates near the equilibrium, and (2) molecular geometries are much better known experimentally than FCs. However, the experimental geometry must be chosen judiciously, since uncertainties in interatomic distances (particularly in hydrogen compounds) can significantly affect stretching FCs. For systematic calculations (*12*, *91*, *92*) a good compromise is to use theoretical geometries corrected for deviations due to the quantum chemical method used.

9-8. FORCE CONSTANTS IN POLYATOMIC MOLECULES

9-8-1. General Comments

Besides allowing the prediction and interpretation of vibrational spectra, polyatomic potential functions are indispensable for conformational analysis of strained molecules. Most measurements of molecular properties (e.g., geometry, dipole moment) yield averages over vibrational states, but present experimental accuracy requires that vibrational effects be accounted for. Since in many cases nuclear displacements from reference geometry are small, and large-amplitude motions are restricted to a few degrees of freedom, the potential function can be characterized by the coefficients of its power-series expansion.

The main difficulty in the experimental determination of FCs is that the data are usually insufficient to determine complete force fields, even in the harmonic approximation. For large molecules the only answer seems to be consistent force fields, i.e., transferable force fields for a class of (organic) molecules which reproduce vibrational frequencies, geometries, heats of formation, etc. for the whole class. This approach has been very successful for, e.g., nonconjugated hydrocarbons (*55*, *100*), but its extension to wider classes of compounds is difficult (*37*) because of uncertainties in vibrational assignments for such complex molecules.

9-8-2. Ab Initio Work

Quantum mechanical calculations provide a way out of the difficulties associated with empirical determination of force fields. The only ab initio method routinely used is the (closed-shell) Hartree-Fock method with a finite basis set. Its usefulness depends on two questions. (1) How strongly does electron correlation influence FCs? (2) How sophisticated are the basis sets necessary to obtain results close to the Hartree-Fock limit?

The first question can be answered from several near-Hartree-Fock results on FCs, e.g., water (*34*, *38*, *62*, *95*), methane (*60*), ammonia (*83*), formaldehyde (*63*), hydrogen cyanide (*53*, *111*), acetylene (*3*), carbon dioxide (*77*), and ethylene (*42*). Regarding the harmonic part of the potential function, one can infer that (1) coupling FCs are reproduced remarkably accurately, with errors of the order of ± 0.05 aJ/Å^2 or $\pm 15\%$, whichever is larger; experimental values are frequently less accurate. (2) Diagonal FCs have systematic errors and are usually overestimated. The deviation is small for CH stretches, ~10% for CC stretches and 10-20% for deformations, except for out-of-plane deformations in doubly and triply bonded systems, where it is ~30% and ~45%, respectively. Thus, the accuracy of diagonal FCs does not reach experimental accuracy, although small differences between diagonal FCs of the same type are generally estimated correctly. Seemingly the best way is to combine quantum mechanical force fields with experimental information, either through scaling or by taking theoretical values for experimentally uncertain coupling constants. Bartell's (*5*) suggestion of including theoretical FCs in a carefully weighted least-squares fitting procedure is being pursued with, e.g., the force field of benzene (*91*, *92*). For large molecules one should also include van der Waals terms (cf. consistent force fields, see Ref. 4).

It is difficult to assess the accuracy of coupling constants from a comparison of near-Hartree-Fock and experimental FCs, because of various uncertainties in the latter. Comparing the results from highly correlated wavefunctions with near-Hartree-Fock ones is more conclusive (Tables 9-3 to 9-5). These tables also give some indication about basis-set effects on force fields. From systematic and exhaustive investigations on methane (*60*) and ammonia (*83*), as well as numerous other calculations reported in literature (*4*, *14*, *109*), the following conclusions can be drawn for Gaussian basis sets: (1) For satisfactory FCs, the valence shell requires at least three different elementary Gaussian exponents. There is little change with four exponents, showing that the limiting radial flexibility is easy to reach. Any decent description of atomic cores is satisfactory, at least for first-row atoms. (2) The valence shell should have at least

Table 9-3. Harmonic Force Constants of Methane.[a]

Force Constant	HF					
	C: 4-21[b] / H: 2-1 / CH: –	7s3p[c] / 3s / 1s	9s5p1d[d] / 5s1p / –	near-HF[e]	CEPA[f]	exp.[g]
A_1: F_{11} (CH)	5.722	5.888	5.673	5.668	5.562	(5.503)[g]
E: F_{22} (def)	0.699	0.708	0.646	0.642	0.587	0.581
F_2: F_{33} (CH)	5.546	5.744	5.472	5.472	5.402	5.383
F_2: F_{34} (CH, def)	0.261	0.235	0.220	0.221	0.221	0.225
F_{44} (def)	0.674	0.668	0.619	0.612	0.560	0.547
$F_{rr} = (F_{11} - F_{33})/4$	0.043	0.036	0.050	0.049	0.040	(0.03)
$F_{\alpha\alpha} = F_{22} - F_{44}$	0.023	0.040	0.027	0.030	0.027	0.034

[a]Units: F_{11} and F_{33} in aJ/Å^2; F_{22} and F_{44} in aJ/rad^2; F_{34}, in aJ/Å-rad. All theoretical FCs have been corrected for a common reference geometry, 1.090 Å. For the coordinates see, e.g., Ref. 60. The curly braces mark off the basis sets for the different HF calculations.
[b]Ref. 91.
[c]Ref. 81.
[d]Ref. 89.
[e]Ref. 60.
[f]Highly correlated wavefunction (*63*).
[g]Ref. 31. As shown by Meyer (*61*), the anharmonic correction in the A_1 species used by Duncan and Mills (*31*) is probably incorrect. The most probable value of the harmonic frequency was determined by Meyer as 3044 cm^{-1}. The experimental F_{11}, given in parentheses, is based on this frequency.

Table 9-4. Harmonic Force Constants of Water.[a]

Force Constant	HF[b]	near-HF[c]	SD[c]	SDQ[c]	exp.[d]
F_{RR}	8.72	8.69	8.60	8.54	8.45
$F_{\alpha\alpha}$	0.918	0.783	0.741	0.734	0.697
$F_{RR'}$	-0.081	-0.052	-0.079	-0.085	-0.101
$F_{R\alpha}$	0.381	0.240	0.257	0.262	0.218

[a]R, R', and α denote the two OH bond lengths and the HOH angle, respectively. The FCs are evaluated at the following reference geometry: R = 0.9572 Å; α = 104.520°.
[b]Hartree–Fock results using a small double-zeta Gaussian basis (*91*).
[c]Ref. 95. The harmonic FCs have been transformed to the present reference geometry by means of the cubic and quartic constants. Near-HF is the Hartree–Fock result with an extended Slater-type basis set; SD refers to a CI calculation including all single and double substitutions with respect to the Hartree–Fock determinant; in the calculation designated SDQ, quadruple substitutions (unlinked cluster contributions) have been taken into account approximately.
[d]Ref. 48.

Table 9-5. Harmonic Force Constants of Hydrogen Cyanide.[a]

Force Constant	HF 4-21[b]	HF 7s3p1b[c]	near-HF[d]	SD[d]	exp.[e]
F_{11} (CH)	6.52	6.57	6.57	6.52	6.25
F_{22} (CN)	21.16	20.99	20.5	20.2	18.70
F_{12} (CH, CN)	-0.326	-0.219	-0.18	-0.16	-0.200
F_{33} (bending)	0.398	0.410	0.375	0.296	0.260

[a]Energy in aJ, bond lengths in Å, the angle in radians. All theoretical FCs have been transformed to a common reference geometry, the experimental equilibrium (r_{CH} = 1.0657 Å, r_{CN} = 1.1530 Å), using the theoretical cubic and quartic FCs.
[b]For the 4-21 basis, see Ref. 91.
[c]Hartree-Fock calculation using a small but polarized Gaussian basis (*87*). Note the importance of the polarization functions for the constant F_{12}.
[d]Near-Hartree-Fock and correlated (all singles plus doubles) results (*111*).
[e]Ref. 110.

Table 9-6. Ab Initio Hartree-Fock Force-Constant Calculations on Polyatomic Molecules with Small Double-Zeta (Split-Shell) Basis Sets.[a,b]

List of Molecules in Alphabetic Order

Acetamide,[c] acrolein,[c] benzene (*92*D), BH_3 (*91*D), BH_4^- (*81*A), BO (*23*A), butadiene,[c] CH_4 (*81*A, *101*A, *91*D), C_2H_2[c] (*22*A), C_2H_4[c] (*13*C, *82*A, B, *103*C), C_2H_6[c] (*11*C. *13*C, *103*C), C_3H_8 (*11*C, *13*C), CH_3CN (*32*C, D), CH_3COH,[c] CH_3Cl (*104*C), CH_3F (*16*C, *104*C), CH_2F_2 (*16*C), CHF_3 (*16*C), CH_3NC,[c] CH_3NH_2[c] (*86*A, *104*C), CH_3OCH_3 (*15*C), CH_3OH (*12*C, *20*A, B, *104*C), CH_3SH (*104*C), C_2N_2 (*87*A), CO_2,[c] cyclopropane (*11*C, *13*C, *103*C), FCN (*87*A), FCO (*9*A), formamide,[c] glyoxal,[c] HBN (*23*A), HBO (*23*A), HCCCl (*22*A), HCCF (*22*A), HCN[c] (*18*A, B, *87*A), HCO (*17*A), $HCOOCH_3$,[c] HCOOH,[c] H_2CO[c] (*63*A, B, *103*C), HCP (*18*A, B), H_2CS (*103*C), HCl (*101*A), HF (*81*A, *91*D, *101*A), HNB (*23*A), HNBH (*23*A), HNC (*21*A), $HNCH_2$ (*21*A), HNCO (*21*A), HNO (*19*A), HNO_2 (*108*A), H_2O (*12*C, *80*A, B, *91*D, *101*A), H_2O_2 (*20*A, B), HOB (*23*A), HOF (*20*A, B), H_2S (*101*A), H_2SiCH_2 (*105*C), isoprene,[c] NF_3 (*99*A), NH_3 (*81*A, *91*D, *101*A), NH_2OH (*20*A, B), *N*-methyl acetamide,[c] *N*-methyl formamide,[c] NO_2F (*71*A), ONF (*99*A), PH_3 (*101*A), propene (*13*C), SiH_4 (*101*A).

[a]Calculations are included only if complete or nearly complete harmonic force fields have been determined, and if the calculations show systematic nature.
[b]Gaussian basis sets employed are denoted by one of four letters after reference numbers. A–7*s*3*p*/3*s*, possibly augmented by bond functions; several basis sets of this size are in use; for details, see original references. B–extended basis sets, mostly 9*s*5*p*/4*s*. C–4–31*G* basis set (*30*). D–4–21 basis set (*91*).
[c]Part of a cooperative research program between the University of Texas at Austin and Eötvös Loránd University, Budapest. Publication of the results is in progress. These cal-calculations use the basis set D. See also Ref. 85.

double-zeta (split-shell) accuracy. Basis sets contracted to minimum basis, e.g., Pople's STO-3G (*29*) are poor, especially for stretching FCs. Obviously, "breathing" of the valence shell is important. Releasing the contraction over double zeta has little further effect. (3) The effect of polarization functions is not negligible. But, considering both experimental uncertainties and computational cost, a medium-size, non-polarized, double-zeta basis set seems to be a good compromise, e.g., the STO-4-31G (*30*), the 4-21G (*91*, *92*), and the $7s3p$ sets (*94*). These three sets give similar FCs (Table 9-3). Table 9-6 lists several numerically satisfactory ab initio calculations of FCs by small double-zeta basis sets, using mostly the force method and occasionally the energy surface method. Figures 9-1 to 9-3 illustrate the good prediction of coupling FCs for some molecules which are experimentally well characterized.

In spite of studies on many molecules, so far there is no systematic study of

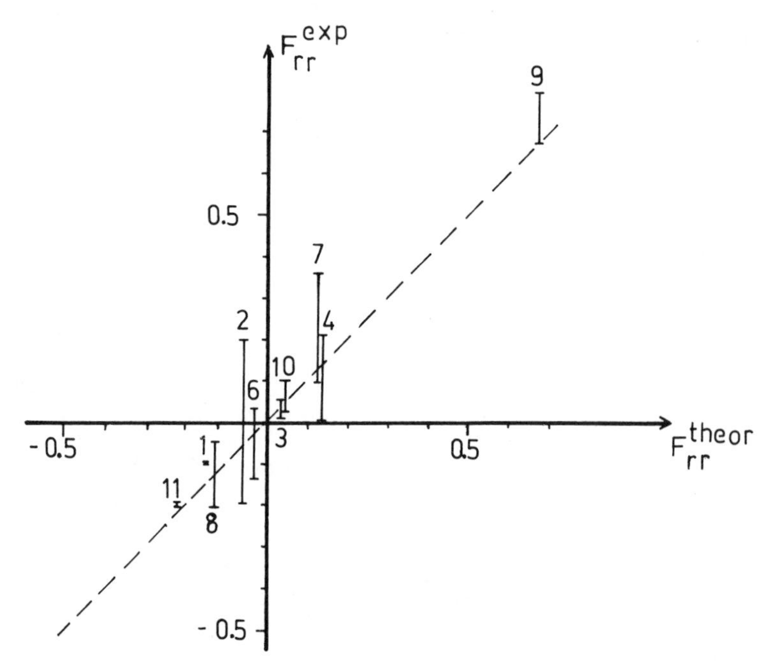

Fig. 9-1. Correlation between experimental and theoretical stretch–stretch coupling force constants. The theoretical values have been obtained from the author's Hartree–Fock calculations, using small Gaussian basis sets. For detailed information, the source of data and coordinates, etc., see the original references in Table 9-6. Units: aJ/Å^2. The vertical bars show the estimated uncertainty of the experimental values. Force constants with large experimental uncertainties have not been included. Key: (1) H_2O, OH–OH′; (2) NH_3, NH–NH′; (3) CH_4, CH–CH′; (4) C_2H_6, CC-sym. CH; (5) C_2H_6, sym. CH-sym. CH′; (6) C_2H_6, as. CH-as. CH′; (7) C_2H_4, CC-sym. CH; (8) C_2H_2, CC–CH; (9) H_2CO, CO-sym. CH; (10) H_2CO, CH–CH′.

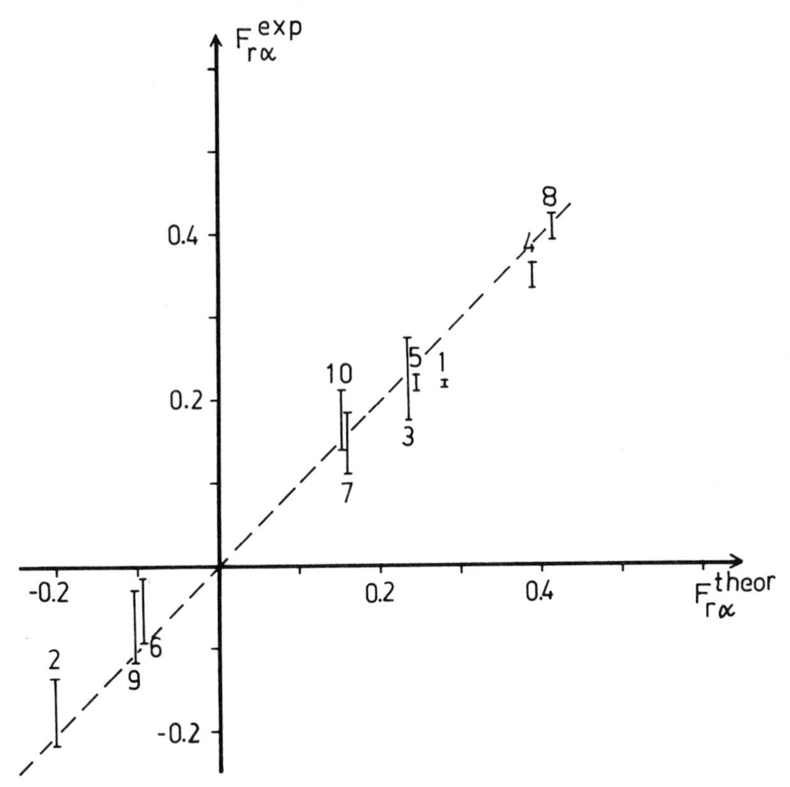

Fig. 9-2. Correlation between experimental and theoretical stretch–deformation coupling force constants. Units: aJ/Å-rad. See Fig. 9-1. Key: (1) H_2O, OH–HOH; (2) NH_3, as. NH–as. def.; (3) CH_4, as. CH–as. def.; (4) C_2H_6, CC–sym. def.; (5) C_2H_4, CC–sym. def.; (6) C_2H_4, sym. CH–sym. def.; (7) C_2H_4, as. CH–rocking; (8) H_2CO, CO–HCH; (9) H_2CO, sym. CH–HCH; (10) H_2CO, as. CH–rocking.

molecular force fields by ab initio methods. Such a study would put our present understanding of vibrational spectra on a much firmer foundation.

9-8-3. Semiempirical Work

Semiempirical calculations are much cheaper than ab initio ones but they do not attain the same accuracy. The widely used CNDO method, for example, chronically overestimates diagonal stretching FCs by a factor of nearly 3. Despite this obvious deficiency, which long discouraged semiempirical FC calculations, the resulting FCs are still useful after a suitable scaling procedure. Couplings between stretchings and deformations are quite well predicted by both CNDO and MINDO; for couplings between deformations, CNDO is superior to MINDO (*52*).

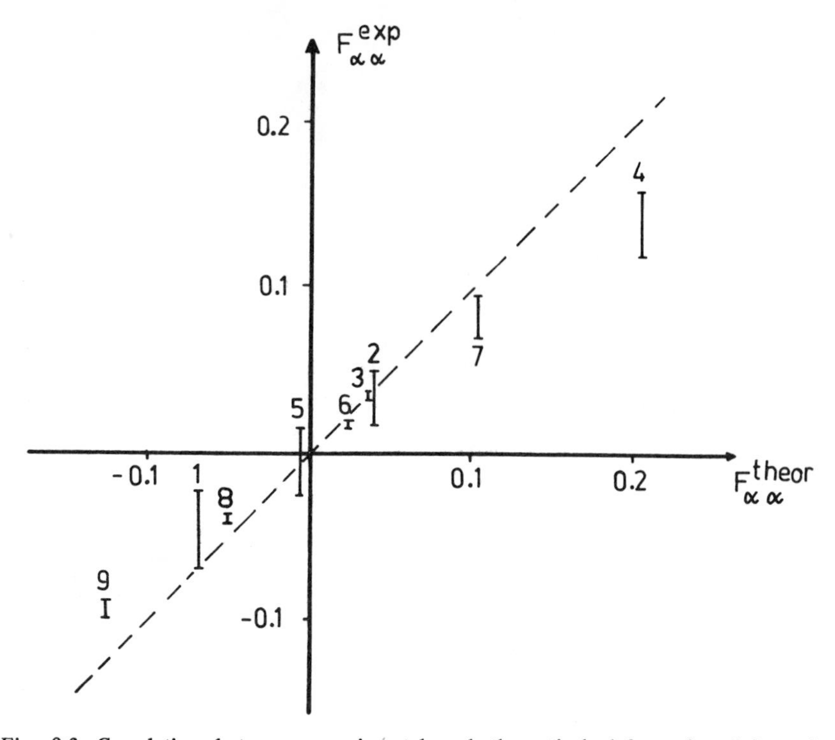

Fig. 9-3. Correlation between experimental and theoretical deformation–deformation coupling force constants. Units: aJ/rad^2. See Fig. 9-1. Key: (1) NH_3, HNH–HNH′; (2) CH_4, F_{22}–F_{44}; (3) C_2H_6, sym. def.–sym. def.′; (4) C_2H_6, rocking–rocking′; (5) C_2H_6, as. def.–as. def.′; (6) C_2H_4, sym. def.–sym. def.′; (7) C_2H_4, rocking–rocking′; (8) C_2H_4, as. def.–as. def.′; (9) C_2H_2, HCC′–CC′H′.

Stretch–stretch couplings are significant only in conjugated systems, and in these cases CNDO was found to perform well, but the small couplings in saturated molecules are not reproduced correctly.

While semiempirical predictions are qualitative in the best case, they may be quite useful in establishing assignments and for estimating the magnitude and direction of FC changes. For example, the changes in the vibrational frequencies between *cis*- and *trans*-butadiene could be correctly predicted by this method (*76*).

9-8-4. Higher Force Constants

Two factors have long hampered the quantum chemical calculation of higher FCs:

1. a general prejudice because of the wrong dissociation behavior of the Hartree-Fock wavefunction;
2. unsatisfactory results obtained for small molecules at the theoretical geometry (Section 9-7).

In reality the reproduction of cubic and quartic FCs seems to be excellent if the calculations are carried out at the experimental geometry. The first systematic calculations of cubic and quartic FCs in polyatomic molecules were performed by Schlegel et al. (*101*). The good reproduction of higher FCs may be rationalized by noting that the bulk of anharmonicity is due to core–core repulsion, which is correctly reproduced in the Hartree–Fock theory. To illustrate this point, Hartree–Fock cubic and quartic FCs in some diatomic molecules are compared to experimental values in Table 9-7.

Complete experimental quartic force fields are known only for a few polyatomic molecules, and many of the higher couplings are very uncertain even in the molecules which are well characterized. Comparison to experiment is thus not very helpful and may be even misleading in deciding how accurate anharmonic couplings are in the Hartree–Fock theory. However, a few calculations have been carried out for anharmonic FCs with CI wavefunctions (*3*, *53*, *63*, *89*, *95*, *111*) and these confirm the expectation that the influence of correlation is small, especially on the coupling terms. Calculation of full quartic force fields is very tedious, even by the force method. Fortunately, most of the anharmonicity is concentrated in the stretching coordinates, and the simple recipe of taking into account only diagonal cubic and quartic stretching FCs (*54*) is probably sufficient in most cases.

Table 9-7. Cubic and Quartic Force Constants for Some Diatomic Molecules.[a]

Molecule	HF	Correlation (CEPA)	HF + corr.	exp.
CH F_{rrr}	-26.61	+0.04	-26.57	-26.6
F_{rrrr}	136.6	-0.6	136.0	138
NH F_{rrr}	-39.30	+0.06	-39.24	-39.2
F_{rrrr}	224.2	-0.6	223.6	221
HF F_{rrr}	-71.84	-0.23	-72.07	-71.2
F_{rrrr}	481.2	-0.7	480.5	480
N_2 F_{rrr}	-173.9			-169.6
F_{rrrr}	987			1004

[a]Units: aJ/Å^3 for F_{rrr}, aJ/Å^4 for F_{rrrr}. All theoretical values have been evaluated at the experimental equilibrium interatomic distance. For the source of data, see Table 9-2.

9-9. SUMMARY

This chapter deals with the direct analytic calculation of energy gradients with respect to the nuclear coordinates. This procedure, if applicable, is superior to the numerical determination of gradients from energy values, both in terms of computer time and numerical accuracy. In contrast to H-F forces, the error of the exact forces (energy gradients) is of second order in the error parameter. It is shown that calculation of the exact forces is simple for wavefunctions which contain only two kinds of parameters: (1) parameters which depend in a simple prescribed way on the nuclear coordinates, and (2) parameters which minimize the total energy. Calculation of the forces from closed-shell and general SCF wavefunctions is discussed. The formulas for transforming forces and force constants from Cartesian to internal coordinates are given. Determination of molecular geometries, saddle points, and force constants using exact forces is treated in detail. Finally, the results of quantum chemical force constant calculations are summarized, with emphasis on ab initio Hartree-Fock results.

Acknowledgment

Part of this chapter is based on Ref. 88. The author is grateful to Plenum Publishing Corporation and Prof. H. F. Schaefer, for their consent to use that material, and to Prof. W. Meyer for his contributions and critical comments.[9]

Notes

1. There is no closed formula for energy as a function of nuclear coordinates.
2. Since the first-order wavefunction determines energy up to third order, knowledge of the C_k^a enables one to calculate analytically the second and third derivatives of the energy for general SCF wavefunctions (*70*). This procedure, however, requires a large amount of computation.
3. There is normally an infinite manifold of solutions.
4. The author is indebted to W. Meyer for pointing this out to him (*64*).
5. This is actually a generalization of Hurley's definition, which requires that all parameters be optimized. Taken literally, Hurley's definition is not useful, as the number of parameters can be increased without limit for every function by considering wider classes of functions.
6. The sum and torque of the exact forces vanish if the basis set ensures that the energy is invariant to rigid translations and rotations of the nuclear frame. The first condition is easily satisfied by fixing the basis functions to the nuclei. Rotational invariance requires that the basis set does not carry torques; this is satisfied if the basis set is rotationally complete around each center, i.e., application of a rotation around a center does not lead out of the subspace.
7. An analogous lucid treatment (*71*) of up to third analytical derivatives, with respect to orbital exponents, remained somewhat hidden because of the unfortunate title.

8. A reaction path usually connects two energy minima.

9. Since the preparation of this chapter we have come across a number of other significant developments. Perhaps the most important recent achievement in computational quantum chemistry in general is an ingenious method for evaluating integrals over Gaussian basis functions [M. Dupuis, J. Rys, and H. F. King, *J. Chem. Phys.*, **65**, 111 (1976)]. This method is particularly advantageous for the calculation of higher angular momentum functions occurring in gradient evaluation, and leads to a significant reduction of the computational effort [M. Dupuis and H. F. King, *J. Chem. Phys.*, **68**, 3988 (1978)]. An ingenious iterative scheme for the determination of the wavefunction coefficient derivatives (see Section 9-2) has overcome much of the difficulty in the direct analytical calculation of the second derivative of the SCF energy, and the first derivative of the second-order part of the correlation energy (J. A. Pople, R. Krishnan, H. B. Schlegel, and J. S. Binkley, forthcoming). An alternative method for the calculation of the gradient of open-shell restricted Hartree–Fock energies has been recently published [J. D. Goddard, N. C. Handy, and H. F. Schaefer, III, *J. Chem. Phys.*, **71**, 1525 (1979)]. Subsequently, this has been extended to the case of very many configurations (B. R. Brooks, W. D. Laidig, P. Saxe, J. D. Goddard, and H. F. Schaefer, III, forthcoming).

References

1. Allavena, M. and Bratož, S., 1963, *J. Chim. Phys.*, **60**, 1199.
2. ——, 1966, *Theoret. Chim. Acta* (Berlin), **5**, 21.
3. Bagus, P. S., Pacansky, J., and Wahlgren, U., 1977, *J. Chem. Phys.*, **67**, 618.
4. Bartell, L. S., Fitzwater, S., and Hehre, W. J., 1975, *J. Chem. Phys.*, **63**, 4750.
5. ——, 1976, private communication.
6. Benston, M. L. and Kirtman, B., 1966, *J. Chem. Phys.*, **44**, 119.
7. Bishop, D. M. and Randić, M., 1966, *J. Chem. Phys.*, **44**, 2480.
8. —— and Macias, A., 1970, *J. Chem. Phys.*, **53**, 3515.
9. Bleicher, W. and Botschwina, P., 1974, *Mol. Phys.*, **30**, 1029.
10. Bloemer, W. L. and Bruner, B. L., 1973, *J. Chem. Phys.*, **58**, 3735.
11. Blom, C. E., Slingerland, P. J., and Altona, C., 1976, *Mol. Phys.*, **31**, 1359, 1377.
12. ——, Otto, P., and Altona, C., 1976, *Mol. Phys.*, **32**, 1137.
13. —— and Altona, C., 1977, *Mol. Phys.*, **33**, 875.
14. —— and Altona, C., 1977, *Mol. Phys.*, **34**, 177.
15. ——, Oskam, A., and Altona, C., 1977, *Mol. Phys.*, **34**, 557.
16. —— and Müller, A., 1978, *J. Mol. Spectrosc.*, **70**, 449.
17. Botschwina, P., 1974, *Chem. Phys. Lett.*, **29**, 98, 580.
18. ——, Pecul, K., and Preuss, H., 1975, *Z. Naturforsch.*, **A30**, 1015.
19. ——, 1976, *Mol. Phys.*, **32**, 729.
20. ——, Meyer, W., and Semkow, A. M., 1976, *Chem. Phys.*, **15**, 25.
21. ——, Nachbaur, E., and Rode, B. M., 1976, *Chem. Phys. Lett.*, **41**, 486.

22. ——, Srinivasan, K., and Meyer, W., 1978, *Mol. Phys.*, **35**, 1177.
23. ——, 1978, *Chem. Phys.*, **28**, 231.
24. Bratož, S., 1958, *Colloq. Intern. CNRS* (Paris), **82**, 287.
25. —— and Allavena, M., 1962, *J. Chem. Phys.*, **37**, 2138.
26. Cade, P. E. and Huo, W. M., 1967, *J. Chem. Phys.*, **47**, 614.
27. ——, Sales, K. D., and Wahl, A. C., 1966, *J. Chem. Phys.*, **44**, 1973.
28. Chong, D. P., Gagnon, P. A., and Thorhallson, J., 1971, *Can. J. Chem.*, **49**, 1047.
29. Ditchfield, R., Hehre, W. J., and Pople, J. A., 1970, *J. Chem. Phys.*, **52**, 5001.
30. ——, Hehre, W. J., and Pople, J. A., 1971, *J. Chem. Phys.*, **54**, 724.
31. Duncan, J. L. and Mills, I. M., 1964, *Spectrochim. Acta*, **20**, 523.
32. ——, McKean, D. C., Tullini, F., Nivellini, G. D., and Perez Pena, J., 1978, *J. Mol. Spectrosc.*, **69**, 123.
33. Dunning, T. H. and Winter, N. W., 1971, *Chem. Phys. Lett.*, **11**, 194.
34. ——, Pitzer, R. M., and Aung, S., 1972, *J. Chem. Phys.*, **57**, 5044.
35. Dykstra, C. E., 1977, *Chem. Phys. Lett.*, **45**, 466.
36. ——, Schaefer, H. F., and Meyer, W., 1976, *J. Chem. Phys.*, **65**, 2740.
37. Ermer, O., 1976, *Structure and Bonding*, Vol. 27 (Springer-Verlag, Berlin), p. 161.
38. Ermler, W. C. and Kern, C. W., 1971, *J. Chem. Phys.*, **55**, 4851.
39. Fink, W. and Allen, L. C., 1967, *J. Chem. Phys.*, **46**, 3270.
40. Fletcher, R. and Powell, M. J. D., 1963, *Computer J.*, **6**, 163.
41. Flood, E., Pulay, P., and Boggs, J. E., 1977, *J. Am. Chem. Soc.*, **99**, 5570.
42. Fogarasi, G. and Pulay, P., 1980, *Acta Chim. Acad. Sci. Hung.*, forthcoming.
43. Frost, A. A., 1967, *J. Chem. Phys.*, **47**, 3707.
44. Gerratt, J. and Mills, I. M., 1968, *J. Chem. Phys.*, **49**, 1719, 1730.
45. Grimmer, M. and Heidrich, D., 1973, *Z. Chem.*, **13**, 356.
46. Hall, G. G., 1961, *Phil. Mag.*, **6**, 249.
47. Hehre, W. J., Lathan, W. A., Ditchfield, R., Newton, M. D., and Pople, J. A., 1973, GAUSSIAN 70, Quantum Chemistry Program Exchange, Indiana University, Bloomington, Indiana, Program No. 236.
48. Hoy, A. R., Mills, I. M., and Strey, G., 1972, *Mol. Phys.*, **24**, 1265.
49. Huber, H., Čarsky, P., and Zahradnik, R., 1976, *Theoret. Chim. Acta* (Berlin), **41**, 217.
50. Hurley, A. C., 1954, *Proc. Roy. Soc.* (London), **A226**, 170.
51. Kern, C. W. and Karplus, M., 1964, *J. Chem. Phys.*, **40**, 1374.
52. Kozmutza, K. and Pulay, P., 1975, *Theoret. Chim. Acta* (Berlin), **37**, 67.
53. Kraemer, W. P. and Diercksen, G. H. F., 1976, *Astrophys. J. Lett.*, **205**, L 97.
54. Kuchitsu, K. and Morino, Y., 1965, *Bull. Chem. Soc., Jap.*, **38**, 805, 814.
55. Lifson, S. and Warshel, A., 1968, *J. Chem. Phys.*, **49**, 5116.
56. McIver, J. W., Jr. and Komornicki, A., 1971, *Chem. Phys. Lett.*, **10**, 303.
57. ——, 1972, *J. Am. Chem. Soc.*, **94**, 2625.

58. Meyer, W. and Pulay, P., 1969, MOLPRO Program Description, Munich and Stuttgart, Germany (correspondence to Meyer at Munich or Pulay at Budapest).
59. ——, 1969, private communication.
60. —— and Pulay, P., 1972, *J. Chem. Phys.*, **56**, 2109.
61. ——, 1973, *J. Chem. Phys.*, **58**, 1017.
62. ——, 1975, Proceedings of the SRC Atlas Symposium No. 4, *Quantum Chemistry–The State of Art*, Chilton, Berks., England, eds. V. R. Saunders and J. Brown (Science Research Council, Great Britain), p. 97.
63. —— and Pulay, P., 1974, *Theoret. Chim. Acta* (Berlin), **32**, 253.
64. ——, 1975, private communication.
65. —— and Rosmus, P., 1975, *J. Chem. Phys.*, **63**, 2356.
66. ——, 1976, *J. Chem. Phys.*, **64**, 2901.
67. —— and Pulay, P., 1980, forthcoming.
68. Mizushima, M., 1975, *The Theory of Rotating Diatomic Molecules* (John Wiley, New York).
69. Moccia, R., 1967, *Theoret. Chim. Acta* (Berlin), **8**, 8.
70. ——, 1970, *Chem. Phys. Lett.*, **5**, 260.
71. Molt, K., Sawodny, W., Pulay, P., and Fogarasi, G., 1976, *Mol. Phys.*, **32**, 169.
72. Morokuma, K., 1976, lecture at the Second International Congress on Quantum Chemistry, New Orleans, Louisiana, April 1976.
73. Murtagh, B. A. and Sargent, R. W. H., 1970, *Computer J.*, **13**, 185.
74. Nesbet, R. K., 1958, *Phys. Rev.*, **109**, 1632.
75. Neumann, D. B., Basch, H., Kornegay, R. L., Snyder, L. C., Moskowitz, J. W., Hornback, C., and Liebmann, S. P., 1972. POLYATOM 2, Quantum Chemistry Program Exchange, Indiana University, Bloomington, Indiana, Program No. 199.
76. Panchenko, Yu. N., Pulay, P., and Török, F., 1976, *J. Mol. Struc.*, **34**, 283.
77. Pacansky, J., Wahlgren, U., and Bagus, P. S., 1976, *Theoret. Chim. Acta* (Berlin), **41**, 301.
78. Pancíř, J., 1973, *Theoret. Chim. Acta* (Berlin), **29**, 21.
79. Pulay, P., 1969, *Mol. Phys.*, **17**, 197.
80. ——, 1970, *Mol. Phys.*, **18**, 473.
81. ——, 1971, *Mol. Phys.*, **21**, 329.
82. —— and Meyer, W., 1971, *J. Mol. Spectrosc.*, **40**, 59.
83. —— and Meyer, W., 1972, *J. Chem. Phys.*, **57**, 3337.
84. —— and Török, F., 1973, *Mol. Phys.*, **25**, 1153.
85. —— and Meyer, W., 1974, *Mol. Phys.*, **27**, 473.
86. —— and Török, F., 1975, *J. Mol. Struc.*, **29**, 239.
87. ——, Ruoff, A., and Sawodny, W., 1975, *Mol. Phys.*, **30**, 1123.
88. ——, 1977, *Modern Theoretical Chemistry*, vol. 4, ed. H. F. Schaefer, III (Plenum Publishing Corporation, New York), p. 153.
89. ——, Meyer, W., and Boggs, J. E., 1978, *J. Chem. Phys.*, **68**, 5077.
90. ——, 1979, *Theoret. Chim. Acta* (Berlin), **50**, 299.

91. ——, Fogarasi, G., Pang, F., and Boggs, J. E., 1979, *J. Am. Chem. Soc.*, **101**, 2550.
92. ——, Fogarasi, G., and Boggs, J. E., 1980, forthcoming.
93. Rinaldi, D. and Rivail, J. L., 1972, *C. R. Acad. Sci.* (Paris), **274**, 1664.
94. Roos, B. and Siegbahn, P., 1970, *Theoret. Chim. Acta* (Berlin), **17**, 199, 209.
95. Rosenberg, B. J., Ermler, W. C., and Shavitt, I., 1976, *J. Chem. Phys.*, **65**, 4072.
96. Rothenberg, S. and Schaefer, H. F., 1970, *J. Chem. Phys.*, **53**, 3014.
97. Salem, L. and Wilson, E. B., Jr., 1962, *J. Chem. Phys.*, **36**, 3421.
98. —— and Alexander, M., 1963, *J. Chem. Phys.*, **39**, 2994.
99. Sawodny, W. and Pulay, P., 1974, *J. Mol. Spectrosc.*, **51**, 135.
100. Schachtschneider, J. H. and Snyder, R. G., 1963, *Spectrochim. Acta*, **19**, 117.
101. Schlegel, H. B., Wolfe, S., and Bernardi, F., 1975, *J. Chem. Phys.*, **63**, 3632.
102. ——, 1975, Dissertation, Queen's University, Kingston, Ontario, Canada.
103. ——, 1976, private communication.
104. ——, Wolfe, S., and Bernardi, F., 1977, *J. Chem. Phys.*, **67**, 4181, 4194.
105. ——, Wolfe, S., and Mislow, K., 1975, *Chem. Comm.*, 246.
106. Schwendeman, R. H., 1966, *J. Chem. Phys.*, **44**, 2115.
107. Shavitt, I., 1976, private communication.
108. Skaarup, S. and Boggs, J. E., 1976, *J. Mol. Struc.*, **30**, 389.
109. ——, Skancke, P. N., and Boggs, J. E., 1976, *J. Am. Chem. Soc.*, **98**, 6106.
110. Strey, G. and Mills, I. M., 1973, *Mol. Phys.*, **26**, 129.
111. Wahlgren, U., Pacansky, J., and Bagus, P. S., 1975, *J. Chem. Phys.*, **63**, 2874.
112. Wilson, E. B., Jr., Decius, J. C., and Cross, P. C., 1955, *Molecular Vibrations* (McGraw-Hill, New York).

Author Index

Numbers in italics indicate page where detailed references are available.

Molecule Index

Subject Index